REPRODUCIBILITY

REPRODUCIBILITY

Principles, Problems, Practices, and Prospects

Edited by

HARALD ATMANSPACHER
Collegium Helveticum, University and ETH Zurich, Zurich, Switzerland

SABINE MAASEN
Munich Center for Technology in Society, Technical University, Munich, Germany

Published by John Wiley & Sons, Inc., Hoboken, New Jersey

Published simultaneously in Canada

Library of Congress Cataloging-in-Publication Data:

Names: Atmanspacher, Harald. | Maasen, Sabine, 1960-
Title: Reproducibility : principles, problems, practices, and prospects /
edited by Harald Atmanspacher, Sabine Maasen.
Description: Hoboken, New Jersey : John Wiley & Sons, Inc., [2016] | Includes
index.
Identifiers: LCCN 2015036802 | ISBN 9781118864975 (cloth)
Subjects: LCSH: Observation (Scientific method) | Science–Methodology.
Classification: LCC Q175.32.O27 R47 2016 | DDC 001.4/2–dc23 LC record available at
http://lccn.loc.gov/2015036802

Typeset in 11/13.6pt CMSS10 by SPi Global, Chennai, India

Printed in the United States of America

10 9 8 7 6 5 4 3 2 1

Table of Contents

PART III: PHYSICAL SCIENCES

PART IV: LIFE SCIENCES

PART V: SOCIAL SCIENCES

PART VI: WIDER PERSPECTIVES

Contributors

Michael Anderson
Department of Psychology
Franklin and Marshal College
Lancaster PA, USA
mljanderson@gmail.com

Harald Atmanspacher
Collegium Helveticum
University and ETH Zurich
Zurich, Switzerland
atmanspacher@collegium.ethz.ch

Sabine Baier
Collegium Helveticum
University and ETH Zurich
Zurich, Switzerland
baier@collegium.ethz.ch

David H. Bailey
Lawrence Berkeley
National Laboratory
Berkeley CA, USA
david@davidhbailey.com

Ladina Bezzola Lambert
Department of English
University of Basel
Basel, Switzerland
ladina.bezzola@unibas.ch

Jonathan Borwein
School of Mathematical
and Physical Sciences
University of Newcastle
Callaghan NSW, Australia
jon.borwein@gmail.com

Jerome Busemeyer
Department of Psychological
and Brain Sciences
Indiana University
Bloomington IN, USA
jbusemey@indiana.edu

Suyog Chandramouli
Department of Psychological
and Brain Sciences
Indiana University
Bloomington IN, USA
suchandr@indiana.edu

Harry Collins
School of Social Sciences
Cardiff University
Cardiff, UK
CollinsHM@cardiff.ac.uk

Werner Ehm
Heidelberg Institute
for Theoretical Studies
Heidelberg, Germany
wernehm@web.de

Hinderk Emrich
Psychiatric Clinic
Hannover Medical School
Hannover, Germany
emrich.hinderk@mh-hannover.de

Vladimir Estivill-Castro
Department of Information
and Communication Technologies

University Pompeu Fabra
Barcelona, Spain
vestivill@gmail.com

Georg Feulner
Earth System Analysis
Potsdam Institute for
Climate Impact Research
Potsdam, Germany
feulner@pik-potsdam.de

Gerd Folkers
Collegium Helveticum
University and ETH Zurich
Zurich, Switzerland
folkers@collegium.ethz.ch

Martina Franzen
Wissenschaftszentrum
für Sozialforschung
Reichpietschufer 50
Berlin, Germany
martina.franzen@wzb.eu

Holger Kantz
Nonlinear Dynamics
and Time Series Analysis
Max-Planck-Institute for
Physics of Compex Systems
Dresden, Germany
kantz@pks.mpg.de

Marianne Martic-Kehl
Collegium Helveticum
University and ETH Zurich
Zurich, Switzerland
martic@collegium.ethz.ch

Felix Keller
Humanities and Social Sciences
University of St. Gallen
St. Gallen, Switzerland
felix.keller@unisg.ch

Johannes Lengler
Theoretical Computer Science
ETH Zurich
Zurich, Switzerland
johannes.lengler@inf.ethz.ch

Sabine Maasen
Munich Center for Technology in Society
Technical University
Munich, Germany
sabine.maasen@tum.de

Theodore Porter
Department of History
University of California
Los Angeles CA, USA
tporter@history.ucla.edu

Martin Reinhart
Institute for Social Sciences
Humboldt University
Berlin, Germany
martin.reinhart@hu-berlin.de

P. August Schubiger
Collegium Helveticum
University and ETH Zurich
Zurich, Switzerland
schubiger@collegium.ethz.ch

Richard Shiffrin
Department of Psychological
and Brain Sciences
Indiana University
Bloomington IN, USA
shiffrin@indiana.edu

Werner Stahel
Seminar for Statistics
ETH Zurich
Zurich, Switzerland
stahel@stat.math.ethz.ch

Angelika Steger
Theoretical Computer Science
ETH Zurich
Zurich, Switzerland
steger@inf.ethz.ch

Friedrich Steinle
Institute for Philosophy
Technical University
Berlin, Germany
friedrich.steinle@tu-berlin.de

Victoria Stodden
Graduate School of Library
and Information Sciences
University of Illinois
Urbana-Champaign IL, USA
victoria@stodden.net

Holm Tetens
Institute for Philosophy
Free University
Berlin, Germany
tetens@zedat.fu-berlin.de

Kai Vogeley
Department of Psychiatry
University Hospital
Cologne, Germany
kai.vogeley@uk-koeln.de

Zheng Wang
School of Communication
Ohio State University
Columbus OH, USA
wang.1243@osu.edu

Walther C. Zimmerli
Graduate School
Humboldt University
Berlin, Germany
walther.ch.zimmerli@hu-berlin.de

Introduction

Harald Atmanspacher and Sabine Maasen

Reproducibility has become a hot topic both within science and at the interface of science and society. Within science, reproducibility is threatened, among other things, by new tools, technologies, and big data. At the interface of science and society, the media are particularly concerned with phenomena that question good scientific practice. As bad news sell, today problems of reproducibility seem to be ranked right next to fraud. The economy, and especially the biotechnology economy, is interested in innovation based upon novel yet robust knowledge and politics in the so-called knowledge societies seek to base their decisions on best evidence, yet is regularly confronted with competing expertise.

A key step toward increasing attention to deep problems with reproducible findings in science was the paper "Why most published research findings are false" by Ioannidis (2005). One among many recent urging proclamations following it was published in *The Scientist* magazine (Grant 2012):

> The gold standard for science is reproducibility. Ideally, research results are only worthy of attention, publication, and citation if independent researchers can reproduce them using a particular study's methods and materials. But for much of the scientific literature, results are not reproducible at all. The reasons and remedies for this state of affairs was the topic of a recent panel discussion titled "Sense and Reproducibility", held at the annual meeting of the American Society for Cell Biology in San Francisco, California. ... The panel offered suggestions, such as raising journals' publication standards, establishing the use of electronic lab notebooks at research facilities, and helping laboratory supervisors provide improved supervision by reducing the size of labs.

Since about a decade voices abound – both in academia and in the media – that lament lacking reproducibility of scientific results and urgently call for better practice. Given that scientific achievements ultimately rest upon an effective division of labor, it is of paramount importance that we can trust in each other's findings. In principle, they should be reproducible – as a matter of course; however, we often simply rely on the evidence as published and proceed from there. What is more, current publication practices systematically discourage replication, for it is novelty that is associated with prestige. Consequently, the career image of scientists involved with cutting-edge research typically does not include a strong focus on the problems of reproducing previous results.

And, clearly, there are problems. In areas as diverse as social psychology (Nosek 2012), biomedical sciences (Huang and Gottardo 2013), computational sciences (Peng 2011), or environmental studies (Santer *et al.* 2011),[1] serious flaws in reproducing published results have been and keep being detected. Initiatives have been launched to counter what is regarded as dramatically undermining scientific credibility. Whether due to simple error, misrepresented data, or sheer fraud, irreproducibility corrupts both intra-academic interaction based on truth and the science–society link based on the trustworthiness of scientific evidence.

Among the initiatives introduced to improve the current state of affairs we find workshops, roundtables, and special issues addressing the topic, e.g., in *Nature* (Schooler 2011, Baker *et al.* 2012). The journal *Biostatistics* changed its policy with a focus on reproducible results in an editorial by Peng (2009). The journal *Science* devoted a special issue to the topic in December 2011, and later revised its publication guidelines concerning the issue of reproducibility (McNutt 2014). Three prominent psychology journals jointly established a "reproducibility project" recently,[2] and the journal *PLOS ONE* launched a "reproducibility initiative" in 2012. The *European Journal of Personality* published recommendations for reproducible research (Asendorpf *et al.* 2013) as the result of an expert meeting on "reducing non-replicable findings in personality research."

Funding agencies have also joined forces: e.g., the National Science Foundation of the United States created the "Sustainable Digital Data Preservation and Access Network Partners" (DataNet) program to provide an infrastructure for data-driven research in 2007. And, very recently, the National Academy of Sciences of the United States hosted an internal symposium titled "Protecting the Integrity of Science" (Alberts *et al.* 2015). An extensive report on reproducible research as an ethical issue is due to Thompson and Burnett (2012).

In sum, these examples point to an increased attention toward reproducibility as a topic *sui generis*. They testify to an increasing interest in reproducibility as a scientific ethos that needs to be upheld – even more so, as new tools and technologies, massive amounts of "big data," inter- and transdisciplinary efforts and projects, and the complexity of research questions genuinely challenge and complicate the conduction of reproducible research. Many of them call for methods, techniques (including their epistemic and ontological underpinnings),

[1]These references are a tiny subset of the existing literature on problems with reproducibility. Many more examples will be addressed in the main body of this volume.

[2]The project is a large-scale, open collaboration currently involving more than 150 scientists from around the world. The investigation is currently sampling from the 2008 issues of the *Journal of Personality and Social Psychology*, *Psychological Science*, and *Journal of Experimental Psychology: Learning, Memory, and Cognition*; see Open Science Collaboration (2012). The results have been published by the Open Science Collaboration (2015); see also `https://osf.io/ezcuj/`.

and/or best practices that are intended to improve reproducibility and safeguard against irreproducibility.

The challenges of sound reproducible research have moved into the focus of interest in an increasing number of fields. This handbook is the first comprehensive collection of articles concerning the most significant aspects of the principles and problems, the practices and prospects of achieving reproducible results in contemporary research across disciplines. The areas concerned range from natural sciences and computational sciences to life sciences and social sciences, philosophy, and science studies.

Accordingly, the handbook consists of six parts. Each of them will be introduced by separate remarks concerning the background and context of aspects and issues specific to it. These introductory remarks will also contain brief summaries of the chapters in it and highlight particularly interesting or challenging features.

Part I covers contextual background that illuminates the roots of the concept of reproducibility in the philosophy of science and of technology (Tetens, Zimmerli), and addresses pertinent historical and sociological traces of how reproducibility came to be practiced (Steinle, Collins). Part II frames the indispensable role that statistics and probability theory play in order to assess and secure reproducibility. Basic statistical concepts (Stahel), new ideas on model selection and comparison (Shiffrin and Chandramouli), the difficult methodology of meta-analysis (Ehm), and the novel area of data mining and knowledge discovery in big-data science (Estivill-Castro) are covered.

Parts III–V are devoted to three main areas of contemporary science: physical sciences, life sciences, and social sciences. Part III includes the viewpoints of computational physics (Bailey, Borwein, and Stodden), severe novel problems with reproducibility in complex systems (Atmanspacher and Demmel), the field of extreme and rare events (Kantz), and reproducibility in climate research (Feulner). Part IV moves to the life sciences, with articles on drug discovery and development (Martic-Kehl and Schubiger, Folkers and Baier), the neurobiological study of cortical networks (Lengler and Steger), cognitive neuroscience (Anderson) and social neuroscience (Vogeley).

Part V offers material from the social sciences: a critical look at the reduction of complex processes to numbers that statistics seems to render unavoidable (Porter), innovative strategies to explore question order effects in surveys and polls (Wang and Busemeyer), original views on public opinion research (Keller), issues of reproducibility as indicated in the "blogosphere" (Reinhart), and an in-depth study of notorious problems with reproducibility in scholarly communication (Franzen).

Part VI widens the perspective from reproducibility as a problem in scientific disciplines (in the narrow sense) to literature and literature studies (Bezzola Lambert) and psychopathology and psychoanalysis (Emrich). There is a clear shift in viewpoint here from the attempt to repeat experiments and reproduce their results to an analysis of why strict repetition is not only impossible but also undesirable. Another article (Atmanspacher) proposes that the way reproducibility is studied needs to be adapted to the granularity of the description of the system considered. The final contribution (Maasen) leads us back to the science–society link and the impact of extrascientific forces on research that often remains underrepresented or even disregarded.

This volume investigates the principles, problems, and practices that are connected with the concept of reproducibility, but there is a fourth "p-word" in addition: *prospects*. In some of the chapters the point is not only to understand principles, address problems, or scrutinize practices – there can also be a strongly constructive dimension in research questions that, on the surface, suffer from a lack of reproducibility. One pertinent example in this volume is the paper by Anderson, who builds on the limited reproducibility of neural correlates of mental states and proposes new ways of interpreting them coherently. Another example is the radical shift in theories of decision making proposed by Wang and Busemeyer, which furthers our understanding of order effects in sequential decisions for which no systematic and consistent modeling framework was available until recently.

While the handbook as a whole is dedicated to explore "reproducibilities" on cognitive and technical levels, its other goal is to scrutinize the notorious difficulties with producing reproducibility in a reflexive manner. Such reflexive perspectives are scattered throughout the individual contributions to this book, addressing, e.g., challenges enforced by information technology. However, parts I, V, and VI in particular inquire into philosophical, historical, social, and political contexts and their interaction with notions and practices of reproducibility. In this way, they elucidate the manifold conditions and consequences of the quest for reproducibility induced by those interactions.

From these perspectives, science is regarded as both a socio-epistemic endeavor and a highly specialized, yet integral part of society. Science and its particular modes of producing knowledge thus cannot be understood without considering them as historically evolved practices and as objects of societal expectations: Most importantly, science is expected to produce reliable knowledge. While peers within the same scientific discipline – under the considerations and within the limits outlined in the chapters of this volume – apply established means to verify their research, things become markedly problematic in interdisciplinary research with differing standards, and they may become impossible in extra-scientific contexts.

To put this in simple terms: Society cannot but trust in science and its internal procedures to produce knowledge that is both true and, hence, trustworthy. Therefore, the recently perceived lack of reproducibility is not only an intra-scientific affair but is also critically noticed by societal actors such as the mass media (e.g., Zimmer 2012). Moreover, it is addressed by science political actors (Maasen, this volume) or editors of scholarly journals (Franzen, this volume) and by concrete measures such as codes of conduct to improve scientific practice.

Finally, a few remarks concerning terminology are in order. Although we chose to use the notion of *reproducibility* to characterize the topic of this volume, there is another term that is often used interchangeably: *replicability*. This is visible in the surprisingly few volumes or reviews on the topic that can be found in the literature (e.g., Sidman 1960, Smith 1970, Krathwohl 1985, Schmidt 2009). And it also shows in the contributions to this volume – some authors prefer one, some the other notion, and (to the best of our knowledge) there is no authoritative delineation between them.

One feature, however, seems significant for "replication" and is very rarely addressed in terms of "reproduction": As the literature shows (e.g., Charron-Bost 2009), replication is mostly referred to if one focuses on how data are *copied* (rather than reproduced by repeated observations or measurements). This is particularly evident in information technology communities (distributed systems, databases), but of course also in genetic (DNA) replication.

A third related notion is *repeatability*. It is primarily adopted if an observational act (measurement) or a methodological procedure (data analysis) is repeated, disregarding the replication or the reproduction of the result of that act or procedure. An experiment may be repeated within the same laboratory or across laboratories. This difference is sometimes addressed as the difference between "internal" and "external" repetitions. Whether or not its results and conclusions from them are compatible would be a matter of reproducibility.

Whatever the most appropriate notion for "reproducibility" may be, this volume shows that it would be wrong to think that it can be universally stipulated. Depending on context, one may want to reproduce system properties characterizable by single values or distributions of such values. One may be interested in patterns to be detected in data, or one may try to reproduce models inferred from data. Or reproducibility may not relate to quantitative measures at all. All these "reproducibilities" unfold an enormous interdisciplinary tension which is inspiring and challenging at the same time.

As reflexive interactions between science and society take place, novel discourses and means of control emerge which are ultimately designed to enforce the accomplishment of truly reliable knowledge. All this happens in addition to the variety of reproducibilities as well as in view of ever-more contexts regarded as relevant and ever-changing (technical) conditions. One prospect for reproducibility seems to be clear: Challenges from within science will certainly continue to meet with those from the outside and jointly leave their traces on the ways in which science and technology produce robust knowledge. Reproducibility will remain at the heart of this process.

The nucleus of this volume has been a long-term research project on reproducibility at Collegium Helveticum, an interdisciplinary research institution jointly operated by the University of Zurich and the Swiss Federal Institute of Technology (ETH) Zurich. Our engagement with the project as editors, grounded in the exact sciences and in the area of science studies, emerged from our long lasting affiliation with the Collegium as associate fellows. We appreciate great encouragement and support by the Collegium and its fellows, in particular by Gerd Folkers and Martin Schmid.

An editorial conference at the Munich Center for Technology in Society (Technical University Munich) in fall 2014 with all authors was instrumental for the preparation of the volume in a consistent fashion, with numerous cross-references among the papers. Our thanks go to the staff of the center for organizing this event. Various colloquia, workshops, and symposia on reproducibility at Zurich and Munich have been influential all along the way.

But most of all, we want to express our gratitude to the contributors to this volume. For none of them, thinking and writing on reproducibility is their regular day job. Nevertheless, we realized so much enthusiasm about this project that any possible grain of doubt concerning its success dissolved rapidly. As all reviewers of the proposal for the project emphasized, the volume is of utmost timeliness and significance, and this spirit pervaded all our conversations and correspondences with its authors. Without their deep commitment, and without the support of Susanne Steitz-Filler and Sari Friedman at Wiley, this book would not have become reality.

References

Alberts, B., Cicerone, R.J., Fienberg, S.E., Kamb, A., McNutt, M., Nerem, R.M., Schekman, R., Shiffrin, R., Stodden, V., Suresh, S., Zuber, M.T., Kline Pope,

B., and Hall Jamieson, K. (2015): Self-correction in science at work. *Science* **348**, no. 6242, 1420–1422.

Asendorpf, J., Connor, M., de Fruyt, F., de Houwer, J., Denissen, J.J.A., Fiedler, K., Fiedler, S., Funder, D.C., Kliegl, R., Nosek, B.A., Perugini, M., Roberts, B.W., Schmitt, M., Vanaken, M.A.G., Weber, H., and Wicherts, J.M. (2013): Recommendations for increasing replicability in psychology. *European Journal of Personality* **27**, 108–119.

Baker, D., Lidster, K., Sottomayor, A., and Amor, S. (2012): Research-reporting standards fall short. *Nature* **492**, 41.

Charron-Bost, B., Pedone, F., and Schiper, A., eds. (2009): *Replication – Theory and Practice*, Springer, Berlin.

Grant, B. (2012): Science's reproducibility problem. *The Scientist*, 18 December 2012.

Huang, Y., and Gottardo, R. (2013): Comparability and reproducibility of biomedical data. *Briefings in Bioinformatics* **14**, 391–401.

Ioannidis, J. (2005): Why most published research findings are false. *PLOS Medicine* **2**(8), e124.

Krathwohl, D.R. (1985): *Social and Behavioral Science Research*, Jossey-Bass, San Francisco.

McNutt, M. (2014): Reproducibility. *Science* **343**, 229.

Nosek, B. (2012): An open, large-scale, collaborative effort to estimate the reproducibility of psychological science. *Perspectives on Psychological Science* **7**, 657–660.

Open Science Collaboration (2012): An open, large-scale, collaborative effort to estimate the reproducibility of psychological science. *Perspectives on Psychological Science* **7**, 657–660.

Open Science Collaboration (2015): Estimating the reproducibility of psychological science. *Science* **349**, No. 6251, aac4716/1–8.

Peng, R. (2009): Reproducible research and biostatistics. *Biostatistics* **10**, 405–408.

Peng, R. (2011): Reproducible research in computational science. *Science* **334**, 1226–1227.

Santer, B.D., Wigley, T.M.L., and Taylor, K.E. (2011): The reproducibility of observational estimates of surface and atmospheric temperature change. *Science* **334**, 1232–1233.

Schmidt, S. (2009): Shall we really do it again? The powerful concept of replication is neglected in the social sciences. *Review of General Psychology* **13**(2), 90–100.

Schooler, J. (2011): Unpublished results hide the decline effect. *Nature* **470**, 437.

Sidman, M. (1960): *Scientific Research*, Basic Books, New York.

Smith, N.G. jr. (1970): Replication research: A neglected aspect of psychological research. *American Psychologist* **25**, 970–975.

Thompson, P.A., and Burnett, A. (2012): Reproducible research. *CORE Issues in Professional and Research Ethics* **1**, 6.

Zimmer, C. (2012): A sharp rise in retractions prompts calls for reform. *New York Times*, April 17, D1.

PART I
CONTEXTUAL BACKGROUNDS

PART I: CONTEXTUAL BACKGROUNDS

Introductory Remarks

Harald Atmanspacher

The reproducibility of results is considered one of the basic methodological pillars of "good science." Why? As always, normative rules and other firm beliefs of this kind hardly derive from scientific research *per se* (although they may be informed by it) – they are typically motivated by or founded on reasons external to science. In this sense, the concept of reproducibility is a paradigm case for historical and philosophical analyses, which are the framework of the contributions to this part of the volume.

One of these extra-scientific reasons refers to a very basic *ontological* commitment shared by many (if not most) scientists – that there are stable fundamental structures of our universe. In contrast to sense data or introspective data, these ontic structures are assumed to be universal rather than particular. Their stability underlies the assumption that there are laws of nature (such as Maxwell's equations) and universal constants (such as the speed of light) which are valid irrespective of where, when, and by whom they are investigated. If a law of nature is assumed to hold, its empirical predictions must hold with the same sense of stability. And this means that a reproduction of an experiment under the same conditions must lead to the same result.[1]

In the contribution by *Holm Tetens at the Institute of Philosophy, Free University Berlin*, the ontological dimension of science is expressed by its claim to unveil *objective* truths. But such truths themselves are not directly accessible. Instead, if empirical results are reproducible across laboratories, *intersubjective* agreement can be generated about those results so that they become part of the accepted body of scientific knowledge. As a supplement to this intersubjective dimension, Tetens indicates the technological aspect that every experiment is a potential prototype of an artificial device that can be controlled and manipulated

[1] In fact, this sounds much more innocent than it is. For instance, the notion of a result can refer to values of observables, to their distributions, to patterns of behavior, or even to entirely qualitative issues. This will become clearer in the more specialized parts of this volume. Moreover, only *relevant* conditions must be the preserved for a valid reproduction of a result. If time is relevant (e.g., in memory, learning, or other path-dependent processes), then strict reproducibility is impossible since the initial conditions for any follow-up study depend on the results of the preceding study.

for particular purposes. The history of philosophy presents this as a special truth criterion as well: Giambattista Vico's *verum-factum* principle says that "the true and the made are interchangeable" (see, e.g., Miner 1998).

A notion that expresses the idea of stability in mathematical terms is *invariance*. A property is invariant under a transformation if it does not change when the transformation is applied. The so-called conservation laws in physics are well-known examples, such as the conservation of momentum or energy. Closely related to invariance principles are *symmetry* principles: In the basic sciences, symmetries are major driving forces for theoretical progress (see, e.g., the classic by Weyl 1952). Of course, this is much less prominent as we move to more complex structures, or to living systems, where most if not all symmetries that govern physics are broken. And as a consequence of broken symmetries, the significance of universal laws and constants declines naturally.

Walther Ch. Zimmerli writes as a philosopher of mind and of technology at the Graduate School of Humboldt University Berlin and amplifies the theme of reproduction in its relation to technology. He advocates that technology is more than just the result of applied science, a position strongly defended by Bunge (1966). Zimmerli's approach comes closer to more recent tenets of "science and technology studies," which emerged in the 1980s and focus on how social, political, economic, and cultural values affect scientific research and technological innovation. See Mitcham (1994) for an overview of the wide variety of perspectives in this field.

Zimmerli offers a guided tour through a number of topics related to reproducibility from various angles: production, continuity, rationality, prediction, and prognosis. His starting point is a critique of the myths of a strict distinction between sciences and humanities, of the idea that knowledge is a purely mental product, and of the assumption of something like a unity of the sciences. One of his main conclusions is that the reproducibility of results in science and the (re-)production of products in modern knowledge societies will never be free from ignorance and uncertainty.

Friedrich Steinle, historian of science at the Technical University Berlin, gives an in-depth introduction into the history of the concept of reproducibility in the empirical sciences, with a strong emphasis on experimentation in physics. The first time the criterion of reproducibility was made systematic use of was provably due to the Arab scholar Alhazen in the early 11th century. Galileo Galilei, Francesco Redi, Felice Fontana in Italy, Robert Boyle in England, Claude Bernard in France, and Otto von Guericke in Germany reported studies highly relevant for reproducible research in the Early Modern period. Many of Steinle's cases, however, refer to the 19th and 20th centuries, from Ørsted, Faraday and

Coulomb to gravitational wave detection (Weber), high-temperature superconductivity (Bednorz and Müller) and "cold fusion" (Pons and Fleischmann).

The case studies in this chapter show that the reproducibility of experimental results has many faces indeed. Sometimes results get accepted without any replication; sometimes they get only accepted after many detailed replications (sometimes even without any theoretical underpinning); sometimes attempts at replication discredit results.[2] This indicates that replication alone has never been the only key to accepted knowledge. Various kinds of contexts play major roles, such as mainstream trends, economic factors, skills and status of experimenters. All these contexts together with the norm of reproducibility provide the ground from which the stability of an experimental result ultimately emerges.

This theme is highlighted in greater detail by *Harry Collins, sociologist of science at Cardiff University.* Collins outlines a concept that he introduced into the debate about reproducibility as *experimenters' regress* (Collins 1985): To know that an experiment is well conducted one needs to know that its outcome is correct, and to know that the outcome is correct, the experiment must be well conducted. This vicious circle clearly implies problems if experiments are intended to resolve scientific controversies. Collins argues and illustrates that, typically, such a resolution cannot be achieved by "sound science" alone but depends crucially on social processes.

Another focus of his chapter is what he calls a "statistical uncertainty principle," one of the many instances of the experimenters' regress. Based on the example of gravitational wave detection, Collins discusses the problem of properly assessing *statistical* significance, and balancing it with *substantive* significance: How do we decide what a result obtained by subtle experimentation and involved methods of data processing actually *means* after all? The numerous ways in which data analysis and noise reduction are continuously refined in large research projects may easily affect one another and obfuscate the results obtained. Collins suggests a view on how to understand this, which exploits analogies with the uncertainties of quantum theory[3] – quite speculative but certainly stimulating.

The papers in part I of this volume provide background material to the more specialized parts to follow. To keep things as simple as possible, the examples and illustrations they use are mainly taken from the so-called exact sciences –

---

[2]See also Franklin and Perovic (2015) and references therein for more details about the many faces of experiments in physics.

[3]Applying the mathematical framework of quantum theory outside physics is a new field of research that has proven to produce interesting novel insights, e.g., in cognitive science (for an introdution see Wang *et al.* 2013).

and yet one realizes pretty quickly how difficult it is "to keep things simple" even in a restricted physics-informed framework. The later parts of the volume will make it clear that the increasing complexity we meet in the life sciences and social sciences entails additional sophistications not apparent in physics. However, the chapters in part I make it clear that even in physics and its close derivatives, extra-scientific background often plays a role that cannot and must not be neglected.

References

Bunge, M. (1966): Technology as applied science. *Technology and Culture* **7**, 329–347.

Collins, H.M. (1985): *Changing Order: Replication and Induction in Scientific Practice*, Sage, London.

Franklin, A., and Perovic, S. (2015): Experiment in physics. In *Stanford Encyclopedia of Philosophy*, ed. by E.N. Zalta. Available at `http://plato.stanford.edu/entries/physics-experiment/`.

Miner, R.C. (1998): Verum-factum and practical wisdom in the early writings of Giambattista Vico. *Journal of the History of Ideas* **59**(1), 53–73.

Mitcham, C. (1994): *Thinking Through Technology: The Path between Engineering and Philosophy*, University of Chicago Press, Chicago.

Wang, Z., Busemeyer, J., Atmanspacher, H., and Pothos, E. (2013): The potential of quantum theory to build models of cognition. *Topics in Cognitive Science* **5**, 672–688.

Weyl, H. (1952): *Symmetry*, Princeton University Press, Princeton.

1
Reproducibility, Objectivity, Invariance

Holm Tetens

Abstract. The independent reproducibility of experiments and their results, for instance, the determination of fundamental constants of nature, belongs to the most important conditions for the possibility of nomological empirical science. Two aspects are particularly significant: (1) laws of nature could not be empirically tested and confirmed if experiments and measurements were not reproducible, and (2) the independent reproducibility of experiments and measurements guarantees the objectivity of scientific results that otherwise would amount to mere subjective beliefs of individual researchers, neither comprehensible nor verifiable by others.
The nomological character and the intersubjectivity of propositions in science share a common basic root, namely, the technical applicability of research – and thereby control over nature as one of the most significant goals of modern occidental science. This goal hinges on the requirement of reproducibility. It will be discussed which structural features of our theories, e.g., conservation laws, actually are consequences of the demand that research results be reproducible.

1.1 Introduction

Results of scientific inverstigations must be reproducible. This is what scientists expect from their findings. It is also what laypeople expect, however vaguely. How can this requirement and expectation be justified?

To demand that scientific results should be reproducible is not to demand something minor. The reproducibility of scientific results is at the core of every scientific inquiry. Science lays claim to truth.[1] Science cannot give up this claim to truth without giving up itself. However, the concept of truth is to be understood in science or in the individual sciences, their claim to truth would be amiss if every researcher in one and the same scientific discipline would take something different to be true. Instead of the objective truth of scientific propositions, we would merely have subjective opinions of individual scientists. Without sufficient intersubjective consensus among the representatives of a scientific discipline, it would be impossible to uphold a claim to objective truth in science.

What is a sufficient intersubjective consensus in the context of philosophy of science? Science distinguishes itself by trying to justify its claims about what

[1]For the following considerations compare Tetens (2013), especially pp. 17–28.

Reproducibility: Principles, Problems, Practices, and Prospects, First Edition. Edited by Harald Atmanspacher and Sabine Maasen.

is the case in the world. The methods and procedures of justification can differ substantially across individual scientific disciplines. This plurality of methods of justification is grounded in the fact that the different sciences deal with vastly different segments of reality (cf. Tetens 2013, pp. 34–38). For the sciences, in a broad sense of the term, also include the social sciences and even many of the humanities.

But all methods of justification share the feature that different scientists or scientific teams can apply a method of justification to the pertinent matters of their discipline over and over again. And here, only one principle is constitutive: *If scientists or scientific teams independently and correctly apply the same method of justification to the same subject matter, then they must reach the same conclusions about this subject matter.* This is what the intersubjective reproducibility of scientific results means in general. In order to hold its results to be true, a scientific discipline must fulfil this principle categorically.

Aren't scientists often in disagreement? Isn't the principle of the intersubjective reproducibility of scientific results often unfulfilled? This is certainly the case, in some scientific disciplines to a greater, in others to a smaller extent. But as long as scientists are in disagreement, as long as scientific propositions are disputed because researchers cannot comprehend their justifications or because they achieve different results, one cannot reasonably raise a claim to scientific truth. Scientists will then try to make their propositions more precise, to modify and improve their methods, and to scrupulously control the correct application of these methods to reach an unanimously accepted result.

Where this does not work out, they will at least have to agree that they disagree, where they disagree and why they disagree – especially whether the disagreement has a foundation in the subject matter itself and in the difficulties of approaching it with the established methods. But this shows: The principle of the intersubjective reproducibility of scientific results is generally accepted, and the lack of its fulfilment must remain an exception. Even the exceptions (e.g., anomalies, see Atmanspacher 2009), always annoying and problematic for science, are dealt with in a way that itself tries to live up to the principle of reproducibility.

1.2 Reproducibility in the Empirical Sciences

As mentioned above, the principle of the intersubjective reproducibility of scientific results must certainly be spelled out differently for the different sciences and their different methods of justification. And it should turn out that the principle can be redeemed more or less strictly in the different sciences. It looks very different in mathematics compared with molecular genetics, and again very different compared with history. But however limited the intersubjective reproducibility of the results of a particular science may be, no discipline can wholly renounce the reproducibility of its results without seeing its claim to objective truth dissolve and, as a consequence, disappearing as a science.

Since there is not much more to be said *in general* about the principle of intersubjective reproducibility, let us turn to a special class of sciences that, however, count among the most important ones: the empirical laboratory sciences. Here, the demand of the reproducibility of results takes a special and an especially important form. First, natural processes are experimented with, and, second, characteristics of natural processes are measured in experiments. I assume that readers are sufficiently familiar with the fundamental structural features of experiments and measurements, and will therefore not comment on those here.[2]

But there is one question concerning the empirical sciences which I must briefly touch upon. Why has the measuring experiment become such an important method in the sciences? There is a general answer to this question which can be well supported by the history of science (see Tetens 2013, pp. 28–34). What happens in a measuring experiment is structurally the same as in our efforts to technically control natural processes and thereby to redesign nature with regard to human purposes. If we have experimentally investigated a natural process successfully, then we know how we can technically create a corresponding process with the aid of artifical devices. Every successful scientific experiment provides a prototype of such a device, with which we can change and manipulate the process concerned with respect to certain parameters.

We can only act directedly and purposefully if we can successfully plan our actions in advance. To do so, we must know in advance under which circumstances we generate which effects with which action. Such knowledge is produced, among other things, in scientific experiments. Every kind of experiment has the goal to make sure that, under the same boundary and initial conditions, executing the same action leads to the same results. In other words: The principle of the reproducibility of scientific results is indispensable from the point of view of our technical control over nature. Insofar as we want to learn how certain natural processes can be technically controlled, we have to investigate them experimentally in the laboratory (see also Zimmerli, this volume).

Tailoring the principle of the intersubjective reproducibility of scientific results to the empirical laboratory sciences means that scientists must be able to independently reproduce measurements and their results in the context of exper-

[2] See Tetens (1987) and Tetens (2006) for more extensive and detailed discussion.

iments. We can, therefore, formulate the general principle of the intersubjective reproducibility of scientific results as follows: *Under the same boundary and initial conditions, an experiment must always proceed in the same way with respect to the quantitative parameters measured within its measurement accuracy.*

The principle of the intersubjective reproducibility of scientific results only gives a necessary condition for empirical laboratory science. If it were not satisfied, then laboratory science would be impossible with objective results, so it would be impossible at all. Therefore, we know *a priori* that the principle of reproducibility is mandatory for any successful empirical laboratory science.

1.3 Objectivity

The principle of the intersubjective reproducibility of scientific results has a truth-theoretic aspect that aims at the *objectivity* of scientific results.[3] This truth-theoretic aspect is supplemented by the technological aspect mentioned above. This supplementation means that both aspects can be distinguished, but do not exclude each other. Instead, both are significant features in empirical laboratory science.[4] This will be the topic of this section.

What follows from the fulfilment of the principle of reproducibility, i.e., if experiments and their results can be successfully reproduced? Wherever this principle can be redeemed for a specific experiment or a class of experiments, this experimental practice corresponds to true propositions of the form: Whenever boundary and initial conditions of the kind U fix certain results of a measurement by particular values at an initial time and location, then after some spatial and/or temporal shift, one measures particular (other) values as the results of a measurement. Such general propositions can be called experimental-scientific laws of nature. Now we see that the reproducibility of experiments in science and empirically adequate experimental laws of nature are merely two different sides of one and the same scientific effort.[5]

At this point, we have to turn to the fact that experiments do not always succeed in reproducing certain values for measurement results. This holds es-

[3]For a comprehensive account of the concept of scientific objectivity, how it should be defined, whether it is desirable, and to what extent scientists can achieve it, see Reiss and Sprenger (2014).

[4]Compare Giambattista Vico's famous statement that the true and the made are convertible: *verum et factum convertuntur.* See Miner (1998) for more discussion.

[5]A proposition L which involves an experimental law of nature is empirically adequate if and only if, when $B_1, ..., B_n$ are true observation statements that entail, together with L, the observation statement B, then B is also true.

pecially for a large class of experiments in quantum mechanics, where one can only reproduce the relative frequencies of different measurements (though with impressive reliability). If not even relative frequencies were reproducible, this would be the end of physics as we know it. Furthermore, there are three things to note in the context of quantum mechanics.

1. If physicists could choose between experiments that reproduce fixed measurements and those that only reproduce relative frequencies of measurements reliably, they would choose experiments of the first kind. The reason is obvious given the goal of controlling nature.
2. Quantum mechanics also fulfils the general principle of the intersubjective reproducibility of scientific results, yet in a slightly modified form: *Under the same boundary and initial conditions, the probabilities of the different measurement results are always the same. The probabilities of the measurement results are reproducible.*
3. The principle of the reproducibility of quantum experiments and their results depends on the strict reproducibility of the boundary and initial conditions under which they are repeatedly executed in order to gradually stabilize the relative frequencies.[6]

1.4 Invariance and Symmetry

After this excursion into quantum mechanics, let us return to more general considerations. Every experiment is defined by certain boundary and initial conditions that experimenters can reproduce in principle. Certain realizations of a measurement $m_1, ..., m_n$ yield certain reproducible values depending on each other, on time t, and on location x, y, and z.[7] If the course of an experiment can be reproduced under the boundary and initial conditions U_i, then it can in principle be described by an equation of the form $f_i(m_1, ..., m_n, x, y, z, t) = 0$, where f characterizes the state of the system.

[6]This is an aspect which especially the proponents of the so-called Copenhagen interpretation of quantum mechanics have been stressing. Quantum mechanics relies on experiments in which certain values of *classical* measurements are registered using devices that have been and must have been reliably built and reproduced in the framework of classical physics.

[7]In quantum mechanics, the probabilities of the values of measurement results must be the same. This entails that in long series of tests, the relative frequencies of the different measurements deviate the less from the probabilities, the longer one continues the series of tests. The distribution converges in the limit of infinitely many tests (see also Stahel, this volume). In what follows, we shall leave the peculiarities of quantum mechanics aside.

A further aim in science is to subsume as many different experiments as possible by deducing the respective functions $f_i(m_1, ..., m_n, x, y, z, t) = 0$, the state evolution, from the same differential equation or system of differential equations. The different boundary and initial conditions U_i classifying the experiments provide the parameters for determining the special function describing the state evolution from the general family of solutions of the differential equations. In this way, reproducible experiments correspond to (systems of) differential equations, the solutions of which quantitatively describe the courses and the results of the experiments.[8]

Is the reproducibility of experiments reflected in corresponding laws which, as just elaborated, are to be expressed as differential equations? Well, regardless of when and where one experiments, whenever the relevant boundary and initial conditions are created, an experiment always proceeds in the same way. This is what it means that experiments can be reproduced. But this also means that the differential equations, the solutions of which describe the processes of the reproducible experiments, do not change under temporal translations, movements of the spatial coordinate system or rotations of its axes. Technically speaking: The differential equations must be invariant with respect to temporal and spatial translations and spatial rotations.

At this point, a much more general connection between the reproducibility of scientific results and their objectivity emerges. For the reproduction of an experiment, it must not matter where and when it is performed. One can also express this as follows: The result of an experiment must not depend on where and when the experimenters are situated. More generally: The scientific results must not differ arbitrarily from experimenter to experimenter, from observer to observer, from scientist to scientist. Otherwise, the results would lose their objective liability. Put the other way around: *The less scientific results depend on and are influenced by differences between researchers, the more objective are they and the better can they be reproduced intersubjectively.*

In the empirical sciences, most distinctly in physics, this is reflected by *invariance principles.* Most basic among them is the invariance of physical laws under temporal translations and spatial translations and rotations. But physics does not stop here. It is not only the time and location of observers that ought not influence the results but also the state of motion of the observers. This independence from the state of motion leads to the different versions of the

[8] It would take us too far afield to elaborate more closely on the overwhelming success in describing different reproducible experiments and their processes uniformly by deducing the process functions from one and the same differential equation or system of differential equations. For the case of classical mechanics, this is elaborated more extensively in Tetens (1987).

principle of relativity, which plays a crucial role in classical mechanics and in the special and general theory of relativity.[9]

Another example is the cosmological principle in astrophysics, according to which the universe appears the same from every reference point. And there are many other invariances and symmetries without which physics would not be the same. From a mathematical point of view, they are all connected to corresponding conservation laws. The close connection between invariances, conservation laws, and symmetries can be formulated mathematically: A relation R, defined on a set S of objects, is an equivalence relation (cf. Atmanspacher, this volume) if there is a mathematical group G of transformations $T : S \rightarrow S$ from the set S onto itself such that two elements $x, y \in S$ are mapped onto one another by T. If a property stays the same (is invariant) under certain changes (transformations), this is called a *symmetry*.[10]

We have seen that some important symmetries or invariance principles can be derived immediately from the principle of the reproducibility of experimental results. Other symmetries are at least indirectly connected with the fundamental requirement of reproducibility. Generally speaking, symmetries can be identified by equivalence relations R, which leave a property unchanged under the transformation of some variable.

If certain symmetries hold for processes in nature, then certain changes do not affect these processes and do not have to be considered, controlled and eliminated in the technical–empirical generation of these processes. The less circumstances must be considered, controlled, or, if necessary, eliminated in order to technically generate a process, the easier it can be reproduced technically–empirically.

1.5 Summary

A natural science aimed at objectivity one the one hand and at technical applications on the other is an empirical laboratory science.

If it is successful, then, for methodological reasons, it necessarily fulfils the fundamental principle of the reproducibility of its results: *Under the same boundary and initital conditions, an experiment must always proceed in the same way*

[9] In the special theory of relativity, this is the principle: If the laws of relativistic mechanics and Maxwell's equations in electrodynamics apply in a reference frame I, and I' is a reference frame which is in rectilinear and uniform motion with respect to I, then the same laws of relativistic mechanics and Maxwell's equations in electrodynamics apply in I'.

[10] For some more background about symmetries and invariance principles in physics compare Scheibe (2001, Part VII), Wigner (1979, Secs. 1–5), or Weyl (1952).

with respect to the quantitative parameters measured within its measurement accuracy.

Fulfilling this principle, the experimental investigation of nature in the laboratory will in the long run produce theories in which natural processes are described by differential equations that satisfy symmetry and invariance principles.

Some of these invariance principles can be directly deduced from the fundamental demand of the reproducibility of scientific experiments, while others support the reproducibility of experiments at least indirectly.

References

Atmanspacher, H. (2009): Scientific research between orthodoxy and anomaly. *Journal of Scientific Exploration* **23**, 273–298.

Miner, R.C. (1998): Verum-factum and practical wisdom in the early writings of Giambattista Vico. *Journal of the History of Ideas* **59**(1), 53–73.

Reiss, J., and Sprenger, J. (2014): Scientific objectivity. In *Stanford Encyclopedia of Philosophy*, ed. by E.N. Zalta. Available at plato.stanford.edu/entries/scientific-objectivity/.

Scheibe, E. (2001): *Between Rationalism and Empiricism*, ed. by B. Falkenburg, Springer, Berlin.

Tetens, H. (1987): *Experimentelle Erfahrung. Eine wissenschaftstheoretische Studie über die Rolle des Experiments in der Begriffs- und Theoriebildung der Physik*, Meiner, Hamburg.

Tetens, H. (2006): Das Labor als Grenze der exakten Naturforschung. *Philosophia Naturalis* **43**(1), 31–48.

Tetens, H. (2013): *Wissenschaftstheorie. Eine Einführung*, Beck, München.

Weyl, H. (1952): *Symmetry*, Princeton University Press, Princeton.

Wigner, E. (1979): *Symmetries and Reflections*, Ox Bow Press, Woodbridge.

2
Reproducibility between Production and Prognosis

Walther Ch. Zimmerli

Abstract. It is a prejudice widely held that technology is to be understood as an application of science, the latter being some kind of Platonic result of pure theoretical intuition. The heuristic starting point of this chapter is a different hypothesis: that a modern scientific criterion like reproducibility – a crucial element of a highly sophisticated system of abstractions with respect to technology – is better understood if seen against and in the light of this background.
The chapter attempts to demonstrate this by scrutinizing the relations between reproducibility, production, continuity, scientific rationality, prognosis, and prediction, thus reexamining and reevaluating some of the most prominent topics of traditional philosophy of both science and technology.

2.1 Preliminary Remarks: Three Myths

In a post-enlightenment culture like our European civilization it is both a vice and a virtue to take nothing for granted – since Francis Bacon's (1857) theory of idols, criticism of prejudices has become a fundamental characteristics of occidental reasoning. But since Gadamer's (1989) rehabilitation of prejudice we know that we always have to be aware of the fact that criticism of prejudices tends to become self-referential: One of the most detrimental prejudices is the prejudice against prejudices. Therefore, it is probably safe to adopt a somewhat hypothetical critical attitude in identifying those prejudices (or myths) that have already turned out to be obstacles for an understanding of the problem in question: in our case the problem of reproducibility.

In order to critically discuss this problem it is, therefore, necessary (although not sufficient) to first identify the most powerful prejudices to either confirm or reject them. Some assumptions have become so strong that to most of us they do not even have the character of hypotheses any more but seem to be self-evident. Seemingly self-evident assumptions, however, could develop into veritable myths. The myths I would like to critically discuss to begin with are the *myth of the two cultures, the myth of knowledge as pure mental product,* and *the myth of the unified science.*

Reproducibility: Principles, Problems, Practices, and Prospects, First Edition. Edited by Harald Atmanspacher and Sabine Maasen.

Trying to avoid the pitfalls of these three myths, this chapter will critically reflect on their different aspects by discussing five approaches closely related to each other: First of all, the relation between reproducibility and production will be examined, and from there we will proceed to the relation between production and continuity. This will lead us to the core question as to the nature of scientific rationality, and from there to the nature of scientific prognosis. Finally we close the circle with the relation between scientific prognosis and reproducibility.

2.1.1 The Myth of the "Two Cultures"

Drawing on a long modern history originating in Cartesian dualism, Hegel's dichotomy of the philosophy of nature and the philosophy of spirit as well as its Neo-Kantian version of the methodological distinction between explaining and understanding, Snow (1959) coined the formula of the "two cultures." Its bottom line is the assumption that the occidental academic system is based on two strictly opposed or even hostile cultures: the natural sciences and the humanities.

It does not even need a second glance, however, to see that there are more than just two academic or scientific cultures: Lepenies (1985) already wrote about "the three cultures" – and in addition to the natural sciences and the humanities, the social sciences and the engineering sciences have been established as increasingly powerful scientific cultures since the second half of the 19th century (Zimmerli 1997, 2006).

2.1.2 The Myth of Knowledge as a Purely Mental Product

Closely related to the myth of the two cultures is the not less influential myth of knowledge as a purely mental product. In order to compensate for the weaknesses of the seemingly intuitive inductive verificationism of the neo-positivist Vienna Circle (Kraft 1997), regarding theories as generalized statements about observations, Popper (1959) claimed the opposite position of deductive falsificationism. Theories are mentally produced generalized propositions that cannot possibly be verified, but only falsified by other propositions about observations. The consequences of this stunning logical insight into the asymmetry of verification and falsification resulted, however, in a detrimental Platonism according to which the main drivers of scientific dynamics are primarily mental products (ideas) and the attempts to falsify them.

This assumption has developed into a very powerful myth which, I think, is doomed to collapse as soon as one realizes that knowledge, especially scientific knowledge, originates from technico-pragmatic behavior in and against the world.

2.1.3 The Myth of a Unified Science

Underlying these two assumptions a third one has developed into a dominant myth as well – the myth of a unified science, more exactly: of the unity of the sciences. This idea is considered to bridge the – although erroneously – diagnosed abyss between the two cultures by introducing a unifying instance. This could be the "language of physics as universal language of science" (Carnap 1931), the "anthropic principle" (Barrow and Tipler 1988) or the concept of "self-organization" (Nicolis and Prigogine 1977).

The main motivation behind these ideas is an – again Platonic – attempt to reduce plurality and diversity to unity, something like an "Esperanto mistake": The attempt to overcome the plurality and diversity of natural languages by introducing a unifying artificial language was doomed to failure because it neglected the fact that plurality and diversity are the core characteristics of natural languages. The only way to unlock this potential is to learn foreign languages. And pretty much the same applies to the many different scientific cultures and their languages. You have to try to learn to at least passively participate in as many languages as possible that are relevant for the topic in question. This is a key part of the operational definition of *transdisciplinarity*.

The topic in question here is the notion of "reproducibility." It is most feasible for a transdisciplinary critical approach because it does itself cover the full range from a characteristic requirement of (natural) sciences to a universal criterion for scientific activity in general. The reproducibility of a crucial experiment is considered to be the decisive difference between the corroboration or even proof of a hypothesis and some arbitrary "one-time effect."

Furthermore, reproducibility has become an earmark or even a "shibboleth" of science proper in discussions between the different scientific cultures. It is probably not too far-fetched to suspect that reproducibility is in itself running the risk of being hypostatized from a descriptive quality into a prescriptive criterion for any scientific endeavour whatsoever – in short, into a myth. This suspicion is supported by the fact that reproducibility has even become part of the discourse in the humanities.[1]

[1]See Benjamin (1968). It would be a tempting task to submit Benjamin's essay to an extensive critical analysis to find out whether his position is falling victim to the mythical aura of the concept of reproducibility. For other approaches to reproducibility from the perspective of the humanities see the contributions in the later parts of this volume.

2.2 How Does Reproducibility Connect with Production?

2.2.1 Knowledge Production

Theoretical knowledge – as is well known since ancient Greek philosophy – is characterized by contemplation. The Greek notions *eidenai* and *theorein* both mean as much as "having personally seen," *theoros* being the observing eye witness. This background in combination with the Platonic idea of the immediate intellectual view of the ideal forms could result in the misleading assumption that knowledge was a purely spiritual or mental product.

There is, however, plenty of evidence even in Plato's own dialogues that the role model for the production of knowledge is rather handicraft than contemplation. The divine "demiurgos" is modeled after a craftsman, and the Platonic Socrates distinguishes craftsmen, in contrast to the so-called experts, as those who know what they are doing. Yet there has been so much of a smoke screen of idealistic neo-Platonic interpretation, especially in medieval times, that it took centuries until this idea was explicitly reformulated at the beginning of modernity.

It was the arch-rationalist René Descartes who insisted that the knowledge provided by the – at that time just emerging – new modern science was not some kind of contemplative theory, but an endeavor to produce knowledge that is useful for the people (Descartes 2001). And even before that the arch-empiricist Francis Bacon is said to have formulated his famous line "knowledge is power."[2]

In combination with another famous quotation from another famous author, Giambattista Vico, published more than a century later, this line makes even more sense. Vico's axiom *verum ipsum factum* (Vico 1979) can now be interpreted in a pragmatist way: Knowledge is not just theoretical contemplation, but on the one hand, it is produced, i.e., fabricated, on the other hand, the – technical and practical – producibility is its criterion, its *signum veritatis*. In brief: The background of the quest for reproducibility in science is technical or practical producibility (see also Tetens, this volume).

[2] For quite a time it has been doubted that Bacon was the source of this line because it could nnot be found in his major writings. And indeed, Bacon expressed it a little different, and not in his main works, and not in English. What he explicitly said in his *Meditationes Sacrae* of 1597 was that not just knowledge but science itself is power: *ipsa scientia potestas est* (Bacon 1864). Everybody knowledgeable about Bacon will be aware that talking about science for him means equivalently talking about technology.

2.2.2 Manufacture and Industrial Production

At the dawn of industrialization in the first half of the 19th century, this insight became even more sophisticated. Karl Marx demonstrated in numerous excerpts and finally in the famous Chapter 13 of the first volume of the "Kapital" (Marx 1867) the connection between the development of knowledge, industrial machine technology, economy, and society.

Like many deep insights this one is stunningly simple as well: Following the writings of Babbage (1986) and Ure (1835), Marx analyzed the transition from craftsmanship and manufacture into industrial production of goods by machines. In this context he distinguished the three necessary elements of each machine: the motive part, the transmission, and the tool machine. As soon as the motive part became independent of its location, i.e., as soon as the steam engine was invented (or rather reinvented) by Watt and Newcomen, the whole "big machinery" (in Marx's earlier terminology, the "Atelier") also became independent of its location. Neither water nor wind was required any more to run the mills or the production plants. This task was now fulfilled by steam engines and later by the electricity generated by them.

2.2.3 Repetition and Mass Production

But there is much more included: The big machinery renders repetitive processes and thus the mass production of identical products possible. This, however, has implications for both the process of fabrication and the organization of labor. From the simple analysis of the necessary elements in each machine it is not very far to division of labor, Taylorism, and the modern design of production plants. The decisive next step in this development is the introduction of "intelligent" computer-aided technology, from the early computer-numerical-control machines to modern robots (Zimmerli 2014).

At the beginning of the industrial revolution, however, not much of this was happening yet, except for an additional effect with considerable impact on industrial production: By the dissection of all processes into small identical individual activities at the tool machine, the tendency of deskilling got ever more dominant. Craftsmanship and skills became less demanded, and the wages of the unskilled assembly-line workers decreased as they became but parts of a machine.

The driving force behind this whole process is – and here Karl Marx was quite right – the economy, in this case the economy of scale. To produce more identical products in the same time span at barely higher costs results in producing less expensive goods, and this again increases the profit of the entrepreneurs.

Within this historical context, the conjecture becomes very plausible that the requirement of reproducibility in science reflects features of industrial production. If this is the case, then it raises the suspicion that we are caught in an outdated and misleading normative requirement of scientific activity in the same way as we still stick to an outdated and misleading idea of industrial production.

2.3 How Does Production Connect with Continuity?

2.3.1 "Natura Non Facit Saltus"

Following this heuristic hypothesis for one more step, the inverse question arises: If the conditions for the possibility to apply the concept of reproducibility originate from production technology, what does this imply for relations within science? Usually we take for granted that nature as we usually refer to it is continuous – nature does not make jumps, or *natura non facit saltus*. Although this phrase is familiar since ancient times, the first explicit source is much more recent.[3]

And the farther our understanding of nature differs from a purely physicalist conception toward a more life-science-oriented perspective, the more surprising is it that we – despite all contradicting evidence – still seem to be convinced that nature makes no jumps but behaves continuously. "L'homme machine" of the French materialists (de la Mettrie 1912) is just one aspect of "la nature machine". Although today we know that nature in fact "jumps" a lot and is discontinuous in many respects, our epistemological picture is still grossly dominated by the technomorphous idea of continuity.

2.3.2 Increasing Knowledge by Ignorance

A proof for this claim can be given *ex negativo*: The continuity hypothesis requires that, under the same conditions, from the same premise p always the same conclusion q follows. In logical notation this reads $(x)\ px \rightarrow qx$, and this is the logical form of a lawlike proposition. But now it seems that "in nature," cases that do not satisfy this logical rule are much more common than those that do. And as we – true believers in (scientific) truth – are used to accept as true (scientific) knowledge only what satisfies the requirement of universal lawlikeness,

[3]The idea of continuity has been present, in different versions, throughout the history of the philosophy of nature from Aristotle via Leibniz and Newton to Darwin. Its literal phrasing *natura non facit saltus* is due to Linné (1751, Chap. III, par. 77, 27).

these other cases are typically categorized as cases of lacking knowledge or even of ignorance.

Ignorance, however, is one of the main drivers of knowledge, and even more so if considered from the perspective of continuity. Probability calculus and, consequently, statistics is an example for the transformation of lacking knowledge into a different kind of – statistical – knowledge. If, like in quantum mechanics, it is impossible to simultaneously determine location and momentum of a particle, we consider the whole population of particles and their statistical distribution.

We do something very similar if we do not know how an individual will behave in a ballot or an election. The statistical distribution of a population of individuals is considered likely based on previous experiences under particular conditions. And we go even one step further, presupposing that the statistical distribution is continuous (in the limit of infinitely many individuals) and can be related to probabilities for the behavior of individuals. That this often induces misunderstandings becomes especially obvious in cases like the statistical success rate of clinical surgery or various forms of disease risk.

Thus, our technology-induced metaphysical presupposition of continuity seems to be applicable even where we scientifically know it is not valid. This could be called "statistical continuity as a tool for coping with ignorance" – in logical notation $(\%x)\ px \to qx$.

2.3.3 Deficiency and Innovation

Another equally venerable source of wisdom teaches us that there is nothing new under the sun (preacher Salomo). Taken literally, this would be an obvious logical consequence of a concept of nature as a machine operating continuously. But again, there is plenty of empirical evidence opposing this assumption. Today we live in a period of civilization characterized by a radically positive prejudice in favor of novelty, manifesting itself in an encompassing, almost religious belief in the power of innovation (Gimmler *et al.* 2010).

So, under the undisputed assumption of continuity and the conviction that there is nothing new under the sun, the question is: How could the belief in novelty originate in the history of science? How was it possible in the 19th century for both thermodynamics and the theory of evolution to be developed?

And again the metaphysical assumption of continuity proves unavoidable: The "new" does not really irritate the continuity assumption much – it rather expresses exceptions to the rules governing the manufacture of continuity. Already in the elaborated version of Darwin's theory, the so-called "modern synthesis" (Huxley 2010), small variations ("more of the same") are possible and larger mutations ("something different") are not just permitted, but even desirable if they prove helpful for survival.

In short, biological evolution became a theory that legitimizes irregularities and flaws in the weaving of the continuous fabric of nature. With respect to innovations, however, we have to keep in mind that not every irregularity is innovative, even though all innovations are based on or even caused by irregularities.

2.4 How Does Continuity Connect with Scientific Rationality?

2.4.1 The Myth of the Two Cultures Revisited

In his Rede lecture on May 7, 1959, at the University of Cambridge, Snow (1959) had claimed that the culture of science and the culture of the humanities are divided by an insurmountable abyss and strictly opposed or even hostile against each other. Since then, the literary intelligentsia is considered ignorant or even hostile *vis-à-vis* science and technology by many representatives of the natural and engineering sciences. The converse position, that scientists and engineers are ignorant with respect to broader cultural issues, has not been explicitly formulated by Snow. Nevertheless, his claim supports the cheap prejudice that a culturally educated person can be distinguished by their ignorance with respect to the second law of thermodynamics.

One aspect of our culture, however, has been – intuitively but quite correctly – identified by Snow: The occidental canon of learned education contains many more mediocre novels than seminal scientific theorems. For our argument it is important that the very theorem in question, the second law of thermodynamics, in actual fact is indeed a fundamental scientific theorem of physics. However, it originated from a basic technological insight, generated by the engineering science inventing the steam engine (Carnot *et al.* 1890) – to know is to make.

Today Snow's thesis of the two cultures is, in the words of Weinrich (1993), "snow of yesterday." It looks so anachronistic and outdated because the new cultural knowledge technology is *ipso facto* already bridging the gap between different scientific cultures. It re-introduces continuity as knowledge/information technology and re-defines standards of scientific rationality.

2.4.2 Three Types of Theories

A crucial element of scientific rationality is defined by scientific theories that need to be corroborated by reproducible results. But what are scientific theories?

From the perspective of a nonreductive philosophy of science I would like to distinguish at least three different types of theories: explanatory, narrative, and descriptive.

By an *explanatory theory* we understand a certain number of interconnected generalized law-like propositions (including their statistical variants discussed above) from which, together with the relevant boundary conditions, a proposition about a particular state of affairs can be logically deduced. This deduction is then called an "explanation." Since neo-Kantian philosophy and Dilthey, this type is methodologically associated with the "explaining" natural sciences as opposed to the "understanding" humanities (Wright 1971).

Since Danto (1965) the dominant type of theory connected to the method of "understanding" is called *narrative theory*. Instead of deducing a proposition about a particular state of affairs from some generalized lawlike proposition, propositions about a particular state of affairs are arranged in terms of a narrative. The plausibility of such a theory (which is called "hermeneutical") is generated by the systematic narrative, e.g., of historical developments or literary connections.

The third type is a *descriptive theory* without any law-like propositions or narratives. Its plausibility is generated by the utmost precision of the recording and description of data and their recurrent connection. Examples are theoretically underpinned demoscopic inquiries or data collections concerning longitudinal weather developments.

In all three types the status of scientific rationality depends on the assumed continuity of nature, of history, of large data sets, or of the behavior of social groups. Briefly speaking, scientific rationality, as represented in these different types of theories, depends on the supposition of continuity. This applies even to theories about discontinuous, discrete, or nonlinear processes as in the theory of deterministic chaos.

2.4.3 The "Covering-Law" Model

In analytical philosophy of science, the analysis of an explanatory theory is by far the most advanced. This might be one of the reasons for the widespread opinion that only scientific endeavors based on theories of the explanatory type are really rational.

In the late 1940s Hempel and Oppenheim (1948) proposed a theory of scientific explanation, which has been further developed by numerous authors, including Popper (1959). Their proposal is known under various names, such as "H-O scheme," "D-N (for deductive–nomological) model," "subsumption model," "Hempel–Popper scheme," or "covering-law model." According to this model, an event or a state of affairs is explained by syllogistically deducing propositions about it from generalized propositions ("covering laws") plus propositions about the relevant boundary conditions.

While other types of theory allow for a less restrictive interpretation, the "covering-law model" expresses best the rigidity of the narrow conception of science as favored by the logical empiricism of the neo-positivist Vienna Circle and its followers. It presupposes not just logical but also diachronic continuity. This is why a narrow and strict generalization is necessary, which in turn implies the dogma of the "symmetry of explanation and prediction" (Hanson 1959; for the most recent debate on the "covering law" model in the social sciences cf. Opp 2013 and Ylikoski 2013).

2.5 How Does Scientific Rationality Connect with Prognosis?

2.5.1 Symmetry of Explanation and Prognosis

In order to better understand why and how scientific rationality is connected to prognosis we have to recall that in our interpretation, science and its characteristics are to be seen in the light of technology. The purpose of science is not only to produce knowledge for knowledge's sake but also knowledge useful for human beings. This implies that not only science but also technology produces relevant insights for the future.

Talking about the importance of insights for the future we leave the realm of pure epistemology and enter the theory of time.[4] Although scientific propositions, and especially lawlike ones, seem to be applicable to each and every event in time, a closer examination results in a different picture. We usually distinguish between three different extensions of time: the past, the present, and the future, and at first glance, they seem to be exactly just this: extensions of the one, ever-flowing linear time.

But it has been noticed as early as in the fourth century B.C. by Aristotle (1949) in his famous sea-battle-argument, that there is a decisive (logical) difference between propositions concerning the past and the present on the one hand, and propositions concerning the future on the other hand. Strictly speaking, only propositions about the past and the present express a truth value, whereas

[4]For more details about the following argumentation on the direction of time compare Zimmerli (1997, 2015).

propositions about the future are neither true nor false unless they have been "caught up with" by real time, and have thus been transformed into propositions about the present or past.

Analyzing this even further we see that it corresponds with yet another categorical difference with respect to both modality and quantity. While in present and past tense we speak about present and past *reality*, sentences in future tense refer to *possibility*. As far as quantity is concerned, sentences in future tense, although grammatically pretending to speak about the one future, are in actual fact about different (possible) *futures*; otherwise they would not talk about possibilities.

This means that the dogma of the symmetry of explanation and prediction does not relate to genuinely future events at all. Scientific predictions are deductions of propositions from generalized lawlike propositions. They do not express anything about time as they do not belong to the temporal but to the logical realm. It is helpful to turn this into a terminological difference between logically deduced propositions as *predictions* and propositions referring to future events as *prognoses*.[5] Thus, there is indeed a symmetry between explanation and prediction, but not between explanation and prognosis.

2.5.2 Nature Does Not Have a Future

This has serious implications regarding the ontology of time and of nature. If the target of scientific explanation is "nature," and if there is a fundamental asymmetry between explaining and prognosticating the future(s), then two questions arise: (1) Does this entail that nature itself has no future? (2) Can we still rely on the continuity of nature? The first question could even be generalized by referring not to future only, but to the past–present–future type of tensed time in general. This position has been defended by McTaggart (1908) with his notion of the "unreality of time."

One thing, however, has to be made quite clear: By claiming that extrahuman nature does not have a future I am neither denying nor even disputing the direction (or "arrow") of time (Popper 1974). One of the key principles of modern physics, the second law of thermodynamics expresses the directedness of time by the unavoidability of entropy increase. But this law is, as already mentioned, not a principle of theoretical physics but an abstraction and generalization of a technological insight into the functioning of a steam engine (cf. Brush 1967).

[5]This terminological distinction does neither perfectly match our ordinary language understanding of these two notions nor their scientific use. We usually speak of prognoses, in the sense of "prognostics," mainly with respect to medical forecasts.

If, according to the asymmetry of explanation and prognosis, there isn't anything like future in nature, then science would not make much sense with respect to the purpose of (modern) science, to generate knowledge useful for people. In order to transform logical predictions into useful prognoses, human beings must interfere not just cognitively but actively and, in many cases, even technologically. To relate this closer to the topic of this chapter: The ultimate proof of reproducibility is reproduction.[6]

2.5.3 Three Types of Prognosis

Given these results of our analysis, future (or even tensed time as a whole) does not come into existence without the cooperation of "timing" by human beings. If the task of modern science is to prove useful by producing prognoses concerning future events, there is but one conclusion: Prognoses within the realm of technological modern science are future events coproduced by both extra-human nature and human impact. This becomes especially obvious if we pay attention to the degree to which the predicted event needs human involvement to be transformed from prediction to prognosis.

Among the vast number of different types of prognoses let me comment on just three of them: the explanation type, the description type, and the simulation type. Much has already been said about the explanation type, insofar as it coincides with the "covering-law model" of explanation (Section 2.4.3). But there is one decisive additional point. Transforming an explanatory prediction logically deduced from a covering law into a prognosis requires the co-productive synchronization of a given point on the arrow of time with the observer's point in time.[7] This point in time is then called the "present," while all earlier time instances are called "past" and all later time instances are called "future."'

The co-productive aspect becomes even more evident in the other two types of prognosis. The description type is quite common, not just in science, but also in the lifeworld. It does not need a covering law, not even a law-like proposition, but just large amounts of data and experiences connected with expectations of probability. Poll predictions, e.g., are typically not based on any kind of theory, but on a longitudinal probability, i.e., the experienced relation of the data gathered in previous studies to the result of the poll (see Keller, this volume).

[6]This coincides with the result of linguistic analysis of the term "reproducible" as a so-called disposition predicate. Such predicates are characterized by the fact that a proposition like, say, "x is breakable" is true or false depending on a proposition like "x is broken."

[7]This synchronizing act enforces the breakdown of time-translation invariance, introducing a distinguished point along the time axis. Ignoring this important detail explains why prognosis and prediction are often confused with one another.

The logical structure applied here is the logic of analogy (Hofstadter and Sander 2013). The coproductive role of the observer consists in not just identifying the present but also presupposing that individual decisions at the ballot are less important than the statistical distribution of their decision behavior.

The more data there are to be processed, the more urgent becomes the development of powerful data processing tools. Therefore, with the increasing speed and numerical capacity of data processing technology, a new type of prognosis has emerged not just in science, but also in the lifeworld: the simulation type. Here, a system is algorithmically modeled, and within it the relevant parameters are simulated, resulting in different scenarios for optional futures.[8] In this case, the coproductive character of the transformation of predictions into prognoses becomes even more obvious: the very notion of "optional" and the plural "futures" emphasize the human impact of evaluating different possible situations.

2.6 How Do Prediction and Prognosis Connect with Reproducibility?

2.6.1 A-Series and B-Series

In the preceding section, I argued that for predictions to become prognoses, they need analytical, descriptive, and evaluative coproduction by human beings. To reach a more in-depth understanding of this it might be helpful to look at still another aspect of these relations. As human beings share the ability to express themselves in language as well as reflect on it, we might ask the question of how the co-productions mentioned above are represented in language.

Such a linguistic analysis shows a result stunningly coherent with the distinctions discussed so far. As McTaggart pointed out, we talk about temporal sequences of events in two different ways by distinguishing a sequence which he calls "B-series" from another sequence which he calls "A-series" (McTaggart 1908, p. 458):

> Each position is Earlier than some, and Later than some, of the other positions. And each position is either Past, Present, or Future. ... For the sake of brevity I shall speak of the series of positions running from the far past through the near past to the present, and then from the present to the near future and to the far future, as the A-series. The series of positions which runs from earlier to later I shall call the B-series.

[8] See also Bailey *et al.*, this volume, for technical details and Feulner, this volume, for applications in climate research.

The B-series obviously pretends to identify the relation of events independent of an observer, whereas the A-series needs an observer or a participant to identify the "present" in order to then distinguish between "past" and "future" events, relative to the identified present. Propositions in past and present tense address the one reality, whereas propositions in future tense address multiple possibilities. So in order to transform predictions into prognoses, one has to transform propositions from the B-series to propositions in the A-series.

2.6.2 The World as a System of "Merton-Sensible" Systems

There is yet another aspect to be included if the consequences for the problem of reproducibility are to be understood: the self-referential effect that predictions have on the system they refer to as they are transformed into prognoses. This aspect has been highlighted in the work of the sociologist Robert Merton.

Merton focused on the fact that the coproductive transformation of predictions into prognoses within social systems not only concerns those human beings performing them but also those whose behavior is subject to the prediction. This might be of increasing importance up to the point where a prediction may be overthrown. In order to express this dialectical character of predictions he talked of "self-fulfilling" and "self-destroying prophecies" (Merton 1949, pp. 421–436). We know an analogous effect as the "indeterminacy principle" from quantum mechanics (Bynum *et al.* 1981, p. 202): "in measuring one of a pair of conjugate physical quantities characterizing a micro-system (e.g. position-momentum or energy-time), the experimental act necessarily destroys the possibility of measuring the other to arbitrary accuracy."[9]

To better understand this analogy let us have a look at the distribution of the accuracy of predictions from microscopic over mesoscopic to macroscopic scales. Such a distribution looks roughly like a Gauss curve with two wings where the probability that a prediction is correct is relatively low and a central part where this probability is rather high. In quantum mechanics, e.g., arbitrarily accurate predictions in the microcosm of very small particles and short periods of time are impossible. And something similar is – although for different reasons – true in the macrocosm with respect to very large systems and very long periods of time. In the range in between, however, which is described rather well by Newtonian mechanics, the probability of accurate predictions is rather high.

If we now take a look at the humanities and social sciences, the distribu-

[9]This parallel with quantum theory is also addressed by Collins, this volume, who relates it to what he calls "experimenter's regress." A concrete example, order effects in sequential decisions, is described in detail by Wang and Busemeyer, this volume.

tion of accuracy has a similar shape, although much steeper and narrower. This means that in the mesocosm, where the probability of accurate predictions is rather high, the distribution of accuracy is much narrower than in the natural sciences. It is (almost) impossible to predict short-time behavior of individuals or long-term developments in history and society, but in the mesocosm of behavior, e.g., in a ballot, at least statistical predictions for ensembles of agents are possible. The same applies to predictions based on simulation. One of the reasons for this analogy is that in physics, the social sciences, and the humanities, there is a "Vico effect" at work. True in the strict sense of the word is only what we can do or make (or more pointedly, what we did or made).

2.6.3 Knowledge Technology

The development of production, as analyzed in Section 2.2, culminated in the introduction of data-processing "intelligent" technologies. Something quite similar happened in the development of different kinds of prognosis that culminated in the simulation type. Simulation has all the characteristics of a mental experiment or of producing a virtual product. In this way it creates a space that is both real and virtual. However, it has all the advantages of not having to bring into technological existence products with unknown consequences and implications.

This very advantage of the virtual character of the computer-aided simulation type of prognosis, however convincing and promising it may be, does not have this seemingly positive effect only. The reason is that one does not just play with a virtual copy of relevant structures of a sector of reality, but one is reproducing structures of the cognitive simulation itself. And due to the vast amount of data ("big data") and the incredibly high velocity of data processing, one knows that one cannot possibly control this simulation and the prognoses its predictions are transformed into.

What is at stake here is a self-referential amplification of the problem of technology assessment. It is well known that such assessment is absolutely necessary – for the same reason as it is known to be impossible. It will never be possible to know all the consequences and problems implied in a (new) technology because, on the one hand, each solution of one of those problems will create more problems and, on the other hand, every agent acts within a Merton-sensible system where what is predicted might turn into self-fulfillment or self-destruction.

2.7 Concluding Remarks

Once we free ourselves of the myths of the "two cultures," of "knowledge as a purely mental product," and of the "unity of science," we realize that modern science from its very beginnings is directed toward producing knowledge practically useful for human beings: technological knowledge.

The notion of the reproducibility of scientific results is formulated according to the requirements of (industrial) mass production of identical products. The possibility of mass-producing identical products depends on the presupposition of a natural continuum and thus of continuity.

The well-corroborated expectation that the same production process will produce identical products is mirrored in the scientific abstraction called prognosis. A logical analysis of different types of scientific prognoses results in the insight that predictions are being transformed into prognoses as soon as they become co-products of the theory (and/or the data) invested and the human beings deducing them.

As the world can be understood as a system of Merton-sensible systems, it becomes quite obvious that and why predictions, once transformed into prognoses, necessarily become to some extent indeterminate, both in the (natural and social) sciences and in the humanities.

Although in technology indeterminate situations are ruled out by definition, indeterminacy recurs as soon as technology is embedded in a system of Merton-sensible systems. This applies in an especially obvious way to modern technology which is permeated by and rooted in net-based knowledge technology and predictions of the simulation type.

Unveiling the analogy with the problem of technology assessment goes far beyond the analogy between scientific reproducibility and technical production. Technology is socially, environmentally, and historically embedded. Even in its seemingly well-defined and rigidly closed system, it is impossible to rule out uncertainty and ignorance.

References

Aristotle (1949): *De Interpretatione.* In *Aristotelis Categoriae at Liber de Interpretatione*, ed. by L. Minio-Paluello, Oxford University Press, Oxford.

Babbage, Ch. (1986): *On the Economy of Machinery and Manufactures*, McGraw-Hill, New York. Originally published in 1831.

Bacon, F. (1857): *Novum Organum.* In *The Works of Francis Bacon Vol. 1*, ed. by J. Spedding, R.L. Ellis, and D.D. Heath, Houghton, Mifflin and Co., Boston.

Originally published in 1620.

Bacon, F. (1864): *Meditationes Sacrae*. In *The Works of Francis Bacon Vol. 14*, ed. by J. Spedding, R.L. Ellis, and D.D. Heath, Houghton, Mifflin and Co., Boston. Originally published in 1597.

Barrow, J.D., and Tipler, F.J. (1988): *The Anthropic Cosmological Principle*, Oxford University Press, Oxford.

Benjamin, W. (1968): The work of art in the age of mechanical reproduction. In *Illuminations*, transl. by H. Zohn, ed. by H. Arendt, Fontana London, pp. 214–218. Originally published in 1936 in German.

Brush, S.G. (1967): *The Kind of Motion We Call Heat. A History of the Kinetic Theory of the Gases in the 19th Century, 2 Vols.*, North-Holland, Amsterdam.

Bynum, W.F., Browne, E.J., and Porter, R., eds. (1981): *Dictionary of the History of Science*, Princeton University Press, Princeton.

Carnap, R. (1931): Die physikalische Sprache als Universalsprache der Wissenschaft. *Erkenntnis* **2**, 432–465.

Carnot, S., Carnot, H., and Thomson, W. Baron Kelvin (1890): *Reflections on the Motive Power of Heat and on Machines*, Wiley, New York.

Danto, A.C. (1965): *Analytical Philosophy of History*, Cambridge University Press, Cambridge.

de la Mettrie, J.O. (1912): *Man a Machine*, English transl. by G.C. Bussey, Open Court, Chicago. Originally published in 1748 in French.

Descartes, R. (2001): *Discourse on the Method of Rightly Conducting the Reason and Seeking the Truth in the Sciences*, Bartleby, New York. Originally published in 1637 in French.

Fellmann, F. (1978): *Das Vico-Axiom: Der Mensch macht die Geschichte*, Alber, Freiburg.

Gadamer, H.-G. (1989): *Truth and Method*, transl. and ed. by J. Weinsheimer and D.G. Marshall, Sheed and Ward, London. Originally published in 1960 in German.

Gimmler, A., Holzinger, M., and Knopp, L., eds. (2010): *Vernunft und Innovation. Über das alte Vorurteil für das Neue*, Fink, München.

Hanson, N.R. (1959): On the symmetry between explanation and prediction. *Philosophical Review* **68**, 349–358.

Hempel, C.G., and Oppenheim, P. (1948): Studies in the logic of explanation. *Philosophy of Science* **15**(2), 135–175.

Hofstadter, D., and Sander, F. (2013): *Surfaces and Essences. Analogy as the Fuel and Fire of Thinking*, Basic Books, New York.

Huxley, J. (2010): *Evolution – The Modern Synthesis*, MIT Press, Cambridge.

Kraft, V. (1997): *Der Wiener Kreis. Der Ursprung des Neopositivismus*, Springer, Wien.

Lepenies, W. (1985): *Die drei Kulturen. Soziologie zwischen Literatur und Wissenschaft*, Hanser, München.

Linné, C. von (1751): *Philosophia Botanica*, Gottfried Kiesewetter, Stockholm.

Marx, K. (1867): *Das Kapital, Buch 1*, Meiner, Hamburg.

McTaggart, J.M.E. (1908): The unreality of time. *Mind* **17**, 457–474.

Merton, R. (1949): *Social Theory and Social Structure*, Free Press, New York.

Nicolis, G., and Prigogine, I. (1977): *Self-Organization in Nonequilibrium Systems: From Dissipative Structures to Order through Fluctuations*, Wiley, New York.

Opp, K.-D. (2013): What is analytical sociology? Strengths and weaknesses of a new sociological research program. *Social Science Information* **52**, 329–360.

Popper, K.R. (1959): *The Logic of Scientific Discovery*, Routledge, London. Originally published in 1934 in German as *Logik der Forschung*.

Popper, K.R. (1974): Boltzmann and the arrow of time. In *Unended Quest. An Intellectual Autobiography*, Routledge, London, pp. 181–184.

Snow, C.P. (1959): *The Two Cultures and the Scientific Revolution*, Cambridge University Press, Cambridge.

Ure, A. (1835): *The Philosophy of Manufactures: An Exposition of the Scientific, Moral and Commercial Economy of the Factory System of Great Britain*, Charles Knight, London.

Vico, G. (1979): *Liber metaphysicus (De antiquissima Italorum sapientia liber primus)*, ed. by S. Otto and H. Viechtbauer, Fink, München. Originally published in 1710 in Latin.

Weinrich, H. (1993): Diskussionsbemerkung. In *Einheit der Wissenschaft. Wider die Trennung von Natur und Geist, Kunst und Wissenschaften*, ed. by H. Mainusch and R. Toellner, Westdeutscher Verlag, Opladen.

Wright, G.H. von (1971): *Explanation and Understanding*, Cornell University Press, Ithaca.

Ylikoski, P. (2013): The (hopefully) last stand of the covering-law theory: A reply to Opp. *Social Science Information* **52**, 383–393.

Zimmerli, W.Ch. (1997): Zeit als Zukunft. In *Die Wiederentdeckung der Zeit. Reflexionen – Analysen – Konzepte*, ed. by A. Gimmler, M. Sandbothe, and W.Ch. Zimmerli, Wissenschaftliche Buchgesellschaft, Darmstadt, pp. 126–174.

Zimmerli, W.Ch. (2006): *Die Zukunft denkt anders. Wege aus dem Bildungsnotstand*, Huber, Frauenfeld.

Zimmerli, W.Ch. (2014): From fiction to science: A German-Japanese era project. In *Robotics in Germany and Japan. Philosophical and Technical Perspectives*, ed. by M. Funk and B. Irgang, Lang, Frankfurt, pp. 11–25.

Zimmerli, W.Ch. (2015): Human responsibility for extra-human nature. An ethical approach to technofutures. In *Technofutures*, ed. by C. Deane-Drummond, S. Bergmann, and B. Szerszynski, Ashgate, Farnham, pp. 17–30.

3
Stability and Replication of Experimental Results: A Historical Perspective

Friedrich Steinle

Abstract. The debate about the replication and replicability of experimental results has a long and diverse history since the Early Modern period. In my contribution, I shall highlight some important points of that history. Such historical analysis shows, moreover, important aspects and differentiations of the debate that might enrich contemporary discussion.

3.1 Experiments and Their Reproduction in the Development of Science

Experimental outcomes, sometimes called "experimental facts," have been used, from early on, for supporting all sorts of claims in science, be it about the existence of certain entities (like electrons), about regularities or laws (like Hooke's law of force and extension of a spring, or Mendel's laws of heredity), about physico-mathematical theories (like Newton's theory of different refrangibility), about specific processes (like combustion or photosynthesis), or about the causes of natural processes (like invisible electric fluids).

Prominent examples reach back to antiquity: Galen, in the second century AD, performed cruel experiments with living animals to demonstrate the pathway of urine from the kidneys into the bladder; Alhazen attempted to prove, in the 11th century, the rectilinear propagation of light by experiments with an apparatus that later would be called *camera obscura* (Schramm 1998, Lloyd 1991); and Roger Bacon and Pierre de Maricourt, in the 13th century, emphasized the importance of experiments for natural philosophy and produced astounding results in optics and magnetism (Crombie 1961). Alchemists, from the late mediaeval period on, supported their claims about cleaning and transmuting metals by experiments – indeed, alchemy formed, from early on, an experimental research field *par excellence* (Newman 2006).

As those cases make clear, experiments have also been used to prove the manageability of processes and to enable or optimize the production of materials (pure metals or strong magnets), of items (spirits, solubles, ethereal oils,

---

Reproducibility: Principles, Problems, Practices, and Prospects, First Edition. Edited by Harald Atmanspacher and Sabine Maasen.

philosophers' stone, compass needles, etc.), or of processes (optimized distillation, calcinations, etc.). Moreover, in the case of alchemy, it is clear that manageability and philosophical claims were often not separated or separable (Newman 1998).

The experiment as an intervening procedure to create knowledge about nature had already a long history when protagonists of a "new" science in the 17th century started to highlight its importance and sometimes claimed to have invented it. These protagonists – such as Gilbert, Galileo, Bacon, or Descartes – declared experiments as the royal road to science. Indeed, they extended the use of experiments in the process of generating knowledge far wider than before, to domains like floating bodies, falling projectiles, electric attraction, and heated bodies.

Moreover, they made the use of experiments – and this was new indeed – an explicit topic in their published reflections of how natural knowledge should be gained. Gilbert, e.g., highlighted, in his *De Magnete* (1600), every experiment he mentioned by an asterisk in the margin, larger or smaller, according to the importance he attributed to it. Bacon, in his programmatic *Novum Organum* (1620) and his utopian *Nova Atlantis* (1626), described far ranging experimental activity as the main source of empirical knowledge.

Given that broad discussion (or sometimes propaganda) about experiments as central means of knowledge generation, it comes as no surprise that also further aspects of experiment were first addressed explicitly in that context. Among those, the question of how to secure experimental results, why to believe other experimenters' claims, and how to convince others of one's own results played a major role. Possible responses to that question were sometimes explicitly addressed, but more often just practiced without explicit emphasis.

The issue of the possible roles of the reproduction of experiments, and even the very concept of reproducibility, has been subject of philosophical, historical, and sociological discussion from the 1980s on, starting with a provocative analysis by Harry Collins that reduced reproducibility to the social process it involves and, hence, supported social constructivism (see Schickore 2010, pp. 568ff, for a brief account of that history and the recent philosophical literature).

In this chapter, I shall discuss the question of reproduction and reproducibility from a historical perspective. I do not have in mind the (impossible) task to develop a comprehensive view. Rather I shall look at some historical cases and episodes with specific questions in mind and hereby explore the spectrum of possible responses that history offers. At the end of my chapter, I shall consider some wider implications that might be suggested by that analysis. Most of my historical episodes are taken from the physical and chemical sciences up to the

20th century, with all the restrictions that come with such a focus. Experiments that involve a large variety inherent to its subjects, as in physiology, are not in my focus, nor are experiments that centrally work with statistical approaches. To what degree my general observations hold also for the history of experimenting in the life sciences, psychology, or the social sciences is, hence, a matter that needs further exploration.

Before dealing with historical material, it is necessary to address some fundamental distinctions that are instrumental in selecting, organizing, and analyzing the material and have been proposed in the philosophy of science. First, it is one thing for an experimenter to assure an experimental result for her- or himself, and another thing to convince others of it. In more recent times, individual experimenters have become rare in favor of larger working groups. To have a shorthand terminology for that difference, I shall speak of the *repetition* of experiments when dealing with the first case (i.e., repetition of the experiment by the experimenter or the working group that made it in the first place). I shall use the notion of a *replication* when it comes to redoing the experiment (or at least reproducing the experimental outcomes, which is not necessarily the same, see Bogen 2001) by other, possibly competing experimenters or working groups. I use *reproduction* as an umbrella term for both cases. Such a distinction is much in line with what other analysts have proposed (Schickore 2010 in particular), and the main difference between the two cases is the point of making the result public. Even while borders between "inside" and "outside" of the research group might be blurred or shifted in settings with large numbers of researchers involved in one project (as in the case of the Human Genome Project or of particle physics), the main point of difference, *mutatis mutandis*, remains valid.

My second observation concerns variations. Strictly speaking, it is simply impossible to repeat or replicate an experiment without varying some parameters. Every repetition or replication necessarily involves at least a change of time or space or both, and of the specific item (object, material, specimen, etc.) to be worked with. Even if the same item is used, it has become older and might have undergone (probably uncontrolled and/or unnoticed) changes as compared with its first use. This was already stated by Collins (1985) as a starting point of his general approach to degrade replication as criterion for experimental stability. I see it as a welcome occasion to distinguish between various types of variations to be found in replications, much in the way Schickore (2011, pp. 344f) proposed. The hope to reproduce the same experiment by repetition or replication presupposes the idea that those inevitable variations do not affect the result – an idea that might look innocent and usually is taken for granted in experiments with non-living items such as in physics and chemistry. But it leads to complications in the realm of living beings such as plants and animals: Living beings grow older in time, and no living item (such as a frog leg) is exactly the same as another one of the same type.[1] It is partly for this reason that even those reproductions of experiments that keep variations as limited as possible will often lead to variations in outcomes. Experimenters have, from early on, been faced with that difficulty and have developed methods to cope with inevitable variations.

A different type of variation in repetition or replication is the deliberate variation of apparatus, of materials, or of procedures. Those variations might have different goals: showing what often is called the robustness of an experimental result, i.e., showing that the same result can be achieved by different apparatuses or methods, but also using those variations to explore more broadly the causal factors that contribute to an effect. These are epistemic goals that go further than the basic goal of showing the stability of just one specific experimental result. However, they presuppose already that for every single one among the varying experimental arrangements such experimental stability can be obtained. Such stability, with as few as possible variations, is the most basic requirement for all further experimental approaches pursued and questions to be asked. It is in the center of my considerations in this article: How was experimental stability in this sense achieved, and how did experimenters manage to convince others of the validity of their individual results?

Third, it is important to keep in mind the possible difference between how experimenters actually proceed and how they explicitly reflect about their procedures (if they do so at all). Of course, dealing with those two aspects historically requires different historiographical approaches and types of reconstruction, and there have been recent studies of numerous cases that yield insights into both aspects (Holmes *et al.* 2001, among others). Here I shall deal both with practices and reflections, while staying aware of their specific powers and limits.

3.2 Repetition of Experiments

Before an experimenter (or a research group) goes public with an experiment, they typically have convinced themselves of the validity of their results – a point that has, in response to what he regarded as an overemphasis of social aspects, very much been highlighted by Holmes (1992). As a rule, this conviction is usually reached by repeating the experiment. This is mentioned (sometimes in passing) in many historical cases, from the 11th century onwards.

[1]The issue of variability is also addressed by Shiffrin and Chandramouli (this volume) from the perspective of statistical modeling. Questions as to the relevance of varying contexts at different descriptive levels are discussed by Atmanspacher (this volume).

One of the first scholars to make systematic use of experiments in arguing for general claims was the Arab scholar Ibn-al-Haytham, living in Cairo in the early 11th century and called Alhazen in Latin Europe. His *Book on Optics* ("Kitab al Manazir") was of utmost impact on the later Latin European development (Ibn-al-Haitham 1989b or Schramm 1998). Already here we find hints to successful repetition, even under variation, of experiments. Often Ibn-al-Haytham ends the description of an experiment with the remark that "this is always found to be so, with no variation or change" (e.g., Ibn-al-Haitham 1989a, book I, chap. 2, §8, p. 8; or book I, chap. 2, §10, p. 9). And typically it was only after that confirming remark that he went on to spell out the finding in general terms. At a later place, Alhazen summarized the finding of the previous chapters, and then supported them by a general remark on the experiments: "When this state of affairs is examined at all times it is found to be uniform, suffering no variation or change" (Ibn-al-Haitham 1989a, book I, chap. 3, §3, p. 13). Repeatability was presented here as a central argument to trust in the validity of an experimental finding.

Moving to the Early Modern period, we find Galileo claiming to have repeated his experiments on the inclined plane "a full hundred times." While Galileo did not explain why he did so, historian Peter Dear interprets this as an attempt to connect experimental research to the Aristotelian notion of common experience as the fundament of natural philosophy, of course with the subversive aspect that now a very narrow segment of experience was given the status that previously was attributed only to broad and common experience (Dear 2001, p. 133).

Such an aspect might also have been important for Alhazen. But there might be a more general point in the background of those cases: It is hard to imagine that experimenters who wanted to base firm philosophical or mathematical claims on experiments did not quickly become aware of how easily accidental circumstances could change the outcome of experiments. Hence, to make an experimental outcome as firm as possible (and thereby give its philosophical or mathematical implications a firm base), they had to cope with that situation. Repetition was perhaps one of the ways close at hand since it was (and is) so close to everyday practice.

Whether or not this general consideration holds for those early cases, we find it made explicit by Francesco Redi, Early Modern physician, naturalist, and prolific experimenter, among others, on the question of how maggots came into meat. As Schickore has shown in detail, he emphasized explicitly, in his study of snake venoms, the need of repetition of experiments and criticized those who had, in his view, not done so sufficiently. He even gave a reason why this was so important: He saw repetitions as a means to make sure that the obtained outcome was not just the result of some "unknown or unseen hindrance" present at the moment of the experiment (Schickore 2011, pp. 330f). With a similar argument, Felice Fontana, 18th century microscopist and physiologist, emphasized in his studies of the effects of viper venom that he had repeated his experiments oftentimes, with or without voluntary variations (Schickore 2010, 2011).

Remarks about the necessity of repetition and replication are also found in the writings of Robert Boyle, one of the main promoters of what he called "experimental philosophy" in the 17th century, and founding member of the Royal Society. He recommended repetition as a way to ensure that the experimental result was not a one-time chance occurrence, and went so far as to regard experiments that could not be successfully repeated as failures, as unsuccessful altogether (Sargent 1995, pp. 176ff; see also Boyle's own remarks (1999, pp. 37, 58, etc.)).

A somewhat different argument has been provided by Claude Bernard, famous physiologist and promoter of experimental medicine in the 19th century. In his influential *Introduction à l'étude de la médicine expérimentale* (1865), he emphasized the importance of repeating experiments oftentimes. Since natural processes are absolutely and sharply determined by their conditions, he argued that a repetition of the same experiment must necessarily lead to exactly the same result. Conversely, if the result of the repetition comes out to be different, one has to conclude there was some change in the conditions. Hence, repetitions first of all make the experimenter aware of those changes, and can then serve to identify those changes (Bernard 1865, pp. 121f). Bernard suggested that such an enterprise was "relatively very easy" when dealing with inorganic bodies but became very difficult when dealing with living material, in particular warm-blooded animals. Nevertheless, he maintained that procedure as the central procedure of all experimental research (Bernard 1865, p. 202). Repetition for him had a central epistemic function.

These examples indicate that repetition might have been an omnipresent feature of experimental research for a very long time. The stability of an individual experimental result was and is indeed defined as stable repeatability. It might be indicative of the degree to which that idea belonged to the self-understanding of experimenters that it has typically been mentioned only in passing (if at all). And it might be likewise indicative that someone like Goethe, who entered experimental research from the "outside" as a new practitioner in the 1790s and, as a sharp methodological analyst, highlighted that feature (Steinle 2014).

A particular aspect concerns those experiments that are not publicized by text or pictures, but by performance. This includes in particular experiments that have been designed for public show, to make money, or to gain reputation. Most intriguing cases are provided by public vacuum experiments in the 17th century or of the 18th century culture of learned "salons," of performances at court or on the marketplace. This has been done most intensely in the fields of electricity, of pneumatics, and of optics with large lenses and mirrors. Showmanship of spectacular experiments continues up to this day: Term-opening lectures in chemical or high-voltage electric engineering departments have often the character of show and entertainment rather than of academic teaching, not to speak of those that are deliberately addressed at a wider public.

While those experiments did and do not have the goal to convince other (possibly competing) researchers of experimental outcomes, they have, nevertheless, the essential requirement of experimental stability: no showman, medical doctor, engineer, or craftsman will have a chance to succeed if their experiments run serious risk to fail at the moment of the show, treatment, or performance. Nobody would announce an experimental show, ask spectators for money, or invite nobility (with the aim of financial or moral support) without being sure that the experiments to be shown would work with certainty. Stability has always been essential here, and has constantly been achieved and assured by repetition. Repetition of experiments forms a characteristic of the procedure, regardless of whether the experiments belong to academic research or aim at other purposes.

3.3 The Power of Replicability

Convincing others of the validity of experimental results is, by definition, a social process. Experimenters of all times and branches have been aware of this point, and the historical development of the sciences shows a plethora of strategies to achieve credibility. They include hinting to the *authority* and trustworthiness of the experimenter, listing up *witnesses*, again alluding to their trustworthiness, and describing the experimental setup in so much detail as to reach, even without real replication, something that has been called *virtual witnessing*.

The foremost strategy we find in historical records, however, again has been based on repetition and replication. On the one hand, we often find remarks and emphasis on the successful repetition by the experimenter, but on the other hand, the thrust of most arguments is directed at the attempt to enable or facilitate *replication*. Many experimenters have been aware of the difficulties of replication and, hence, pursued various strategies to facilitate it for their readers: describing the apparatus and the procedure in much detail, illustrating the description with figures, sending around materials or sample apparatuses, or even directly sending specialist personnel to those places where replication was attempted.

Which of these strategies have been pursued in particular cases depended much on the specific constellation concerning character and availability of materials, apparatus, and skills, but also status of the experimenter within the community (once such a thing existed). It also depended on the character of the experimental claim as related to the general conceptual and theoretical framework of the time and the expectations implied by it and, hence, on the possible consensual or controversial character of the experimental claims. What also forms a most important part of the specific constellation is the possible use of the experimental result in practical fields, i.e., its possible technical and economical importance. I shall illustrate these general points by some historical examples that include both practices and explicit reflections.

It is not by chance that we find many explicit reflections on experiment in general and replication in particular in the Early Modern period, in which experiment was declared the central procedure of natural research. One of the first institutions conducting systematic and well-organized experiment in a group was the Accademia del Cimento, active in Florence under the Medici from 1657 for about 10 years (Boschiero 2007, Beretta *et al.* 2009). The topics of the Academy's research were wide, spanning from heat and cold to magnetism, from acoustics to light, and from compression of water to colors of liquids. Among its members were Vincenzo Viviani, former collaborator of Galileo, and the already mentioned Francesco Redi, court physician to the Medici. The thrust of the Academy's approach laid in the invention and development of new instruments and their proper use. In its first and only publication, the *Saggi di naturali esperienze*, published in 1667 and presenting a synthesis of a decade of experimental work, the Academy explained its program (translation from Knowles Middleton 1971, pp. 90f):

> Yet besides trying new experiments, it is not less useful to search among those already made, in case any might be found that might in any way have counterfeited the pure face of Truth. So it has been the aim of our Academy ... to make them ... on those things that have been done or written about by others, as unfortunately we see errors being established by things that go by the name of experiments. ... These prudent precepts ... embraced by the Academy ... have mainly been intended to encourage other people to repeat the same experiments with the greatest rigor.

What was strongly encouraged here was the replication of experiments. It was recommended as a means to assure experimental findings and to detect possible errors, not only of the ancients (to which it was probably mainly directed), but also of present day experiments around and in the Academy. Also the Academy's famous motto "provando e riprovando" points into that direction, whichever of the various attempts to translation one might prefer.[2] The Academy not only had made replication a central part of its own workings (Findlen 1993) but also encouraged others to replicate the experiments it had conducted itself – a significant and in its explicitness most rare sign of critical attitude against own experimental activity, an attitude that shows a high awareness of the numerous possibilities of error in doing and interpreting experiments.

The Royal Society of London, founded in the early 1660s, partly along the model of the Accademia del Cimento, had experimental research as one of its key elements from its beginning. As Thomas Sprat, early historian and apologist of the Royal Society, put it in his *History of the Royal Society* (published as early as after five years of its existence), the Society's primary activity was aimed at "Directing, Judging, Conjecturing, Improving and Discoursing upon Experiments," with judgements to be based on "the matter of fact and repeated experiments" (quoted by Shapiro 2000, p. 113).

Replication was at the center of this activity. The Society received many reports of observations and experiments, and on the latter, a specific procedure was established. Once the report was read at the meetings of the Society, it was discussed, and a decision was made whether to pursue it further or not. In the first case, the Society's "curator of experiments" (a position held for decades by Robert Hooke who provides a rich source of historical information, see Shapin 1988) was given the task to prepare the experiment and to perform it at one of the next meetings in front of the assembled members.

The central point was to establish what the Society called "matters of fact," a new concept that was formed directly after the model of Common Law and introduced to point to undisputable experimental results and, hence, a stable empirical basis for science. After all, "receiving all credible accounts" might result in some "hazard and uncertainty," as Sprat put it. This danger was removed, he continued, since the Society reduced "such matters of hearsay and information, into real, and impartial Trials, perform'd by their Experiments," capable of "exactness, variation, and accurate repetition" (Shapiro 2000, pp. 119f).

As Shapiro excellently showed, the procedure of judging in a committee had indeed much in common with the procedures of Common Law, even with respect to who was admitted to the committee. However, there was one essential difference: the jury must not only rely on reports, but were made eyewitnesses to a procedure of replicating the experiment in question. The procedure was as much about witnessing as it was about replication. While the ideal of replication had no doubt been highly visible at the Accademia del Cimento (to which the

[2] Schickore (2010, pp. 577f) provided a convincing analysis of the motto, pointing both to its historical roots in Dante and to the novel twist the Academy gave it.

Royal Society had a close eye), it was at the Royal Society that the ideal found a formal installation for the first time.

Perhaps the one to publish the most detailed reflections on the value of experiment and on the details of experimental work was Robert Boyle. Boyle did a large number of experiments himself or had them done by his assistants whom he could pay out of his own pocket. His experimental work encompassed a vast area of domains, from chemistry (certainly, his most intense research field) to pneumatics, from mineral waters to hydrostatics, from heat and cold to light and colors. Boyle also published extended reflections on the very procedure of experiments: He defined a special type of experimentation that was essentially devoted to an understanding of the instrument, called it "probatory experiments" and distinguished it from "exploratory experiments" that aimed at gaining knowledge of the world by means of those instruments (Sargent 1995, Chap. 7). He also published essays that dealt with the *Unsuccessfulness of Experimental Philosophy* (Boyle 1999) in which he made clear that the failure of repetition and replication was one way to make experiments unsuccessful.

When the young and unknown Isaac Newton published his first paper on light and colors in 1672, he proposed the bold claim of having proven, in a strong sense, that all previous accounts of the generation of colors in refraction were wrong, and that his "new theory" was demonstrated beyond any doubt by experiments, just as strong as a mathematical demonstration. Given the fundamental character of his claims, and the extraordinary weight he put on individual experiments – particularly the *experimentum crucis* with two prisms – it is no surprise that those who were interested in the topic tried to replicate the experiments. Indeed, in the early reactions to his paper, there were numerous debates on how to replicate the experimental results. Since replication did not work out easily and very often failed, the experimental result was put in doubt, and with it the doctrine itself.

As a recent debate among historians has brought out, it was a combination of increasing experimental replicability and the general rise of Newton's authority that led, over the course of some decades, to an increasing acceptance of both his experimental results and his claims about the nature of light and colors (Schaffer 1989, Shapiro 1996). While no consensus has been reached among historians about how to estimate the relative weight of replicability and authority, there is no doubt that missing reproducibility was an essential factor in making Newton's contemporaries suspicious in the first place. Much of the early debate focused exactly on that point, and without successful replication in the end, Newton's claims would not have found acceptance.

My next case study is taken from the 19th century. In 1820, the Copenhagen physics professor Hans Christian Ørsted announced his discovery of an action of electricity on magnetism, an effect that had been sought-for for decades and, due to the failure of those attempts and some theoretical considerations, declared impossible at important centers of research such as Paris. Not only was Ørsted's discovery unexpected for most, but it was also incompatible with some authoritative theoretical approaches, such as the theory of imponderable fluids, developed and cherished at Paris academy.

Ørsted, while well established in Copenhagen, was certainly not an authority on European scale, rather regarded with suspicion due to his connections to the later so-called "romantic" researchers in German countries. Hence, he expected with good reasons that the validity of his experimental result would be a key point, and he combined various strategies to establish it. Besides highlighting, in his text, the numerous witnesses of his experiments, he did much to facilitate replicability: He gave detailed descriptions of the setup and relatively uncomplicated procedures in which he referred to standard equipment of the time. As a consequence of these efforts, replication worked out very well nearly everywhere it was tried. And even at Paris, the place with strongest theoretical resistance and with most personal reservation against Ørsted (indeed, he was suspected to promote another "reverie allemande"), it was the successful replication before the academy that muted all doubts immediately and triggered feverish activity to cope with it (for a comprehensive account, see Steinle 2005, Chap. 2).

In the field of electromagnetism, another case can be observed only a year later when the unknown laboratory assistant Michael Faraday announced his finding of an electromagnetic rotation in 1821 – again a completely new effect that at first sight contradicted the electromagnetic theory that had been proposed by the well-established Paris mathematics professor Ampère. In contrast to Ørsted's experiment, Faraday's was not easy to conduct since it involved a new type of apparatus, with delicate arrangements and adjustments.

Faraday, well aware of the challenging nature of his finding, of the complicated nature of the experiment, and of the striking difference of social status between him and most other researchers in the field, did everything to facilitate replication. He not only gave extensive descriptions in his published texts, he also added illustrations with detailed extra figures on sensitive parts of the apparatus. And, finally, he designed a small and handy version of his apparatus, made numerous copies of it, and sent it around to researchers all over Europe. Probably as a result of these efforts, the experimental result was quickly and widely accepted (Steinle 2005, Chap. 6).

My final case in this subsection deals with superconductivity, the effect that some materials lose completely their electric resistance below a critical temperature and, hence, transport electric current without loss. Ever since its discovery in 1911, the effect had been subject of technological visions of cheap energy transport, but these visions were not developed since the critical temperatures were extremely low (about the temperature of liquid helium, i.e., 4 Kelvin), so that the effort to reach and keep those temperatures widely surpassed any gain to be expected. Moreover, the first and still valid theory of superconductivity, formulated in 1957 by Bardeen, Cooper, and Schrieffer, explained among other things why the critical temperatures had to be that low, and distorted all hope to find materials with higher critical temperature.

Given that background, it was a spectacular event when, in 1986, physicists Bednorz and Müller, working at the IBM laboratory close to Zurich, announced to have found a material with a critical temperature of 35 Kelvin. This was definitely higher than anything allowed by theory. No surprise, researchers all over the world joined in with efforts to replicate the effect, and since Bednorz and Müller had communicated experimental details, in particular about the complex composition of the material, replication was largely successful.

Very rapidly the new research field of high-temperature superconductivity (HTSC) became established and developed with dramatic speed: Within a few years, critical temperatures of above 77 Kelvin were reached, which could be obtained with liquid nitrogen, a material much cheaper than liquid helium and widely used in industrial applications. New perspectives of technical use were certainly one of the strongest driving forces of that development. The spectacular experimental success is characterized by high experimental stability that centrally involves stable replicability, to such a degree that in May 2014, the first large-scale commercial use of HTSC was launched at Essen (Germany).[3]

This is all the more striking as the theoretical situation is still unsolved. There is no theory available that would explain the effects (di Bucchianico 2014). HTSC offers a striking case in which experimental stability is so strong as to keep the research field going even while fundamental theoretical questions remain unsolved. In technical domains, like aircraft wing design or, further back, steam engine technology, such constellations have not been unknown. What makes HTSC specific is that it has come out of experimental research within physics, and still (regardless of its technical applications) remains there up to this day. Stability and reproducibility of experiments are one of the main factors that are important here (Leggett 1995, Holton *et al.* 1996).

As a side remark, it might be noted that there is a particular area of science in which replicability has gained an absolutely central place: modern metrology. While formerly standards have been fixed by material items, such as the prototype

[3]For more details see `www.dw.de/180-tage-supraleiter-in-essen/a-17904074`.

meter or kilogram in Paris, it has become standard nearly in all cases now to define them by a measuring procedure. This, of course, presupposes the idea that all those measurements, whenever and wherever in the world they are carried out, lead to the same outcome with a very high degree of precision. The switch from prototypes to measuring procedures for high precision standardization thus indicates a very high trust in the possibility of very precise replication.

3.4 Cases of Failed Replication

While all the above cases are characterized by successful replication, there are, of course, also cases of a different type – in which replication was tried, but constantly failed. To illustrate their variety, I shall sketch three episodes.

Johann Wilhelm Ritter, working in Jena in Thuringia in the early 19th century, was one of the most prolific researchers in the then new fields of galvanism and electrochemistry. Among many others things, he developed two new forms of the Voltaic pile – a dry pile that allowed much easier handling and, even more important, a chargeable pile that could, after being exhausted, be reactivated by electricity, which would avoid the arduous task of disassembling the whole pile, cleaning all metal plates, and reassembling it. These achievements were spectacular, and he had good reason to hope for the annual Volta prize announced by the Paris academy.

When his friend Ørsted, who was travelling in Europe, was setting off from Jena to Paris in 1803, Ritter commissioned him to present his discoveries at the Paris academy. Ritter was well aware that he could not count on an unbiased audience there – it was well known that Paris academicians had certain reservations against the type of research done in the context of German *Naturphilosophie* to which Ritter belonged. But he was confident that the demonstrations and replications before the academy would speak for themselves. He added, however, a third discovery to Ørsted's package: He had noted that an "open" pile received some charge when it was placed in certain compass directions for some time, an effect that Ritter interpreted as an indication of electric poles of the earth, additional to its magnetic poles.

At Paris, Ørsted was invited to present the experiments at the academy. While the two new types of the pile worked out well, the experiment regarding electric poles could not be replicated, even after intense efforts of academy members such as Biot and Coulomb. That failure of replication led not only to the dismissal of Ritter's claim (and his retraction as a candidate for the Volta prize), but damaged both Ritter's and Ørsted's reputations dramatically and fueled the academicians' general doubts about the soundness of experimental research impregnated by German *Naturphilosophie* (see Meyer 1920, p. xxxi, or Christensen 1995, pp. 164f).

Among more recent episodes, the case of US physicist Joseph Weber on gravitational waves stands out. His research in the 1960s was part of a general search for gravitational waves that had been proposed by Einstein's theory of gravitation, but not been detected so far. Weber developed his own detecting instruments ("antennas") that, like all proposed procedures up to this day, required not only extremely sensitive physical apparatus, but also very complex ways of processing the signals, i.e., increasing the signal-to-noise ratio.

At some point, Weber announced to have detected gravitational waves with his detectors. His claim was both spectacular and bold, since the intensity of the waves he claimed to have discovered was far higher than the intensity that was to be expected by theory. The history of how his claims were received, checked, and finally widely dismissed, has been treated in controversial ways by historians and sociologists (Collins 1985, 1994, Franklin 1994). I can only give a most abbreviated and selective sketch of what I see as their minimal consensus, which is already instructive enough.

Since the apparatus used was very expensive, and its operation required long time spans and sophisticated and skilled handling, direct replication of the observations was difficult, if not impossible. Instead, physicists who wanted to check the result, asked Weber for the raw data and for insight into the procedures of data reduction and evaluation. This can well be regarded as partial replication: while the immediate output of the apparatus itself had to be taken for granted, the highly complex procedure of computationally detecting signals out of a supposedly intense noisy background could be scrutinized and checked for stability.

In Weber's case, those attempts failed: researchers could not reproduce his results out of the raw data, but instead became aware of problems in the algorithms. Even in discussions with Weber, those points could not be clarified, all while Weber upheld his claims and confirmed them even with revised procedures. As a result of a constellation in which (partial) replication by others constantly remained unsuccessful, an increasing part of the community tended to doubt the results, and some physicists even started campaigns to finally dismiss them and close the discussion. While Weber still is seen a pioneer in experimental techniques of gravitational wave detectors, his results are no longer discussed.

My third episode has striking similarities and differences to the HTSC case. In March 1989, the physical chemists Stanley Pons and Martin Fleischmann at the University of Utah announced a spectacular phenomenon: They had observed effects that they interpreted as indicating the fusion of hydrogen nuclei to helium – a nuclear fusion process otherwise occurring at temperatures of millions of Kelvins in stars, including our sun, and exposed most dramatically in hydrogen bombs. It is a process that is capable of delivering endless energy from an omnipresent chemical element – if it can be successfully controlled.

Attempts to generate this process in fusion reactors have been made, with huge international efforts, since the 1950s and are still ongoing. To have that process realized in small table-top experiments under normal laboratory conditions (at room temperature!), as Pons and Fleischmann claimed, opened dramatic perspectives. If it worked, it would have the potential to solve global energy challenges in completely unexpected ways. The technical promise was even more immense than for HTSC but, as in that case, any theoretical explanation was not only far away, but there were serious doubts from the side of theory about the very possibility of such an effect.

Researchers all over the world rushed to work, all the more since the equipment involved was mainly standard and widely available. The new research field of "Cold Fusion" was created. Replication very quickly came out to be problematic, however, since Pons and Fleischmann had not provided much experimental details in their announcements (probably also for reasons of commercial promise). Hence, the reception took a very different pathway than for HTSC, with long discussions and many attempts at replication, all of which failed at the central point of the experiment, the generation of excess heat. Despite the immense technical promise of the effect, research on the field has nearly disappeared.[4]

One core factor for that development certainly was that all attempts of replication by others failed in the end. Hence, and despite Pons and Fleischmann kept up their claims, the trust in the existence of the effect vanished. While for both HTSC and Cold Fusion immense technological and economical interest was visible and kept research going, and while in both cases support from theory was lacking, the development of the two fields went very different ways. One of the decisive factors of that difference (perhaps not the only one, given the complexity of the two histories) is certainly the strikingly different outcome of attempts at replication. Replicability made perhaps not all, but certainly a lot of the difference (Lewenstein 1992, Huizenga 1994).

The above cases show that replication and replicability can provide central arguments for the acceptance of experimental results. They worked for or against them also in those cases in which neither the existing theoretical or conceptual framework nor the constellation of authority within a community favored the

[4]A less pessimistic view about excess heat is due to Beaudette (2002). That the entire controversy is not completely extinct is also shown by novel discussion forums under the acronym LENR for "low energy nuclear reactions."

acceptance of the experimental result. The cases illustrate, on the one hand, how successful replication can establish a result even without the resources necessary to situate it in a theoretical context. They indicate, on the other hand, that in cases in which replication failed, this failure led to increasing doubts and disbelief in the experimental claims in the long run. The cases also illustrate the variety of strategies by the part of the experimenters to enable replication by others.

3.5 Doing Science without Replication and Replicability

But history also offers cases of a different kind, in which replicability was not a main point in communicating experimental results and was not the main argument in the process of its acceptance.

Coming back to Alhazen, with his explicit emphasis on repetition, it is significant to note that there was no attempt to enable replication by others – at Alhazen's time that idea was unfamiliar and he had no reason to think that others would attempt to replicate his experiments. Rather he emphasized the stability, the stable repeatability, of his own experiments. The ideal of gaining acceptance by replication, deeply rooted as it might appear to us, is not necessarily bound to experimental research. As indicated above, expectations as to replication and replicability were fully present not later than in the 17th century. Nevertheless, we find numerous cases in which replication was not a goal.

The first vacuum pump was designed and put in operation by the Magdeburg mayor and former technical advisor Otto von Guericke in the 1650s. While vacuum had been produced earlier by way of barometers in small spaces at the upper end of glass tubes, Guericke's device was the first to visibly suck the air out of a larger space, and even to let it refill again. Hence, his arrangement was of high novelty, and he did a lot to attract public interest. Although he had quite specific, if not idiosyncratic, ideas about what those experiments tell us about cosmic space, even the experiments as such were spectacular.

As is well known, he performed various experiments with the pump in front of noble audience in Magdeburg, Berlin, and not the least at the Regensburg assembly of the emperor in 1653/1654 – scenes of those experiments have been often depicted. Guericke's apparatus was unique, complicated, expensive, and difficult to handle. It is unclear whether Guericke expected others to replicate the experiments. In his publication of 1672 (to which Guericke was pushed by the Jesuit father Gaspar Schott who helped write the text), he gave a detailed description of the apparatus and its components, enriched by illustrations, but it is more than doubtful whether one could have built the complicated apparatus on the basis of those descriptions (Guericke 1672, book 3, Chap. 4).

In any case, it is clear that replication was not an issue for Guericke; but even without replication, no doubts were raised about his results. This had probably to do with his public performance that could be witnessed by a large number of participants. Even more drastically does such a disinterest in replication show up in Guericke's experiments with a sulphur globe that gave rise to curious effects of attraction and repulsion of small bodies – Guericke hesitated to call them electric effects. Here, his description of how to manufacture the globe was that short that he could definitely not expect someone to replicate the globe nor the (rather complicated) experiments on that base (Guericke 1672, book 4, Chap. 15).

Having heard about Guericke's apparatus, the English nobleman Robert Boyle, most interested in natural philosophy and experiments, designed an air pump together with Robert Hooke and made it built in the 1660s (for an overview of early airpumps, see van Helden 1991). He introduced a decisive new element: instead of having the receiver made of metal globes, he switched to glass, which created some problems (since it could break) but also allowed to investigate, for the first time, the behavior of various things brought within the receiver and exposed to the vacuum.

Again the apparatus was unique, designed and produced for the first time. Not only was it very expensive, but it also required techniques and skills that Boyle and his helpers had to develop from scratch when they started to build and operate the machinery step by step. Hence, Boyle knew it would have been unrealistic to expect others to attempt or even achieve replication, at least on broad scale. At the same time, he did not give public performances in the way Guericke had done. Rather, he invited selected persons to his private house where he had installed the apparatus. Being well aware of the need to convince a larger audience of the reliability of his numerous experiments (that led him to significant philosophical consequences), he pursued a different strategy: he described, in his published account of his experiments, the apparatus, its component, its operation, and the process and results of the experiments, in very minute detail, and complemented the text by detailed illustrations. Historians Shapin and Schaffer (1985) have analyzed that strategy in detail, labeled it "virtual witnessing," and argued convincingly that Boyle used it deliberately to compensate missing replicability and, nevertheless, increase trust in his experiments and their results.

In the late 18th century there is again a striking case in which replicability played no role: the first direct measurement of the law of electric attraction and repulsion. That those forces decrease with distance had been taken for granted anyway, and the analogy to gravitation suggested a $1/r^2$ dependence. The problem of measuring those forces was that they were extremely weak, and the first to attempt to measure them directly was the former civil engineer Coulomb.

He had worked on measuring magnetic intensity before, and developed arrangements with a torsion wire that were apt to react to very weak forces. Transferring those experiences to electric forces, he designed and built a torsion balance for electric forces and carried out experimental series with it. As a result, he announced in 1785 to have found a $1/r^2$ law for electric repulsion and, with a different apparatus, also for electric attraction. The torsion balance was just built for those experiments. It was extremely sensitive to all sorts of disturbances, and the procedures to operate it were very refined.

Historians who have tried to replicate the result recently have realized how difficult replication must have been (Heering 1994, Martínez 2006). While Coulomb described his apparatus and procedures in some detail in his recollections, and also added some illustrations, it is unclear whether he expected others to attempt replication. And, as far as we know, there were no attempts to replicate the experiment for decades to come.

Nevertheless, and even without specific rhetoric strategies on the side of Coulomb, the result as such was nowhere put in question, but immediately accepted and praised as success of precision measurement. This strikingly unproblematic acceptance had certainly to do with the authority Coulomb had gained as a master of measurement, with the rising enthusiasm of the period for precision measurement, but probably mostly with the fact that a $1/r^2$ force law for electrical attraction and repulsion was well expected and fit perfectly into the general framework. There was no theoretical or conceptual reason to raise any doubt (Heering 1994).

It is important to note that such an attitude is much more common than it might appear at first sight. Even in our days, a great lot – I guess indeed the vast majority – of published experiments are not subjected to replication. In our competitive science system, replications of other experiments do not count as scientific achievements, and papers that just report successful replication (or even failure of replication) will scarcely be published (see, e.g., Tsou *et al.* 2014). In most cases, there is no incentive for replication, and acceptance of results rather relies on trust. The rare cases in which replication actually is attempted are those that either are central for theory development (e.g., by being incompatible with existing theory) or promise broad attention due to major economical perspectives. Despite the formal ideal of replicability, we do not live in a culture of replication.

Moreover, there are fields of modern science in which replication is no longer feasible for practical and monetary reasons. The devices of particle physics, both accelerators and detectors, have grown larger and larger in the after-war period. Until recently there used to be more than one machine of the same kind around on the globe, and results achieved at one place could, and usually were, placed under scrutiny at some other laboratory.

This situation has changed now. In the run to ever higher energies, and exploding costs of those machines, there is only one laboratory left for highest particle energies: CERN at Geneva has constructed, with support by many nations in the world, the large hadron collider (LHC) with corresponding detectors. This is now the only laboratory that can reach particle energies in the tera-electron volt range (1 TeV = 10^{12} electron volt). A first long series of experiments and longish evaluations has led, in 2012 and 2013, to the discovery of the Higgs particle that has been postulated by theory. That discovery, to be sure, involved immense amounts of statistical evaluation, all the more since the Higgs particle cannot be detected directly (it is too short-lived) but only retrospectively inferred from the analysis of complex decay schemes of other particles.

It is to be expected that those experiments will not be replicated elsewhere, and be it just for monetary reasons (the costs are counted in billions of Euros). Well aware of the problem of lacking external replication, the laboratory has put not only one, but two large detectors at different spots of the LHC accelerator. While they have fundamentally different design, they still work with the same particle beam. The large groups of researchers here and there work quite independently, so as to come close to the ideal of at least partially independent experiments. And the fact that, in the Higgs experiment, they have largely agreed on the result, provides indeed considerable support for its validity.

Similar constellations will arise in other large-scale experimental and observational projects, such as the "Atacama Large Millimeter/Submillimeter Array" (ALMA) observatory, the "IceCube South Pole Neutrino Observatory," or new gravitational wave detectors. One way to compensate, at least partially, the lack of replicability, is certainly to make the raw data and evaluation procedures openly accessible, and this is what usually is done in those cases now. The awareness of measures to be taken at that point makes clear how deeply replicability (or the lack thereof) is taken to be central for experimental research. Many experimenters see a certain lacuna when replicability is not possible for whatever reasons, and try to compensate it by other means. The situation resembles the case of Boyle who, knowing that replicability of his air-pump experiments was not to be expected for practical reasons, was very much concerned with measures to include his readers as tightly as possible.

These cases, taken from very different historical periods and contexts, show a striking variety of attitudes toward replicability. They make clear that replicability was and is by no means the only way to make experimental results accepted. Sometimes it is not required since the result fits expectations so well that no doubt remains. Sometimes it is just not possible for reasons of resources, and in earlier times – think of Alhazen – replication has just not been the standard of experimental research. Replicability may be a sufficient but is certainly not a necessary condition for accepting experimental results.

Going even further, there is another important aspect to be highlighted – an aspect that concerns experiments for practical purposes, e.g., performance for possible sponsors or for a paying public or for medical applications. As emphasized above, it is essential for these experiments that they always work successfully and repeatably.

But it is important to note that this does not imply aiming at replicability by others. Often enough the experimenter has to make a living of successful performances within a competing environment and, hence, it is not in his interest to enable others to perform the same experiments. This marks an obvious tension between the ideal of scientific knowledge as openly accessible in all its aspects, and commercial and technological uses of experiments that often want to keep the way to obtain the results in house. As a consequence, experimenters aim at repeatability, but not at replicability.

That tension has a long history, reaching back to the early days of alchemy in which experiments have typically been communicated in a way that veils rather than enables replicability. In periods and constellations when doing research with the ideal of open scientific knowledge and performing experiments for practical purposes (like show or healing) have been closely aligned, that tension has been constantly visible. An illustrative case is provided by 18th-century electricity, when the field was developing quickly both in the academic realm and in public space where it provided an income to many performers (Hochadel 2003, Chap. 5).

The tension has not vanished – a modern variant can be found in all fields in which research is conducted or funded by industry, and in which there is a permanent negotiation of which details of experiments might be published or kept hidden. Nevertheless, in many of these cases, there is no doubt about the validity and scientific interest of the experimental results. Hence, we should be cautious not to discard too easily cases in which we see the tendency to weaken rather than to enable replicability right away as pseudo-science or fraud, as has so commonly been done, in particular in the case of alchemy.

3.6 What Can We Learn from History?

Questions of replication and its use, necessity, possibility, and reliability in experimental science came into the focus of history and sociology of science in the 1980s. Harry Collins, emphasizing social interaction, claimed the omnipresence of what he called the "experimenters' regress," a vicious circle between the "proper outcome" of an experiment and the "proper functioning" of its apparatus – a circle that can, according to Collins, be broken only by a social process.

Concomitant with that claim he put the claim of the impossibility to establish experimental results in science by replication alone. His claims have been strongly contested, and fierce debates arose, in particular between Collins and Franklin. Looking back, we see that the debate was fruitful in many ways and brought up many issues, among others, the question of what types of historical sources to use and what types of claims to base on them. I take it as a most welcome sign of the vivid life of philosophical–sociological–historical reflections on science that positions have developed further, learnt from each other, even to such a degree that Collins and Franklin now agree on the sources of their disagreement and find much common ground for their current approaches to science studies (Franklin and Collins 2015).

In my historical outlook onto the question of replication and replicability I wanted to provide material for a differentiated view. I attempted to identify various types of constellations of replication, replicability, and the acceptance of experimental results. Since my case studies are taken from very different periods and historical settings, putting them together under one common perspective raises immediately the question of whether that perspective itself can be fruitfully applied over such varieties in time and situation.

As becomes indeed clear in the Alhazen case, that assumption has its limits: We see that the expectation to convince others of the validity of experimental results is not addressed at all – there was just another culture of trust at work. Only from the 17th century on we find these issues addressed regularly. But from then on we find indeed a long continuity of discussion about the role of replication and replicability. Let me summarize some general points.

- The repetition of experiments by the experimenters (or the experimental group) themselves seems to be an omnipresent and essential element in experimental practice in order to achieve experimental stability.
- There are cases in which replicability did all the job, i.e., experimenters became convinced of other experimenters' results just by replication, even while giving no credit to the report and its author, and even while it was clear that the result was not compatible with existing theory.
- There are cases in which an experimental result was stabilized on the long run by successful replication alone while resisting theoretical explanation. For HTSC, e.g., a stable phenomenon created in the laboratory even led to technical applications without theoretical foundation.

- There are cases in which a confirmation of experimental results was of utmost theoretical, technical, or economical importance so that the pressure of successful replication was particularly high. However, replication failed consistently, and the results finally have been discredited.
- There are many cases in which experimental results were accepted without any demand for replication, i.e., where replicability was not an issue at all.

These cases bring various points to light regarding the role of replication and replicability in searching acceptance for the validity of experimental results. First, an asymmetric constellation becomes visible. Acceptance of experimental results has been achieved by different means, among which replication certainly played an outstanding role, but was not always required. Replicability has not been taken as a necessary condition of acceptance. By contrast, however, it obviously formed a sufficient condition: It will be hard to find an experimental result that was constantly replicable but not widely accepted. On the other hand, in cases where replication was seriously and repeatedly tried but constantly failed, this failure would weaken the experimental claim and put it in doubt.

As a consequence, it becomes clear that we should split the question about the role of replication in accepting an experimental result into at least two issues: Was replication aimed at, or attempted, at all? And if it was, has it been successful, and which consequences arose from it?

Historical analysis shows that replicability of experiments has never been a value *per se*, but was always embedded in a set of contexts. The specific settings of each case have to be considered, and one result of the analysis given above has been to identify some of those contexts that must be taken into account in particular. They include at least the type of equipment and skills needed for the experiment (from standard to extreme novelty); the relation of the experimental claim to the theoretical and conceptual situation of the period (from perfect fit to completely incompatibility); the status and trustworthiness of the researcher within the community; the time, effort, and expenses to be invested in attempts at replication; and the expected chances to achieve a replication with given resources in appropriate time.

The question of replication and replicability is much more complex than easy textbook accounts make us believe. Historical analysis shows that there are no easy and clear-cut answers and that it is not useful to have simple general claims such as "it's all about replicability" or "replicability does not decide anything." A more differentiated view helps us to develop a more refined perspective that one should keep in mind when estimating the role and value of replication and replicability in experimental science.

Acknowledgments

I owe a lot to the papers by and discussions with Jutta Schickore – many thanks to her. I also thank Arianna Borrelli for helpful comments and Teresa Hollerbach for assistance in preparing this article.

References

Beaudette, C.G. (2002): *Excess Heat: Why Cold Fusion Research Prevailed*, Oak Grove Press, South Bristol.

Beretta, M., Clericuzio, A., and Principe, L.M., eds. (2009): *The Accademia del cimento and Its European Context*, Science History Publications, Saggamore Beach.

Bernard, C. (1865): *Introduction l'étude de la médecine expérimentale*, Baillière, Paris.

Bogen, J. (2001): "Two as good as a hundred": Poorly replicated evidence in some nineteenth-century neuroscientific research. *Studies in History and Philosophy of Biological and Biomedical Sciences* **32**, 491–533.

Boschiero, L. (2007): *Experiment and Natural Philosophy in Seventeenth-Century Tuscany. The History of the Accademia del Cimento*, Springer, Dordrecht.

Boyle, R. (1999): Two essays concerning the unsuccessfulness of experiments. Containing diverse admonitions and observations (chiefly Chymical) touching that Subject. In *The Works of Robert Boyle*, ed. by M. Hunter and E.B. Davis, Pickering and Chatto, London. Originally published in 1661.

Christensen, D.C. (1995): The Oersted-Ritter partnership and the birth of Romantic natural philosophy. *Annals of Science* **52**, 153–185.

Collins, H.M. (1985): *Changing Order. Replication and Induction in Scientific Practice*, Sage Publications, Beverly Hills.

Collins, H.M. (1994): A strong confirmation of the experimenters' regress. *Studies in History and Philosophy of Science* **25**, 493–503.

Crombie, A.C. (1961): *Augustine to Galileo: Science in the Later Middle Ages and Early Modern Times, XIII–XVII Centuries*, Heinemann, London.

Dear, P. (2001): *Revolutionizing the Sciences. European Knowledge and Its Ambitions, 1500–1700*, Palgrave, Basingstoke.

di Bucchianico, M. (2014): A matter of phronesis: Experiment and virtue in physics, a case study. In *Virtue Epistemology Naturalized: Bridges Between Virtue Epistemology and Philosophy of Science*, ed. by A. Fairweather, Springer, New York, pp. 291–312.

Findlen, P. (1993): Controlling the experiment: Rhetoric, court patronage and the experimental method of Francesco Redi. *History of Science* **31**, 35–64.

Franklin, A. (1994): How to avoid the experimenters' regress. *Studies in History and Philosophy of Science* **25**, 463–491.

Franklin, A., and Collins, H. (2015): Two kinds of case studies and a new agreement. In *The Philosophy of Historical Case Studies*, ed. by T. Sauer and R. Scholl, Springer, Berlin, in press.

Guericke, O. von (1672): *Experimenta nova (ut vocantur) Magdeburgica de vacuo spatio*, Waesberge, Amsterdam.

Heering, P. (1994): The replication of the torsion balance experiment. The inverse square law and its refutation by early 19th century German physicists. In *Restaging Coulomb: Usages, controverses et réplications autour de la balance de torsion.*, ed. by C. Blondel and M. Dörries, Leo S. Olschki, Florence, pp. 47–66.

Hochadel, O. (2003): *Öffentliche Wissenschaft. Elektrizität in der deutschen Aufklärung*, Wallstein, Göttingen.

Holmes, F.L., Renn, J., and Rheinberger, H.-J., eds. (2001): *Reworking the Bench: Laboratory Notebooks in the History of Science*, Springer, Berlin.

Holmes, F.L. (1992): Do we understand historically how experimental knowledge is acquired? *History of Science* **30**, 119–136.

Holton, G., Chang, H., and Jurkowitz, E. (1996): How a scientific discovery is made: A case history. *American Scientist* **84**, 364–375.

Huizenga, J.R. (1994): *Cold Fusion. The Scientific Fiasco of the Century*, Oxford University Press, Oxford.

Ibn-al-Haitham (1989a): Translation. In *The Optics of Ibn al-Haytham, Books I–III, On Direct Vision*, transl. with introd. and comm. by A.I. Sabra, Warburg Institute, London.

Ibn-al-Haitham (1989b): Introduction, commentary, glossaries, concordance, indices. In *The Optics of Ibn al-Haytham, Books I–III, On Direct Vision*, transl. with introd. and comm. by A.I. Sabra, Warburg Institute, London.

Knowles Middleton, W.E. (1971): *The Experimenters. A Study of the Accademia del Cimento*, Johns Hopkins Press, Baltimore.

Lewenstein, B.V. (1992): Cold fusion and hot history. *Osiris* **7**, 135–163.

Leggett, A.J. (1995): Superfluids and superconductors. In *Twentieth Century Physics, Vol. II*, ed. by L.M. Brown, A. Pais and A.B. Pippard, AIP Press, New York, pp. 913–966.

Lloyd, G.E.R. (1991): Experiment in early Greek philosophy and medicine. In *Methods and Problems in Greek Science*, ed. by G.E.R. Lloyd, Cambridge University Press, Cambridge, pp. 70–99.

Martínez, A.A. (2006): Replication of Coulomb's torsion balance experiment. *Archive for History of Exact Sciences* **60**, 517–563.

Meyer, K., ed. (1920): *H.C. Oersted. Naturvidenskabelige Skrifter – Scientific Papers, 3 Vols.*, Hoest & Soen, Copenhagen.

Newman, W. (1998): The place of alchemy in the current literature on experiment. In *Experimental Essays – Versuche zum Experiment*, ed. by M. Heidelberger and F. Steinle, Nomos, Baden-Baden, pp. 9–33.

Newman, W. (2006): *Atoms and Alchemy. Chymistry and the Experimental Origins of the Scientific Revolution*, University of Chicago Press, Chicago.

Sargent, R.-M. (1995): *The Diffident Naturalist. Robert Boyle and the Philosophy of Experiment*, University of Chicago Press, Chicago.

Schaffer, S. (1989): Glass works. Newton's prisms and the uses of experiment. In *The Uses of Experiment: Studies in the Natural Sciences*, ed. by D.C. Gooding, T. Pinch, and S. Schaffer, Cambridge University Press, Cambridge, pp. 67–104.

Schickore, J. (2010): Trying again and again. Multiple repetitions in early modern reports of experiments on snake bites. *Early Science and Medicine* **15**, 567–617.

Schickore, J. (2011): The significance of re-doing experiments: A contribution to historically informed methodology. *Erkenntnis* **75**, 325–347.

Schramm, M. (1998): Experiment in Altertum und Mittelalter. In *Experimental Essays – Versuche zum Experiment*, ed. by M. Heidelberger and F. Steinle, Nomos, Baden-Baden, pp. 34–67.

Shapin, S. (1988): The house of experiment in seventeenth-century England. *Isis* **79**, 373–404.

Shapin, S., and Schaffer, S. (1985): *Leviathan and the Air-Pump. Hobbes, Boyle and the Experimental Life*, Princeton University Press, Princeton.

Shapiro, A.E. (1996): The gradual acceptance of Newton's theory of light and color, 1672–1727. *Perspectives on Science: Historical, Philosophical, Social* **4**, 59–140.

Shapiro, B.J. (2000): *A culture of fact. England, 1550–1720*, Cornell University Press, Ithaca.

Steinle, F. (2005): *Explorative Experimente. Ampère, Faraday und die Ursprünge der Elektrodynamik*, Steiner, Stuttgart. An English version is forthcoming under the title *Exploratory Experiments. Ampère, Faraday, and the Origins of Electrodynamics*, Pittsburgh University Press, Pittsburgh 2016.

Steinle, F. (2014): "Erfahrung der höhern Art". Goethe, die experimentelle Methode und die französische Aufklärung. In *Heikle Balancen. Die Weimarer Klassik im Prozess der Moderne*, ed. by T. Valk, Wallstein, Göttingen, pp. 221–249.

Tsou, A., Schickore, J., and Sugimoto, C.R. (2014): Unpublishable research: Examining and organizing the "file drawer". *Learned Publishing* **27**(4), 253–267.

van Helden, A.C. (1991): The age of the air pump. In *Tractrix. Yearbook for the History of Science, Medicine, Technology and Mathematics* **3**, pp. 149–172.

4
Reproducibility of Experiments: Experimenters' Regress, Statistical Uncertainty Principle, and the Replication Imperative

Harry Collins

Abstract. In the first of three parts I describe the experimenters' regress – a problem that scientists face when they try to replicate experiments. It shows itself where repeatability is used to settle scientific controversies. In the second part I redescribe some of the problems that experimental or observational findings face based on statements of statistical confidence drawing an analogy between the enigmas of statistical confidence statements and the uncertainties of quantum theory. I argue that statistical meta-analysis should be seen as always in need of meta-meta-analysis. The third part argues that even though replication is beset by the kinds of problems described in the first two parts, we must, nevertheless, stick to it as the fundamental way of showing that our scientific results are secure. Science is a matter of having the right aspirations even if they cannot be fulfilled.

4.1 Introduction

This paper has three parts corresponding to the three elements of the subtitle. In the first part (Section 4.2) I describe the experimenters' regress for those who do not know about it. It is a problem that scientists face when they try to replicate experiments, and it shows itself where repeatability is used to settle scientific controversies.

In the second part (Section 4.3) I redescribe some of the problems that experimental or observational findings face based on statements of statistical confidence. I draw an analogy between the enigmas of statistical confidence statements and the uncertainties of quantum theory. In this light statistical meta-analysis should be seen not as definitive but only the best we can do in practice and always in need of meta-meta-analysis.

The third part (Section 4.4) argues that even though replication is beset by the kinds of problems described in the first two parts, we must, nevertheless, stick to it as the fundamental way of showing that our scientific results are secure. Science is a matter of having the right aspirations even if they cannot be fulfilled.

Reproducibility: Principles, Problems, Practices, and Prospects, First Edition. Edited by Harald Atmanspacher and Sabine Maasen.

A contribution to a handbook would normally pull together a wide body of existing and established work. My chapter departs from the normal convention in that the second and third parts include novel interpretations based almost entirely on the work of the author.

4.2 The Experimenter's Regress

Experiments are difficult. One reason is that much experimental skill is tacit (Collins 2010). Therefore, an experiment cannot be fully described in print even with the best will in the world. That the conventions of scientific writing and publishing prevent the promulgation of intimate details of experimental trials, errors, and fudges, only makes things worse. Thus, an experimenter who wishes to test another's findings by replicating the experiment finds it hard to know whether any failure to find the same result is a consequence of the result not being robust or the replication being unlike the original in some crucial way.

In practice we normally test the competence of experiments and experimenters by checking the outcome: we know the experiment has been done right if it produces the right outcome, and we know it has not been done right if it does not produce the right outcome. But where the outcome is at stake, this criterion cannot be used. There are other ways of trying to find out whether an experiment has been carried out competently, such as calibrating it by having it produce some surrogate phenomenon, but since a calibration is itself an experiment, it too can be challenged. And it almost certainly will be challenged where the controversy runs deep.

We can express the problem as a kind of logical regress: To know whether an experiment has been well conducted, one needs to know whether it gives rise to the correct outcome. But to know what the correct outcome is, one needs to do a well-conducted experiment. But to know whether the experiment has been well conducted ... ! This is called the experimenters' regress.[1]

It has been argued that the experimenters' regress can be seen as a modern form of a long established paradox with general application. The following two quotes from Sextus Empiricus and Montaigne demonstrate this.[2] First, Sextus

[1]Experimenter's regress has been extensively illustrated for the case of gravitational wave detection and other examples. See, e.g., Collins (1975) – before the notion was coined – or Collins (1985, 1993, 2004) for a thorough history. Once the phenomenon and its correlates are recognized, they can nearly always be seen in operation wherever there is a scientific controversy.

[2]Sextus Empiricus was writing in the second or third century and Montaigne in the 1500s. Both quotations are taken from Godin and Gingras (2002), who first related the experimenters' regress to more general forms of skepticism. For a response see Collins (2002).

Empiricus (1933, Vol. II, p. 20):

> In order for the dispute that has arisen about standards to be decided, one must possess an agreed standard through which we can judge it; and in order to possess an agreed standard, the dispute about standards must already have been decided.

And here is the quote from Montaigne (1938–1942, Vol. II, p. 322):

> To judge the appearances that we receive from objects, we would need a judicatory instrument; to verify this instrument, we need a demonstration; to verify this demonstration, an instrument: there we are in a circle.

There are forerunners to the experimenters' regress, but these were set out before the notion of controlled experiments was invented. Working scientists did not find themselves troubled by these logical conundrums. The experimenters' regress shows, however, that such problems are not circumvented by the experimental method.

It also shows how the problem works out in the day-to-day practice of scientists: Disputants tend to form groups who believe a (e.g., positive) result of one set of experiments that has been competently performed or another (e.g., negative) result of another set of experiments that has been competently performed. If both groups are determined, disputes between them cannot be settled simply by doing more experiments since each new result is subject to the questions posed by the regress.[3]

Reaching complete consensus in such cases may take many decades while the potential to reopen them never goes away. If the controversy cuts deep, "nonscientific" factors, such as a group's or an individual's track record or institutional affiliation, will play a part in the consolidation of the lasting view. Of course, it does not have to be this way. Sometimes one of the groups of scientists might feel that the balance of scientific credibility has moved so far against them that it is time to give up. Sometimes a group's work will have no outlet but fringe journals or blogs.

The importance of scientific results and "non-scientific" social factors will differ from case to case but the experimenters' regress shows that experiment alone cannot force a scientist to accept a view that they are determined to resist. Whatever, the eventual collective decision about what constitutes the

[3]As mentioned above, trying to break the regress by calibrating an experiment with a surrogate phenomenon simply pushes the question back to the adequacy of the calibration. Franklin (1994) and Collins (1994) disagreed about this previously but have since reached a considerable measure of agreement (Franklin and Collins 2015).

right result will be coextensive with the collective decision about which are the competently performed experiments. It is the outcome of such experiments that sets the criterion for a well-performed experiment once closure is reached.

The research that led to the idea of the experimenters' regress encouraged related studies. Early cases include those by Pickering (1981), Pinch (1981), and Travis (1981), but there are many more. The idea has also been adapted to a number of similar situations. Kennefick (2000) talks of the "theoretician's regress"; Miller (2006) of the "patent specifier's regress"; Gelfert (2011) of the "simulator's regress"; and Stephens *et al.* (2013) of the "regulatory regress." For other versions I recommend a Google search for "experimenters' regress."

4.3 The Statistical Uncertainty Principle

Now I turn to the replicability of experiments or observations that are supported by statistical significance tests. Needless to say, these too are subject to the experimenters' regress, but I want to look at some features of such work from a more narrow perspective. I will base the argument on a chapter that describes and discusses the statistics of gravitational wave detection.[4] That chapter concludes: "A statistical test is like a used car. It may sparkle, but how reliable it is depends on the number of owners and how they drove it – and you can never know for sure."

I am going to go somewhat further and suggest that, looked at in a certain way, the results of significance tests approach the weirdness of quantum theory. This means we should look at the replication of statistical experiments and at meta-analysis in a different way. Though the analysis starts with gravitational wave physics, the argument applies to all uses of statistical confidence statements across the sciences.

4.3.1 Some Selected Examples

Gravitational wave physicists have to use statistical means to distinguish signals from noise. A central problem is what physicists refer to as the *trials factor* or, sometimes, the *look elsewhere effect*.[5] Gravitational wave detection is a

[4]See Collins (2011, Chap. 5). The second edition of this book, entitled *Gravity's Ghost and Big Dog*, deals with two case studies and adds a further section of methodological reflection.

[5]The term "look elsewhere effect" seems to have been coined by Lyons (2008, 2010). He is a member of the statistics committee of the collider detector at Fermilab and has published a textbook on statistics for high-energy physics (Lyons 1986). He argues that when one is searching for similar experiments that need to be taken into account when calculating a trials

pioneering science that has been trying and failing to detect the exquisitely weak waves for around half a century. The detectors "on air" in the late 2000s were enormous laser interferometers with 4 kilometer long arms. A passing gravitational wave, as it might be caused by the final inspiraling and coalescence at the end of the life of a binary neutron star system, would manifest itself by relative length changes in the two arms of a detector of around 1,000th of the diameter of an atomic nucleus. Since such changes are regularly present in the detectors due to various noise sources, a "detection" consists of a statement that a certain pattern of disturbances is unlikely to have been caused by chance. Nowadays, the standard adopted for such a claim is an unlikelihood equivalent to the signal being five standard deviations distant from the mean of a normal distribution – the "5σ" standard. This is also the contemporary requirement for a discovery claim in high-energy physics.[6]

The history of gravitational wave physics has been beset by claims that did not stand the test of time. The most notorious ones were Joseph Weber's claims to have detected gravitational waves with a relatively cheap resonating bar. The high point of these claims was the early 1970s, but by around 1975 they had lost credibility. Some believe that Weber "manufactured" his results through inadvertent statistical massage, notably trying many different ways of cutting up his data and reporting only the successful ones. For instance, one respondent said (Collins 2004, pp. 394–395):

> Joe would come into the laboratory – he'd twist all the knobs until he finally got a signal. And then he'd take data. And then he would analyze the data: he would define what he would call a threshold. And he'd try different values for the thresholds. He would have algorithms for a signal – maybe you square the amplitude, maybe you multiply things ... he would have twelve different ways of creating something. And then thresholding it twenty different ways. And then go over the same data set. And in

factor – when one "looks elsewhere" to see what else has been done – there is a range of possibilities. I was unaware of his work when I wrote Chapter 5 of *Gravity's Ghost*, upon which this section is based, but his and my treatment are analogous. An interesting discussion of the problem, including Lyons's contribution, can also be found in the CERN Courier of October 26, 2010 `cerncourier.com/cws/article/cern/44115`. See also Gross and Vitells (2010). Lyons (2008) accepts that statistical choices in physics are conventional: "Physicists and journal editors do like a defined rule rather than a flexible criterion, so this bolsters the 5σ standard. The general attitude is that, in the absence of a case for special pleading, 5σ is a reasonable requirement." The interest of physicists, of course, is to establish the right convention, whereas it is my intention to point to the indeterminate nature of statistical calculations.

[6]This convention too has changed over the years from 3σ in the 1960s (Collins 2011). For a more detailed discussion of changing experimental practices, including reporting conventions, see Franklin (2013).

> the end, out of these thousands of combinations there would be a peak that would appear and he would say, "Ahawe've found something." And [someone] knowing statistics from nuclear physics would say, "Joe – this is not a Gaussian process – this is not normal – when you say there's a three-standard-deviation effect, that's not right, because you've gone through the data so many times." And Joe would say, "But – What do you mean? When I was working, trying to find a radar signal in the Second World War, anything was legal, we could try any trick so long as we could grab a signal."

We do not need to agree that this was truly how Weber proceeded so long as we can agree that if he had proceeded in that way, and chose his setting retrospectively according to which gave the best signal, his reported statistical confidence levels would not have been sound. Every time you take a new "cut" of the data without finding a positive result, you reduce the statistical confidence you have in the first cut. This problem still besets gravitational wave research (as it besets every science with detection problems of this kind).

Notice, however, that in principle Webers procedure would have been sound in logic – if not in practice – so long as he chose the settings he would report according to criteria that were external to and unaffected by the strength of the signal to which they gave rise. It is likely to be the case in a pioneering science that the best statistical algorithm will take a long time to work out – it requires ingenuity to find the optimum way to separate noise from signal. And if all those knob twistings had been driven by some kind of theoretical program of deeper and deeper understanding of the statistics, then the procedure would be sound in principle.

Thus, understanding what a statistical outcome means depends on knowing what the analyst had in mind. It is rather as though someone asks you for your birthday and, when you report June 13th, they say "what an amazing coincidence: that is exactly the date I had guessed; the odds against that were 365 to 1. Those are the odds indeed, but only if the "someone" really did have that date in mind before you said "June 13th." Otherwise the "guess" is entirely unremarkable.

In the gravitational wave detection business such a problem arises in principle for Weber's procedure and it arose in practice, more recently, when an Italian group found an unusual result that has to do with the direction of the source of certain signals found by their detector. They claimed that the signal was bunching in an hour of the day, which indicated that the direction it was coming from was our galaxy, and that this was unlikely to have happened by chance. They claimed that the odds against this periodicity were around 75 to 1: enough to make the phenomenon interesting.

An American analyst calculated, however, that the odds of a chance signal bunching in 1 "bin" out of the 24 bins representing the hours of the day were only 4 to 1 – which was completely uninteresting (see Astone *et al.* 2002 and Finn 2003). The argument, which became heated, depended, then, on what the Italian group had in mind prior to their analysis. If they had the direction of the galaxy in mind, then it was interesting; if they noticed this only in retrospect, it was not. In practice, such an internal state is very hard to determine to the satisfaction of everyone, especially since we do not have complete control over, or knowledge of, even our own internal states (which is the rationale for "blinded" experiments; see Collins 2011, Chap. 5).

Contemporary gravitational wave physicists are acutely aware of this kind of problem and try to resolve it with a strict protocol that allows the development of statistical procedures – "pipelines" as they are known – by retrospective analysis, but only of an unimportant subset of the data. When this exploratory work is completed the statistical procedures must be frozen and only a single, agreed and finalized, pipeline is applied to the entire body of data. To use the terminology, any kind of analysis is allowed on the "playground data" but the analysis must be frozen "before the box is opened" on the real data.

There can still be arguments about whether an actual analysis meets these criteria.[7] Thus in the late 2000s, one group of scientists claimed to have discovered a particularly ingenious way of analyzing interferometer data showing that a certain signal would only have occurred by chance once every 300 years.[8] They were immediately set upon by the rest of the community who said that the group had tried 12 different ways of analyzing their data. If their probability was divided by 12, the significance of the event became unremarkable. Thus, their claims were dismissed even though their different cuts were not "independent" but developments of the same procedure, and even though they had not made their choices retrospectively (Collins 2011, pp. 102f).

The problem is exacerbated because the statistical analysis techniques – the pipelines – required for a pioneering science like gravitational wave detection are being continually developed, with competing teams inventing new ways of data analysis and noise reduction. Furthermore, there are four basic kinds of signals for which four different groups of analysts are responsible. The "burst group" looks for sudden bursts of energy without a known waveform such as might indicate an exploding star; the "inspiral group" looks for signals from the

---

[7] One can, in fact, argue whether the box-freezing protocol is usable in practice, see Collins (2011, pp. 27–33).

[8] This is not a level of significance that would be accepted as indicating a discovery, but it would encourage the scientists to look more closely at that method of analysis.

final few seconds of the life of binary systems by matching the signals with a bank of many tens of thousands of preformed templates[9]; the "continuous wave group" looks for a steady stream of signals with a fixed waveform, as might be emitted by the rotation of an asymmetrical neutron star such as the Crab pulsar; and the "stochastic group" looks for the random background hiss of primordial gravitational waves equivalent to the electromagnetic cosmic background radiation.

With four groups, yet another problem shows itself. One member of the gravitational wave community, who preferred a highly conservative interpretation of the statistics, argued (Collins 2011, p. 101):

> I just want to say that, if you look across the collaboration as a whole, we have four different search groups, each of which is running a handful of different analyses, so there's something like ten different analyses running – more than ten analyses running across the collaboration – so if you want to say something like we want to have below a 1 percent mistake rate in our collaboration, that means, right away, that you need false alarm rates of something like one in every ten-thousand years if you want to be sure. It's a factor of ten because there's at least ten different analyses running across the collaboration.

This scientist was claiming that the work of the four groups should all be seen as occupying the same experimental space such that a result from any one group affected the meaning of the statistical conclusions of the other three groups. Whether the argument is right or wrong, it raises the question of what counts as the "experiment space" for the calculation of aggregate statistics.

4.3.2 Quantum Analogies

Common to all these problems is that the outcomes of statistical calculations are not independent, but each new calculation affects the meaning of the others. The weirdness of this, I now suggest, is something that needs to be given more attention. It receives practical attention as a strange kind of anomaly (Collins 2011, p. 102):

> That's the other problem, there's a bunch of [different] numbers associated with this. ... Somebody ... wrote a pipeline ... that has never been looked at or studied, which says once in three hundred years. I could write a pipeline tomorrow, I'll bet, which could not see the Equinox Event, thereby degrading its significance. And all I have to do is write a really bad pipeline.

---

[9]It will be hundreds of thousands of templates in the more sensitive detectors soon to come on air.

> In fact I can write eight bad pipelines that are bad in different ways so they're all uncorrelated, and when I do that all eight will miss that event, that will downgrade its significance immediately. So if I wanted to I could kill that event because of the trials factor.

The way different results affect each other has something of the strangeness of quantum theory. With the repetition of experiments, the original and the replication are joined by a kind of correlation reminding us of quantum entanglement. The reasons are that the very act of conducting a second, or third, or nth, experiment or observation changes the meaning of the original, and therefore, we are always uncertain about what a statistical conclusion really means.

Exactly the same is true if someone else in another institution carried out the second experiment. And exactly the same is true if someone else did the experiment but didn't tell anyone. So even if I have done nothing but sit in my armchair since I published my positive result, the true statistical significance of that published finding could be fluctuating all over the place without the printed characters on the page altering in any way, and without anyone knowing.

When you read a paper that presents a claim backed up with a statement of statistical significance, though you cannot see it, the value of that result may be shifting before your very eyes – it has no fixed value. Furthermore, the value of the result may shift some time in the future without you having any influence on it or knowing that it is going to happen or knowing when it does happen. You thought you had reported a measurement but it turns out that all such measurements are linked to all the other similar measurements past and future, and they are all jiggling about in a kind of fuzzy network. It is fuzzy because we are not sure which experiments are in the same "experiment space."

Worse, what a single experiment contributes to the vibrating network depends on the internal states of the experimenters. This is analogous, in a distant way, to the input of the measurement instruments in quantum measurement, where there is no observation without an observer. So the measuring instruments (or, in a more radical interpretation, the consciousness of observers) play their part in turning microscopic indeterminateness into a macroscopically defined value.

Now all this is only a distant *analogy* with quantum theory, not an *application* of it. One difference is that the fluctuations of statistical results are caused not by the fundamental indeterminateness of the universe but by the effects of classical "hidden variables." In principle it is possible to provide a complete account of every contributing node in the network and its causal effects on all other nodes *so long as we have all the information we need.* But this seems completely impossible – firstly, because we do not know what is going on in every laboratory throughout the world; secondly, because we cannot define for sure which experiments are the same and which are different; and thirdly, because of the problem of knowing the internal states of statisticians. All these points are a matter of practice, not of theoretical principle. Nevertheless, the quantum analogy helps to get us to think about certain procedures in the right way.[10]

4.3.3 Meta-Analysis

Meta-analysis is supposed to take care of at least some of the problems discussed above by taking the "hidden variables" into account and working out their joint significance. But meta-analysis should assemble all the experiments done everywhere. A well-known problem is that null results, being thought of as uninteresting, are often not published and remain in the laboratory's file drawers – hence, the notion "file-drawer" effect. Authors do not submit them or editors reject them.

As a consequence, there are now calls for special journals or websites to publish negative or null results. This might improve the situation for meta-analysis. The obvious problem has been hinted at by my respondent above: shoddy experimentation can produce negative results, and if many of these are carried out and reported, meta-analysis will produce a preponderance of false negatives. Much harder problems are (1) knowing which experiments belong to the same "experiment space", and (2) knowing what was in the mind of the experimenters.

It may be thought that the practice is far less difficult than this overly refined analysis would suggest. That is why I started with the experience of gravitational wave physicists and showed how it is affected by problems (1) and (2) and, potentially, by the problem of having one's statistics affected by someone else's poor pipeline design.

There is a whole area of empirical studies – parapsychology – that is sometimes said to owe its existence to the "file-drawer" effect, i.e., the non-publication of negative results (see also Ehm, this volume). There are publications reporting marginal but, over large numbers, highly statistically significant results of experiments where subjects have to guess cards without seeing them or influence random events. Some critics say that these are entirely discounted by what lies unreported in file-drawers all over the world.[11]

---

[10]For more direct applications of, at least, the mathematics of quantum theory to other topics see Wang *et al.* (2013) and Wang and Busemeyer, this volume. Zimmerli, this volume, argues that what he calls "Merton-sensible systems" also show analogies with quantum behavior.

[11]It is necessary to say that the discussion of parapsychology should not be taken as an

Meta-analysis is based on the collection of the results of many replicated scientific experiments, observations, and the aggregation of their results. It is sometimes talked about as though it was a definitive procedure. Instead, we should think of meta-analysis as dipping into, or haphazardly sampling, the jiggling, fuzzy-edged, essentially subjective, network of interlinked statistical activities of scientists. It is probably the best we can do, but each time it is applied there should be a meta-meta-analysis of how likely it is to be correct.[12]

How many unknown statistical activities might not have been gathered? There are probably few in the case of the Higgs boson, but many in the case of parapsychology and, perhaps, many in the case of, say, the investigation of simple medical procedures. How fuzzy are the boundaries of the experiment space? They are probably not very fuzzy in the case of testing a new drug, but surprisingly fuzzy in the case of gravitational waves. How much do results depend on the intentions of the experimenters? These are nice problems.

4.4 The Replication Imperative

4.4.1 Physics as a Social System

There are also simpler problems in the way of replicating experimental findings and observations. There are sometimes singular observations and the only hope for something approaching replication is that they are seen more than once (see Kantz, this volume, for the study of rare events). There is the huge expense of high-energy physics experiments and the like where only one apparatus can be built but something that approaches replicability is accomplished by having more than one team, and perhaps more than one detector, probing the same stream of particles.

In the social sciences there are said to be many obstacles to replicability but these are often misplaced or can be circumvented. Firstly, the aim of much social science, however local the investigation, is to understand general social processes that should reappear in many different social locations. For example, this paper began with a description of the experimenters' regress that, what-

endorsement of its findings but a purely neutral sociological reflection, informed by the principle of methodological relativism (Collins 1981), upon how the claims of parapsychology are analyzed by the wider community.

[12]Such a kind of second-order thinking is found in the simple example of assessing the complexity of a system, where the modeling framework needs to be taken into account explicitly (Atmanspacher and Demmel, this volume). For more along the same lines see Atmanspacher and Jahn (2003).

ever its similarity to previous formulations, emerged from the observation of the particular scientific community of gravitational wave physicists.[13]

Yet the phenomenon is a general one – it is meant to apply to all deeply held scientific controversies in any field of science and if it failed to reveal itself in some other deeply controversial experimental field, it would represent a challenge to the original claim. Again, the experimenters' regress is founded in the fact that scientific experimentation is thoroughly invested with tacit knowledge, and this was demonstrated through the study of a particular group of scientists trying to build transversely excited, atmospheric-pressure CO_2 lasers (TEA-lasers).[14]

Nevertheless, the point is meant to apply not only to TEA-lasers but to all experimentation, and it can be tested by others looking at entirely different sciences. I make the same kind of claim for most of my sociological research – my findings can be tested by others investigating different parts of the social world. But does social science sometimes stray from this aim? The study of gravitational wave physics in Collins (2004) presented many such challenges so I will use it as a way to lever the argument forward.

The development of gravitational wave detection over 40 years (Collins 2004) contains passages that, at first sight, are historical and context bound. But even this kind of research can be aimed at the generation of findings with much more general applications. The long passage above where a physicist described what he believes were Weber's statistical practices recounts, on the one hand, what was believed to be a particular historical event but, on the other, it is an event that is used to reveal a general principle – that the propriety of a statistical procedure depends on what is in the mind of the analyst.

To draw an analytic conclusion, all we have to do is to agree that this is something that could have happened, not that it did happen. The social analyst working with unfolding events is tempted to uncover the particular truth of the matter but this should be avoided since it is the, extremely difficult, job of courtrooms and the like to uncover motives. Social analysts can do their job through revealing the vocabulary of motives available to actors rather than the particular motives that drove any particular individual's actions. This is

[13]For an account of the origins of the experimenters' regress in sociological fieldwork see Collins (2009). For the original scientific claims that gave rise to my study of replication in gravitational wave physics see Weber (1961, 1968, 1969). For Weber's continuing claims see Weber *et al.* (1973) and Weber and Radak (1996). A far more complete list of technical references for the early gravitational wave work will be found in Collins (2004). For a technical entree to more recent work on the detection of gravitational waves using interferometers see `en.wikipedia.org/wiki/LIGO` and Collins (2011).

[14]See Collins (1974, 1985). For an entree to more technical accounts of the TEA-laser see `en.wikipedia.org/wiki/TEA_laser`.

the "antiforensic principle," useful to keep in mind when trying to avoid the temptation of becoming a detective rather than a social analyst (Collins 2004, pp. 412–413).

A central feature is the shift of gravitational wave physics from a small science to a big science and the associated traumas. Again this is a particular historical shift but the analysis includes setting out of the general characteristics of the two kinds of science and the general nature of such transitions. My claim is that however much the account I produced rested on my personal involvement with the field – the quintessentially "subjective" participant comprehension, success in which was tested by exposure to an imitation game – any other observer with similar skills would have seen the same transition in the same way.[15]

4.4.2 Actors and Analysts

Moving further along the particularizing spectrum, there are more specific historical details describing the research and career trajectories of certain individuals who found themselves uncomfortable as the shift to big science practices unfolded. An example is the eventual separation from the Laser Interferometer Gravitational-Wave Observatory (LIGO) of Ron Drever, who invented a number of the crucial innovations that make the modern science possible. Drever eventually found himself banned from entering the principle experimental building on the CalTech campus that contained the apparatus he had designed.

But now I have to admit of a complication for my claim that all social science, including the most particularistic accounts, is aimed at general conclusions and, therefore, replicable in principle. The complication is that there is feedback between the accounts and the verification by others – the quasi-replication. The descriptions of both the transition from small to big science as a whole, and of the way certain personalities interacted with it, are not aimed solely at reflecting what scientists experienced but also at contributing to their understanding of the change and their roles within it. That is to say, if descriptions and explanatory generalizations are effective, they will have at least marginally changed the perspectives of the actors in a direction that will make it more likely that they will agree that the accounts were satisfactory.

It has to be said that the process that led respondents to agreeing with my accounts was helped along by considerable to-and-fro discussion, especially on those occasions when my perspective was not readily accepted. Thus, as we move from the general to the particular, the replicability we seek is not quite

[15]For participant comprehension see Collins (1984), and for my success at passing as a gravitational wave physicist see Giles (2006).

so distant from the research process itself, a criterion of successful research of this type being that the actors come to understand their world in a new and productive way and thus their world changes even as they endorse (replicate) the sociological account.[16]

Throughout this description of various types of scientific investigation I have tried to draw out the replicability or potential replicability with a slightly embarrassed admission of feedback in the last couple of instances. This is because I am talking about social *science* and replicability is a cornerstone of science. And this is for a very simple reason: Science deals with observation, and observations must be stable (cf. Steinle, this volume) and therefore repeatable.[17] A *scientific* social science must stress the repeatability of observations. This is in spite of the fact that when social science is applied to the sciences, it reveals all the difficulties in the way of replicating an experiment and the fact that, in difficult areas, any agreement that a result has been replicated or that an observation has stability will be a matter of social agreement.

Social *scientists* cannot make sense of their world without the degree of compartmentalization that enables them to maintain replicability as an aspiration. The job of the social sciences is not to show that replication is futile or impossible, but to show how to pursue replicability in the face of its recalcitrance. This is possible because science is a set of aspirations for how to try to observe in the best possible way, not a set of rules without exceptions. This position is known as "elective modernism" (Collins and Evans 2015b).

---

[16]For the complexity of the relationship between actors' and analysts' categories, see Collins (2013, Chap. 17) and Collins and Evans (2015a). Keller (this volume) says that this kind of feedback happens with social survey results, too. But it is a mistake to think that this is a feature of the social sciences only – the only unique thing about the social sciences in this respect is that the feedback is mediated by language. Turning to the natural sciences, we can well imagine that capturing animals or birds and fitting transmitters to them so as to monitor their movements might well affect them to alter those movements. And things change all by themselves in the natural sciences just as they change in society. We can imagine that making measurements of the numbers of creatures found in a certain ecological system and returning the following year to find that those numbers are entirely different because the food supply or the relationship between species has changed. The change becomes the phenomenon rather than the single measurement so there is nothing special about the social sciences here. It is just a matter of what is being investigated – stable features or change. But even the scientific investigation of change is generally set within a stable, and testable by replication, theory of change.

[17]Collins (1985, Chap. 1) addresses this point. The argument reflects Popper's (1957) demarcation between unique events and generalizable scientific laws. Popper argued that there could be no "theory of history" because history was a unique unfolding, whereas science must be about what is common to historical events. Popper, of course, raised doubts against the supposed scientific status of the theory of history enshrined in Marxism.

But there is also a humanities approach to the social world that licenses much more context-bound and unique experience-type claims. Reluctantly, I have strayed into the edges of it with the admission of the feedback issue addressed above: there was an aspect of uniqueness even in the generalizability of the accounts of individual personalities and how they reacted to the changes going on around them. This is an aspect of replicability that is outside the pure objectivity of the aspiration. There is a little of this going on wherever respondents change their understanding of their own world in response to the analysis presented to them – their agreement makes them complicit in the "replication" and it makes the replication less of an affirmation of the stability of the observation. There is no equivalent to this process within the natural sciences.

But the humanities approach to social analysis celebrates such things rather than feels them to be a cause of difficulty. Certain kinds of ethnographic account seem like this: "thick description" of a unique social group is the aim with no attempt made to relate that group to other groups. Appealing and colorful descriptions of the world are put in place of analysis, and performance is stressed over and above verisimilitude.[18] Persuasion – the creation of new ways to experience the world – becomes the aim; persuasive novelty, not stability, becomes the aspiration.

The crucial point is that the same fieldwork can be analyzed in different ways, depending on what the analyst prefers to stress. I have argued that most social science need be no different to natural science in its aspiration to give rise to repeatable findings even though sometimes this is hard. That language provides the framework for social life certainly does mean that the social sciences have recourse to certain methods that do not apply to the natural sciences.

For example, the essentially "subjective" participant comprehension has no counterpart in the natural science. But even this does not mean that there is difference in principle in respect of replicability between the natural and social sciences. The outcomes of the "subjective" participant comprehension can be replicated by others if the method is treated as "scientific." Still, it may be that, in the end, non-repeatability is a more pervasive feature of the social

---

[18]The humanities take their template from the arts where brilliant performances of old works are applauded and where innovation is celebrated for its own sake. It is entirely appropriate for an artist to engender a unique experience or a unique interpretation of even a multiplicity of interpretations of a work rather than the one true and, therefore, replicable interpretation. The consumers and assessor of art are the people or their representatives – the media critics – whereas in science it is thought that expertise is required to produce competent assessments of a piece of work. Collins and Evans (2007) talk of the varying "locus of legitimate interpretation" in science and art. For an account of the otherwise bizarrely successful "actor network theory" that draws on some of these notions see Collins (2012).

sciences than the natural sciences because language is central to social scientists' activities. Language forms the world that humans inhabit even as it describes it.

That feedback complicates replicability may be a more common feature of the social sciences than other sciences.[19] That feedback can change the viewpoint of respondents to help them accept an analyst's account of their actions does, however, begin to stray away from the scientific ideal of replication and move in the direction of the humanities. This seems inevitable if our ambition is to change the world as well as analyze it. Finally, should one choose the humanities instead of the sciences as a guide to action, then performances and colorful description of local particulars will replace replicability as the aspiration. The first duty of the social analyst is to declare whether their aspiration is to do science or art.

References

Astone, P., Babusci, D., Bassan, M., Bonifazi, P., Carelli, P., Cavallari, G., Coccia, E., Cosmelli, C., D'Antonio, S., Fafone, V., Federici, G., *et al.* (2002): Study of the coincidences between the gravitational wave detectors EXPLORER and NAUTILUS in 2001. *Classical and Quantum Gravity* **19**, 5449–5465.

Atmanspacher, H., and Jahn, R.G. (2003): Problems of reproducibility in complex mind-matter systems. *Journal of Scientific Exploration* **17**, 243–270.

Collins, H. (1974): The TEA set: Tacit knowledge and scientific networks. *Science Studies* **4**, 165–186.

Collins, H. (1975): The seven sexes: A study in the sociology of a phenomenon, or the replication of experiments in physics. *Sociology* **9**(2), 205–224.

Collins, H. (1981): What is TRASP: The radical programme as a methodological imperative. *Philosophy of the Social Sciences* **11**, 215–224.

Collins, H. (1984): Concepts and methods of participatory fieldwork. In *Social Researching*, ed. by C. Bell and H. Roberts, Routledge, Henley-on-Thames, pp. 54–69.

Collins, H. (1985): *Changing Order: Replication and Induction in Scientific Practice*, Sage Publications, London.

Collins, H. (1994): A strong confirmation of the experimenters' regress. *Studies in History and Philosophy of Science* **25**, 493–503.

Collins, H. (2002): The experimenter's regress as philosophical sociology. *Studies in History and Philosophy of Science* **33**, 153–160.

[19]The social science examples include the remarks by Keller (this volume) and Zimmerli (this volume) about opinion surveys as well as Merton's idea of "self-fulfilling (and self-denying) prophecy." But bear in mind that there is feedback in the natural sciences too.

Collins, H. (2004): *Gravity's Shadow: The Search for Gravitational Waves*, University of Chicago Press, Chicago.

Collins, H. (2009): Walking the talk: Doing gravity's shadow. In *Ethnographies Revisited: Constructing Theory in the Field*, ed. by A.J. Puddephatt, W. Shaffir, and S.W. Kleinknecht, Routledge, London, pp. 289–304.

Collins, H. (2011): *Gravity's Ghost: Scientific Discovery in the Twenty-First Century*, University of Chicago Press, Chicago.

Collins, H. (2012): Performances and arguments. *Metascience* **21**, 409–418.

Collins, H., and Evans, R. (2007): *Rethinking Expertise*, University of Chicago Press, Chicago.

Collins, H., and Evans, R. (2015a): Actor and analyst: A response to Coopmans and Button. *Social Studies of Science* **44**, 786–792.

Collins, H., and Evans, R. (2015b): Elective Modernism. Preprint.

Collins, H., and Pinch, T. (1993): *The Golem: What Everyone Should Know About Science*, Cambridge University Press, Cambridge.

Finn, L.S. (2003): No statistical excess in EXPLORER/NAUTILUS observations in the year 2001. *Classical and Quantum Gravity* **20**, L37–L44.

Franklin, A. (2013): *Shifting Standards: Experiments in Particle Physics in the Twentieth Century*, Pittsburgh University Press, Pittsburgh.

Franklin, A., and Collins, H. (2015): Two kinds of case study and a new agreement. In *The Philosophy of Historical Case Studies, Boston Studies in the Philosophy of Science*, ed. by T. Sauer and R. Scholl, Springer, Dordrecht, in press.

Gelfert, A. (2011): Scientific models, simulation, and the experimenter's regress. In *Models, Simulations, and Representations*, ed. by P. Humphreys and C. Imbert, Routledge, London, pp. 145–167.

Giles, J. (2006): Sociologist fools physics judges. *Nature* **442**, 8.

Godin, B., and Gingras, Y. (2002): The experimenters' regress: From skepticism to argumentation. *Studies in History and Philosophy of Science* **33**, 137–152.

Gross, E., and Vitells, O. (2010): Trial factors for the look elsewhere effect in high energy physics. *European Physical Journal C* **70**, 525–530.

Kennefick, D. (2000): Star crushing: Theoretical practice and the theoreticians' regress. *Social Studies of Science* **30**, 5–40.

Lyons, L. (1986): *Statistics for Nuclear and Particle Physics*, Cambridge University Press, Cambridge.

Lyons, L. (2008): Open statistical issues in particle physics. *Annals of Applied Statistics* **2**, 887–915.

Lyons, L. (2010): Comments on "look elsewhere effect". Unpublished manuscript, available at `www.physics.ox.ac.uk/Users/lyons/LEE_feb7_2010.pdf`.

Merton, R.K. (1948): The self fulfilling prophecy. *Antioch Review* **8**(2), 193.

Miller, D.P. (2006): Watt in court: Specifying steam engines and classifying engineers in the patent trials. *History of Technology* **27**, 43–76.

Montaigne, M. de (1938): *Essays*, J.M. Dent, London. Originally published in 1580.

Pickering, A. (1981): Constraints on controversy: The case of the magnetic monopole. *Social Studies of Science* **11**, 63–93.

Pinch, T.J. (1981): The sun-set: The presentation of certainty in scientific life. *Social Studies of Science* **1**(11), 131–158.

Popper, K. (1957): *The Poverty of Historicism*, Routledge and Kegan, Paul, London.

Sextus Empiricus (1933): *Sextus Empiricus, Vol. 1: Outlines of Pyrrhonism*, transl. by R.G. Bury, Harvard University Press, Cambridge.

Stephens, N., Lewis, J., and Atkinson, P. (2013): Closing the regulatory regress: GMP accreditation in stem cell laboratories. *Sociology of Health and Illness* **35**, 345–360.

Travis, G.D. (1981): Replicating replication? Aspects of the social construction of learning in planarian worms. *Social Studies of Science* **11**, 11–32.

Weber, J. (1961): *General Relativity and Gravitational Waves*, Wiley, New York. 688.

Weber, J. (1968): Gravitational waves. *Physics Today* **21**(4), 34–39.

Weber, J. (1969): Evidence for the discovery of gravitational radiation. *Physical Review Letters* **22**, 1320–1324.

Weber, J., Lee, M., Gretz, D.J., Rydbeck, G., Trimble, V.L., and Steppel, S. (1973): New gravitational radiation experiments. *Physical Review Letters* **31**, 779–783.

Weber, J., and Radak, B. (1996): Search for correlations of gamma-ray bursts with gravitational-radiation antenna pulses. *Nuovo Cimento B* **111**, 687–692.

Wang, Z., Busemeyer, J.R., Atmanspacher, H., and Pothos, E.M. (2013): The potential of using quantum theory to build models of cognition. *Topics in Cognitive Sciences* **5**, 672–688.

PART II
STATISTICAL ISSUES

PART II: STATISTICAL ISSUES
Introductory Remarks

Harald Atmanspacher

A handbook on reproducibility without a section on statistics is an impossibility. Although much of the attention which the notion of reproducibility received in recent years arose from problems to reproduce results in specific scientific disciplines (cf. parts III–V), we cannot properly understand most of these problems without sound knowledge about the statistical tools used to analyze data, test hypotheses, formulate statistical models, and so on.

A basic notion for a sensible discussion of reproducibility from a statistical point of view is the notion of *stability*. A most obvious way in which stability features in statistics is visible in so-called "limit theorems" (cf. Petrov 1995), for instance, the *law of large numbers*. It expresses that the expectation value of a stochastic object (e.g., a random variable) exists in the limit of infinitely many independent (uncorrelated!) instantiations.

From a broader (and deeper) perspective, the *stability of statistical models* of data is crucial to assess the reproducibility of observations. An initial statistical model is reproduced if the data of a replication study are "compatible" (under certain criteria) with that model. Models in this general sense can result from theoretical frameworks (e.g., in theoretical physics) or they may be inferred from data (e.g., in complex systems). In the latter case, the selection of proper models becomes a serious issue. The variety of available model classes includes regression models, Markov models, Bayesian models, neural networks, and so on (cf., e.g., Davison 2008).

Basic concepts and issues concerning reproducibility in terms of the stability of models are introduced and expanded in the chapter by *Werner Stahel, former head of the Statistical Consulting Services at ETH Zurich*. He discusses a number of statistical issues that are important for the validation of reproduced results in repeated experiments. But he also focuses on what he calls "data challenges" – how regression analyses can help to achieve credibility for a model rather than for a replication of data. This becomes especially interesting if the conditions under which a model is to be reproduced are heterogeneized in order to test its generalizability. Yet another step is the idea of *conceptual replication*, where the conclusions of a study are examined with different quantifications of concepts.

Stahel also emphasizes a number of specific *caveats*: First of all, basic probabilistic models typically assume the statistical independence of observations or measurements. This is often violated in practice, and more realistic models alllowing for correlations are often challenging. Second, the assumption of normal (Gaussian) distributions is frequently questionable and should be avoided. Third, the conclusions of statistical analyses are frequently formulated in terms of statistical significance, but this formulation is ill-suited for reproducibility,[1] and confidence intervals should be communicated instead.

Empirical science strives to obtain knowledge from data, and this often means to develop models from observations. The difficulty to do this in ways that are both creative and reproducible is addressed in the contribution by *Richard Shiffrin and Suyog Chandramouli, mathematical psychologists at Indiana University at Bloomington*. This issue of *inductive inference* is particularly relevant in the life sciences and social sciences where "true" values usually have variability and their distributions cannot be derived from "first principles" of an underlying theory. Modeling such distributions is a highly non-trivial task, and the chapter discusses the factors governing the variability of the distribution, termed *replication variance*, and the ways in which those factors influence both model comparison and reproducibility.

The authors present an extension of standard Bayesian model selection – a methodology representing an alternative to hypothesis testing in the style of frequentist analyses by conditioning probabilities for particular outcomes upon prior probabilities, for instance, "prior beliefs" (see Raftery 1995 for a good introduction). Their approach infers the posterior probabilities that a given model instance predicts a data distribution that is the best match to the "true" data distribution. The posterior probabilities can be used to produce a predicted distribution for a statistic defined on the data. Comparing this statistic with that of a replication study yields a quantitative assessment of the quality of the replication. A simple toy model at the end of the chapter illustrates in detail how the approach works in practice.

If many studies have been performed on the same (or, more typically, closely related) research question(s), one is often interested in the overall effect across all those studies. This is particularly relevant if individual studies did not yield a clear and unambiguous result. In such cases, one can lump together all individual analyses in one so-called *meta-analysis* (for a clear and thorough introductory

[1]Statistical significance is typically expressed by a p-value, roughly indicating the probability that a result could be due to pure chance. However, as Porter (part V) argues, a statistically significant quantity can be irrelevant for a given situation if it is not also *substantively significant*. See also Atmanspacher, part VI, or Keiding (2010).

textbook see Borenstein *et al.* 2009). *Werner Ehm at the Heidelberg Institute for Theoretical Studies* introduces this field of research in his contribution.

There are several things that need consideration in meta-analyses – two most significant ones among them are highlighted by Ehm. The first one concerns random-effects modeling, which accounts for the possibility that an effect need not be a fixed quantity but may itself exhibit variation due to unknown factors. The second is about selection biases, most notably the bias resulting from the fact that positive results of a study are published more often than negative or null results.[2] For an illustrative empirical example from mind-matter research, Ehm shows how "corrections meta-analysis" can be applied to assess both random effects and selection biases *together* in one model, and to check the validity of related statistical inferences.

Vladimir Estivill-Castro in the Department of Information and Communication Technologies at the University Pompeu Fabra at Barcelona addresses a novel area of research that rapidly developed momentum in recent years under the notion of *data mining and knowledge discovery*.[3] Trying to infer knowledge from empirical data is nothing new in principle, of course. But modern technologies offer so many extremely efficient tools of collecting huge amounts of data (so-called "big data," such as in high-energy physics and astrophysics and in various areas of biomedical research) that pattern detection and cluster analysis have become virtually impossible by visual inspection alone.

For these reasons, numerous clustering algorithms have been proposed and are being employed to automatize the task. Since the patterns looked for are unknown to begin with, all kinds of "machine learning" are generally *unsupervised* rather than supervised. Clustering algorithms require assumptions about and choices of many parameters that are not prescribed and could be chosen otherwise. This leads to the question of how valid their results are – in other words, how reproducible (how stable) the obtained clusters are under perturbations.

Estivill-Castro discusses novel clustering-quality measures based on what he calls "instance easiness," which are faster and more reliable than many alternatives. Instance easiness links the validity of a given clustering to whether it constitutes an easy instance for *supervised* learning. He illustrates this proce-

---

[2]A *Journal for Negative Results in BioMedicine* with its editorial office at Harvard Medical School was launched in 2002 to counter this well-known deficit in publication policy. They encourage the submission of experimental and clinical research falling short of demonstrating an improvement over current state of the art or failures to reproduce previous results.

[3]See the handbook by Maimon and Rokach (2010) or the textbook by Cios *et al.* (2007). Moreover, two journals exclusively devoted to the topic have been launched by Springer and by Wiley in recent years.

dure and its successful application with synthetic data as well as with empirical gene-expression data for different types of leukemia.

This part of the volume is inevitably its most formal part. Although mathematical formalism usually enhances the contents-to-length ratio of articles and helps to keep them short, the reader will find that most of the contributions in this part are in the long tail of the distribution. The reason is that authors, particularly in this part, tried hard to explain formal expressions as much as possible and to illustrate their topics by concrete examples, so that the key messages of their chapters are as accessible as possible to non-mathematical readers.

References

Borenstein, M., Hedges, L.V., Higgins, J.P.T., and Rothstein, H.R. (2009): *Introduction to Meta-Analysis*, Wiley, New York.

Cios, K.J., Pedrycz, W., Swiniarski, R.W., and Kurgan, L. (2007): *Data Mining: A Knowledge Discovery Approach*, Springer, Berlin.

Davison, A.C. (2008): *Statistical Models*, Cambridge University Press, Cambridge.

Keiding, N. (2010): Reproducible research and the substantive context. *Biostatistics* **11**, 376–378.

Maimon, O., and Rokach, L., eds. (2010): *Data Mining and Knowledge Discovery Handbook*, Springer, Berlin.

Petrov, V.V. (1995): *Limit Theorems of Probability Theory*, Clarendon, Oxford.

Raftery, A. (1995): Bayesian model selection in social research. *Sociological Methodology* **25**, 111–163.

5
Statistical Issues in Reproducibility

Werner A. Stahel

Abstract. Reproducibility of quantitative results is a theme that calls for probabilistic models and corresponding statistical methods. The simplest model is a random sample of normally distributed observations for both the original and the replication data. It will often lead to the conclusion of a failed reproduction because there is stochastic variation between observation campaigns, or unmodeled correlations lead to optimistic measures of precision. More realistic models include variance components and/or describe correlations in time or space.
Since getting the "same circumstances" again, as required for reproducibility, is often not possible or not desirable, we discuss how regression models can help to achieve credibility by what we call a "data challenge" of a model rather than a replication of data. When model development is part of a study, reproducibility is a more delicate problem. More generally, widespread use of exploratory data analysis in some fields leads to unreliable conclusions and to a crisis of trust in published results. The role of models even entails philosophical issues.

5.1 Introduction

Empirical science is about extracting, from a phenomenon or process, features that are relevant for similar situations elsewhere or in the future. On the basis of data obtained from an experiment or of a set of observations, a model is determined that corresponds to the ideas we have about the generation of the data and allows for drawing conclusions concerning the posed questions.

Frequently, the model comprises a structural part, expressed by a formula, and a random part. Both parts may include constants, called parameters of the model, that represent the features of interest, on which inference is drawn from the data. The simplest instance is the model of a "random sample" from a distributions whose expected value is the focus of interest.

The basic task of data analysis is then to determine the parameter(s) that make the model the best description of the data in some clearly defined sense. The core business of statistics is to supplement such an "estimate" with a measure of precision, usually in the form of an interval in which the "true value" should be contained with high probability.

This leads to the basic theme of the present volume: If new data becomes available by reproducing an experiment or observation under circumstances as

Reproducibility: Principles, Problems, Practices, and Prospects, First Edition. Edited by Harald Atmanspacher and Sabine Maasen.

similar as possible, the new results should be compatible with the earlier ones within the precision that goes along with them. Clearly, then, reliable precision measures are essential for judging the success of such "reproductions."

The measures of precision are based on the assessment of variability in the data. Probability theory provides the link that leads from a measure of data variability to the precision of the estimated value of the parameter. For example, it is well known that the variance of the mean of n observations is the variance of the observations' distribution, divided by n if the observations are independent. We emphasize in this chapter that the assumption of independence is quite often inadequate in practice and, therefore, statistical results appear more precise than they are, and precision indicators fail to describe adequately what is to be expected in replication studies. In such cases, models that incorporate structures of variation or correlation lead to appropriate adjustments of precision measures.

Many research questions concern the relationship between some given input or "explanatory" variables and one or more output or "response" variables. They are formalized through regression models. These models even allow for relating studies that do not meet the requirement of "equal circumstances" that are necessary for testing reproducibility. We will show how such comparisons may still yield a kind of confirmation of the results of an original study and thereby generalize the idea of reproducibility to what we call *data challenges*.

These considerations presume that the result of a study is expressed as a model for the data and that quantitative information is the focus of the reproduction project. However, a more basic and more fascinating step in science is to develop or extend such a model for a new situation. Since the process of developing models is not sufficiently formalized, considerations on the reproducibility of such developments also remain somewhat vague. We will, nevertheless, address this topic, which leads to the question of *what* is to be reproduced.

In this chapter, we assume that the reader knows the basic notions of statistics. For the sake of establishing the notation and recalling some basic concepts, we give a gentle introduction in Section 5.2. Since variation is the critical issue for the assessment of precision, we discuss structures of variation in Section 5.3. Regression models and their use for reproducibility assessment is explained in Section 5.4. Section 5.5 covers the issue of model development and consequences for reproducibility. Section 5.6 addresses issues of the analysis of large datasets. Comments on Bayesian statistics and reproducibility follow in Section 5.7. Some general conclusions are drawn in the final Section 5.8.[1]

[1]While this contribution treats reproducibility in the sense of conducting a new measurement campaign and comparing its results to the original one, the aspect of reproducing the analysis

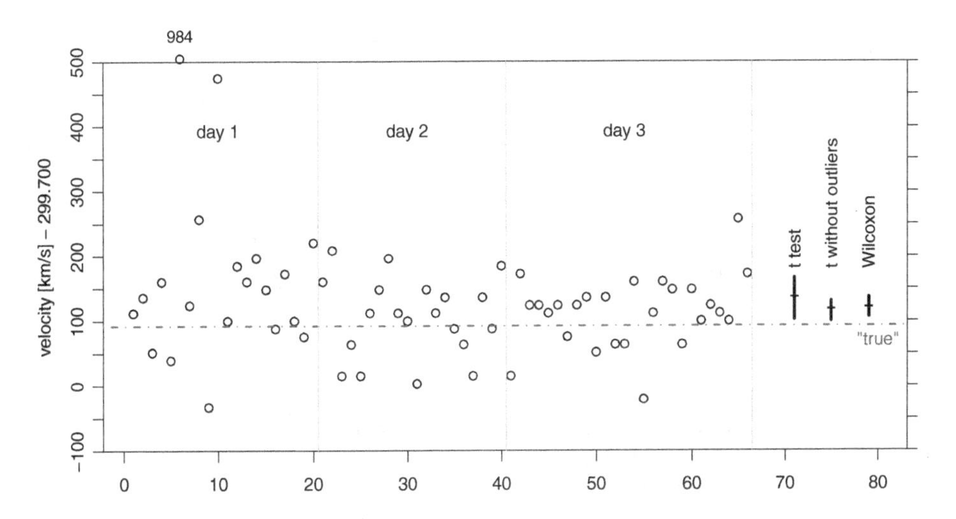

Figure 5.1: Measurements of the velocity of light by Newcomb in 1882. Three confidence intervals are shown on the right.

5.2 A Random Sample

5.2.1 Simple Inference for a Random Sample

The generic statistical problem consists of determining a constant from a set of observations that are all obtained under the "same circumstances" and are supposed to "measure" the constant in some sense. We begin with a classic scientific example: the determination of the velocity of light by Newcomb in 1882. The data set comprises 66 measurements taken on three days, see Fig. 5.1.

The probability model that is used in such a context is that of a random variable X with a supposed distribution, most commonly the normal distribution. It is given by a density f or, more generally, by the ("theoretical") cumulative distribution function $F(x) = P(X \leq x)$. The distribution is characterized by parameters: the expected value μ and the standard deviation σ (in case of the normal distribution). For the posed problem, μ is the parameter of interest, and σ is often called a nuisance parameter, since the problem would be easier if it were not needed for a reasonable description of the data. We will use θ to denote a general parameter and $\underline{\theta}$ to denote the vector of all the parameters of the model. The distribution function will be denoted by $F_{\underline{\theta}}(x)$.

---

of the data is left to Bailey *et al.* (this volume). The tools advocated by many statisticians to support this aim are the open source statistical system R and the documentation tools `Sweave` and `knitr` (Leisch 2002, Xie 2013, Stodden *et al.* 2014).

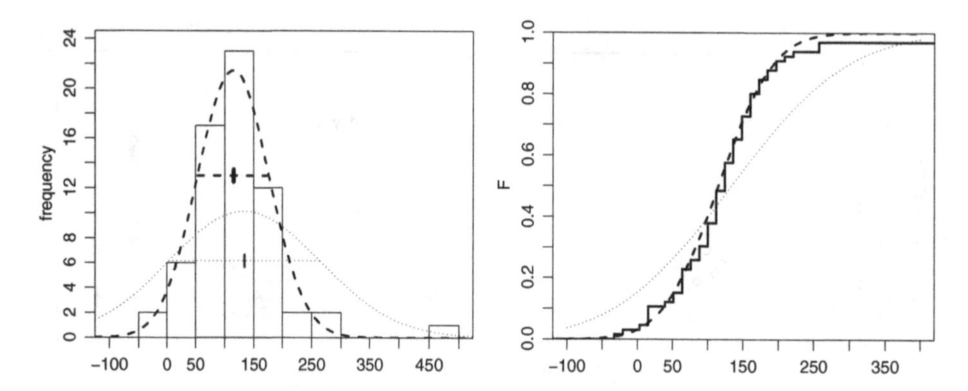

Figure 5.2: Histogram (left) and empirical cumulative distribution function (right) for the Newcomb data. The dashed and dotted lines represent normal distributions obtained from fits to the data without and with two outliers, respectively.

If the model is suitable, the histogram describing the sample values $x_1, x_2, ..., x_n$ will be similar to the density function, and the empirical cumulative distribution function $\widehat{F}(x) = \#(x_i \leq x)/n$ will be like the theoretical one (Fig. 5.2). For small samples, the similarity is not very close, but by the *law of large numbers*, it will increase with n and approach identity for $n \to \infty$. Probability theory determines in mathematical terms how close the similarity will be for a given n.

The question asked above concerns the mean of the random quantity. It is not the mean of the sample that is of interest, but the expected value of the random variable. It should be close to the sample mean, and again, probability theory shows how close the mean must be to the expected value. It thereby gives us the key to drawing statistical inferences about the model from the sample of data at hand – here, for inferences about the parameter θ.

The basic questions Q and answers A in statistical inference about a parameter of a model are the following.

Q1 Which value of the parameter θ (or parameters $\underline{\theta}$) is most plausible in the light of the data $x_1, x_2, ..., x_n$?

A1 The most plausible value is given by a function $\widehat{\theta}(x_1, x_2, ..., x_n)$ (or $\widehat{\underline{\theta}}(...)$) of the data, called an *estimator*.

Q2 Is a given value θ_0 of a parameter plausible in the light of the data?

A2 The answer is given by a *statistical test*. A test statistic $T(x_1, x_2, ..., x_n; \theta_0)$ is calculated, and the answer is "yes" if its value is small enough. The threshold for what is considered "small" is determined by a preselected probability α (called the *level* of the test, usually 5%) of falsely answering "no" (called the *error of the first kind*). Usually, the test statistic is a function of the estimator $\underline{\widehat{\theta}}$ of all model parameters and of the parameter θ_0. If the answer is "no," one says that the test is (statistically) *significant* or, more explicitly, that the *null hypothesis* H_0 that the parameter θ equals the supposed value θ_0 is *rejected.* Otherwise, H_0 is *not rejected* – but *not proven either*; therefore, it may be misleading to say that H_0 is "accepted."

The test statistic is commonly transformed to the so-called *p-value*, which follows the uniform distribution between 0 and 1 if the null hypothesis is true. The test is significant if the p-value is below α.

Q3 Which values for θ are plausible in the light of the data?

A3 The plausible values form the *confidence interval.* It is obtained by determining which values of θ are plausible, in the sense of Q2. Once the confidence interval is available, Q2 can, therefore, be answered by examining whether θ_0 is contained in it.

5.2.2 The Variance of a Mean

The basic model for n random observations is called the *simple random sample.* It consists of n stochastically *independent, identically distributed random variables.* It is mathematically convenient to choose a normal distribution for these variables, which is characterized by its expected value μ and its variance σ^2 and denoted by $\mathcal{N}(\mu, \sigma^2)$. If we want to draw an inference about the expected value μ, the best way to *estimate* it from the data is to use the mean $\bar{x} = \sum_{i=1}^{n} x_i \,/\, n$. This answers Q1 above.

For answering the *test* question Q2, we use the result that under the assumptions made,

$$\bar{X} \sim \mathcal{N}(\mu, \sigma^2/n) ,$$

where the notation means that $\bar{x}$ is distributed as normal with expected value μ and variance σ^2/n. We first postulate that σ is known – as may be the case when the variation of the observations can be assumed to be the same as in many previous experiments. Then,

$$P\left(\mu - 1.96\,\mathrm{se}_{\bar{X}} < \bar{X} < \mu + 1.96\,\mathrm{se}_{\bar{X}}\right) = 0.05 ,$$

where $\mathrm{se}_{\bar{X}} = \sigma/\sqrt{n}$ is called the *standard error of the sample mean,* and 1.96 is the 0.975-quantile of the normal distribution. This is the basis of answering Q2 by saying that μ_0 is "plausible at the level $\alpha = 0.05$" if $\bar{x}$ is contained in the interval $\mu_0 - 1.96\,\mathrm{se}_{\bar{X}} < \bar{x} < \mu_0 + 1.96\,\mathrm{se}_{\bar{X}}$. Otherwise, the null hypothesis $H_0 : \mu = \mu_0$ is rejected. Finally, question Q3 is answered by the confidence interval, which is written as $[\bar{X} - 1.96\,\text{se}_{\bar{X}},\ \bar{X} + 1.96\,\text{se}_{\bar{X}}]$ or, equivalently, as $\bar{X} \pm 1.96\,\text{se}_{\bar{X}}$.

In the more realistic case that σ is not known, it must be estimated from the data. This leads to modifications of the test and the confidence interval known under the notion of *t-test*. We refer to standard textbooks. For the example of the measurements of the velocity of light above we get $\bar{X} = 134$, leading to a velocity of $299,834$ km/s and a confidence interval of $[102, 166]$.

The assumption of a normal distribution for the observations is not very crucial for the reliability of the test and confidence interval. The *central limit theorem* states that the sample mean is approximately normally distributed even if the distribution of the observations is non-normal, possibly skewed, and moderately long-tailed (as long as the observations have a well-defined variance).

For heavily long-tailed but still symmetric distributions, there is a nice model called the *stable distribution*. It contains the normal distribution as the limiting case. Except for this case, stable distributions have no variance (or an infinite one), but still a scale parameter. The mean of such observations has again a stable distribution (whence the name), but its scale parameter is proportional to $n^{-\gamma}$ with $\gamma < 1/2$ (where $1/2$ corresponds to the normal distribution, see above). This means that there is a realistic model for which the standard error of the mean goes to 0 at a slower rate than under the normal model. Nevertheless, if standard methods are applied, levels of tests and confidence intervals are still reasonably valid, with deviations to the "safe side."

Long tails lead to "extreme" observations, well known under the term *outlier*. The Newcomb dataset contains (arguably) two such outliers. Long tails are so widespread that the normal distribution for observations is often not more than wishful thinking. Although this does not affect the validity of the methods based on normal theory too much, it is preferable to avoid means, but estimate or test parameters by alternative methods.

It is often tempting to disregard outliers and analyze the data without them. In the example above, this results in the confidence interval $[100, 130]$. However, dropping outliers without any corrective action leads to confidence intervals that are too short. Methods that deal adequately with restricting the undue influence of outliers (and other deviations from assumptions) form the theme of *robust statistics*. These methods are statistically more efficient than those based on the mean, in particular for pronounced long tails. An interval based on robust methods for our example yields $[101, 132]$.

For simple random samples, there are methods that avoid assuming any specific distribution, known as *nonparametric procedures*. These methods are robust and often quite efficient. Therefore, the t-test and its confidence interval should not be applied for simple random samples. In such cases, a Wilcoxon signed-rank test and the respective confidence interval (associated with the "Hodges–Lehman estimator") should be preferred. Our example yields an estimate of 118 and a confidence interval of $[106, 136]$. The interval is much shorter than the t-interval for the whole dataset. It is as long as, but different from, the interval obtained without outliers or the robust interval.

The value corresponding to today's knowledge, 92, is not contained in either one of the intervals. We are lead to concluding that the experiments of Newcomb had a systematic "error" – a very small one, we should say.[2]

5.2.3 General Parameters and Reproducibility

Using the mean for inferences about the expected value of a distribution is the generic problem that forms the basis for inferences about a parameter in any parametric probability model. For a general parameter, the principle of *maximum likelihood* leads to an estimator. The central limit theorem yields an approximate distribution for it, allowing for approximate tests and confidence intervals. It relies on a linearization of the estimator. Refinements are needed for small sample sizes, robustness issues, and some special cases.

As to the issue of reproducibility, let us note that the confidence interval is supposed to cover the "true value" of the parameter, not its estimate (e.g., the mean) for the new sample. For *testing the success of a reproduction*, the difference of the parameter estimates should be examined. It has an approximate normal distribution with variance equal to the sum of the variances of the parameter estimates. This yields a confidence interval for the true difference, which should contain difference zero for a successful reproduction.

5.2.4 Reproducibility of Test Results and the Significance Controversy

A statistical test gives a "yes" (accept) or a "no" (reject) for the null hypothesis H_0 as its result. Reproducibility would then mean that a "yes" in an original study entails a "yes" in its reproduction; and a "no" entails another "no" in the reproduction.

Table 5.1 collects the four possible cases determined by the validity of the

---

[2]Since 1983, the velocity of light is *defined* by the definition of the second and the meter, according to resolution 1 of the 17th Conférence Générale des Poids et Mesures, `www.bipm.org/en/CGPM/db/17/1/`. As a consequence, the velocity of light is exactly 299,792,458 m/s and has become a *defined* constant in the SI system of units.

hypothesis	non-significant result	significant result
H_0 correct	$P = 95\%$	$P = 5\%$ *
H_A correct	$P = 1-$power *	P = power

Table 5.1: The probabilities of getting the same result in the replication study, for the four possible cases of correct null hypothesis H_0 and alternative hypothesis H_A and significance of test results in the original study. * Replication of wrong results is undesirable.

hypothesis and (non-significant or significant) test results in the original study. Assume first that the null hypothesis H_0 is correct. Then, the probability for its acceptance is large (equal to $1 - \alpha$, usually 95%) in both studies. Given that acceptance was reached in the first study, the probability of a successful replication is $1 - \alpha$. However, since H_0 can never be "proven," one merely obtains a further indication that there may be "no effect." If the null hypothesis is rejected even though it is correct, we do not really want this to be replicated, and the respective probability is as low as α. Since one rarely wants to replicate a null result, such cases are less important. If the null hypothesis is wrong, the probability of getting rejection is equal to the power of the test and depends, among other things, on the sample size. Only in the case of good power for both studies, we obtain the same, correct result with high probability. Thus, test results are generally not reproducible.

These considerations are related to a major flaw of hypothesis testing: A potential effect that is subject to research will rarely be precisely zero, but at least a tiny, possibly indirect and possibly irrelevant true effect should always be expected. Now, even small effects must eventually become significant if the sample size is increased according to the law of large numbers. Therefore, a significance test indeed examines if the sample size is large enough to detect the effect, instead of making a clear statement about the effect itself.

Such arguments have led to a controversy about "null hypothesis significance testing" in the psychology literature of the 1990s. It culminated in a task force discussing to ban the use of significance testing in psychology journals (Wilkinson and Task Force on Statistical Inference 1999). The committee regrettably failed to do so, but gave, among fundamental guidelines for empirical research, the recommendation to replace significance tests by *effect size* estimates with confidence intervals. Krantz (1999) presents a thorough discussion, and Rodgers (2010) and Gelman (2014) provide more recent contributions to this issue.

In conclusion, in spite of its widespread application, null hypothesis significance testing does not really provide useful results. The important role of a statistical test is that it forms the basis for determining a confidence interval (see Section 5.2.1). *Results should be given in terms of confidence intervals for parameters of interest.* The statistical test for no difference of the quantity of interest between the replication and the original study may formally answer whether the reproduction was successful. But even here it is preferrable to argue on the basis of a confidence interval for this difference.

The null hypothesis test has been justified as a filter against too many wild speculations. Effects that do not pass the filter are not considered worth mentioning in publications. The main problem with this rule is its tendency to pervert itself: Whatever effect is found statistically significant is regarded as publishable, and research is perverted to a hunt for statistically significant results. This may indeed have been an obstacle to progress in some fields of research. The replacement of tests and their p-values by confidence intervals is only a weak lever against this perversion. At least, it encourages some thought about whether or not an effect is relevant for the investigated research question and may counteract the vast body of published spurious "effects."

A second problem of the "filter" is that it explicitly leads to the publication and selection bias of results, which is in the focus of this book as a mechanism to produce a vast body of spurious published "effects" or relationships, see Section 5.5.4 and Ehm (this volume, Section 7.2.4).[3]

5.3 Structures of Variation

5.3.1 Hierarchical Levels of Variation

Even if measurement procedures are strictly followed, ample experience from formal replications shows that the test mentioned in Sectiom 5.2.1 too often shows significant differences between the estimated values of parameters of interest.

This experience is formally documented and treated in *interlaboratory studies*, in which different laboratories get samples from a homogeneous material and are asked to perform two or more independent measurements of a given property. The model says that the measurements within a laboratory h follow a (normal) distribution with expectation m_h and variance σ^2, where the latter is assumed to be the same for all laboratories. The expectations m_h are modeled

[3]Killeen (2005) proposed a *measure of replication* p_{rep} to overcome the problems with null hypothesis testing. However, since this measure is a monotonic function of the traditional p-value, it does not overcome the problems associated with significance testing.

as random variables from a (normal) distribution with the "true" value μ as its expected value and a variance σ_m^2.

If there was no "laboratory effect," σ_m would be zero. Then, the precision of the mean of n measurements obtained from a single laboratory would be given by $\sigma/\sqrt{n}$. If $\sigma_m > 0$, the deviation of such a mean from μ has variance $\sigma_m^2 + \sigma^2/n$ and obtaining more measurements from the same laboratory cannot reduce this variance below σ_m^2.

When looking at an interlaboratory study as a whole, measurements from the same laboratory are more similar than those coming from different laboratories. This is called the *within laboratories (or within groups) correlation.* The term *reproducibility* is used in this context with a well-defined quantitative meaning: It measures the length of an interval that contains the difference of two measurements from *different* laboratories with probability 95%. The analogous quantity for measurements from the *same* laboratory is called *repeatability.*

Interlaboratory studies form a generic example for other situations, where groups of observations show a within-group and a between-group variance, such as production lots, measurement campaigns, regions for localized measurements, and genetic strains of animals. Then the variances σ^2 and σ_m^2 are called the *variance components.* There may, of course, be more than two levels of a grouping hierarchy, such as leaves within trees within orchards within regions, or students within classes within schools within districts within countries.

5.3.2 Serial and Spatial Correlations

When observations are obtained at different points in time, it is often intuitive to suppose that subsequent measurements are more similar to each other than measurements with a long time lag between them. The idea may lead to the notion of *intra-day correlations*, for example. In other situations, it may be formalized as a *serial auto-correlation* ρ_h, the correlation between observations X_t at time t and a later one, X_{t+h}, at time $t+h$. (This correlation is assumed not to depend on t.)

The chemist and statistician Gosset made a reality check of the fundamental assumption of independence and published the results under the pseudonym Student (1927). Between 1903 and 1926, he carefully measured the nitrogen content in pure crystalline aspartic acid once a week. This is supposedly an ideal example for the model of independent observations scattering around the true concentration. Figure 5.3 shows the 131 measurements from 1924 to 1926. There appears to be an ascending and slowly oscillating trend. The confidence interval based on the whole dataset does not overlap with the one obtained for

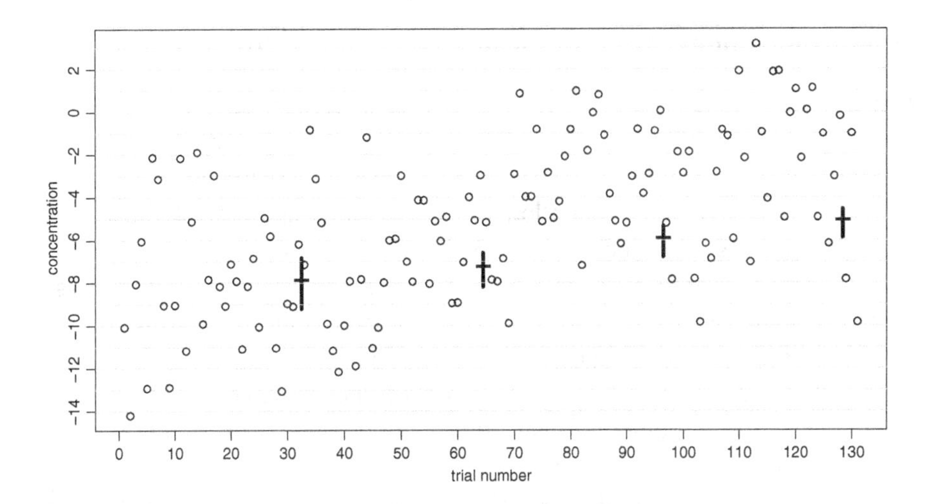

Figure 5.3: Student's measurements of the concentration of nitrogen in aspartic acid. The bars represent confidence intervals based on the t-test for the first 32, 64, 96, and 128 of 131 observations.

the first fourth of it. This indicates that the confidence intervals are too short – because independence is assumed but clearly violated.

If there is a positive auto-correlation, the variance of the mean as well as other estimators obtained from n observations is larger than the variance calculated under the assumption of independence. Therefore, tests and confidence intervals calculated under this assumption are liberal (true probabilities of the error of the first kind are higher, intervals are too short) to an extent that depends on the auto-correlation and increases with the sample size n.

The situation of a series of observations can be modeled by *time series models* or, more generally, by probabilistic process models. If an appropriate model can be found, adequate tests and confidence intervals can again be produced. In practice, auto-regressive models are often used to describe correlations between random deviations, and they entail methods for adequate inference. Note, however, that these models assume that auto-correlations go to zero rather quickly as the lag increases. There is evidence that, in reality, so-called long-range dependence, describing slow decay, occurs amazingly often. Respective models are less well known and more difficult to fit (Künsch *et al.* 1993).

In analogy to correlations in time, one can study correlations between close-by measurements at different locations in space. (If locations occur in groups or clusters, they lead to variance components; see Section 5.3.1.) *Spatial correlations* may be described and treated by the models of *spatial statistics*.

5.3.3 Consequences for Reproducibility and Experimental Design

What are the consequences of within-group, temporal, or spatial correlations on reproducibility? Firstly, they often give an explanation for undesired significant differences between original and reproduced results. Secondly, if they can be adequately modeled, they may allow for a refined assessment of the validity of the original results. Specifically, if an estimate of the variance components due to "measurement campaigns" is available, it may be used to conduct a refined test of the compatibility of original and reproduced results. When many studies assessing the same parameter, for instance, many clinical trials for the same treatment, are available, then an estimate of the "between studies" variance component can be obtained. This leads to the possibility of a "meta-analysis", see Ehm (this volume).

The disturbing effect of correlations as mentioned above may be circumvented or at least reduced by an adequate design of experiments. The goal of an experiment usually is to assess the effect of some "input variables" (often called "parameters") on a "target variable" – like the influence of a "treatment" on the yield in an industrial production. For such a goal, only *differences* of values of the target variable are relevant.

This fact allows for avoiding disturbing effects of variance components by *blocking*: A block is composed of experimental units that are as homogeneous as possible with respect to nuisance factors that, while influencing the response variable, are not in the focus of the study, like measurement devices, laboratories, environmental conditions, subjects, time, or space.

The treatments or other variables whose effects are to be assessed should be varied within each block. Differences within a block will then be unaffected by the conditions used for blocking and lead to adequate and efficient estimates of treatment effects. Using as many blocks as possible will allow for combining these estimates and thereby gaining precision in a joint model for all blocks. We come back to these considerations in the next section. In order to allow for a sound *generalization of results*, blocks should differ in the values of the nuisance factors.

Analogous ideas can be applied to situations with temporal or spatial correlations. Künsch *et al.* (1993) discussed this topic in connection with long-range dependence.

If adequate blocking is used, a replication study can be conceived as the realization of more blocks, in addition to those of the original study, and a joint model may be used to check whether the new results support the original ones (see Section 5.4.2).

5.4 Regression Models

5.4.1 The Structure of Models

The vast majority of well-posed scientific questions to be studied by analyzing data lead to regression models. The variables that are observed or measured fall into three categories:

- *Target variables.* These variables describe the outcome, or response of a system to an input: yield, quality, population size, health indicator, survival, test score, and the like.
- *Explanatory variables.* They influence the target variable as inputs, or stimuli. This influence is the focus of the scientific question of the study. Often, these variables can be manipulated in order to achieve a change in the target variable. Otherwise, there may be no causal relation, and the terms "explanatory variable" and "influence" may be misleading. In this case, the model simply describes non-causal relations.
- *Nuisance variables.* They also have a potential influence on the target variable, but their effect is not of interest and rather disturbs the examination of the influences of the explanatory variables.

Explanatory and nuisance variables will appear in the model in the same form, and we will call both of them *input variables.*

A regression model describes the target variable as a function of the input variables up to random deviations. We will call its ingredients the *regression function* or *structural part* and the *error distribution* or *random deviations part* of the model. The regression function includes parameters that are called regression *coefficients.* The coefficients may be *fixed,* unknown numbers or random variables called *random coefficients* or *random effects.*

Note that all variables may be quantitative, binary, categorical, or of a more involved nature. A rather comprehensive zoo of models have been developed in the last decades to cover the cases resulting from combining different types of target variables, fixed and random coefficents, kinds of functions (linear or nonlinear in the coefficients), as well as distributions and correlations of the random deviations.

As an example, consider a clinical study of a medication. The target variable may be blood pressure (quantitative), disappearance of a symptom (binary, yes or no), time to recovery (quantitative, often censored), or appearance and severity of side effects (ordered categorical). The primary explanatory variable is the treatment with the medication (binary). Several other variables may be of interest or perceived as nuisance variables: age and gender, body mass index, fitness, severity of disease, and compliance in taking the medication. If several health care units (hospitals or physicians) are involved, they give rise to a categorical variable with random effects.

Many models combine all effects of input variables into a *linear predictor*, which is then related to the target variable via a *link function*. The linear predictor is a linear function of unknown coefficients, but need not be linear in the input variables. In fact, input variables U_k may be transformed and combined to form the *terms* of the linear predictor $\sum_j \beta_j X_j(\underline{U})$. If only one U_k is involved in a term ($X_j = U_j$ or $= g(U_j)$), we call it a *main effect*, otherwise an *interaction*. The simplest form of an interaction term involves a product, $\beta_{jk} U_j U_k$.

The structures discussed in Section 5.3.1 can be included into such a model as a random effects term, treating the grouping as a categorical variable with random coefficients. Combined with fixed effects, such models are called *mixed (effects) linear models*, and a commonly used subclass are the *hierarchical regression models*, also called *multilevel models*. Correlation structures of the random deviations can also be built into regression models as a joint multivariate normal distribution with correlations given by a time series or a spatial model (Section 5.3.2).

5.4.2 Incorporating Reproducibility and Data Challenge

Regression models allow for prediction in a sense that refers to the idea of reproducibility: They produce a joint distribution of the values of the target variable to be obtained for given values of the input variables (those with fixed effects). Thus, they describe the expectations for the data to be obtained in a reproducibility study.

Testing the success of a reproduction for a general parameter has been introduced in Section 5.2.3 by generalizing the situation of a simple random sample in both the original and the reproduction studies. In the simple case, the observations of both studies can be joined, and a toy regression model for all the data can be set up: The observations play the role of the target variable and the only explanatory variable is the binary indicator of the replication versus original study. The test for failure of the replication is equivalent with the test for the coefficient of the study indicator to be zero.

This basic model can be enhanced with many additional features:

- Different precision of the observations in the two studies lead to a heteroscedastic model and weighted regression for its analysis.

- Random effects and correlation terms can be incorporated to take care of the structures of variation discussed in Sections 5.3.1 and 5.3.2.

- A random study effect can be added if an estimate for the respective variance component is available (Section 5.3.1).
- Additional explanatory and nuisance variables can be introduced. For each explanatory variable, an interaction term with the study indicator should be included to test if its effect on the target variable is different in the two studies.

The idea of a replication study leads to values of the explanatory variables that are selected to be the same as in the original study, or at least drawn from the same distribution. This is an unnecessary restriction: One could select different values for the explanatory variables on purpose and still test if the *model* of the original study (including the parameters) is compatible with the new data. Since such an approach differs from a replication study, we will refer to it by the notion of a *data challenge* of the model.

The notion of data challenge applies not only to regression models but also *to any theory that allows for predictions of data to be expected under given circumstances.* Within the context of regression, prediction is often the main goal of the modeling effort, from the development of technical processes to regression estimators in survey sampling.

It is easy to see, and well known, that predictions for values of explanatory and nuisance variables outside the range of those used for the modeling are often unreliable. Such "extrapolation" is less of a problem in our context. We will rather examine how large the differences of model coefficients between the original and the new study are – in the form of the study indicator variable and its interactions with all relevant explanatory variables – and hope that zero difference lies within the confidence intervals. Unreliable "extrapolated" predictions will be stabilized by fitting a joint model for both studies. If the new study does not incorporate similar values of explanatory variables as the original one, this adaptation may go too far, and the data challenge will not really be a challenge.

The concept of data challenge extends the notion of reproducibility in a general and useful sense: Rather than requiring new experiments or observation campaigns under the same circumstances, it also allows for double-checking knowledge under different circumstances. This will often be more motivating than a simple reproduction, since it may add to the generalizability of a model. Such studies are common in many research areas. Referring to them as "data challenge" may express that they strengthen the reliability of models (or theories).

5.5 Model Development and Selection Bias

5.5.1 Multiple Comparisons and Multiple Testing

When data are graphically displayed, our perception is typically attracted by the most extraordinary, most exceptional pattern. It is tempting to focus on such a pattern and test, with suitable procedures, if it can be produced by random variation in the data. Since our eye has selected the most salient pattern, such a test will yield significance more often than permitted by the testing level α, even if it is purely random.

In well-defined setups, such test procedures can be formalized. The simplest case is the comparison of g groups. If we examine all pairwise differences among $g = 7$ means by a statistical test, we will perform $k = 21$ tests. If the significance level is $\alpha = 0.05$ as usual and if all observations in all groups came from the same distribution, we should expect one (wrongly) significant test among the 21 that are performed. Such considerations have long been pursued and lead to appropriate tests for the largest difference of means among g groups (and other procedures), which feature as *multiple comparisons*.

In more general situations, this theme is known as *multiple testing*. If statistical tests on k null hypotheses are performed and all hypotheses are correct (i.e., there is no real effect at all), the *family-wise error rate* of a testing procedure is the probability that any null hypothesis is rejected. A rate bounded by α can be obtained by performing the k individual tests at a lower level. The safest and simplest choice is the Bonferroni procedure, testing all hypotheses at the level α/k and rejecting those that turn out significant at this lower level.

These considerations formalize the concern mentioned at the beginning of this subsection, except that in an informal exploration of data it is usually unclear how many hypotheses could have caught our attention. Therefore, k is unknown. Hence, any salient pattern should be interpreted with great caution, even if it is formally significant (at level α). It should be *reproduced* with a new, independent dataset before it can be taken seriously.[4]

Why have we gone into lengthy arguments about the significance of tests after having dismissed their use in Section 5.2.4? These thoughts translate directly into conclusions about estimated effects and confidence intervals: Those that are most distant from zero will be more so than they should, and their coverage probabilities will be wrong (unless corrected adequately).

If an effect is highly significant (if zero is far from the confidence interval) and

[4]Collins (this volume) discusses a related issue in gravitational wave physics, where an extensive search resulted in the "detection" of spurious effects.

has a plausible interpretation, we tend to regard it as a fact without replicating it. A strict scientific rule of conduct would require, however, that an effect can only be regarded as a fact, and its precision (expressed by the confidence interval) is only reliable if derived from data that was obtained with the specific goal of testing or estimating that specific effect. Only such a setup justifies the interpretation of the testing level as the probability of an error of the first kind, or the level of a confidence interval as a coverage probability. The statistical analysis of the data obtained in such a study was called *confirmatory analysis* by John Tukey in the 1970s. But Tukey also emphasized the value of *exploratory data analysis*: analyzing data for unexpected patterns, with the necessary caution concerning their interpretation.

5.5.2 Consequences for Model Development

In many statistical studies, the relationship of one or more target variable(s) with explanatory and nuisance variables is to be examined, and the model is not fixed beforehand. Methods of model development are then used to select

- the explanatory and nuisance variables that enter the model at all,
- the functional form in which the variables enter the model (monotonic transformations, or more flexible forms like polynomials and splines, may often be adequate),
- the function that combines the effects of the resulting transformed variables (nonlinear, linear) and relates the combination to the target variable (via a "link function"),
- interaction terms describing the joint influence of two or more variables in a more refined way than the usual, additive form,
- any correlation structure of the random deviations if applicable.

These selections should be made with the goal of developing a model that fits the data well in the sense that there are no clear violations of the assumptions. This relates to the *exploratory mode* of analysis mentioned in the preceding subsection. It would be misleading to use estimated effects, tests for a lack of effect, or confidence intervals based on such a model without any adjustments.

Developing a model that describes the data well is a task that can be as fascinating as a computer game. Strategies for doing so should be formalized and programmed. They would form procedures for the "estimation of a model" that can be evaluated. Activities in this direction have been restricted to the limited task of selecting variables from a pre-defined set of potential variables, known as *model* or *variable selection* procedures. The next task would be to formalize the uncertainty of these procedures. In analogy to the confidence interval, one might obtain a set of models that are compatible with the data in the sense of a statistical test for the validity of a model.

Whether formalized or not, doing too good a job for this task leads to *overfitting* (cf. Bailey *et al.*, this volume): The part that should remain random sneaks into the structural part. This hampers *reproducibility,* since the phony ingredients of the structural part of the model will be inadequate for the new study.

An alternative to developing a model along these lines is to use a very flexible class of models that adapts to data easily. Two prominent types of such classes are *neural networks* and *vector support machines.* Such model classes have been proposed mainly for the purpose of *predicting* the target variable on the basis of given values of the input variables. Care is needed also for these methods to avoid overfitting. The fitted models are hard to interpret in the sense of effects of the explanatory variables and, therefore, generally of less value for science.

An important part of model development, for which formalized procedures are available, is the task of variable selection. In order to make the problem well defined, a *sparsity condition* is assumed for the correct model: Among the potential regressors X_j, $j = 1, ..., p$, only a subset $S = \{j_1, j_2, ..., j_q\}$ has coefficients $\beta_j \neq 0$, and these are not too small. Thus, the true linear predictor is $\sum_k \beta_{j_k} X_{j_k}$. We have argued in Section 5.2.4 that such a condition is unrealistic in general, since at least tiny effects should be expected for all variables. An important field for which the condition is plausible is "omics" research: In genomics, regression models explain different concentrations of a protein by differences in the genome, and in proteomics, a disease is related to the activities of proteins. Supposedly, only a small number of genes or proteins, respectively, influence the outcome, and the problem is to find them among the many candidates.

A model selection procedure leads to a subset $\widehat{S}$. In analogy to a classification rule, the quality of such a method is characterized by the number of regressors in $\widehat{S}$ that are not contained in S, the "false discoveries," and the number of regressors in the true model S that are missed by $\widehat{S}$, the "false negatives," or, equivalently, the number of "true discoveries," $\#(\widehat{S} \cap S)$. The family-wise error rate introduced in Section 5.5.1 is the probability of getting at least one false discovery. A model selection procedure is *consistent* if it identifies the correct model with probability going to one for $n \to \infty$. Consistent variable selection methods include methods based on the Bayesian information criterion, the lasso and its extensions with suitable selection criteria, and a method introduced by Meinshausen *et al.* (2009), briefly characterized in Section 5.5.3. These concepts and methods are treated in depth by Bühlmann and van de Geer (2011).

In a reproducibility study, any variable selection procedure will most likely produce a different model. This raises the question of *which aspects of the results may be reproducible.* Three levels may be distinguished: (1) Assuming that the procedure provides a measure of importance of the individual input variables, the more important ones should again be found in the reproduction study. A formalization of this thought seems to be lacking. (2) A less ambitious requirement would ask for the selected model to fit the new data without a significant lack of fit. More specifically, adding variables that have been dropped by the variable selection in the original study should not improve the fit significantly (taking care of multiplicity!). (3) If the focus of the study is on prediction, a similar size of the variation of the random deviations may be enough.

An important point that is often overlooked in this context is that the effect of a variable X_j on the target variable *depends crucially on the set of other variables* in the model – as far as they are correlated with X_j. In (generalized) linear models, the coefficient β_j of X_j measures the effect of changing variable X_j while keeping all other variables constant. Such a change may be trivial to perform or, in other applications, unrealistic or impossible, if any change of X_j leads to a change in other input variables. In fact, the *scientific question* must be explicit about this aspect: What should be kept constant for specifying the effect of X_j to be examined? A similar difficulty arises with *interaction* terms: They imply that the effect of X_j depends on the value of another variable, and it must be clear which effect of X_j is to be estimated.

Let us use the distinction between explanatory variables, whose effects are of primary interest in the planning phase of the study, and nuisance variables, which are used to obtain a more precise model. If the nuisance variables are roughly orthogonal to (uncorrelated with) the explanatory variables, this provides freedom to include or exclude the latter in the model. Thus, those having little influence on the target variable can be safely dropped. Adjustments to the inference about the explanatory variables are, nevertheless, necessary, as Berk *et al.* (2013) showed, naming the problem "postselection inference." The role of nuisance variables that are correlated with explanatory variables must be determined without reference to the data. If a nuisance variable shows a significant, unanticipated effect, this should be interpreted with great caution.

The desired independence of the explanatory and nuisance variables can be obtained by randomization if the former can be manipulated. Thus, when the effect of a treatment is to be assessed, the assignment of individuals to the treatment or control group must be based on random numbers or on so-called orthogonal designs. The block designs mentioned in Section 5.3.3 also keep treatment and nuisance effects orthogonal.

With these rules, the effects of explanatory variables should be reproducible in the sense of Section 5.2.3, whereas inference on effects of nuisance variables is less reliable.

5.5.3 Internal Replication

The idea of replicating results on new data has inspired procedures that alleviate the problem of overfitting in model selection.

The choice among competing models should be based on a suitable performance measure, like a mean squared prediction error or misclassification rate. A straightforward estimation of such measures uses residuals as prediction errors or counts misclassifications of the observations of the data set used to fit the model. These estimates, however, are biased, since the estimated model fits the data better than the correct model does. In order to assess the performance correctly, a promising approach is to split the dataset into two parts, a "training set" used to estimate the model and a "test set" for which the performance is evaluated. Clearly, this mimics an original and a replication study.

Unless the number of available observations is large, there is an obvious trade-off when choosing the relative sizes of these two parts. A small training set leads to poor estimation of the model parameters, whereas a small test set entails an imprecise performance estimate. This trade-off is avoided by keeping the test set small, but splitting the dataset many times and averaging the performance estimate over all splits. The extreme choice is using a single observation as the test set. These procedures are called *cross validation*.

Many model selection techniques lead, in a first step, to a sequence of models with increasing size – number of terms in a regression model. The unbiased measures of performance are used to select the final model among them.

A different approach was proposed by Meinshausen *et al.* (2009). They split the sample into halves, use the first part to select variables and the second to calculate a significance measure. Repeating this step for many random splits, they derived p-values for the effects of all the variables that are asymptotically correct in the sense of multiple testing and thereby get a final variable selection.

5.5.4 Publication and Selection Bias

The problem of multiplicity (Section 5.5.1) is also relevant for the publication of scientific results. Since results are hard to publish if there is no statistically significant effect (cf. Section 5.2.4), those that are published will be a selection of the studies with highest effect estimates among all conducted studies. In addition, in some fields, the effects to be examined are first selected within a study among many possible effects that can appear to be significant by choosing different target variables and exploiting the possibilities of model selection listed in Section 5.5.2.

These choices have been called "researcher degrees of freedom" (Simmons *et al.* 2011) or "forking paths" (Gelman and Loken 2014). The published results will, therefore, include too high a number of "false positives," that is rejections of "correct" null hypotheses or confidence intervals excluding the zero effect value. This was prominently discussed in the paper by Ioannidis (2005), "Why most published research findings are false," and has led to a considerable number of follow-up articles; see also Ehm (this volume).

Notably, the analysis by Ioannidis (2005) relies on significance testing, discussed critically in Section 5.2.4, where we questioned the plausibility of an exactly zero effect. Therefore, "correct effects" would need to be defined as those that are larger than a "relevance threshold." Then, an estimated effect would need to be significantly larger than the threshold to be regarded as a valid result. This would reduce the number of effects declared as significant, but would not rule out selection bias.

In the delicate field of drug development, the problem would lead to useless medications which by chance appeared effective in any study. This is avoided through strict regulations requiring that *clinical trials* be planned in detail and announced before being carried out. Then, the decision is based on the results of that study, and no selection effect of the kind discussed above is possible. Such pre-registration has been proposed as a general remedy to fight the problem of publication and selection bias (Ioannidis 2014).

While these lines of thought consider the "correctness" of an effect obtained as significant, let us briefly discuss the consequences for reproducibility. If the effect to be tested by reproduction has been subject to selection bias in the original paper, the distribution of the estimate is no longer what has been used in Section 5.2.3, but its expected value will be smaller (in absolute value) and its variance larger than assumed. Since the biases can rarely be quantified, a success of the replication should then be defined as reaching the same conclusion, that is obtaining a confidence interval that again lies entirely on the same side of zero (or of the relevance threshold) as in the original study.

5.6 Big and High-Dimensional Data

Datasets in traditional scientific studies comprise a few dozen to several thousand observations of a few to several dozen variables. Larger datasets have appeared for a while in large databases of insurance companies and banks as well as automated measuring instruments in the natural sciences, and the data that can be collected from the world wide web today leads into even larger orders of magnitude. If organized in rectangular data matrices, traditional modeling treats clients, time points, and records as "observations", and their characteristics, concentrations of substances, and fields in records as "variables."

Usually, the number of observations grows much more rapidly than the number of variables, leading to the problem of handling *big data*. The large number of observations allows for fitting extremely complicated models, and it is to be expected that no simple model fits exactly, in the sense of showing non-significant results for goodness-of-fit tests. Nevertheless, it is often more useful to aim for simple models that describe approximate relationships between variables. Other typical problems deal with searching patterns, like outliers hinting to possible fraud or other extraordinary behavior, or clusters to obtain groups of clients to be treated in the same way (see Estivill-Castro, this volume).

A useful strategy to develop reasonable models for big data is to start with a limited random subset, for which traditional methods can be used, and then validate and refine them using further, larger subsets. In principle, all aspects of reproducibility checks can then easily be applied within the same source of data, possibly for studying its stability over time. However, when descriptive methods are applied that do not intend to fit a parametric model (as is typical for cluster analysis), it may be unclear how the results of their application to subsets or different studies should be compared.

In genomics and proteomics, the situation is reversed as compared to "big data": measurements of thousands of "variables" are collected on chips, which are used on a few dozen or a few hundred subjects. This leads to *high-dimensional data*. In this situation, traditional methods typically fail. Regression models cannot be fit by the popular least-squares principle if the number of observations is smaller than the number of coefficients to be estimated. The basic problem here is variable selection, for which reproducibility has a special twist, as discussed in Section 5.5.2.

5.7 Bayesian Statistics

In the introduction, the core business of statistics was presented as providing a way to quantitatively express uncertainty, and confidence intervals were proposed as the tools of choice. Another intuitive way to describe uncertainty is by probabilities. The parameter to be assessed is then perceived as a random variable whose distribution expresses the uncertainty, and it does so in a much more comprehensive way than a simple confidence interval does.

This is the paradigm underlying Bayesian statistics, where the distribution of the parameter expresses the belief about its plausible values. This allows for formalizing, by the *prior distribution*, what a scientist believes (knows) about a parameter before obtaining data from which to draw inferences about it. When data becomes available, the belief is updated and expressed again as a distribution of the parameter, called the *posterior distribution*.

The general rule by which updating is performed is the formula of Bayes connecting conditional probabilities – whence the name of the paradigm. The idea of Bayesian statistics is to describe in a rational way the process of *learning* from a study or, by repeated application, from a sequence of studies. Note that the idea still is that there is a "true value" of the parameter, which would eventually be found if one could observe an infinite amount of data. The distribution of the parameter describes the available knowledge about it and, hopefully, will have a high density for the true value.[5]

It is useful to distinguish the two kinds of probabilities involved:

- The prior distribution expresses a belief and is, as such, not objective. It may be purely subjective or based on the consent of experts. Such probabilities describe *epistemic* knowledge and their use contradicts the objective status of a science which is supposed to describe nature. This is not too severe if we accept that objectivity in this sense is an idealization which cannot be practically achieved.
  Furthermore, if little is known *a priori*, this may be expressed as a "flat" or "uninformative" prior distribution. This leads us back to an analysis that does not depend on a subjective choice of a prior distribution.
- The probability model that describes the distribution of the observations for given values of the parameters is of a different nature. The corresponding probabilities are called *frequentist* or *aleatory*. This is the model used in the main body of this paper. Note that the choice of the model often includes subjective elements and cannot be reproduced "objectively," as discussed in Section 5.5.

The paradigm of Bayesian statistics is in some way contrary to the idea of *reproducibility*, since it describes the process of learning from data. For a replication study, the posterior distribution resulting from the original work would be used as a prior, and the new data would lead to a new, sharper posterior distribution of the parameter.

On the other hand, judgments about the success of a replication attempt can be formalized by quantifying the evidence from the new data for the hypotheses

[5]For more details on Bayesian modeling approaches, see the chapter by Shiffrin and Chandramouli, this volume.

of a zero and of a nonzero effect. When a prior distribution for the effect parameter is fixed, the (integrated) likelihood can be calculated for the nonzero effect hypothesis, and compared to the likelihood under the null hypothesis. The ratio of the two likelihoods is called the *Bayes factor.*

Verhagen and Wagenmakers (2014) explained this idea and proposed, in the replication context, to use the posterior distribution of the original study as the prior within the non-zero hypothesis for this evaluation. This leads to a "point estimate" for the Bayes factor and leaves the judgement whether the evidence expressed by it should suggest a success or a failure of replication open. Reasonable boundaries for such a decision would depend on the amount of information in the data of the original and the replication studies (and in the prior of the original study, which may be standardized to be non-informative).

Gelman (2014) advocated the use of Bayesian statistics as an alternative to hypothesis testing, cf. Section 5.2.4. On the one hand, he suggested to boost the insufficient precision of studies with small sample sizes by using an informative prior distribution. We strongly prefer to keep the information expressing prior knowledge separated from the information coming from the data for the sake of transparency, especially for readers less familiar with the Bayesian machinery. On the other hand, Gelman advocated the use of multilevel models, which are often formulated in Bayesian style. We agree with this recommendation but prefer the frequentist approach as described in Section 5.3.1.

5.8 Conclusions

5.8.1 Successful Replication

Reproducibility requires that a scientific study should be amenable to replication and, if performed again, should yield "the same" results within some accuracy. Since data is usually subject to randomness, statistical considerations are needed. Probability models are suitable to distinguish between a *structural part* that should remain the same for the replication study, and *random deviations* that will be different, but still subject to a supposed probability distribution.

This is an idealized paradigm. It relies on the assumption that the "circumstances" of the new study are "the same" in some sense of *stability* (cf. Tetens and Steinle, both in this volume). Stochastic models may include components to characterize the extent of such stability in a realistic manner: Usually, a variance component that describes the deviations from experiment to experiment or period to period will be appropriate. The lack of such a component may be a candidate to explain the common experience that reproducibility fails too often.

One can formalize a successful replication as one for which the difference from the original study in the estimate of the parameter of interest θ is not significantly different from zero. This may lead to different situations for the conclusion about the effect measured by θ itself. As an extreme example, one might have a significantly positive estimate $\widehat{\theta}$ in the original study and a (non-significant) negative estimate in the replication and still get an insignificant difference so that the replication would look successful. Thus, it is useful to distinguish two aspects:

- *Statistical compatibility.* Is the data obtained in the replication compatible with the data from the original study in the light of the model used to draw inference? If not, this may point to a flaw in the model, like neglecting a source of variability, to a selection bias, to manipulation of data, or to a failure to reproduce the same circumstances.
- *Confirmation of conclusions.* Does the replication study lead to the same conclusions as the original one? If a relevant effect has been found in the original study, one would hope to find it again (as statistically significant) in the replication. The result may then be indecisive, if statistical compatibility is fulfilled, but significance fails in the repeated study.

5.8.2 Validation and Generalization

In designing studies for empirical modeling, there is a trade-off between keeping the circumstances as homogeneous as possible to get clearly interpretable results and making them as different as possible to get *generalizability*, i.e., to allow for more generally applicable conclusions. The latter suggests to repeat an experiment under different conditions, as discussed above, using block designs (Section 5.3.3), or including nuisance variables that characterize different conditions.

This may also suggest an *extension of the notion of reproducibility* to include studies that explore different circumstances with the expectation to reproduce the essential *conclusions* from an original study rather than reproducing the data. We introduced the concept of *data challenge* for such studies (Section 5.4.2). However, it is not clear if possible discrepancies should be interpreted as a failed replication or as a variation due to altered circumstances. A most rewarding way to conduct validation studies will be to design a *replication part* geared at reproducing the results under the same circumstances and an additional *generalization part*, varying them to yield the potential for novel insight.

Another, related direction of extending the idea of reproducibility is the *conceptual replication*, in which the essential conclusions of a study are examined, intentionally using different methods than in the original study. In psychology,

	same follow-up study features	different follow-up study features	level of validation
repetition, repeatability	all settings, experimenters	—	all data features; compatible model and estimated effects
replication, reproducibility	all settings, procedures	experimenters, institution	compatible model and estimated effects
data challenge	model	settings of explanatory and nuisance variables	model fits both studies; conclusions
replication of concepts	concepts (constructs)	methods (instruments)	stable concepts and relations; conclusions

Table 2: Levels of validation for same and different features (middle columns) in the original and follow-up study. The different terms in the left column are related to different types of validation in the right column.

the personality characteristics, called "traits" or "constructs," cannot be measured directly, but are derived from the analysis of questionnaire data or from the performance of subjects in test tasks. Their supposed relationships (like "Is 'emotional intelligence' related to 'work memory'?") should be independent of variations of questionnaires ("instruments") that intend to measure the same two constructs. This principle is often examined in so-called "multitrait-multimethod" analyses. The idea of conceptual replication could and should be transferred to other fields of science.

As a summary, Table 5.2 collects the various levels of validation considered in this chapter. They differ in the features of the follow-up study that are intended to be identical or different from the original one and refer to different levels of validation and generalization of the results. The upper two rows in the table aim at confirming quantitative effects. Since these are only defined for a given – possibly quite trivial – model, this model must be compatible in the first place. The lower two rows in the table aim at generalizing conclusions, which may be quantitative or qualitative. They deserve more discussion and recognition as tools for achieving related, but broader goals than reproduction in a narrow sense do.

Indeed, the purpose of science is to discover models that generalize to novel, different circumstances. Such models earn the honorary title of a *theory* if they predict and explain outcomes for new situations and either provide interpretable relationships or, even better, have a *causal interpretation*. It is well known that causal relations cannot be proven by empirical modeling alone but need either experimentation or a connection to a plausible mechanism that implies causality. However, while causality is a fundamental concept in the so-called exact sciences, many questions in the life sciences and social sciences can only be answered by correlative phenomena.

5.8.3 Scope of Reproducibility

In the *social sciences,* reproducibility may play a less rigorous role than in the exact and life sciences. This is partly due to the difficulty of experimenting or getting "similar circumstances," like in macro-economics or sociology, and partly due to diverse valuable goals of these sciences. Gelman and Loken (2014) and Gelman (2014) discussed the role that is and should be assigned to reproducibility in these fields.

There are ongoing efforts to introduce quantitative concepts in psychology and to establish relations among them and with variables that may influence them. Standardized questionnaires, thoroughly tested for reproducibility, are used to measure concepts. This push towards measurable concepts entails the risk of *deprecating those questions that resist quantification and replicability* as non-scientific (as expressed by Collins, this volume). Nevertheless, *models* may well play a role for making ideas more transparent, even if they cannot be empirically verified and reproducibility of quantitative results is not feasible.

In the discussion of model development in Section 5.5.2 we mentioned the tradeoff between *flexibility*, which helps to describe the data well, and *simplicity*, which tends to increase generalizability. That simple models are more likely generalizable is implicitly based on experience, although it would be difficult to justify by empirical data. It is related to the principle of *Occam's razor*, relying on a belief that reality can be described by "simple," general laws. Why are such models possible in the natural sciences? This final question leads us way beyond exact and social sciences into *philosophy and religion*.

Acknowledgments

I thank many colleagues for helpful discussions, including Christian Hennig, Andreas Buja, Carolin Strobl, and Ted Anderson. An editorial workshop at Munich in October 2014 has been very fruitful for me to become familiar with the subject and its implications for the many fields of science represented in this volume.

References

Berk, R., Brown, L., Buja, A., Zhang, K., and Zhao, L. (2013): Valid post-selection inference. *Annals of Statistics*, **41**, 802–837.

Bühlmann, P., and van de Geer, S. (2011): *Statistics for High-Dimensional Data. Methods, Theory and Applications*, Springer, Heidelberg.

Gelman, A. (2014): The connection between varying treatment effects and the crisis of unreplicable research: A Bayesian perspective. *Journal of Management* **20**(10), 1–12.

Gelman, A., and Loken, E. (2014): The statistical crisis in science. *American Scientist* **102**, 460–465.

Ioannidis, J. (2005): Why most published research findings are false. *PLOS Medicine* **2**, 696–701.

Ioannidis, J. (2014): How to make more published research true. *PLOS Medicine* **11**(10): e1001747.

Killeen, P.R. (2005): An alternative to null-hypothesis significance tests. *Psychological Science* **16**(5), 345–353.

Krantz, D.H. (1999): The null hypothesis testing controversy in psychology. *Journal of the American Statistical Association* **94**, 1372–1381.

Künsch, H.R., Beran, J., and Hampel, F.R. (1993): Contrasts under long-range correlations. *Annals of Statistics* **21**(2), 943–964.

Leisch, F. (2002): Sweave: Dynamic generation of statistical reports using literate data analysis. In *Compstat 2002 – Proceedings in Computational Statistics*, ed. by W. Härdle and B. Rönz, Springer, Berlin, pp. 575–580.

Meinshausen, N., Meier, L., and Bühlmann, P. (2009): P-values for high-dimensional regression. *Journal of the American Statistical Association* **104**, 1671–1681.

Rodgers, J.L. (2010): The epistemology of mathematical and statistical modeling. *American Psychologist* **65**, 1–12.

Simmons, J.P., Nelson, L.D., and Simonsohn, U. (2011): False-positive psychology: Undisclosed exibility in data collection and analysis allows presenting anything as significant. *Psychological Science* **22**, 1359–1366.

Stodden, V., Leisch, F., and Peng, R.D. (2014): *Implementing Reproducible Research*, CRC Press, London.

Student (1927): Errors of routine analysis. *Biometrika* **19**, 151–164.

Verhagen, J., and Wagenmakers, E.J. (2014). Bayesian tests to quantify the result of a replication attempt. *Journal of Experimental Psychology: General* **143**, 1457–1475.

Wilkinson, L., and Task Force on Statistical Inference (1999): Statistical methods in psychology: Guidelines and explanations. *American Psychologist* **54**, 594–604.

Xie, Y. (2013). *Dynamic Documents with R and knitr*, CRC Press, London.

6
Model Selection, Data Distributions, and Reproducibility

Richard Shiffrin and Suyog Chandramouli

Abstract Models offering insights into the infinitely complex universe in which we reside are always wrong but vary along many dimensions of usefulness. It is most often useful in science to prefer models that are simultaneously a good representation of the "truth" and simple (a form of Occam's Razor). However, we cannot match "truth" to models directly. Thus in a limited experimental domain we represent "truth" by a distribution of possible experimental outcomes. Inference is carried out by assuming the observed data are a sample from that unknown data distribution. This chapter discusses the factors that govern the variability of that distribution, termed "replication variance," and how those factors do and should influence both model comparison and reproducibility. We present an extension of Bayesian model selection (BMS) that infers posterior probabilities that a given model instance predicts a data distribution that is the best match to the "true" data distribution. We point out close similarities to the other chief method for model comparison, minimum description length (MDL). Finally we show how posterior probabilities for the data distributions can be used to produce a predicted distribution for a statistic S defined on the data: Reproducibility of S can be assessed by comparing the value of S in the replication to the predicted distribution.

6.1 Introduction

This note provides background, technical definitions, and analytic derivations needed to explicate the issue of replication variance – its sources, its importance in model selection, and its utility in estimating reproducibility.

Section 6.2 lays out Bayesian model selection (BMS) in a form that highlights its close relation to the other main method for model comparison, minimum description length (MDL), approximated by a form of normalized maximum likelihood (NML). All sensible model comparison methods, including these two, strike a balance between good fit and complexity. BMS assumes the true state of the world[1] is represented by some instance of the model classes being compared.

---

[1] In this chapter we use "true" in the sense of "real": neither measurements nor models can *exactly and exhaustively* represent the "real" state of the world. We will abstain from any philosophical discussion about the simplicity or complexity of the notion of "truth."

Reproducibility: Principles, Problems, Practices, and Prospects, First Edition. Edited by Harald Atmanspacher and Sabine Maasen.

Section 6.3 extends BMS by changing this assumption and inferring the posterior probabilities, given the observed data, that each model instance provides the best *approximation* to the true state of the world. In these first two sections the true state of the world is represented by a distribution of possible outcomes of the experiment providing the observed data.

Section 6.4 presents the reasons why the true state of the world is represented by a distribution of possible outcomes, outlines some factors that determine the variability of that distribution, and discusses the way the assumptions underlying replication variance govern both model comparison and reproducibility.

6.2 Bayesian Model Selection and Relation to Minimum Description Length

We first fix some terminology. Model classes are collections of model instances, usually characterized by parameters with values to be specified. Each instance has specific values assigned to all parameters. For convenience and to highlight the fact that inference should not extend past the precision and power of the data to discriminate, these model instances, each possible outcome of an experiment, and all probabilities are discretized.

Let M_0 and M_1 denote two model classes to be compared, with prior probabilities $p_0(M_0)$ and $p_0(M_1)$ that sum to 1. Each class is made of model class instances that will be denoted θ_i and λ_i, respectively. Let $p_0(M_0) = \sum_i p_0(\theta_i, M_0)$ and $p_0(M_1) = \sum_i p_0(\lambda_i, M_1)$. The notation $p_0(\lambda_i, M_1)$ is used to emphasize the fact that the prior probabilities sum to one across both classes.

A given instance has all parameter values specified and, therefore, predicts exactly one distribution, $\psi_k(\theta_i)$ (or $\psi_k(\lambda_i)$), of data outcomes. Each ψ assigns a probability to each possible data outcome, x (including the observed data outcome, y): $p_0(x|\theta_i)$ or $p_0(x|\lambda_i)$ for the prior probabilities, and $p(x|\theta_i, y)$ or $p(x|\lambda_i, y)$ for the posterior probabilities (the probabilities of the instances *after* inference based on the observed data).

6.2.1 Bayesian Inference and Bayesian Model Selection (BMS)

Assume to begin the discussion that M_0 and M_1 share no instances that predict very similar distributions of data outcomes. That is, there is no case where $\mathbf{E}(\theta_i) \approx \mathbf{E}(\lambda_i)$, where $\approx$ means equal up to the precision of the ability to measure in some experimental setting. We show next that using this assumption in combination with Bayesian inference makes posterior inference concerning instances and posterior inference about classes consistent with each other:

1. *Inference by Class.* We use Bayes to calculate the posterior probabilities of M_0 (and of M_1). The posterior probability of M_0 is:

$$p(M_0|y) = \frac{p(y|M_0)\, p_0(M_0)}{p(y|M_0)\, p_0(M_0) + p(y|M_1)\, p_0(M_1)} \tag{6.1}$$

Note that the denominator equals $p(y)$. Let us substitute $p(y)$ for the denominator in the following. Note that the numerator is the joint probability $p(y, M_0)$, which can be written as a sum of joint probabilities over the instances: $p(y, M_0) = \sum_i p(y, \theta_i) = \sum_i p(y|\theta_i, M_0)\, p_0(\theta_i, M_0)$. Note also that $p_0(M_0) = \sum_i p_0(\theta_i, M_0)$. Thus, Eq. (6.1) can be written as:

$$p(M_0|y) = \frac{\sum_i p(y|\theta_i, M_0)\, p_0(\theta_i, M_0)}{p(y)} \tag{6.2}$$

These prior probabilities add to 1 over both classes, but they can be conditioned on the class they come from, leading to the following version of the same equation:

$$p(M_0|y) = \frac{\sum_i p(y|\theta_i, M_0)\, p_0(\theta_i|M_0)\, p_0(M_0)}{p(y)} \tag{6.3}$$

Similar equations can of course be written for $p(M_1|y)$. We take the ratio of these posterior probabilities to connect with the usual way of characterizing classes by the Bayes factor:

$$\begin{aligned}\frac{p(M_0|y)}{p(M_1|y)} &= \frac{1/p(y)\, \sum_i p(y|\theta_i, M_0)\, p_0(\theta_i|M_0)\, p_0(M_0)}{1/p(y)\, \sum_i p(y|\lambda_i, M_1)\, p_0(\lambda_i|M_1)\, p_0(M_1)} \\ &= \frac{\sum_i p(y|\theta_i, M_0)\, p_0(\theta_i|M_0)}{\sum_i p(y|\lambda_i, M_1)\, p_0(\lambda_i|M_1)}\, \frac{p_0(M_0)}{p_0(M_1)}\end{aligned} \tag{6.4}$$

The left-hand ratio in Eq. (6.4) is usually termed the Bayes factor, and the right-most term is usually considered the prior odds for the model classes, albeit this is the discrete formulation (see Raftery 1995, Kass and Raftery 1995). Therefore, we define the Bayes factor BF as:

$$\mathrm{BF} = \frac{\sum_i p(y|\theta_i, M_0)\, p_0(\theta_i|M_0)}{\sum_i p(y|\lambda_i, M_1)\, p_0(\lambda_i|M_1)} \tag{6.5}$$

2. *Inference by Instance.* We use Bayes' theorem to calculate the posterior probabilities of the θ_i and λ_i:

$$p(\theta_i|y) = \frac{p(y|\theta_i)\, p_0(\theta_i)}{p(y)} \tag{6.6}$$

We sum these posteriors over all θ_i in M_0 to get the posterior for M_0:

$$p(M_0|y) = \sum_i p(\theta_i|y) = \frac{\sum_i p(y|\theta_i)\, p_0(\theta_i)}{p(y)} \tag{6.6a}$$

Equation (6.6a) is the same as Eq. (6.2), so Eqs. (6.3), (6.4) and (6.5) follow.

It is important to note the stipulation that the instances in one class produce distributions that are all distinguishably different from the instances in the other class. If this is not the case, then one cannot compare classes by summing their instance posteriors. This is most clear when one class is wholly embedded in another, because the sum of instance posteriors must be larger for the larger class.

We shall assume in this chapter that the shared instances in the smaller class are subtracted from the larger class prior to model comparison. The reasons and examples are given in Shiffrin *et al.* (2015). We note here only that in most cases of model comparison in which a full model is compared to one in which some of the parameters are restricted, the larger model is of larger measure than the smaller so such subtraction does not change any of the traditional results.

6.2.2 Occam's Razor and BMS

When comparing model classes, one prefers a model class that has instances that fit the observed data well, but not a model class that has instances that fit all possible data sets well. We prefer a model that has an appropriate balance of good fit and complexity. BMS achieves this by distributing prior probabilities across a class: Recall that $p_0(M_0) = \sum_i p_0(\theta_i, M_0)$ and $\sum_i p_0(\theta_i|M_0) = 1$. Thus, when model instances are not shared across classes, a larger class must have lower conditional prior probabilities, on average, for its instances. When $p_0(M_0) = p_0(M_1)$, then this statement is also true of the unconditional prior probabilities.

6.2.3 An Equivalent Characterization of BMS and Bayes Factor

The Bayesian inference in Eq. (6.5) shows that the Bayes factor is a ratio of sums of the conditional joint probabilities of the observed data and the model instance. We now rewrite this equation to give a characterization that will show us how the usual Bayes factor and NML align, even when the model classes have shared instances.

Because the sum of $p(x|\theta_i, M_0)$ over all data sets x must be 1, the numerator of the Bayes factor (Eq. 6.5) can be written as:

$$\sum_i p(y|\theta_i, M_0)\, p_0(\theta_i|M_0) = \frac{\sum_i p(y|\theta_i, M_0)\, p_0(\theta_i, M_0)}{p_0(M_0)}$$
$$= \frac{\sum_i p(y|\theta_i, M_0)\, p_0(\theta_i, M_0)}{\sum_x \sum_i p(x|\theta_i, M_0)\, p_0(\theta_i, M_0)} \quad (6.7)$$

Replacing $p_0(\theta_i, M_0)$ by $p_0(\theta_i|M_0)\, p_0(M_0)$, in both the numerator and denominator of Eq. (6.7), one notes that the term $p_0(M_0)$ cancels, so Eq. (6.7) can equally well be stated in terms of conditional joint probabilities

$$\frac{\sum_i p(y|\theta_i, M_0)\, p_0(\theta_i|M_0)}{\sum_x \sum_i p(x|\theta_i, M_0)\, p_0(\theta_i|M_0)} \quad (6.7a)$$

The denominator of the Bayes factor can be written similarly, with λ_i, M_1 replacing θ_i, M_0. The Bayes factor is thus seen as a ratio across the two classes: each term in the ratio is the *sum* of the conditional or unconditional joint probabilities of the model instances for the observed data normalized by a sum of like quantities for all data outcomes.

This characterization can be generalized: One can multiply the numerator and denominator of Eq. (6.7a) by any constant without changing the result. If the constant is $1/n_i$, the number of instances in the class, then one can restate the result as a normalized *average* joint probability. The statement of the Bayes factor with the use of Eq. (6.7) highlights similarities to one formulation of MDL, termed NML#. In the next section, we show that NML# can be expressed as a ratio for each class of the conditional (or unconditional) *maximum* joint probabilities.

6.2.4 Minimum Description Length and Normalized Maximum Likelihood

The field has a comprehensive overview of MDL as a principle to guide model selection in the book by Grünwald (2007); see also Barron *et al.* (1998). We give only brief remarks here. The goal of MDL, as with BMS, is inductive inference. At its core it uses the idea that one can characterize the regularities in data by a code that compresses the description to an optimal degree.[2] Model classes can be viewed as defining candidates for such a code, and one should prefer a class that compresses the available data to the greatest degree.

[2] MDL is very much the same as algorithmic complexity, the length of the shortest algorithm that is capable of reproducing the given data. In the study of complex systems, it has been used as a measure of complexity; cf. Atmanspacher and Demmel, this volume.

MDL is typically – like BMS – applied to probabilistic model classes, i.e., model classes whose instances are probability distributions. To make MDL model selection well defined for such model classes one needs a mapping from sets of probability distributions to codes. The generic principle for doing this is to pick the code that minimizes the worst-case coding regret, a method that gives the same result irrespective of any particular parameterization of the model class at hand. "Basic" MDL model selection tells us to associate each model class with its corresponding worst-case regret code and then pick the one minimizing codelength of the data at hand.

Because of a natural interpretation of minimum code length as maximum probability, this turns out to be equivalent to associating each model class with a specific corresponding *distribution* and then picking the distribution maximizing the probability of the data. If one wants to incorporate prior knowledge, one is led to an approximation that we term normalized maximum likelihood#, or NML# (see Zhang 2011): It is a form of what Grünwald (2007) termed *luckiness* NML-2. It is the same as Eq. (6.7), with the sum of joint probabilities replaced by the maximum joint probability:

$$\mathrm{NML\#} = p_{\mathrm{NML}}(y|M_0) = \frac{\max_i\Big(p(y|\theta_i, M_0)\, p_0(\theta_i|M_0)\Big)}{\sum_x \max_i\Big(p(x|\theta_i, M_0)\, p_0(\theta_i|M_0)\Big)} \tag{6.8}$$

The same is done for all models under consideration (e.g., for model 1, we have λ_i, M_1 replacing θ_i, M_0), and the amount of evidence in the observed data y for model M_0 over M_1 is measured as the ratio. Readers familiar with the original version of NML will notice that, if the prior over model instances is uniform, then both numerator and denominator in Eq.(6.7a) share a constant factor $1/n_0$ that cancels from the equation – the resulting equation is then the original NML.

In sum, the Bayes factor can be seen as a ratio across the two classes: each term in the ratio is the sum (or average) of the conditional or unconditional joint probabilities of the model instances for the observed data normalized by a sum of like quantities for all data outcomes. NML# is the same with the sum (or average) replaced by the maximum joint probability. See Shiffrin *et al.* (2015) for details and discussion.

6.3 Extending BMS (and NML#): BMS*

To make the discussion in Section 6.4 clear, it is useful and important to extend BMS by making explicit the idea that all the models are wrong to begin with, and that one should be choosing the model that *best approximates* the truth. To do this, it is necessary to make explicit the fact that the true state of the world is represented by distributions of outcomes, and that the model instances predict distributions of outcomes (see, e.g., Karabatsos 2006).

The equations that determine Bayesian inference, given in Section 6.2, remain unchanged, but the inference is different: Both prior and posterior probabilities refer the probability that a model instance (or class) is the best approximation to the distribution that specifies the true state of the world (what will be the true distribution in Section 6.4).

Best approximation is determined by a function G that is used to match distributions. There are a number of choices for G, such as some form of Kullback–Leibler information. For present purposes we do not need to choose G, but note that the choice should be made in light of one's inference goals. For example, for different goals one might want to choose a G that emphasizes or de-emphasizes extreme values. For a given G, the new system is termed BMS*. Thus, in BMS*, as for BMS:

$$p(\theta_i|y) = \frac{p(y|\theta_i)\, p_0(\theta_i)}{\sum_j p(y|\theta_j)\, p_0(\theta_j)}, \tag{6.9a}$$

but

$$p(y|\theta_j) = \sum_k (y|\psi_k(\theta_j))\, \frac{p_0(\psi_k(\theta_j))}{\sum_l p_0(\psi_l(\theta_j))} \tag{6.9b}$$

where $\psi_k(\theta_j)$ denotes the set of distributions for which $\psi_j(\theta_j)$ (the distribution produced by θ_j) is the best approximation according to G. By the definition of "best approximation," all possible distributions of experimental outcomes are partitioned into disjoint subsets, one subset $\psi_k(\theta_j)$ for each model instance. $p_0(\psi_k(\theta_j)$ is the prior probability of each of those distributions being true (the sum of which is the prior probability of θ_j, or equivalently of $\psi_j(\theta_j)$ being the best approximation).

The probability $p(y|\psi_k(\theta_j))$ denotes the probability of the data given that $\psi_k(\theta_j)$ is the true distribution. The term in the right-hand bracket of Eq. (6.9b) is the prior probability of a given distribution being true among the set of distributions for which θ_j is best, conditional on θ_j being best. It should be noted that Eq. (6.7) remains true in BMS*, so the relation to NML# remains unchanged, if one applies Eqs. (6.9a,b) as given above.

priors of data distributions ψ_i		data outcomes x_j						ψ's best fit by θ_k	model instances θ_k
	x_1	x_2	x_3	...	x_j	...	x_n		
$p_0(\psi_1)$	$p(x_1\|\psi_1)$	$\boldsymbol{p(x_2\|\psi_1)}$	$\boldsymbol{p(x_3\|\psi_1)}$	...	$p(x_j\|\psi_1)$	...	$p(x_n\|\psi_1)$		
$p_0(\psi_2)$	$p(x_1\|\psi_2)$	$\boldsymbol{p(x_2\|\psi_2)}$	$\boldsymbol{p(x_3\|\psi_2)}$	...	$p(x_j\|\psi_2)$	...	$p(x_n\|\psi_2)$	ϕ_1	θ_1
$p_0(\psi_3)$	$p(x_1\|\psi_3)$	$\boldsymbol{p(x_2\|\psi_3)}$	$\boldsymbol{p(x_3\|\psi_3)}$	...	$p(x_j\|\psi_3)$	...	$p(x_n\|\psi_3)$		
$p_0(\psi_4)$	$p(x_1\|\psi_4)$	$\boldsymbol{p(x_2\|\psi_4)}$	$\boldsymbol{p(x_3\|\psi_4)}$	...	$p(x_j\|\psi_4)$	...	$p(x_n\|\psi_4)$		
$p_0(\psi_5)$	$p(x_1\|\psi_5)$	$\boldsymbol{p(x_2\|\psi_5)}$	$\boldsymbol{p(x_3\|\psi_5)}$	...	$p(x_j\|\psi_5)$	...	$p(x_n\|\psi_5)$	ϕ_2	θ_2
...	...	...	...	...	...	...	...		
$p_0(\psi_i)$	$p(x_1\|\psi_i)$	$\boldsymbol{p(x_2\|\psi_i)}$	$\boldsymbol{p(x_3\|\psi_i)}$	...	$p(x_j\|\psi_i)$	...	$p(x_n\|\psi_i)$		
...	...	...	...	...	...	...	...		
$p_0(\psi_m)$	$p(x_1\|\psi_m)$	$\boldsymbol{p(x_2\|\psi_m)}$	$\boldsymbol{p(x_3\|\psi_m)}$	...	$p(x_j\|\psi_m)$	...	$p(x_n\|\psi_m)$	ϕ_l	θ_l
	$p_0(x_1)$	$\boldsymbol{p_0(x_2)}$	$\boldsymbol{p_0(x_3)}$	...	$p_0(x_j)$	...	$p_0(x_n)$		

$$\boldsymbol{S = s_y}$$

Matrix 6.1: Prior probabilities for data distributions (left margin), distributions predicted by model instances (right margin), and model outcomes (bottom margin). A posterior version of the same matrix can be produced where the margins are conditioned on the observed data outcome y. As explained in the text, virtually all the equations and predictions in this chapter can be derived and explained with reference to this matrix. Data outcomes in bolded columns are explained in Section 6.4.2.

Matrix 6.1 clarifies the situation. The left-hand column shows the prior probabilities for each data distribution ψ; these sum to 1. The top row shows the possible experimental data outcomes x. Each row of the table shows the data distribution ψ: $p(x_j|\psi_i)$ gives the probability that outcome x_j will be observed if ψ_i is the "true" data distribution. Each row adds to 1.

All model instances θ in all model classes are listed in the right most column. Each of these predicts a data distribution, so θ_k is shown adjacent to the data distribution it predicts – there are of course far fewer model instances than data distributions. Each model instance is a best match to a disjoint subset ϕ_k of all data distributions. The data distributions are, therefore, organized so that the best matching data distributions for model instance θ_k are listed successively to the left of θ_k.

The posterior probabilities of the ψ_i, given that outcome x_i has been observed, are calculated by multiplying the vector of data distribution priors times the column entries for x_j, and normalizing. That is, for an observed outcome y (the jth column if $y = x_j$), the posterior probability $p(\psi_i|x_j) = p(\psi_i|y)$ is just the matrix entry (the joint prior probability for ψ_i, x_j) normalized by the sum of joint prior probabilities for the entire x_j column.

The posterior probability for model instance θ_k is just the sum of the posterior probabilities for the data distributions in ϕ_k. Assuming that one of the model instances predicts the "true" data distribution (i.e., traditional BMS), then there would be just one distribution for each model instance, but the statements in this paragraph would remain valid. (The data outcomes in the bolded columns are described in Section 6.4.2 when discussing reproducibility).

Note that the emphasis upon data distributions increases the dimensionality of the problem space. This will almost always make analytical derivations impossible and will generally pose computational challenges even larger than those existing for traditional BMS and NML. It is possible that these challenges could be met by sophisticated sampling schemes, the use of simplifying assumptions, or both. For example, Karabatsos (2006) was able to make considerable headway by assuming a class of "stick-breaking" priors. This chapter will not pursue these practicalities but rather lay out the theoretical basis for what we hope will be seen as a coherent approach to model selection and reproducibility.

There are several benefits to the BMS* extension of BMS. Several of these are not the subject of this chapter, but are worthy of brief mention. First, this approach places scientific inference much more squarely on the data, represented by the prior and posterior distributions of data outcomes, than the models. This is desirable, given that the models are wrong, and often very simplified and crude approximations to whatever is the truth.

Second, the way of introducing prior knowledge can be placed on the data rather than models: in BMS*, the prior probabilities can be specified as the probabilities of each data distribution; when these are summed across the distributions for which a given model instance is the best approximation, the result becomes the prior probability of that instance. Such a result makes data priors and model priors consistent while placing the emphasis on data.

Third, the approach provides a natural way to deal with model misspecification. Model misspecification is universal, but sometimes is seriously awry, and causes undesirable inferences when applying traditional BMS (see Grünwald and Langford 2007, Erven *et al.* 2008). The importance of BMS* for this chapter is the emphasis upon representing the true state of the world, and the approximations to that true state, as distributions of data outcomes. This allows inferences about data distributions, and those posterior probabilities can be used as a basis for inference concerning reproducibility. This is the subject of the next section.

6.4 Replication Variance and Reproducibility

It is unclear whether the true (i.e., real) state of the universe is deterministic, perhaps depending on understandings of quantum uncertainty. Given as complete a specification as possible of every factor operating in the universe at a given time and the complete history of all relevant factors, the outcome of some experiment might or might not be deterministic – but if not, it would certainly be close to deterministic at the macro-scale at which experiments are carried out.

Thus a coin flip may appear random, but complete knowledge of all factors and forces such as the air currents and resistance and the linear and rotational acceleration and velocities at which the toss is made, and numerous other factors of this sort, will in principle allow a (close to) deterministic prediction of the result. But the universe is infinitely complex, and we never have or could have knowledge of the values of all the factors that underlie a given experimental outcome – the rare cases (if any) in which the scope of an experiment is so narrow and the relevant variables so few and so well controlled that replications will produce identical outcomes need not be considered in this chapter.

We will, therefore, represent the true state of the universe, as restricted to some experimental setting, to be a distribution of outcomes. A model instance containing probability parameters will predict a deterministic distribution of outcomes, but such a distribution will embody statistical variation, to a degree that could be either smaller or larger than that used to represent the "true" state of the world (as applied to the current study), but the predicted distribution will never match the "true" distribution due to the many unknown and unknowable factors operating in any setting.

6.4.1 Within- and Between-Setting Replication Variance and the True State of the World

Within a given experimental setting we use a distribution of outcomes for the entire experiment to represent the true state of the world as applied to the present setting. However, we cannot generally expect this distribution to be the same even for within-setting repeated trials in a given condition – there will almost always be at least some time dependence of the results. To take one example, in human studies fatigue can set in, or various sorts of learning can take place.

Thus within-setting replication variance is a theoretical construct by which we imagine a replication of the experiment in the same setting, but with different values of the many variables that are unknown, ignored, or unmeasured that affect performance. Of course, good scientific design often allows one to assume that within-setting repeated trials are sampled from the same distribution. This is a useful assumption because Bayesian inference can be applied to a subset of trials in a given condition, and the posterior probabilities from that analysis can be used as priors for inference applied to the remaining trials in that condition.

Note that the "true state of the world" is not some abstract truth applying generally, whatever that could mean, but that applying to the present experimental setting (or a set of such settings if there are more than one). Thus if a study measures the success rate $p(s)$ for a drug, a few of the factors that contribute to within-setting replication variance would be the individuals receiving treatment, positive or negative conscious or unconscious experimental bias and errors or mistakes in programming, design, reporting, and analysis. These and many other factors would contribute to within-setting replication variance to the extent that they might vary in an imagined replication in the same setting at the same time. There is no such thing as a "true" value of $p(s)$ because this will vary with every setting, and every set of values of the host of relevant factors that operate to determine $p(s)$.

It is of course possible to guess at the operation of some of these factors. Such guesses are based on prior knowledge and can be entered into the model selection process in the form of priors assigned to the outcome distributions. The guesses might also be incorporated explicitly in the various model instances, but such model implementations are in a sense independent of the within-setting replication variance: The guesses might be reflected in the representation of replication variance through choices of priors for data outcomes and might also be reflected in the models we are testing, but need not match.

It is not critical that they match because we are only looking to infer the model instance that produces the *best approximation* to the "true" distribution. Thus we might have one model class predict the same value of $p(s)$ for every individual tested and, hence, predict a binomial distribution of data outcomes, and another model class predict two classes of individuals with differing values of $p(s)$, thus predicting bi-modal distributions of data outcomes (a mixture of two binomials). Independently we can specify priors on the data distributions, based on prior knowledge, that try to allow for many other sources of variance due to unknown and guessed at factors, with the result that relatively high priors would be assigned to distributions that are neither binomial nor bimodal but spread fairly widely across the outcome space.

The within-setting replication variance will be represented in the priors assigned to the data distributions, and these will in turn determine the posteriors for the data distributions. As we shall show in the following subsection, the data distribution posteriors will allow prediction of any data statistic for a new "replication" study. Of course, the new replication study will have different values for many of the unknown, unmeasured, and ignored factors – in general, this produces additional variability not represented in the within-setting replication variance.

Thus, if the goal is to show that the new study does have such new values, then one would try to represent as accurately as possible the within-setting replication variance and represent that value in the priors for the data distributions. On the other hand, one might want to admit and assume that additional between-setting replication variance exists, and take this into account when assessing reproducibility. In this case one would want to represent both within-setting replication variance and appropriate amounts of between-setting replication variance in the data distribution priors. The predicted distribution for some data statistic would then have a higher variance and one would need more deviant results in the replication study to conclude that replication failed.

Thus, when assessing reproducibility one must consider carefully one's inference goals. For example, suppose that the prior studies of $p(s)$ for a drug were all carried out by the drug's manufacturer and the new study was carried out by a "neutral" agency. In this case one might want to include only within-setting replication variance in predicting some data statistic, because one expects and wants to show a change. On the other hand, suppose all studies, including the replication, are carried out by drug companies with unknown but variable biases. In this case one might include such variability in the within-setting replication variance and in the corresponding data distribution priors, making it harder to demonstrate a replication failure.

6.4.2 Reproducibility

The concept of reproducibility has several meanings. One can attempt to find a statistically similar total or partial outcome, a repeated qualitative pattern, an analysis that reaches a similar conclusion, or an outcome consistent with the posterior probabilities from the prior study. In addition, one must choose some criterion or metric, qualitative or quantitative, to decide whether some result is sufficiently deviant to be called a failure of reproduction. Some of these reproducibility issues are not easy to specify quantitatively, and are best left in the eye of the beholder, but most assessments of reproducibility can be based on a statistic S calculated from a given experimental data outcome. The approach we suggest operates as follows.

It is possible to use Bayesian induction on the observed data from a prior study to produce posterior probabilities for each outcome distribution, and then use those posterior probabilities to predict various aspects of the new study. Reference to Matrix 6.1 makes this clear: Every data distribution on the left column assigns a probability to the observed data. The set of these, when normalized to add to 1, is the set of posterior probabilities for the data distributions. One can write this in equation form:

$$p(\psi_i|x) = \frac{p(x|\psi_i)\,p_0(\psi_i)}{\sum_j p(x|\psi_j)\,p_0(\psi_j)} \tag{6.10}$$

As an important aside, we must admit that this approach in all but a few toy examples is impractical: There are far too many data distributions for such induction to be carried out and then applied. The same could be said about BMS* used for model selection, but various approximations and/or simplifying assumptions can nonetheless make model selection feasible. Similarly, various approximations and simplifications might allow reproducibility to be assessed. The simplest approximation may be the use of the model instance posteriors as a proxy for the true data distributions: Each model instance predicts one data distribution. One could assign the corresponding data distribution the posterior probability of its instance. This would allow predictions of many quantitative aspects of reproducibility to the extent that one believes the model instances provide a decent account.

However one approximates the posterior probabilities of the data distributions, here is a way to assess reproducibility. Let the posterior estimates be denoted $p(\psi_i|y)$, y being the prior outcome that was used along with the priors to produce the posterior estimates. These posterior estimates can be used to predict the distribution $S(x_j)$ of any statistic S defined on the possible outcomes x_j.

This is easiest to see from Matrix 6.1. Multiplying the posterior probabilities with the entries in the matrix produces posterior joint probabilities that sum to 1 over the entire matrix. Summing down columns of the matrix gives posterior probabilities of each data outcome, $p(x_j|y)$: This is the posterior equivalent of the lowest row in Matrix 6.1. We can then use these to predict the distribution of any statistic $S(x_j)$ defined on the data outcomes; let the probabilities in this distribution be termed $p(S(x_j|y)$.

Suppose we next observe a new set of data, z, in a study intended as a replication of one sort or another. If the outcome space in the new study is the same as the earlier study (or can be restricted to be the same), then the same statistic can be calculated for the outcome z: Let s_v denote the value of that statistic for the new data. To assess reproducibility we want to know where s_v lies on the distribution $S(x)$.

$$p(S = s_v) = \sum_i p(s_v|\psi_i)\, p(\psi_j|y) \tag{6.11a}$$

$$p(s_v|\psi_i) = \sum_x p(S = s_v|x, \psi_j) \tag{6.11b}$$

This is a cumbersome way of saying that a given value of S may be found for more than one outcome, so the probability of that value is a sum across the probabilities of those outcomes. In Matrix 6.1 this situation is represented for one value of S by the entries in bold. The bold entries are data outcomes for which $S(x_j|y)$ equals a particular value s_v. Each value of S will be represented by a different set of columns in Matrix 6.1.

Assessing reproducibility by comparing the new value of S to the predicted distribution for S is easiest to do in the cases when S is one-dimensional: When S is multidimensional, its value in the replication might be near the predicted mean on some dimensions and deviant on others, making it harder to come up with a generally accepted criterion for what constitutes a failure. This approach in general is related to traditional power analyses and even more closely related to a Bayesian version of power analyses (e.g., Kruschke 2014, Fig. 13.2).

It is worth emphasizing again that both model selection and assessments of reproducibility depend critically on the choice of data distribution priors that are then used in turn to produce posteriors. Different priors might well be chosen for these two different purposes, because the goals are different. For instance, one might want to favor a model instance (or class) that can predict a broad range of observed outcomes even when they differ considerably, as occurs when a replication seems to fail.

Where to draw the line in these cases is a delicate matter: In science we do not generally want inference to extend only to a single experimental setting. The findings ought to be robust enough to be found in variously precise replications in other settings.[3] Yet this generality is only useful when the studies in the several settings are all valid.

6.4.3 A Toy Example

A toy example illustrates the system that has been described in the preceding sections. Suppose an experiment is carried out to assess the probability of success, $p(s)$, in some domain. The study has 10 trials. It is assumed that the trials are identical and order independent, so the outcome of the study is the number of successes, n, ranging from 0 to 10 (11 outcomes). These potential outcomes are listed across the top of Table 6.1. Each data distribution thus consists of 11 probabilities that add to 1.

Even at a coarse discretization there are a vast number of potential data distributions, so we have reduced them to a very small number here for the sake of illustration. There are also a vast number of potential models, so we will again consider only a small discretized set of model instances, corresponding to $p(s)$ values of 0.0, 0.1, 0.2, ..., 1.0, each predicting a corresponding binomial distribution of number of successes. These model instances are shown towards the right-hand side of Table 6.1 which is in fact an illustration of Case A of our example: the data distributions are limited to just the 11 binomial distributions that are exactly predicted by the 11 model instances.

On the far left margin of Table 6.1, we list the prior probabilities of the data distributions, and on the far right margin we list the prior probabilities of the model instances. Of course the priors in this case have to match. For the sake of the example we have made the priors geometric, giving higher probabilities to distributions based on low values of $p(s)$.

The set of distributions and the priors assigned to them are plotted in Figs. (6.1a) and (6.1b). We used shading to denote both the model instance ($p(s)$ ranging from 0.0 to 1.0 is steps of 0.1), and that instance's best matching data distribution(s). The data distribution indices conform to Table 6.1. For example, index 1 refers to the binomial distribution with $p(s) = 0$, or both that distribution and its higher variance partner.

[3] This refers to the issue of generalization discussed in more detail by Stahel, this volume.

priors of data distributions ψ_i	data outcomes x_j											model instance θ	priors on θ
	0	1	2	3	4	5	6	7	8	9	10		
$p_0(\psi_1)$=0.500	1	0	0	0	0	0	0	0	0	0	0	0.0	0.500
$p_0(\psi_2)$=0.250	0.349	0.387	0.194	0.057	0.011	0.001	0	0	0	0	0	0.1	0.250
$p_0(\psi_3)$=0.125	0.107	0.268	0.302	0.201	0.088	0.026	0.008	0.001	0	0	0	0.2	0.125
$p_0(\psi_4)$=0.063	0.028	0.121	0.233	0.267	0.200	0.103	0.037	0.009	0.001	0	0	0.3	0.063
$p_0(\psi_5)$=0.031	0.006	0.040	0.121	0.215	0.251	0.201	0.111	0.042	0.0011	0.0002	0	0.4	0.031
$p_0(\psi_6)$=0.016	0.001	0.010	0.044	0.117	0.205	0.246	0.205	0.117	0.044	0.011	0.001	0.5	0.016
$p_0(\psi_7)$=0.008	0	0.002	0.011	0.042	0.111	0.201	0.251	0.215	0.121	0.040	0.005	0.6	0.008
$p_0(\psi_8)$=0.004	0	0	0.001	0.009	0.037	0.103	0.200	0.267	0.233	0.121	0.028	0.7	0.004
$p_0(\psi_9)$=0.002	0	0	0	0.001	0.006	0.026	0.088	0.201	0.302	0.268	0.107	0.8	0.002
$p_0(\psi_{10})$=0.001	0	0	0	0	0	0.001	0.011	0.057	0.194	0.387	0.349	0.9	0.001
$p_0(\psi_{11})$=0.0005	0	0	0	0	0	0	0	0	0	0	1	1.0	0.0005

Table 6.1 follows the form of matrix 6.1, with probabilities corresponding to the priors and data distributions for the toy example, for Case A described in the text. Case A assumes all distributions to be binomial, based on the corresponding value of $p(s)$ (from 0.0 to 1.0 in steps of 0.1).

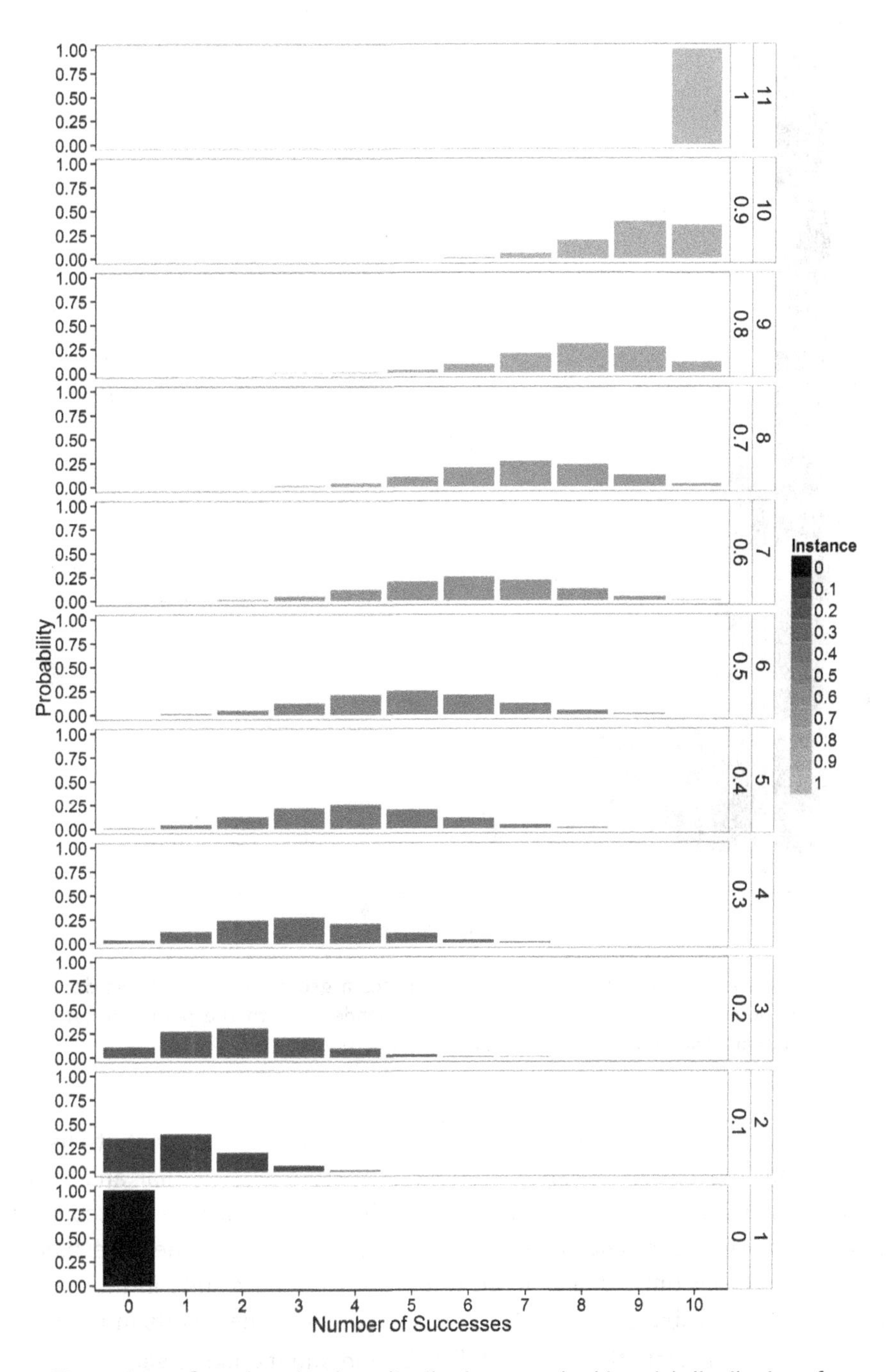

Figure 6.1a: Case A – the data distributions are the binomial distributions for the number of successes in 10 trials predicted by model instances with $p(s)$ in the range from 0.0 to 1.0 in steps of 0.1.

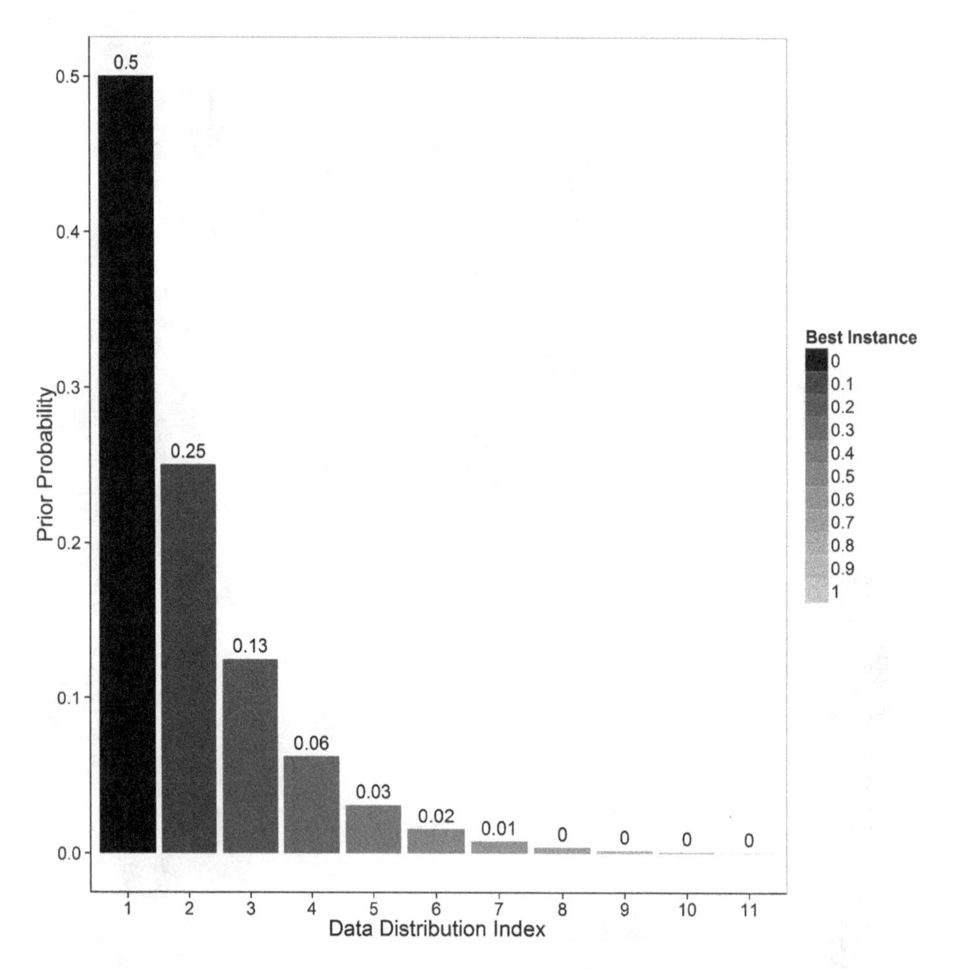

Figure 6.1b: Case A – the toy example imposes a geometric prior on the data distributions. Due to their one-to-one correspondence with the priors on data distributions, the model instances have the same priors.

For Case B we add 11 more data distributions. The extra distributions allow for additional replication variance and are a mixture of binomial and uniform distributions. Based on a given binomial distribution and a given outcome, the probability of that outcome in the new distribution is 0.5 times the binomial probability plus 0.5 times 1/11. If we would present the results in matrix form, it would look like Table 6.1 but with 22 rows. We show the results in Figure 6.2.

Figure 6.2a presents the binomial distribution and its higher variance partner adjacent to each other in a given row. We do this because we assume that the distribution-matching criterion G finds both the binomial distribution and its higher variance partner to be the best matches to a given instance. The

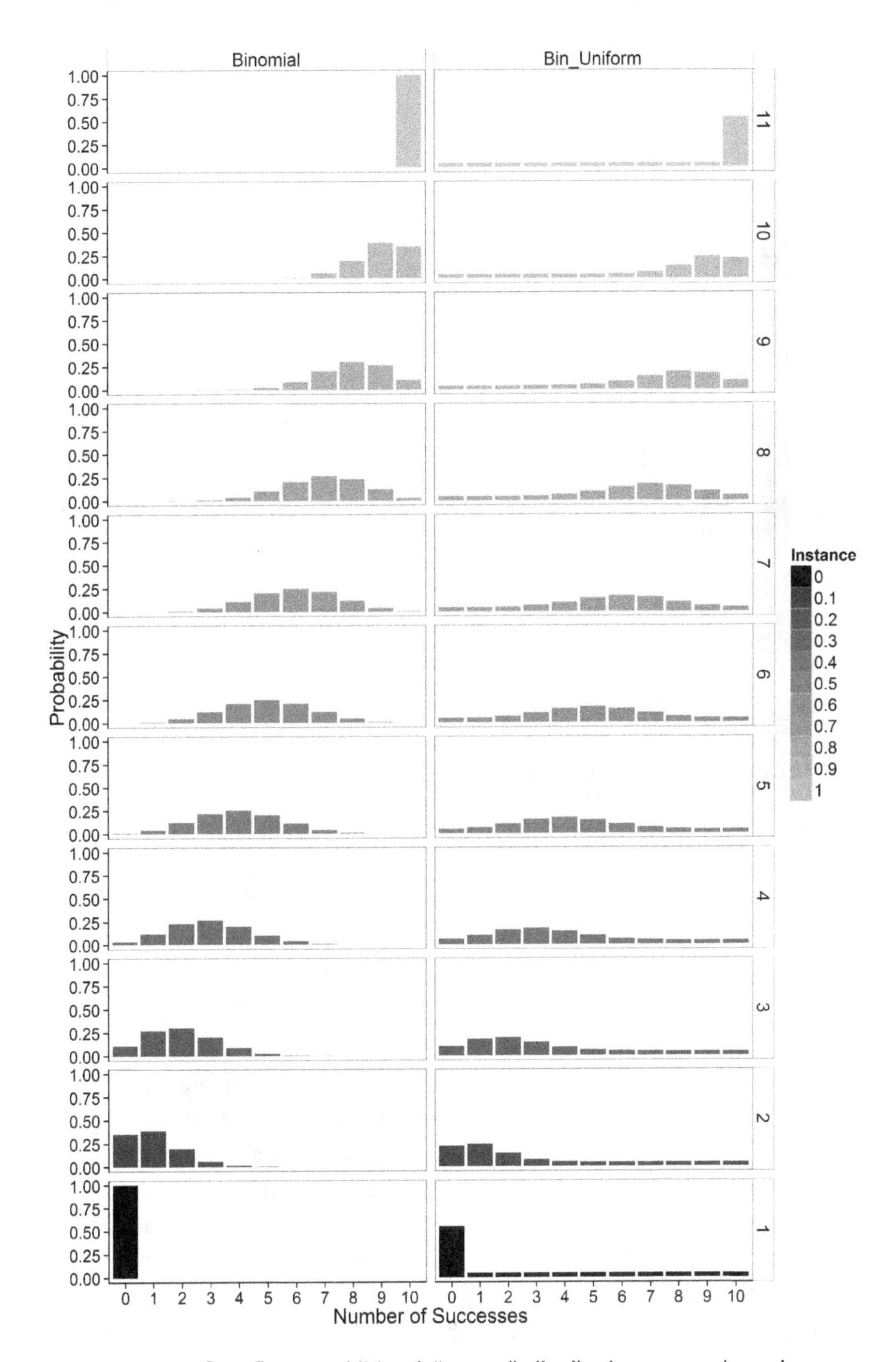

Figure 6.2a: Case B – 11 additional "partner" distributions are each a mixture of one of the binomials with a uniform distribution, and are shown to the right of the binomial partner.

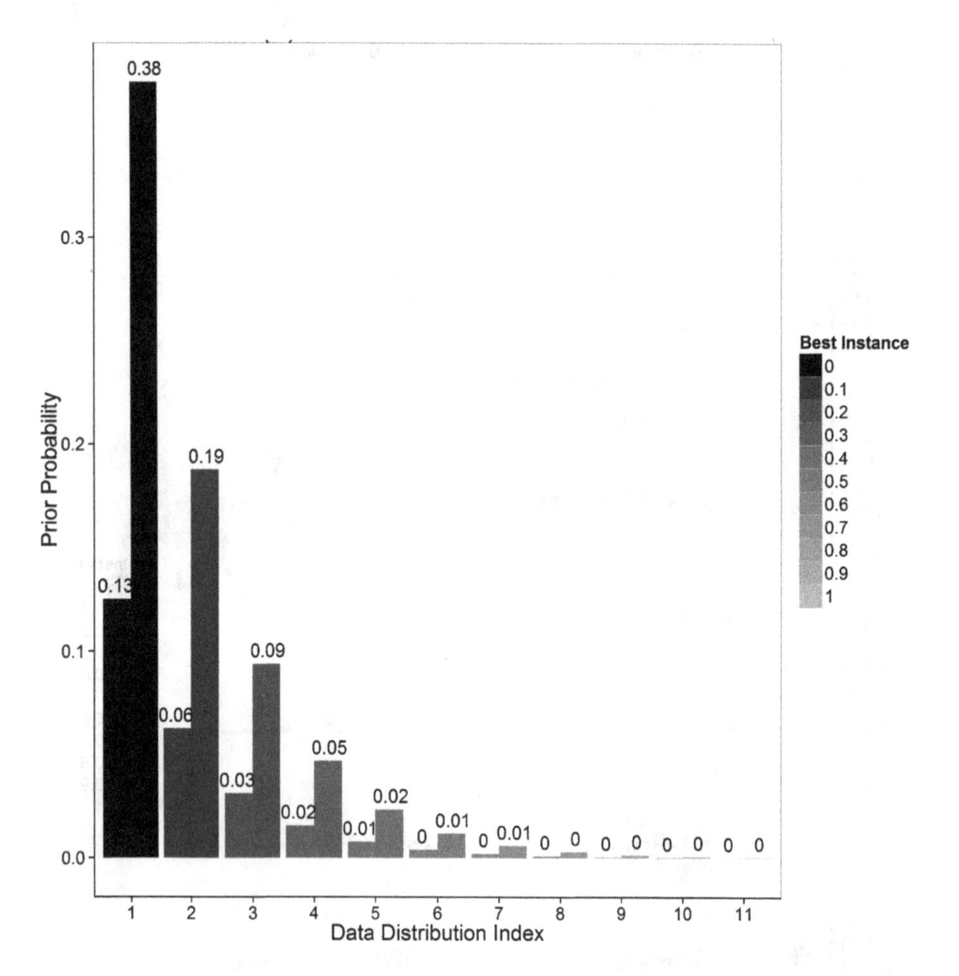

Figure 6.2b: Case B – prior probabilities for the 22 distributions: the binomial distribution is assumed to be three times less likely than its partner.

distributions with added replication variance seem more likely, so we let the priors for the binomial plus uniform distributions be three times as likely as those for the binomial distributions (Fig. 6.2b; the priors retain the preference for low values of $p(s)$). Now, let us assume that "best matches" are defined so that each model instance is a best match for its own distribution, as well as the adjacent partner of that distribution. The model instance priors are then just the sum of the priors for the two partner distributions (Fig. 6.2c).

The posterior probabilities for both the data distributions and the model instances depend on what data are observed. Let us assume that three successes occur. Bayes' theorem in graphical form says we go to the column for three successes, multiply the entries by the priors, and normalize the results so they

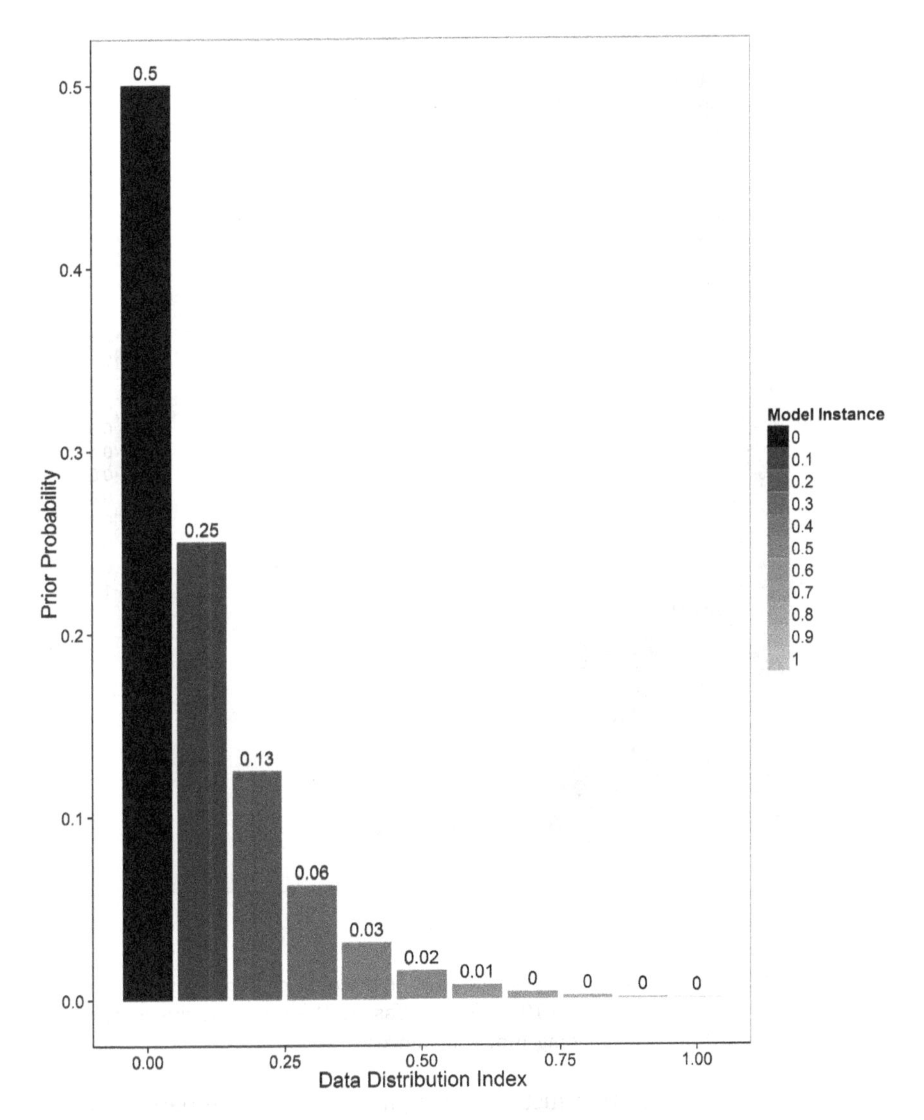

Figure 6.2c: Case B – the prior of each model instance is just the sum of its two best matching partner distributions.

add to 1. We do this for each possible data distribution. We show the resulting posterior for Case A (the 11 binomials) in Fig. 6.3a. The data distributions are listed on the horizontal axis (index 1 is for the binomial with $p(s) = 0$), and the posterior probability for each distribution is plotted on the vertical axis. In this case these are also the posteriors for the model instances.

The posteriors for Case B (11 binomials plus 11 binomials plus uniform) are shown in Fig. 6.3b for all 22 data distributions. These are grouped in pairs with

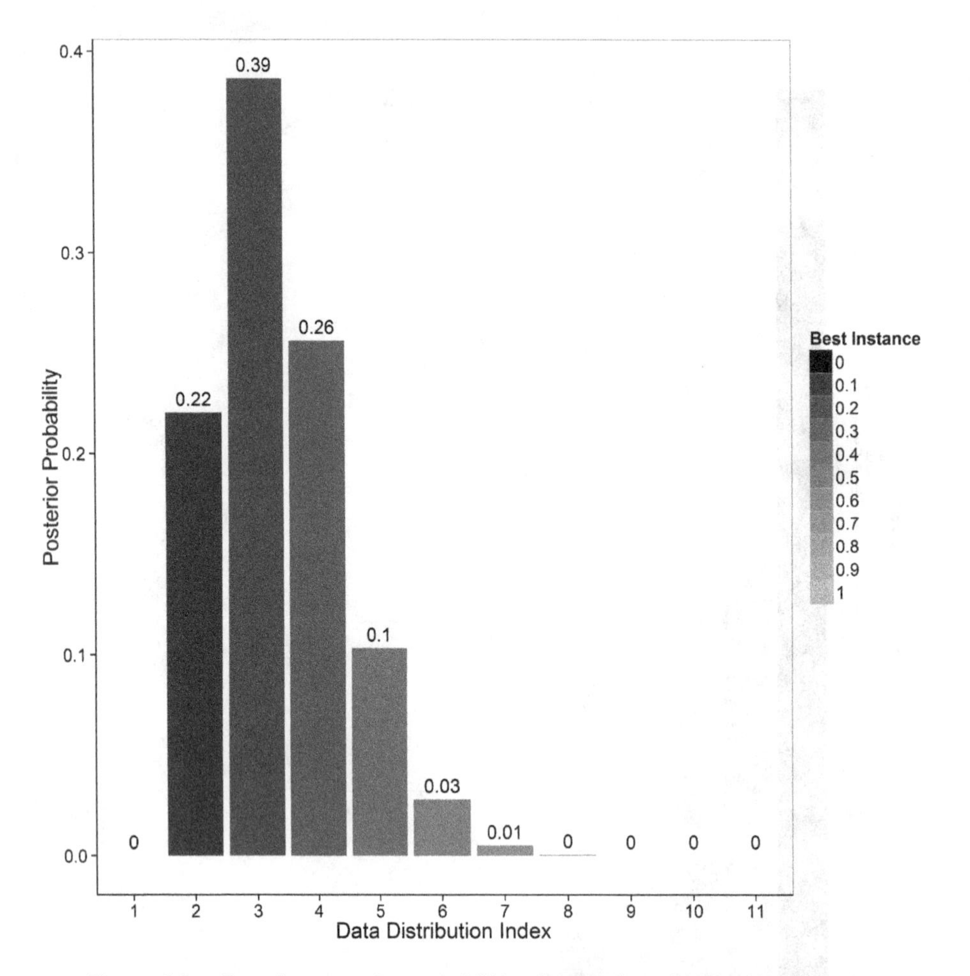

Figure 6.3a: Case A – posterior probabilities for the data distributions following an outcome of three successes in 10 trials. For Case A, these are also the posterior probabilities of the corresponding model instances.

the higher variance distribution just to the right of its binomial partner; thus the binomial for $p(s) = 0$ cannot predict three successes and has zero probability. For the other pairs the binomial has a smaller posterior than its partner largely because it was assigned a lower prior probability. The posterior probabilities for the 11 model instances in Case B are simply a sum of the posteriors for a given binomial and its partner, as shown in Fig. 6.3c. We can see that giving more prior probability to distributions with added sources of replication variance leads to less sharp posterior distributions.

If we then wish to compare model classes that do not share instances, then the posterior probability for each class is just the sum of the posteriors for its instances. Suppose class I consisted of $p(s)$ values from 0.0 to 0.5 and class

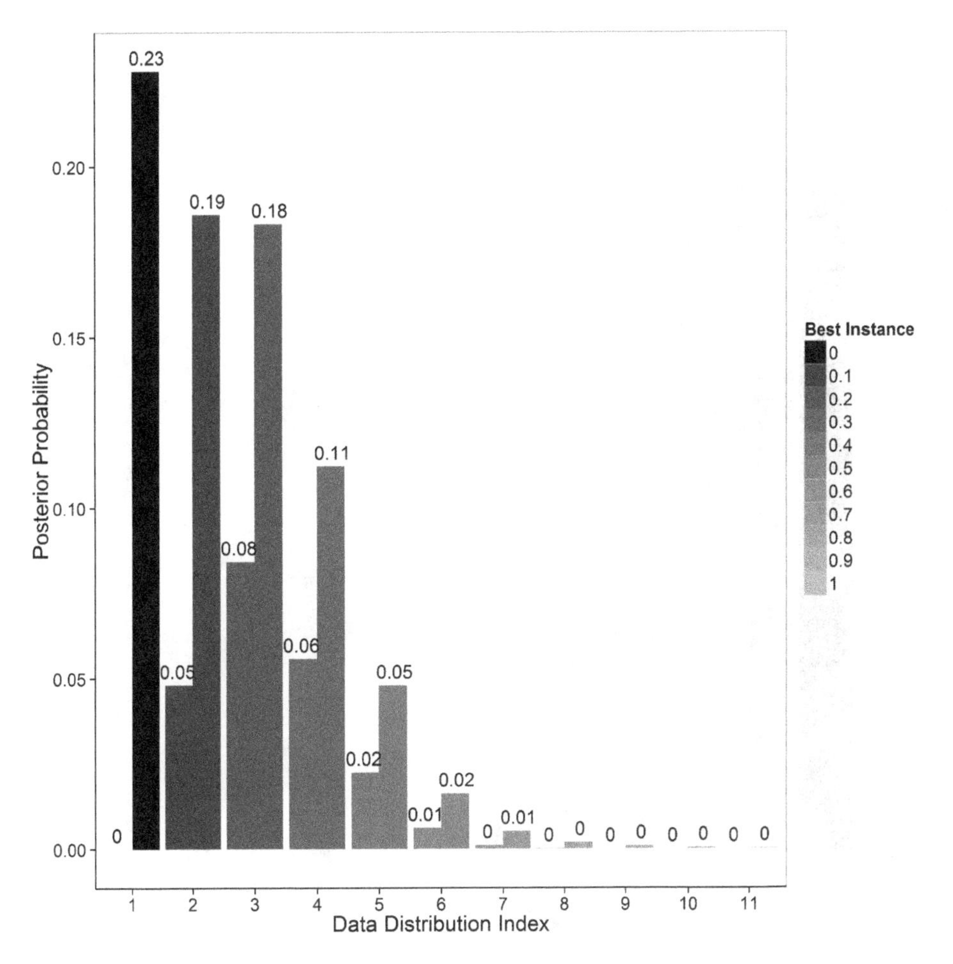

Figure 6.3b: Case B – posterior probabilities for the 22 data distributions, following an outcome of three successes in 10 trials.

II consisted of the higher $p(s)$ values. For Case A, the odds for class I/II are $0.9943/0.0057 \approx 175$, and for Case B we get a value of $0.99/0.0098 \approx 100$. As expected, Case B lessens the odds.

Suppose we wish to assess reproducibility of some statistic S and carry out a new study, with 10 observations. Our toy example has only one outcome, n, so the only statistic that can be checked is the "mean" n. The predicted distribution of n for Case A is shown in Fig. 6.4 – as the dots connected by solid lines for Case A and as the triangles connected by dashed lines for Case B, each showing probabilities for the 11 possible values of n. The value of n in the replication study would be compared to this predicted distribution, and some criterion used to decide whether S had been replicated or if it is too far out in the

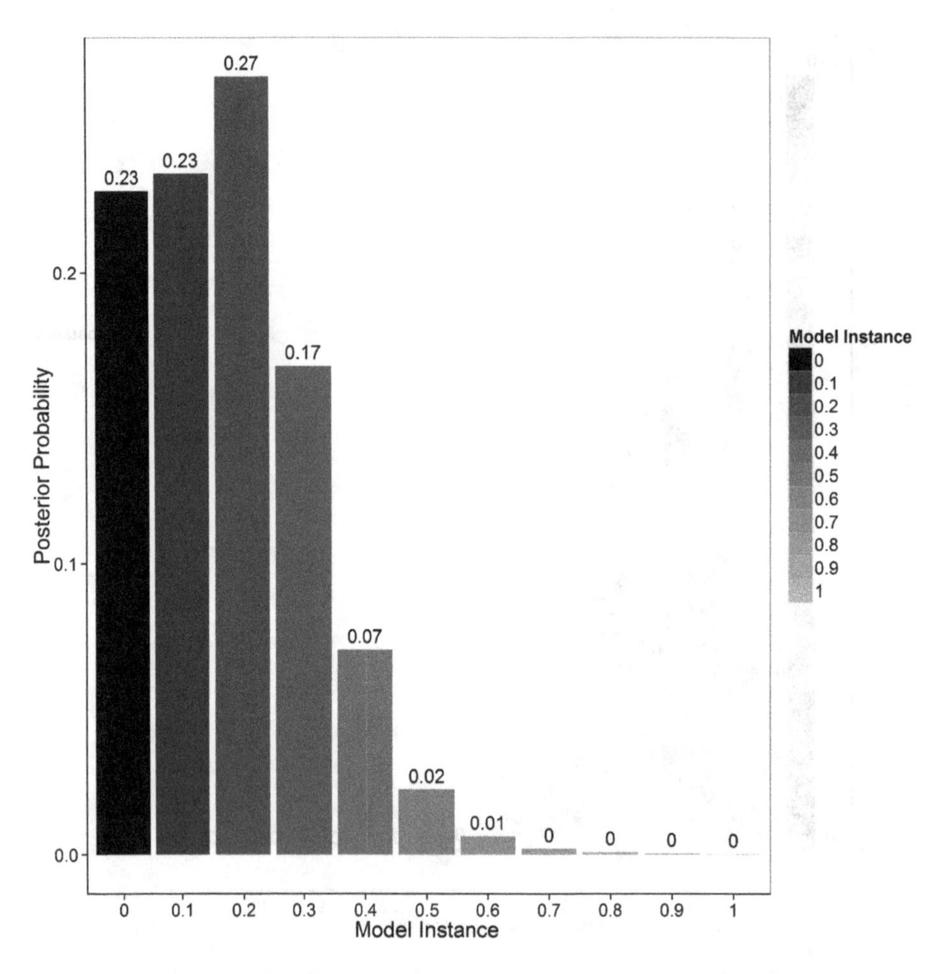

Figure 6.3c: Case B – posterior probabilities for the model instances are the sum of the posterior probabilities for the two partner distributions.

tail. The predicted distribution for n is of course calculated by multiplying the posterior for distribution j by the value of n for that distribution, and summing across all distributions. For instance, for $n = 5$, the predicted probability would be $\sum_j p(\psi_j|n = 3)\, p(5|\psi_j)$. This probability contains 3 as the observed value in the study and gives the prediction that 5 will be observed in the replication.

6.5 Final Remark

Matrix 6.1, and its posterior version, pretty much capture everything about the theory we propose and its applications to model selection and reproducibility.

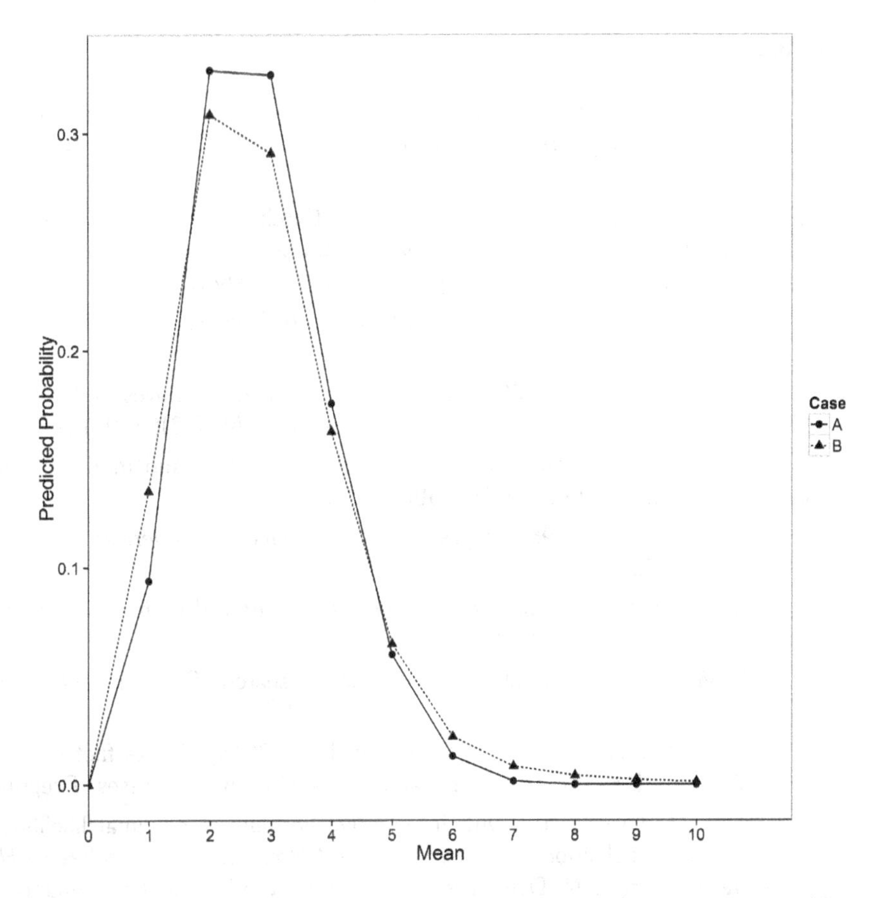

Figure 6.4: Predicted distribution for the mean number of successes expected in ten trials, based on posterior probabilities following an outcome of three observed successes in 10 trials. Solid lines connecting dots are for Case A, and dashed lines connecting triangles are for Case B.

The theory extends BMS, includes its traditional form as a special case, focuses inference on data and data distributions rather than theoretical constructs, aligns model instances with its associated data distributions (thereby also aligning priors and posteriors), evaluates model classes by summing the posteriors for the instances in each class, and produces posterior probabilities for experimental outcomes, thereby allowing reproducibility to be assessed.

The theory is simple, coherent, easy to understand, and consistent with normal scientific practice. Unfortunately the theory is impractical for large databases because the vast number of distributions of the very large number of data outcomes will almost always exceed any possible computational power. Thus approximations and simplifications will be needed to make the approach feasible. This will be a target of future research.

References

Barron, A., Rissanen, J., and Yu, B. (1998): The minimum description length principle in coding and modeling. *IEEE Transactions on Information Theory* **44**, 2743–2760.

Erven, T.V., Grünwald, P., and de Rooij, S. (2012): Catching up faster by switching sooner: A predictive approach to adaptive estimation with an application to the AIC-BIC dilemma. *Journal of the Royal Statistical Society B* **74**, 361–417.

Grünwald, P.D. (2007): *The Minimum Description Length Principle*, MIT Press, Cambridge.

Grünwald, P., and Langford, J. (2007): Suboptimal behavior of Bayes and MDL in classification under misspecification. *Machine Learning* **66**(2-3), 119–149.

Karabatsos, G. (2006): Bayesian nonparametric model selection and model testing. *Journal of Mathematical Psychology* **50**(2), 123–148.

Kass, R.E., and Raftery, A.E. (1995): Bayes factors. *Journal of the American Statistical Association* **90**, 773–795.

Kruschke, J. (2014): *Doing Bayesian Data Analysis: A Tutorial with R, JAGS, and Stan*, Academic Press, New York.

Raftery, A.E. (1995): Bayesian model selection in social research. *Sociological Methodology* **25**, 111–163.

Shiffrin, R.M., Chandramouli, S.H., and Grünwald, P.G. (2015): Bayes factors, relations to Minimum Description Length, and overlapping model classes. Preprint.

Zhang, J. (2011): Model selection with informative normalized maximal likelihood: Data prior and model prior. In *Descriptive and Normative Approaches to Human Behavior*, ed. by E.N. Dzhafarov and L. Perry, World Scientific, Singapore, pp. 303–319.

7
Reproducibility from the Perspective of Meta-Analysis

Werner Ehm

Abstract. Exact replication of experimental results is unattainable in the social and life sciences. Meta-analysis provides formal statistical procedures for dealing with heterogeneity and selection bias, offering a more flexible, quantitative approach to the issue of (ir-)reproducibility. This chapter presents basic methods and problems of meta-analysis, with an emphasis on model building, model checking, and assessing the validity of related statistical inferences.

7.1 Introduction

Recently, much attention has been paid to the observation that empirical studies with similar research questions and experimental protocols often yield substantially differing or even contradictory results. Striking examples from various subject areas were presented by Lehrer (2010). The many instances of experimental findings that afterwards proved irreproducible have raised concerns about the status of "scientific facts," ranging from critical questions (Pashler and Wagenmakers 2012) to resolute skepticism (Ioannidis 2005). Compare Martic-Kehl and Schubiger (this volume) for more examples.

In the physical sciences replicability is achieved to an impressive extent, guaranteeing the reliability of everyday technical devices. Physical materials and systems typically have few degrees of freedom, and boundary conditions can be effectively controlled (see, e.g., Zimmerli, this volume). This is very different with the subjects studied in the social and life sciences. Living organisms exhibit high complexity and are not easily controllable even when carefully prepared to meet certain standardized conditions in an experiment. Unsurprisingly, therefore, perfect replication is not attainable in those areas.

Meta-analysis takes a broader view in that diversity is considered from the outset. In larger surveys some variation of study-specific features is acceptable and even desirable for probing a reasonably broad range of conditions. Results differing in magnitude are, therefore, not unexpected. Even results differing in

---

Reproducibility: Principles, Problems, Practices, and Prospects, First Edition. Edited by Harald Atmanspacher and Sabine Maasen.

sign do not necessarily indicate irreproducibility as they might be due to random fluctuation when sample size is small. However, converse results also may point to a serious problem shaking the trust in a thus far accepted "fact." What may look as if an effect was waning could in fact be due to a bias that vanishes as further evidence is gathered.

Heterogeneity and bias make for two major reasons why meta-analysis should be relevant to the issue of reproducibility. Publication bias and selective reporting are suspected to be involved when initial results cannot be replicated. The file-drawer problem, referring to studies that remain unpublished (and unknown) on whatever reasons, was early addressed in psychological research (Rosenthal 1979) and is now put forward in connection with non-replicable clinical study reports (Doshi *et al.* 2012, Ioannidis 2005). Selective reporting practices, in particular, contribute to a severe overrepresentation of "positive" results that do not survive the test of time.

The dimension of these problems seems to have become clear only recently (Franco *et al.* 2014), although related warnings were issued, and methods for dealing with selection bias have been available much earlier (e.g., Iyengar and Greenhouse 1988, Hedges 1992). Their practical use is not widespread, however, presumably due to the complexity, and risks of going wrong, of some of these techniques, which preclude a routine application.

Meta-analysis emerged in the social sciences as a tool for synthesizing the body of information available about a certain research topic, by systematically searching the relevant literature and combining the individual results in an overall statistical analysis. Its major aims are strengthening of the evidence for, and clarification of the generalizability of, a significant finding observed repeatedly in independent experiments. A review of recent developments in meta-analysis may be found in Sutton and Higgins (2008).

Our purpose here is to give an account of basic meta-analytic methods and problems, with special emphasis on (1) the statistical modeling of heterogeneity and selection bias, and (2) techniques for model checking and for assessing the validity of related statistical inferences. Item (2) is vital because generally little is known about the mechanisms underlying heterogeneity and bias, so that any modeling attempt will involve speculation. Further obstacles ensue from the lack of individual data in meta-analysis, where only a few summary statistics are available for each study. Nevertheless, tools such as the simulation of surrogate data under varying parameter configurations can be useful for avoiding grossly false assumptions and conclusions.

The basic elements of meta-analysis are presented in Section 7.2. Some more special topics are discussed in Section 7.3 using an earlier meta-analysis from mind-matter research as a case study. The chapter ends with a brief summary in Section 7.4.

7.2 Basics of Meta-Analysis

7.2.1 Conceptual Preliminaries

Let us first recall the distinction between an estimate and the parameter it is intended to estimate, which is of fundamental importance to the correct interpretation of any statistical analysis. For example, the relative frequency of heads coming up in 100 tosses of a coin is an estimator of the probability p – the parameter – of observing head in any (future) toss. If the coin is symmetric, one can assume $p = 0.5$; otherwise, p may differ from one half, if only minimally, and is actually unknown. Similarly, if one wants to know the average age a of the inhabitants of a city at a certain time, one may draw a sample and take the average age $\widehat{a}$ of the drawn persons as an estimate of the population average age a.

Both the relative frequency of heads and the average age in the sample are random variables, i.e., their observed value will vary from trial to trial. This variability can be quantified by, e.g., the variance of the random variable. For 100 coin tosses, this variance is $p(1 - p)/100$, hence is known exactly if the coin is symmetric. Otherwise it depends on the unknown p and has to be estimated itself (e.g., by replacing p in the variance formula by the observed relative frequency). In the average age example, the formula for the population variance further depends on the way the sample is drawn. In any case, its value is unknown and has to be estimated, too. This is the generic case.

A second point to be recapitulated is the concept of the *effect size.* Typically, this is the difference or the ratio of two parameters characterizing a certain response under two different conditions. For example, consider two groups of students one of which receives some special training whose merit is assessed by a subsequent test. The effect size, generically denoted η, could be the difference of the population average scores achieved in the trained and untrained groups.

The population depends on how the groups are selected. For instance, one may compare student groups drawn from two "populations": those who received the training and those who did not. Or, interest may be focused on a fixed, given group of students who are randomly assigned to training or no training. The interpretation of the effect size clearly differs. In the second case, randomization (plus blind evaluation of the tests) allows ascribing a non-zero effect to the training. In the first case, a causal interpretation would be doubtful; still, a difference could tell something of interest about teaching strategies (provided the samples are representative).

Another common effect size measure is the odds ratio. Suppose that a certain event occurs with probability p_0 under some well-defined condition C_0. The chances for the event to happen versus not to happen – its "odds" – are then in the relation $\omega_0 = p_0/(1-p_0)$. The odds $\omega_1 = p_1/(1-p_1)$ for the event under another condition C_1 may be changed by a factor α, say, which is taken to represent the effect size. That is, the effect size is conceived as the *odds ratio*, $\eta = \omega_1/\omega_0$, which here assumes the value α. Alternatively, one may pass to the logarithmic scale and consider instead the *log odds ratio*

$$\eta = \log \frac{\omega_1}{\omega_0} = \log \frac{p_1}{1-p_1} - \log \frac{p_0}{1-p_0}, \tag{7.1}$$

thus conceiving of the effect size as the difference of the two log odds. Thus, the effect is measured on an additive rather than multiplicative scale, which is often more convenient.

7.2.2 Systematic Reviews

An important prerequisite of any meta-analysis is a systematic review of the literature about the research question of interest. Data bases and citation indexes such as *Web of Science* or *PubMed* are indispensable in this search task. Ideally, the review should provide an exhaustive list of all relevant studies, published or unpublished. This may be difficult to achieve, and worse, one does not know whether the missing studies are negligible or not. Thus in any case, the possibility of bias has to be taken into consideration.

In the second step the body of literature provided by the systematic review is sifted according to a number of criteria determining whether a study qualifies for inclusion in the meta-analysis. Independent studies almost never address exactly the same problem, use identical experimental designs, pools of test subjects, and so on. Thus expert judgment is inevitable when it comes to decide which problem statements, protocols, quality criteria, and so on are similar enough to allow for meaningful comparisons and joint analyses. Corresponding guidelines have been developed for various subject areas – e.g., by the *Cochrane* collaboration for medical studies – that assist the team of investigators in reaching agreement on those issues. The care applied in the two steps is decisive for the validity and significance of the ensuing research synthesis.

7.2.3 Fixed-Effects and Random-Effects Meta-Analysis

Suppose, then, that a systematic review addressing some particular research question has identified N studies for inclusion in the meta-analysis. Usually, the complete source data is not available, and what can be extracted from the published material essentially consists of two summary measures: an estimate of the effect size, and an estimate of the variance of that estimate. Let us denote the respective quantities for study $\#i$ as

$$\widehat{\eta}_i = \text{estimate of the effect size } \eta_i, \quad \widehat{\sigma}_i^2 = \text{estimate of the variance } \sigma_i^2 \text{ of } \widehat{\eta}_i.$$

Often slightly more information is provided from which the pair $(\widehat{\eta}_i, \widehat{\sigma}_i^2)$ can be derived. As an example we show parts of the first three rows of Table 1 in Carlin (1992), arranged somewhat differently. Entries such as 3 | 36 indicate that in study #1, 3 subjects died and 36 survived in the control group. The entries below $\widehat{\eta}_i$ are the empirical log odds ratios for the treatment / control comparison, e.g., $\widehat{\eta}_2 = \log 7/107 - \log 14/102 = -0.741$. Thus in study #2, the odds for death when treated are estimated as only $e^{-0.741} = 0.477$ times those in the control group. The entries below $\widehat{\sigma}_i$ are the estimates of the corresponding standard deviation σ_i computed as, e.g., $\widehat{\sigma}_2 = \sqrt{1/7 + 1/107 + 1/14 + 1/102} = 0.483$.[1]

study #	control	treated	$\widehat{\eta}_i$	$\widehat{\sigma}_i$
1	3 | 36	3 | 35	0.028	0.850
2	14 | 102	7 | 107	-0.741	0.483
3	11 | 82	5 | 64	-0.541	0.565
...	...	...	...	...

The data stem from a meta-analysis of long-term clinical trials investigating the effects of the prophylactic use of beta-blockers after myocardial infarction (Yusuf *et al.* 1985). A good graphical summary is provided by a so-called *forest plot*, such as shown in Fig. 7.1. There the single studies are listed line by line along with the respective effect size estimates $\widehat{\eta}_i$ and the associated 95% confidence intervals. A negative value of $\widehat{\eta}_i$ indicates that there were less deaths in the treatment than in the control group, while the length of the confidence interval (here approximated as $\widehat{\eta}_i$ plus/minus twice[2] the estimated standard deviation $\widehat{\sigma}_i$) reflects the uncertainty in the estimated log odds ratio.

The slight tendency for negative $\widehat{\eta}_i$s is counterbalanced to some degree by the lengths of the confidence intervals. Many intervals contain positive values

[1]Our values differ insignificantly from those of Carlin (1992) because there an alternative estimation method was used (Mantel–Haenszel).

[2]The factor 2 is an approximation to the exact value 1.96 from the normal distribution.

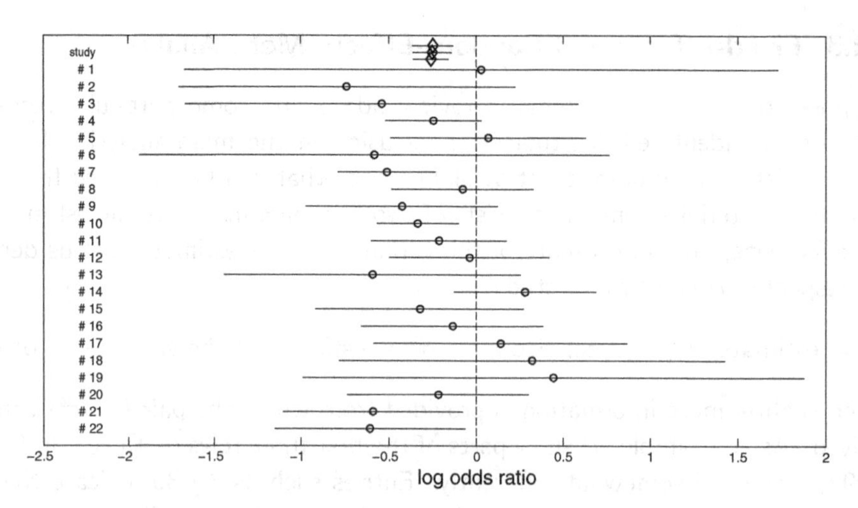

Figure 7.1: Forest plot of the beta-blocker data from Carlin (1992). See text for explanation.

as well, meaning that the corresponding studies are inconclusive in regard to the possible benefits of beta-blockers. Similarly ambiguous results are not uncommon in empirical studies, which raises the question of whether, and how, a combined assessment is feasible in such cases.

Statistical models. Meta-analysis offers essentially two approaches. In *fixed-effects* meta-analysis, it is assumed that the effect size is identical for all studies. The observed variability of the effect size estimates thus is attributed to chance fluctuations of different magnitude across studies. Formally, the fixed-effects model postulates that the effect size estimates $\widehat{\eta}_i$ are normally distributed with a common mean η and possibly different variances σ_i^2; in symbols,

$$\widehat{\eta}_i \sim \mathcal{N}(\eta, \sigma_i^2). \tag{7.2}$$

In *random-effects* meta-analysis the effect sizes themselves are conceived of as being subject to unsystematic variation. This entails a two-stage description: the model supposes that the effect sizes η_i of the single studies are random variables with the same mean η and variance τ^2 and that, conditionally given η_i, the effect size estimate $\widehat{\eta}_i$ is normally distributed with mean η_i and variance σ_i^2, for every i. Symbolically,

$$\eta_i \sim (\eta, \tau^2), \quad \widehat{\eta}_i \,|\, \eta_i \sim \mathcal{N}(\eta_i, \sigma_i^2). \tag{7.3}$$

The modeling is completed by further assuming, for both models, that quantities from different studies are statistically independent. Finally, it is common practice to suppose that the variance estimates are so accurate that it is not necessary to distinguish between the estimate $\widehat{\sigma}_i^2$ and the true value σ_i^2. Accordingly, we often shall write σ_i^2 for both the true value and its estimate.

Combination of individual results. In the fixed-effects model, reducing the individual results to one summary measure appears unproblematic in view of the (hypothetical) existence of a "universal" effect size η that applies likewise to every single study. A weighted average of the individual effect size estimates is a natural candidate for estimating this parameter, and efficiency considerations suggest to choose the weights so as to minimize the estimator's variance. This is achieved by selecting the weights w_i proportional to the inverse variances $1/\sigma_i^2$ of the individual effect size estimates $\widehat{\eta}_i$. The resulting fixed-effects estimate of η then is

$$\widehat{\eta}^{\mathrm{F}} = \sum\nolimits_i w_i^{\mathrm{F}} \, \widehat{\eta}_i \Big/ \sum\nolimits_i w_i^{\mathrm{F}}, \quad \text{where} \quad w_i^{\mathrm{F}} = \sigma_i^{-2}, \tag{7.4}$$

and its variance is

$$\mathrm{Var}(\widehat{\eta}^{\mathrm{F}}) = \left(\sum\nolimits_i w_i^{\mathrm{F}}\right)^{-1}. \tag{7.5}$$

For random-effects meta-analysis matters are less straightforward. The model admits, realistically, that the effect sizes may vary from study to study which, on the other hand, blurs the notion of an "overall" effect size. The pragmatic workaround is to suppose that due to unavoidable differences between studies, the respective effect sizes scatter unsystematically around the (hypothetical) overall effect size η; hence, the assumptions made in Eq. (7.3). Given this conceptualization, one again may consider estimating the parameter η as a weighted average of the single effect size estimates $\widehat{\eta}_i$. The additional random components increase the (conditional) variance of $\widehat{\eta}_i$ by τ^2, $\mathrm{Var}(\widehat{\eta}_i) = \sigma_i^2 + \tau^2$, so the optimal weights here are

$$w_i^{\mathrm{R}} = 1/(\sigma_i^2 + \tau^2). \tag{7.6}$$

Everything else is analogous to Eqs. (7.4) and (7.5) for the fixed-effects model, except that the quantity τ^2 in Eq. (7.6) is unknown, and so are the weights.[3] The standard recipe for overcoming this problem is "plug-in": in all terms involving w_i^{R}, an estimate $\widehat{\tau}^2$ is substituted for τ^2. The classical approach of DerSimonian and Laird (1986) yields a moment estimator of τ^2 in explicit form. Alternatively, one can use a maximum-likelihood-type estimator.

---

[3]Strictly speaking, the fixed-effects weights σ_i^{-2} are unknown, too. There, however, one often can assume that $\widehat{\sigma}_i^2 \doteq \sigma_i^2$; cf. the assumptions made above, and the further discussion below.

With estimates of the overall effect size η and of the variance Var$(\widehat{\eta})$ in hand, an approximate 95% confidence interval for η can, in both cases, be given as $\widehat{\eta} \pm 2\sqrt{\mathrm{Var}(\widehat{\eta})}$. For our data example, three such intervals are shown at the top of Fig. 7.1, the respective estimate $\widehat{\eta}$ being represented by a diamond. The upper two are computed assuming the random-effects model with $\widehat{\tau}^2$ determined by maximum likelihood (top) or DerSimonian–Laird (middle), the third one is derived from the fixed-effects model. Despite minor differences, the confidence intervals unanimously suggest that long-term use of beta-blockers significantly reduces the death rate after myocardial infarction (odds ratio between 0.69 and 0.88), a finding further corroborated by trials carried out later on (Blood Pressure Trialists Collaboration 2003).

Heterogeneity. Close inspection of Fig. 7.1 shows that the three confidence intervals for η are nested, each one containing the one below it. Furthermore, from bottom to top the estimates $\widehat{\eta}$ slightly move toward the zero (no effect) line. Such a pattern can often be seen in meta-analyses, and often more distinctly. It has prompted the notion that random-effects estimates and confidence intervals are more conservative than those assuming fixed effects (Poole and Greenland 1999), in the sense that fixed effects suggest a larger effect size and less uncertainty about its magnitude. As for uncertainty, this indeed holds true because the fixed-effects weights dominate the random-effects weights, $\sigma_i^{-2} \geq 1/(\sigma_i^2 + \tau^2)$, so that by Eq. (7.5) Var$(\widehat{\eta})$ is smaller for fixed than for random effects.

However, there is no generally valid relationship between the effect size estimates assuming random or fixed effects, respectively. Suppose, e.g., study #1 has such a small variance that its fixed-effects weight $1/\sigma_1^2$ dominates all others. Clearly then, $\widehat{\eta}^{\mathrm{F}} \approx \widehat{\eta}_1$. Suppose, furthermore, there is considerable heterogeneity between the single $\widehat{\eta}_i$s. With random effects, these are weighted almost uniformly (because $\widehat{\tau}^2$ is then large), whence $\widehat{\eta}^{\mathrm{R}}$ can assume almost any value depending on the configuration of the $\widehat{\eta}_i$s. In extreme cases it is possible that the fixed and the random-effects confidence intervals entirely lie on opposite sides of the origin, massively contradicting one another.

The above suggests that discrepancies should occur in particular when there is one "mega trial" with a huge number of participants that enters the combined estimate $\widehat{\eta}$ with an overwhelmingly high weight. A case in point are investigations of the effect of intravenous magnesium on mortality in suspected acute myocardial infarction. A meta-analysis of Teo *et al.* (1991) reported clear benefits of magnesium therapy, and similar conclusions were reached in another subsequent trial (LIMIT-2; Woods and Fletcher 1994). These findings were overthrown in 1995 by an extremely large trial with 58,040 patients (ISIS-4 Collaborative Group 1995) which resulted in a small adverse effect.

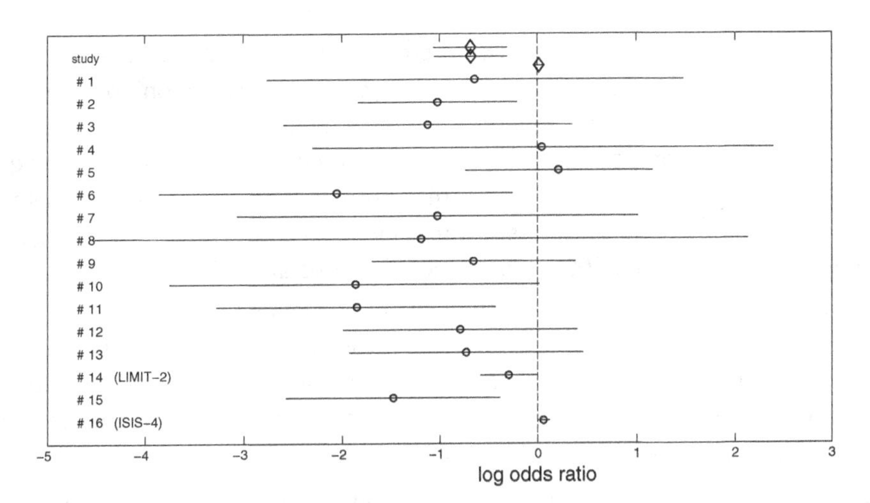

Figure 7.2: Forest plot of the magnesium data from Sterne and Egger (2001). See text for explanation.

The forest plot (Fig. 7.2) for the 16 magnesium trials listed in Sterne and Egger (2001, Tab. 1) demonstrates the above-mentioned discrepancy: the confidence interval for the fixed-effects estimate $\widehat{\eta}^{\mathrm{F}}$ includes zero, implying that there is no substantial treatment effect; the confidence intervals for the two random-effects estimates are both disjoint from the one for $\widehat{\eta}^{\mathrm{F}}$ and include negative values only, implying a significant risk reduction for myocardial infarction. The inconsistency was resolved by taking the high quality of the controlled randomized ISIS-4 trials and their huge sample size into account; magnesium therapy is no more considered as beneficial in suspected myocardial infarction.

Heterogeneity of studies is always a point in question. A test routinely used in fixed-effects meta-analysis rejects the hypothesis of equal effect sizes (presupposed by the model) if

$$Q = \sum\nolimits_i (\widehat{\eta}_i - \widehat{\eta}^{\mathrm{F}})^2 / \sigma_i^2 \tag{7.7}$$

exceeds the 95% quantile of the χ^2 distribution with $N - 1$ degrees of freedom. While a rejection clearly speaks against the fixed-effects model, that does not automatically suggest random effects as a viable alternative, as seen above. In fact, obvious recommendations do not follow from any test result. Even substantial heterogeneity can go undetected due to low test power, so non-rejection does not *per se* support the fixed-effects model.

In view of such imponderables, Higgins and Thompson (2002) proposed considering measures of heterogeneity rather than testing for equal effect sizes. Certainly most desirable would be an identification of the sources of the heterogeneity. Auxiliary information could provide useful hints. For example, there might be a systematic dependence on the year of study, or on some measure of study quality if such is available. Corresponding trends can be uncovered using forest type plots, by reordering the data in chronological order or by quality. Adjusting for the known covariates through a regression model helps to avoid confounding, and it may substantially reduce the residual heterogeneity (Thompson and Higgins 2002, van Houwelingen *et al.* 2002).

To be sure, trying to reduce heterogeneity or dealing with it by, e.g., random-effects modeling is not always appropriate. If the results of the statistical analysis change significantly with the model or method applied, it may be better to simply present the existing evidence "as is" and abstain from further attempts to summarize the research results obtained so far.

Approximations. Technically, a statement like "a 95% confidence interval for the overall odds ratio is $[.69, .88]$" is based on probabilistic calculations which as a rule are not exact but depend on approximations to the distributions of the relevant random variables. These approximations are usually derived by appealing to the standard limit theorems of probability theory – the law of large numbers and the central limit theorem – and pretending that the limit is already attained in the case at hand. This never holds exactly, of course, and it is notoriously difficult to estimate the "gap" between the actual and the limit distribution. If the gap is too big, statistical statements based on the limit distribution can be misleading and invite false conclusions. Accordingly, the suitability of approximations is an issue throughout statistics.

In meta-analysis, the focus ultimately is on the confidence interval for the overall effect size η, so our main concern will be with the standard normal distribution approximation to the pivotal quantitiy

$$T = (\widehat{\mathsf{Var}(\widehat{\eta})})^{-1/2}\,(\widehat{\eta} - \eta)$$

underlying its calculation. Very broadly, the fixed-effects confidence interval for η is expected to be valid if (a) the single effect sizes η_i differ only moderately and (b) the participant groups in each study are sufficiently large and homogeneous. The argument goes as follows. By the above-cited limit theorems and (b), each effect size estimate $\widehat{\eta}_i$ is approximately normally distributed (cf. Eq. (7.2)); moreover, the variance estimates $\widehat{\sigma}_i^2$ can be treated as (non-random) constants, $\widehat{\sigma}_i^2 \doteq \sigma_i^2$. Then $\widehat{V} = \sum_i \widehat{\sigma}_i^{-2}$ represents a good, essentially non-random approximation to the variance of $\sum_i \widehat{\sigma}_i^{-2}(\widehat{\eta}_i - \eta_i)$, whence it follows that the random variable

$$\widehat{V}^{-1/2} \sum_i \widehat{\sigma}_i^{-2}(\widehat{\eta}_i - \eta_i) = \widehat{V}^{1/2}(\widehat{\eta}^{\mathrm{F}} - \eta) - \widehat{V}^{-1/2} \sum_i \widehat{\sigma}_i^{-2}(\eta_i - \eta)$$

is approximately standard normally distributed. If condition (a) is understood to mean that the bias term

$$B = \widehat{V}^{-1/2} \sum_i \widehat{\sigma}_i^{-2} (\eta_i - \eta)$$

is negligible, then $\widehat{V}^{1/2}(\widehat{\eta}^{\mathrm{F}} - \eta) \equiv T$ is approximately $\mathcal{N}(0, 1)$-distributed, too, and the (approximate) validity of the fixed-effects confidence interval follows.

A problem with random-effects confidence intervals is that the estimates $\widehat{w}_i^{\mathrm{R}}$ of the random-effects weights can be very unstable even under condition (b). Hence, they cannot be treated as constants as it would be necessary to transfer approximate normality from the single estimates $\widehat{\eta}_i$ to the combined estimate $\widehat{\eta}^{\mathrm{R}}$. The instability results from the estimate of the unknown variance component τ^2, which is very uncertain particularly when there is marked heterogeneity and the number N of studies is small. Conversely, if N is large, the combination of many independent influences would damp the variability of $\widehat{\tau}^2$, thanks to the law of large numbers.

To summarize, judging the validity of confidence statements about the overall effect size is not straightforward. The fixed-effects confidence intervals tend to be biased and too narrow in the presence of heterogeneity among the η_is. Random-effects constructions, too, are suspected to underestimate the actual variability, and fully Bayesian approaches have been proposed as a remedy (Carlin 1992, Lunn *et al.* 2013). Good agreement of the results obtained with different methods certainly is a good sign. However, Monte Carlo simulations of the distribution of the pivot T will often remain the only means to realistically assess the situation at hand.

7.2.4 Biases in Meta-Analysis

Inappropriate approximations are not the only kind of bias in meta-analysis, and not the worst. Violations of the basic model assumptions that underlie, explicitly or implicitly, the construction of tests, confidence intervals, and so on have the greatest potential to entail false conclusions. One notorious example is dependence among observations that was not accounted for in the model (compare Stahel, this volume).

In meta-analysis, "intrinsic" correlations between study outcomes are unlikely to occur. However, dependence appears in other guises. It may for instance result from different research groups obeying peculiar rules in conducting their trials, recruiting particular subjects for testing, and so forth. This type of dependence is best thought of as covariation with certain characteristics of the study. It can be accounted for by admitting the effect sizes η_i to vary across studies as a function of study features if these are known (meta regression), or in a random fashion if not (random-effects meta-analysis).

Here the focus is on dependence that results from hidden selection processes biasing the publicly accessible information in a certain direction. Selective publication and reporting of studies is a main source of such bias. If "positive results" and significant studies are more likely to be published, correctives from evidence to the contrary are lacking or do not receive appropriate attention.

Journal policies discriminating against negative or nonconclusive results have often been criticized, with modest success. However, imbalanced representation of experimental findings in the literature is not only the responsibility of editors and reviewers. Hedges (1992) argued that reporting biases could have an even greater impact on imbalance. Thus authors may be reluctant to submit research work that did not yield the hypothesized or intended outcome.

Presumably, however, it is common practice in various subject areas of the social and life sciences (not excluding others) to investigate a whole bunch of research questions at the same time that introduces distortions in the first place. The sheer mass of intermediate results obtained from the same source data with diverse analysis methods, choices of "tuning parameters," and so on necessitates decisions about what to include in the paper to be submitted. The ensuing temptations in a "publish-or-perish" scientific community need no commentary.

The potentially dramatic consequences of the various sorts of selection bias call for methods that permit to investigate its presence, extent, and possible impact. In the following we give a sketch of some approaches to and procedures for dealing with this issue.

Funnel plot. A simple but useful graphical method for detecting publication bias in a meta-analysis is the *funnel plot* (Light and Pillemer 1984). It may be regarded as a variant of the forest plot, with a different choice of the ordinate: in the funnel version the data are ordered vertically according to some measure reflecting the (im-)precision of the individual effect size estimates $\widehat{\eta}_i$. Typical choices are the variance or the standard deviation of $\widehat{\eta}_i$, or their inverses. Here we follow a proposal by Sterne and Egger (2001) who compared various possibilities of choosing the axes.

In Fig. 7.3 standard deviations of the estimates (SE) are plotted against the respective log odds ratios, using an inverted scale where SE increases downwards, starting from SE zero at the top. The vertical line indicates the location of the overall effect size estimate $\widehat{\eta}$ on the abscissa, and the two other lines demarcate the endpoints of 95% confidence intervals about $\widehat{\eta}$ as a function of SE. As in Sterne and Egger (2001), the fixed-effects model is assumed for this representation.

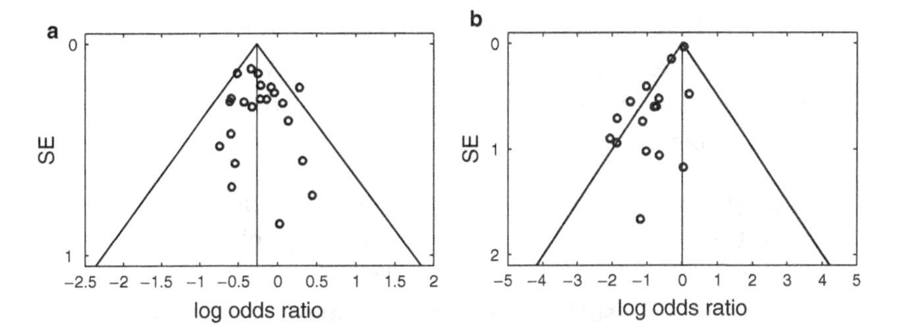

Figure 7.3: Funnel plots of standard deviations of the estimates (SE) versus log odds ratio for the beta blocker (a) and the magnesium data (b) using *fixed-effects* estimates. See text for explanation.

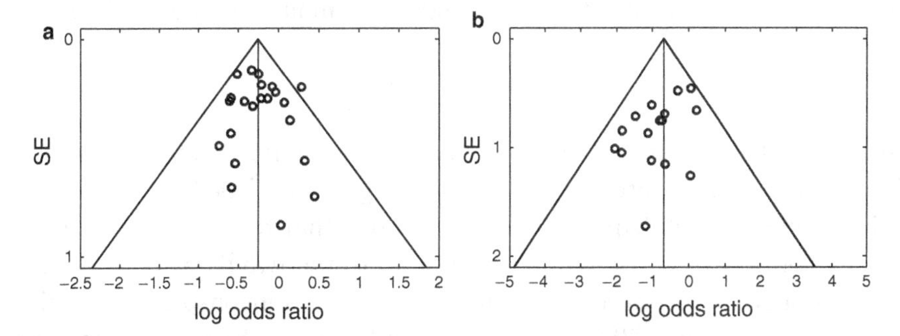

Figure 7.4: Funnel plots of standard deviations of the estimates (SE) versus log odds ratio for the beta blocker (a) and the magnesium data (b) using *random-effects* estimates. See text for explanation.

The distinct asymmetry visible in Fig. 7.3b is taken as an indication of publication bias. The rationale for this interpretation is that large trials require a great effort and are, therefore, more likely published, irrespective of the outcome, than small trials where non-publication of non-significant results might be considered a minor loss. Consequently, one would expect that the effect size estimates with a larger variance, coming from the smaller trials, are biased towards "positive" results. In Fig. 7.3 the positive direction corresponds to negative log odds ratios, and these indeed appear overrepresented in Fig. 7.3b: the data points accumulate in the left half of the plot, and at least if the lowest lying point is ignored, there appears to be a tendency for the deflections to increase with SE. The funnel plot in Fig. 7.3a has a more symmetric appearance suggesting that publication bias is unlikely to be a serious problem in the beta-blocker data.

For the latter, the picture remains nearly unchanged if the funnel plot is constructed using the random-effects model (Fig. 7.4a). For the magnesium data the distinct asymmetry of Fig. 7.3b gets somewhat diluted in Fig. 7.4b which, when inspected on its own, furnishes limited guidance regarding the possible presence of publication bias. This is connected to the discrepancy between the random- and fixed-effects estimates observed in Fig. 7.2. With random effects the single $\widehat{\eta}_i$s are weighted more uniformly, thus damping the influence of large trials and blurring possible asymmetries in the data.

Quantitative approaches. The funnel plot has become popular as a rough and ready check for the presence of publication bias. Its graphical nature leaves much room for subjective interpretation, thus calls for quantitative procedures assisting the assessment. One such method, the so-called *Egger test* (Egger *et al.* 1997), is based on a reformulation of the fixed-effects model, as follows. On dividing by σ_i and putting $y_i = \widehat{\eta}_i/\sigma_i$, one can write Eq. (7.2) in the form of a linear regression through the origin with $x_i = 1/\sigma_i$ as the regressor variable, effect size η as the slope parameter, and standard normally distributed errors ϵ_i,

$$y_i = \eta x_i + \epsilon_i, \qquad \epsilon_i \sim \mathcal{N}(0,1). \tag{7.8}$$

The idea is that under this transformation, asymmetry in the funnel plot would introduce a nonzero intercept in the linear regression. The Egger test then consists in testing the null hypothesis $H_0 : \alpha = 0$ within the augmented linear regression model $y_i = \alpha + \beta x_i + \epsilon_i$. When applied to the two data sets, the test confirms the visual impression from the funnel plots: it is inconspicuous for the beta-blocker data (p-value ≈ 0.5), and highly significant for the magnesium data ($p < 10^{-4}$). Interestingly in view of Fig. 7.4b, it remains marginally significant in the latter case ($p \approx 0.03$) if it is applied with the random-effects estimates.

Other procedures building on an asymmetry of the funnel plot as an indication of publication bias include a nonparametric version (Begg and Mazumdar 1994) of the Egger test and the *Trim and Fill* method (Duval and Tweedie 2000). The latter algorithm iteratively trims the most extreme $\widehat{\eta}_i$s at the "positive" side of the funnel plot and afterwards refills the plot by imputing "likely missing" studies as the "negative counterparts" of those trimmed. This yields a corrected estimate of the overall effect size as well as an estimate of the number of likely missing studies.

Similarly, in Rosenthal's *fail-safe* n method the number of studies "resting in the file drawer" is determined as the number of overlooked zero-effect studies that would be necessary to reduce a significant result to non-significance (Rosenthal 1979).

Explicit selection models. The common premise underlying all statistical models for a hypothetical selection mechanism is that the reason why a study is missing is connected to the outcome of the study – otherwise, missing studies would reduce the available information, but they would not distort it. Iyengar and Greenhouse (1988) proposed to make the connection explicit by assuming that a study happens to be included in the meta-analysis with a probability w that depends on its outcome.

Related work on selection models mainly differs in regard to the kind of assumptions made about the functional dependence of w on the available data. Normally, the only available data is $(\widehat{\eta}_i, \widehat{\sigma}_i)$, whence a common assumption is $w = w(\widehat{\eta}_i, \widehat{\sigma}_i)$. Further reduction is achieved by compressing the pair $(\widehat{\eta}_i, \widehat{\sigma}_i)$ to a "t-value" $t_i = \widehat{\eta}_i / \widehat{\sigma}_i$ and assuming $w = w(t_i)$.

This route has been followed in, e.g., Iyengar and Greenhouse (1988), Dear and Begg (1992), and Hedges (1992). An entirely nonparametric choice of w is infeasible due to data sparseness, so dimensionality-reducing restrictions are necessary. One possibility is to suppose that w has a particular shape tunable by a small number of parameters (Iyengar and Greenhouse 1988). Another option is to specify w as a step function, either with prescribed subdivision points (Hedges 1992), or semiparametrically, with a mild restriction on the number of points (Dear and Begg 1992).

A difficulty with such explicit selection models is that the parameter estimates can be very unstable due to a flat likelihood function. Rather than trying to maximize the full likelihood, it may then be more appropriate to estimate only a subset of the parameters keeping the other (critical) parameter(s) fixed, and to demonstrate graphically how the estimates vary in dependence of the value(s) of the fixed parameter(s). Several variants of such *sensitivity analyses* were proposed by Copas (1999, 2013) in connection with slightly different selection models.

7.3 Meta-Analysis of Mind-Matter Experiments: A Case Study

This section recapitulates work that is special in two respects. First, the research problem addresses a controversial issue, namely, the possible existence of correlations between consciousness and the material world.[4] Secondly, the number of studies in the meta-analysis is unusually large ($N > 500$); moreover, the statistical model of the null hypothesis of no effect is given explicitly.

The precise question is as follows: Are there correlations between the "state of mind" of a conscious subject and the "state of matter" of a physical system? Here, state of matter is identified with the output of a physical random event

[4]The underlying motivation appears to spring from the "observer problem" of quantum theory: very plainly, a quantum system may behave differently "if someone looks."

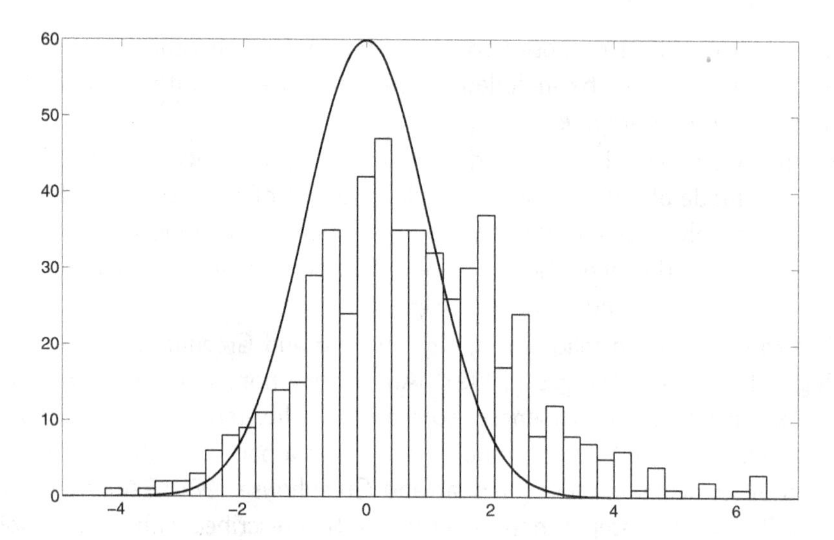

Figure 7.5: Histogram of z-scores under experimental conditions (data from Radin and Nelson (1989)). The bell curve represents the distribution expected under the null hypothesis of no effect.

generator (REG), while state of mind is operationally defined as an intention the subject is instructed to "have" with respect to the physical system.

Empirical evidence for the existence of intention-output correlations was reported in various experiments conducted with REGs. In a meta-analysis of these studies Radin and Nelson (1989) (henceforth "R&N") found a small but highly significant experimental effect: "... it is difficult to avoid the conclusion that under certain circumstances, consciousness interacts with random physical systems."

Indeed, a histogram of the effect size estimates transformed to z-scores (see Eq. (7.12)) exhibits a distinct, "positive" deflection from the distribution expected under the null hypothesis of no interaction (Fig. 7.5) that is extremely significant when tested by, e.g., the Kolmogorov–Smirnov test ($p < 10^{-25}$).

The findings of R&N later were contradicted by a large-scale replication study where the overall summary statistic failed to be statistically significant (Jahn *et al.* 2000). This discrepancy motivated another attempt to analyze the data of R&N using more refined statistical models (Ehm 2005). Our purpose here is to give an account of this latter meta-analysis emphasizing selection effects modeling and model validation.

Experiments with random event generators. Each experiment in R&N's meta-analysis involved some sort of REG. The source of randomness could for instance be a coin, a dice, or thermal or quantum noise. The REG was devised so as to produce a "perfectly random" stream of bits coded as 0 or 1 so that ideally, the output of the REG would mimic a sequence of independent and identically distributed (i.i.d.) 0–1 random variables. Accordingly, under the null hypothesis H_0 describing the control condition the bits $Y_1, Y_2, \ldots$ are regarded as i.i.d. Bernoulli variables with a known "hit probability" $p_0 = Pr[Y_j = 1] = 1 - Pr[Y_j = 0]$.

Under experimental conditions ("under H_1") the REG is run in the presence of a human subject, called agent, who is asked to "have the intention" that "hits" should occur with increased frequency. What counts as a hit depends on the experimental condition: 1 for condition "high," 0 for condition "low." The idea is that an increased frequency of hits under H_1 would lend support to the conjectured existence of mental–material correlations.

Preliminary consideration. The data obtained under experimental conditions is indeed shifted away from the null distribution in Fig. 7.5. Isn't that proof enough of the conjecture? General wisdom negates this: empirical findings not backed up by an elaborate theory may be (tentatively) ascribed to the envisaged grounds only if alternative explanations can be excluded. At least, it should be possible to formulate an adequate statistical model that (F) fits the observed data, and that (I) implements the conjectured relationship in terms of internal properties of the model. Both requirements are essential: without a model fulfilling (I) the interpretation of the data remains vague; and without (F) the model does not relate to the data.[5] The following developments aim at statistical models for the data of R&N that satisfy (F) and (I).

7.3.1 Statistical Modeling

The data available for the i-th study are: the total number n_i of 0/1-bits collected, and the total number $S_i = \sum_{j=1}^{n_i} Y_{ij}$ of hits (under H_1 each); the hit probability $p_{i,0}$ under H_0; and a (row-)vector x_i of q covariates[6] comprising additional information about study $\#i$. Under H_0, S_i has the binomial distribution $\mathcal{B}(n_i, p_{i,0})$, so the statistical model is fully specified. Under H_1, the agent's intention is supposed to increase the relative frequency of hits, a broad assumption leaving ample room for interpretation; it has to be concretized as a statistical

[5]This could be seen as a statistical version of Kant's famous *dictum* in his *Critique of Pure Reason*: "Gedanken ohne Inhalt sind leer, Anschauungen ohne Begriffe sind blind" – think of "Gedanken/Begriffe" as models and of "Inhalt/Anschauung" as data.

[6]As usual, the first component of x_i is taken to be constant, $x_{i1} = 1$, so that there are in fact $q - 1$ proper covariates. Here $q = 6$, and covariates include REG type and a quality score, among others.

model to become verifiable. The *generalized Bernoulli* (GB) scheme adopted in Ehm (2005) yields various models of interest as special cases. It involves the following assumptions referring to the bit level of study $\#i$.

GB1 The Y_{ij} $(j = 1, \ldots, n_i)$ are Bernoulli(p_{ij}) variables that are conditionally independent across bits, given the individual hit probabilities p_{ij}.

GB2 The p_{ij} $(j = 1, \ldots, n_i)$ are themselves random variables, namely, of the form

$$p_{ij} = s\left(\vartheta_{i,0} + \epsilon_{ij}\right), \qquad \epsilon_{ij} = x_i\left(\beta + \rho_{ij}\right), \tag{7.9}$$

where s is a scale function, $\vartheta_{i,0}$ is defined by $s(\vartheta_{i,0}) = p_{i,0}$, $\beta \in \mathbb{R}^q$ is a (column) vector of parameters, the $\rho_{ij} \in \mathbb{R}^q$ are zero mean (column) random vectors detailed later on, and ab denotes the inner product of row vector a and column vector b.

Some explanation is in order. The experimental effect in the GB model is conceived of as resulting from small bit-by-bit disturbances of the pure chance hit probability $p_{i,0}$. These bit-level effects are encoded in the quantities ϵ_{ij}, which measure the (bit-wise) effect size in a certain scale specified by the smooth, strictly increasing function s. The choice $s(t) = 1/(1 + e^{-t})$, e.g., yields the familiar log odds scale where $\vartheta_{i,0} = s^{-1}(p_{i,0}) = \log(p_{i,0}/(1 - p_{i,0}))$, and the effect size ϵ_{ij} is given by the log odds ratio of the perturbed with respect to the unperturbed hit probabilities,

$$\epsilon_{ij} = \log(p_{ij}/(1 - p_{ij})) - \log(p_{i,0}/(1 - p_{i,0})). \tag{7.10}$$

According to Eq. (7.9), the inner products ϵ_{ij} are composed of a systematic, non-random term plus a random perturbation. The first term, $x_i\beta$, is constant across bits and represents the study-specific effect size as a linear function of the respective covariates. The terms $x_i\rho_{ij}$ add unsystematic random variability – "unsystematic" because the ρ_{ij} have expectation zero – that may operate at various stages from the bit level upwards.

From the GB to the RCR model. For the meta-analysis, only the relative frequency $\widehat{p}_i = S_i/n_i$ of the hits is available, so an estimator of the effect size $x_i\beta \equiv \eta_i$ has to be a function of $\widehat{p}_i$. By Eqs. (7.9) and (7.10), a natural estimate is $\widehat{\eta}_i = s^{-1}(\widehat{p}_i) - s^{-1}(p_{i,0})$ (note that $p_{i,0}$ is known). Under mild conditions that are uncritical in our case – e.g., effect sizes are small and n_is are large – it can be shown that $\widehat{\eta}_i$ can be approximated as follows,

$$\widehat{\eta}_i \doteq z_i/\gamma_i \tag{7.11}$$

where

$$z_i = \frac{\sqrt{n_i}\,(\widehat{p}_i - p_{i,0})}{\sqrt{p_{i,0}(1-p_{i,0})}}, \qquad \gamma_i = s'(\vartheta_{i,0})\left(\frac{n_i}{p_{i,0}(1-p_{i,0})}\right)^{1/2}. \tag{7.12}$$

The random variable z_i is a *z-score,* which means that it is approximately standard normally distributed under H_0. Under H_1, i.e., under the GB model, approximate normality is maintained, with different moments,

$$E\, z_i \doteq \gamma_i\, x_i\beta, \tag{7.13}$$

$$\mathsf{Var}(z_i) \doteq 1 + \gamma_i^2\, x_i\Omega_i x_i^T. \tag{7.14}$$

Here $x_i\Omega_i x_i^T$ is a quadratic form involving the dispersion of the random effects via the covariance matrix

$$\Omega_i = \mathsf{Cov}\left(n_i^{-1}\sum\nolimits_j \rho_{ij}\right). \tag{7.15}$$

Ehm's (2005) meta-analysis was not targeted at a synthesis of the single effect size estimates. It rather focused on the distribution of the z-scores z_i across studies, with a view toward model checking and selection effects modeling. The *random coefficients regression* (RCR) model finally adopted for the analysis of the R&N data is specified as follows:

R1 The z-scores z_i are statistically independent across studies and normally distributed as $\mathcal{N}(\mu_i\, \sigma_i^2)$ with

$$\mu_i = \gamma_i\, x_i\beta, \qquad \sigma_i^2 = 1 + \gamma_i^2\, \delta_i^2\, x_i\Omega x_i^T. \tag{7.16}$$

R2 Unknown parameters are the regression coefficients β and the covariance matrix Ω.

R3 Known quantities are the covariate vectors x_i, the γ_i from Eq. (7.12), and certain scaling constants δ_i (to be explained).

As seen above, the RCR model derives from the GB model, except for two additional assumptions. First, a scale function s has been fixed, making the terms $s'(\vartheta_{i,0})$, hence γ_i, known quantities. The choice of s is immaterial if the $p_{i,0}$ are all identical because $s'(\vartheta_{i,0})$ becomes a constant that can be absorbed into the unknowns β and Ω. Here the $p_{i,0}$ cover a wide range of probabilities, and the choice of s can make a difference. On the other hand, for all but very few studies, $p_{i,0}$ was close to 1/2, which damps the influence of the scale function. The reasons for ultimately selecting the log odds scale are detailed in

Ehm (2005). Incidentally, with this choice, γ_i^2 reduces to the familiar binomial variance $n_i p_{i,0}(1 - p_{i,0})$.

The second restriction concerns the covariances of the random-effect averages $\bar{\rho}_i = n_i^{-1} \sum_j \rho_{ij}$, which are assumed to be of the form $\Omega_i = \delta_i^2 \Omega$ with study-dependent scaling factors δ_i and a common covariance matrix Ω. The δ_i account for possibly different orders of magnitude of the $\bar{\rho}_i$. Very briefly, the individual bits Y_{ij} giving rise to $\widehat{p}_i$ are collected across several hierarchically ordered stages (trials, runs, sessions, participants), and random effects may, or may not, accrue at each of those aggregation levels. Random effects accruing at the lowest level are averaged across n_i bits, a random effect at the top (study) level occurs only once. Therefore, $\bar{\rho}_i$ should be of the order $O(n_i^{-1/2})$ in the former case, or $O(1)$ in the latter case (of course with all possibilities in between). The workaround was to try both extremes, $\delta_i = n_i^{-1/2}$ and $\delta_i = 1$, and to further proceed with the better fitting one.

The RCR model represents a general form of meta-regression (van Houwelingen *et al.* 2002) including various submodels as special cases. For instance, putting $\delta_i = 0$ yields a fixed-coefficients regression (FCR) model. If there are no proper covariates ($q = 1$), FCR further reduces to the standard fixed-effects model (Eq. (7.2)), with β_1 representing the scalar fixed effect η. If $q > 1$, the RCR model also allows for random perturbations of the coefficients of the proper covariates. Usually, these are not taken into consideration, nor is allowance made for study-dependent orders of magnitude of the random effects: in Eq. (7.3), the variance component τ^2 is constant across studies. In summary, each term in Eq. (7.16) describes some specific, well-interpretable contribution to the total variability of the z-scores, in compliance with requirement (I) above.

Modeling selection effects – the RCRS model. While biased samples can arise in many ways, all related modeling approaches assume a correlation between the selection of a study for inclusion in the meta-analysis and its outcome. In Ehm (2005), the model of Hedges (1992) was adapted for use with the z-scores. It is put "on top" of the RCR model, hence is referred to as the *random-coefficients regression selection* (RCRS) model.

Let S denote the set of all studies included in the meta-analysis, considered as a subset of the unknown set of all relevant studies. The conditional likelihood of the z_i, given they are included, is

$$L = \prod_{i \in S} \frac{f_{\theta_i}(z_i)\, w(z_i)}{\int f_{\theta_i}(z)\, w(z)\, dz}, \tag{7.17}$$

where $f_{\theta_i}(z) = \varphi((z - \mu_i)/\sigma_i)/\sigma_i$ is the unconditional density of z_i, φ is the standard normal density, and $\theta_i = (\mu_i, \sigma_i^2)$. As in Hedges (1992), the weight function is chosen as a step function with respect to a partition of the real line into $K = 10$ intervals I_k, with the nine (fixed) cutpoints given by familiar $\mathcal{N}(0,1)$-percentiles. One may identify w with the values $w_k \equiv e^{\alpha_k}$ it takes on the intervals I_k, hence parameterize it by vector α. This parameterization is slightly redundant as L is invariant under multiplication of w by a constant. Therefore, α_K, say, can be fixed arbitrarily. Intuitively, the model says that the i-th study is selected with a probability proportional to w_k if z_i happens to fall into I_k, as if this was decided by flipping a (suitably biased) coin.

Statistical inferences about the parameters of the RCRS model can be carried out using the maximum-likelihood method. For example, the null hypothesis $\alpha = 0$ that selection effects are absent is rejected at the 5% level if the likelihood ratio test statistic

$$LR_{sel} = 2\left(\sup_{\beta,\phi,\alpha} \log L(\beta,\phi,\alpha) - \sup_{\beta,\phi} \log L(\beta,\phi,0)\right) \tag{7.18}$$

exceeds the 95% quantile of the χ^2 distribution with 9 degrees of freedom. (Here the factor ϕ in the Cholesky decomposition $\Omega = \phi\phi^T$ is used to parameterize the random components.) Generally, setting certain parameters equal to zero yields a test of the null hypothesis that the corresponding effects are absent. This opens up the possibility of a *corrections meta-analysis* (Copas 1999), namely, to make statistical inferences about the residual experimental effect when adjustment is made for possible selection effects, and *vice versa*.

7.3.2 Analysis of the R&N Data

Model checking. The results of a statistical analysis are difficult to interpret if they rely on a statistical model that does not fit the data at hand; cf. requirement (F) mentioned above. Thus, first and foremost, appropriate checks are necessary to sort out the obviously inadequate models. Among the four models considered, the FCR and the RCR model with $\delta_i = 1$ (RCR1) turned out unsatisfactory, as may be seen from a probability plot of the residuals. In such a plot the standardized residuals should concentrate along the diagonal, with a bit more dispersion allowed in the tails.

Evidently from Fig. 7.6, the FCR model totally misses the scattering of the z-scores visible in Fig. 7.5, and so does the RCR1 model that exhibits further inconsistencies. The plots look more satisfactory for the RCRS and the RCR0 model ($\delta_i = n^{-1/2}$), although close examination of the latter reveals a small but consistent shift to the right. For further assessment, surrogate data were generated under the four models taking the respective (maximum likelihood)

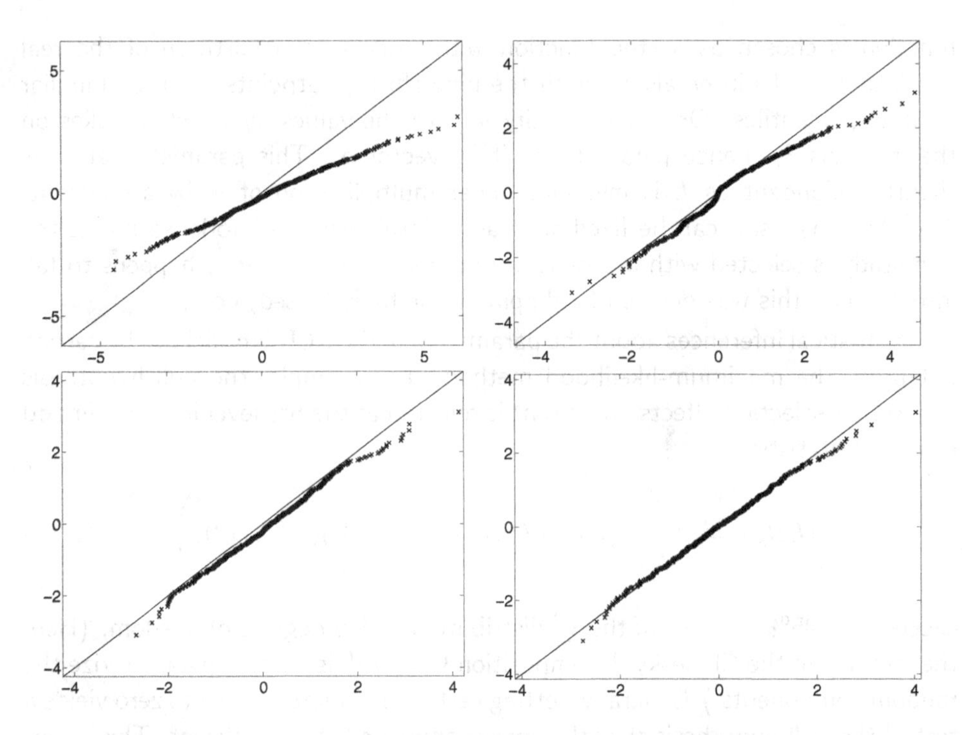

Figure 7.6: Normal probability plots of the standardized residuals for the FCR, RCR1, RCR0, and RCRS model (from upper left to lower right).

estimates as the underlying parameters. Repeated inspection of the simulated data and the respective residual probability plots provides a good impression of the model's capability to produce data such as those observed.

Experimental and selection effects. The following table summarizes the results of three tests of interest carried out under the two remaining models. (Entries are the p-values from the respective likelihood ratio tests.) The hypotheses pertain to the parameter β accounting for systematic experimental effects: β either is unrestricted, or is set to zero ("no effect") or corresponds to a constant mean shift β_1 (on setting the coefficients $\beta_2, \ldots, \beta_q$ of the proper covariates to zero). Note that a (positive) mean shift represents the simplest way to state the

assumed model	"no effects" vs. "unrestricted"	"no effects" vs. "constant shift"	"constant shift" vs. "unrestricted"
RCR0	$8.860 \cdot 10^{-6}$	$1.975 \cdot 10^{-6}$	0.0563
RCRS	0.2484	0.6666	0.1750

premise that the frequency of hits is increased under experimental conditions. Under the RCR0 model the "no effects" hypothesis is rejected at very small test levels, speaking for a highly significant shift effect. The "constant shift" versus "unrestricted" test just fails to be significant at the 5% level, meaning that the covariates contribute at most marginally to the experimental effect.

In stark contrast, none of these tests is even close to being significant under the RCRS model. Could selection be responsible for what the RCR0 model ascribes to an experimental effect? This presumption is corroborated by the test for selection: the null hypothesis of no selection (all $\alpha_k = 0$) is rejected at level $p < 10^{-12}$ within the RCRS model. Still, the meaning of the test results is in question if the tests are biased, i.e., if they fail to have the correct distribution under the null hypothesis.

Test validity. The RCRS model offers two competing explanations for increased z-score values under experimental conditions: (1) shift of the expected z-scores and (2) selection effects. In general, the two are indistinguishable because any translation of a density related to (1) can as well be accomplished by a multiplicative density modification related to (2). However, for the data under study, the RCRS-based likelihood ratio tests proved to be essentially unbiased, i.e., they do not confound experimental and selection effects.

This can be checked by simulating the distribution of the test p-values under the respective null hypothesis (i.e., under the parameter configurations $(\widehat{\beta}, \widehat{\phi}, 0)$ or $(0, \widehat{\phi}, \widehat{\alpha})$, respectively, where $\widehat{\beta}, \widehat{\phi}, \widehat{\alpha}$ denote the maximum-likelihood estimates computed for the full RCRS model). In fact, the simulated p-values were roughly uniformly distributed, as they should; see Fig. 7.7a,b. Likewise, the estimates of the weights w_k unbiasedly reproduce the w-profile underlying the simulations, no matter whether selection is present or not; see Fig. 7c,d. By contrast, the RCR-based likelihood ratio test for an experimental effect turned out to be severely biased in the presence of selection: it rejects the null hypothesis of no experimental effect with high probability even if the latter is true.

Conclusion. Taken together, the analyses suggest that the anomalies in the data of Radin and Nelson (1989) might as well result from some kind of selection effect, and not necessarily from a mind-matter correlation. For a more detailed assessment see Ehm (2005). See also Jahn *et al.* (2000) and Bösch *et al.* (2002).

7.4 Summary

Meta-analysis provides formal tools for dealing with heterogeneity and selection bias in a systematic manner, making it relevant to a quantitative conception of

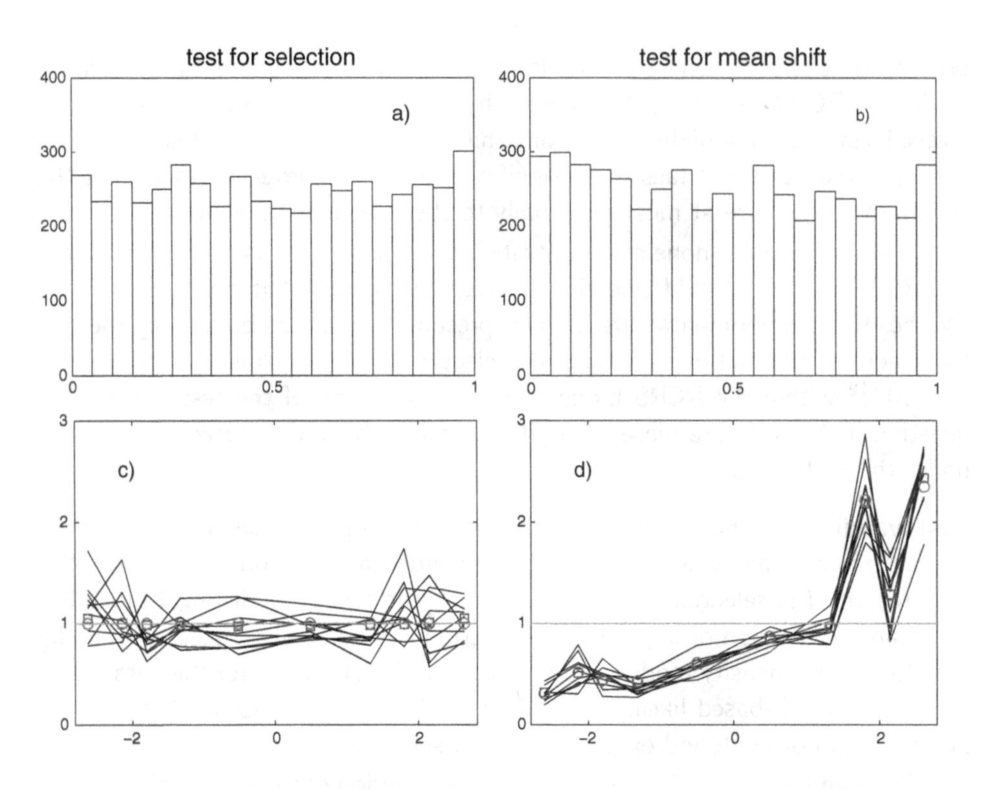

Figure 7.7: (a) and (b) are histograms of 5,000 test p-values simulated under the null hypotheses of no selection and no effect, respectively. (c) and (d) are profiles of weights w_k estimated within the RCRS model, when there is selection (d), and when there is not (c). Squares: averages of estimated w_k-profiles from 5,000 simulated data. Circles: w_k-profiles underlying the simulations. Broken lines: 10 w_k-profiles randomly selected from the 5,000 simulated profiles (providing an impression of the variability to be expected).

reproducibility. Such an approach fits the complexity encountered in the social and life sciences better than the requirement of strict replicability. In Section 7.2 we reviewed basic techniques of meta-analysis, thereby pointing to a number of problems that may arise. Special attention should be paid to possible sources of variability and the way they are incorporated in the statistical model.

An underestimation of the variance of the overall effect size estimate often is a point in question as it invites overly optimistic conclusions. In fact, the very meaning of an overall effect size deserves consideration in the presence of substantial heterogeneity. In such a case, alternative methods of analysis may be called for.

An example of an alternative approach based on the distribution of z-scores has been presented in Section 7.3, with focus on the following points. First, an adequate statistical model is sought that could have produced data such as those observed. Analysis of the residuals and simulation are useful tools for assessing the adequacy of competing models, in particular for sorting out grossly inadequate ones. The danger of overfitting is an issue at this stage (see Bailey *et al.*, this volume). However, it can be diminished by only including model components that have a clear interpretation and by keeping the parametrization sparse in relation to the available data.

Secondly, provision is made for a selection mechanism that might bias the available information. This permits asking for the specific influences of the various factors, for instance, whether a significant experimental effect persists when corrected for possible selection. Unfortunately, such corrections meta-analyses can be seriously misleading due to confounding: a genuine experimental effect may be falsely ascribed to selection, and *vice versa*. Unraveling the two effects may be possible if the number of studies is large and the data is sufficiently diversified through variation of sample sizes and covariates across studies.

Third, in any case it is necessary to assess the validity of the related statistical inferences. This can again be accomplished by simulation, upon turning off either the experimental or the selection effect in the surrogate data generating process, and seeing whether the statistical analysis "detects" an inexistent effect, or finds only those present.

It has to be emphasized that the validity of such analyses is not guaranteed in general. Warnings against an undiscriminating use of corrections meta-analysis are well founded, and more conservative sensitivity analyses are often preferrable (Copas 1999, 2013). On the other hand, if corrections meta-analysis is applicable, it can be a useful tool for more incisive analyses.

Generally, meta-analysis cannot replace replication studies that may, or may not, gradually settle the issue under scrutiny. Taking selection bias into account, it may, however, uncover problems with empirical findings before these are contradicted by a huge trial, and thus safeguard against a premature acceptance of facts or the application of ineffective or harmful treatments.

References

Begg, C.B., and Mazumdar, M. (1994): Operating characteristics of a rank correlation test for publication bias. *Biometrics* **50**, 1088–1099.

Blood Pressure Lowering Treatment Trialists Collaboration (2003): Effects of different blood-pressure-lowering regimens on major cardiovascular events: Results of prospectively-designed overviews of randomised trials. *Lancet* **362**, 1527–1535.

Bösch, H., Steinkamp, F., and Boller, E. (2006): Examining psychokinesis: The interaction of human intention with random number generators – A meta-analysis. *Psychological Bulletin* **132**, 497–523.

Carlin, J.B. (1992): Meta-analysis for 2×2 tables: A Bayesian approach. *Statistics in Medicine* **11**, 141–158.

Copas, J. (1999): What works? Selectivity models and meta-analysis. *Journal of the Royal Statistical Society, Series A* **162**, 95–109.

Copas, J. (2013): A likelihood-based sensitivity analysis for publication bias in meta-analysis. *Journal of the Royal Statistical Society, Series C* **62**, 47–66.

Dear, K.B.G., and Begg, C.B. (1992): An approach for assessing publication bias prior to performing a meta-analysis. *Statistical Science* **7**, 237–245.

DerSimonian, R., and Laird, N. (1986): Meta-analysis in clinical trials. *Controlled Clinical Trials* **7**, 177–188.

Doshi, P., Jefferson, T., and Del Mar, C. (2012): The imperative to share clinical study reports: Recommendations from the Tamiflu experience. *PLOS Medicine* **9**, e1001201.

Duval, S., and Tweedie, R. (2000): A non-parametric "trim and fill" method of assessing publication bias in meta-analysis. *Journal of the American Statistical Association* **95**, 89–98.

Egger, M., Smith, G.D., Schneider, M., and Minder, C. (1997): Bias in meta-analysis detected by a simple, graphical test. *British Medical Journal* **315**, 629–634.

Ehm, W. (2005): Meta-analysis of mind-matter experiments: A statistical modeling perspective. *Mind and Matter* **3**, 85–132.

Franco, A., Malhotra, N., and Simonovits, G. (2014): Publication bias in the social sciences: Unlocking the file drawer. *Science* **345**, 1502–1505.

Hedges, L.V. (1992): Modeling publication selection effects in meta-analysis. *Statistical Science* **7**, 246–255.

Higgins, J.P.T., and Thompson, S.G. (2002): Quantifying heterogeneity in a meta-analysis. *Statistics in Medicine* **21**, 1539–1558.

ISIS-4 Collaborative Group (1995): ISIS-4: A randomised factorial trial assessing early oral captopril, oral mononitrate, and intravenous magnesium sulphate in 58,050 patients with suspected acute myocardial infarction. *Lancet* **345**, 669–687.

Iyengar, S., and Greenhouse, J.B. (1988): Selection models and the file drawer problem (with discussion). *Statistical Science* **3**, 109–135.

Jahn, R.G., Dunne, B., Bradish, G., Dobyns, Y., Lettieri, A., Nelson, R., *et al.* (2000): Mind/machine interaction consortium: PortREG replication experiments. *Journal of Scientific Exploration* **14**, 499–555.

Ioannidis, J.P.A. (2005): Why most published research findings are false. *PLOS Medicine* **2**, e124.

Lehrer, J. (2010): The truth wears off – Is there something wrong with the scientific method? *NewYorker*, December 13, 2010.

Light, R., and Pillemer, D. (1984): *Summing Up: The Science of Reviewing Research*, Harvard University Press, Cambridge.

Lunn, D., Barrett, J., Sweeting, T., and Thompson, S. (2013): Fully Bayesian hierarchical modelling in two stages, with application to meta-analysis. *Journal of the Royal Statistical Society, Series C* **62**, 551–572.

Pashler, H., and Wagenmakers, E.J. (2012): Editors' introduction to the Special Section on replicability in psychological science: A crisis of confidence? *Perspectives on Psychological Science* **7**, 528–530.

Poole, C., and Greenland, S. (1999): Random effects meta-analyses are not always conservative. *American Journal of Epidemiology* **150**, 469–475.

Radin, D.I., and Nelson, R.D. (1989): Evidence for consciousness-related anomalies in random physical systems. *Foundations of Physics* **19**, 1499–1514.

Rosenthal, R. (1979): The "file drawer problem" and tolerance for null results. *Psychological Bulletin* **86**, 638–641.

Sterne, J.A.C., and Egger, M. (2001): Funnel plots for detecting bias in meta-analysis: Guidelines on choice of axis. *Journal of Clinical Epidemiology* **54**, 1046–1055.

Sutton, A.J., and Higgins, J.P.T. (2008): Recent developments in meta-analysis. *Statistics in Medicine* **27**, 625–650.

Teo, K.K., Yusuf, S., Collins, R., Held, P.H., and Peto, R. (1991): Effects of intravenous magnesium in suspected acute myocardial infarction: overview of randomised trials. *British Medical Journal* **303**, 1499–1503.

Thompson, S.G., and Higgins, J.P.T. (2002): How should meta-regression analyses be undertaken and interpreted? *Statistics in Medicine* **21**, 1559–1573.

van Houwelingen, H.C., Arends, L.R., and Stijnen, T. (2002): Advanced methods in meta-analysis: Multivariate approach and meta-regression. *Statistics in Medicine* **21**, 589–624.

Woods, K.L., and Fletcher, S. (1994): Long term outcome after magnesium sulphate in suspected acute myocardial infarction: The Second Leicester Intravenous Magnesium Intervention Trial (LIMIT-2). *Lancet* **343**, 816–819.

Yusuf, S., Peto, R., Lewis, J., Collins, R., and Sleight, P. (1985): Beta blockade during and after myocardial infarction: An overview of the randomized trials. *Progress in Cardiovascular Diseases* **XXVII**, 335–371.

8
Why Are There So Many Clustering Algorithms, and How Valid Are Their Results?

Vladimir Estivill-Castro

Abstract. Validity is a fundamental aspect of any machine learning approach. While for supervised learning there is now a plethora of standard methods, there are significantly fewer tools for unsupervised learning. Moreover, the three types of current validity approaches (external, internal, and relative) all have serious drawbacks and are computationally expensive. We discuss why there are so many proposals for clustering algorithms and why they detach from approaches to validity. This leads to the question of whether and how we can validate the results of clustering algorithms.
We present a new approach that differs radically from the three families of validity approaches. It consists of translating the clustering validity problems to an assessment of the easiness of learning in the resulting supervised learning instances. We show that this idea meets formal principles of cluster quality measures, and thus the intuition inspiring our approach has a solid theoretical foundation. In fact, it relates to the notion of reproducibility. We contrast our suggestion with prediction strength. Finally, we demonstrate that the principle applies to crisp clustering algorithms and fuzzy clustering methods.

8.1 Introduction

8.1.1 Data Mining and Knowledge Discovery

Around 1990 many private and public organizations, scientific institutes, and individual researchers found that computer technology had progressed to the point that collecting and storing large volumes of information was feasible, and rather inexpensive. While the size of these datasets was now beyond the analytical capabilities of individuals, the progress in database technology combined with advances in machine learning gave birth to technologies in data mining and knowledge discovery.

The promise of this somewhat new discipline was that *novel* and *valid* knowledge would be extracted from such large datasets, leading to useful insights and actionable processes that would benefit those interested in the phenomena from which the data was obtained (Fayyad *et al.* 1996). Today, datasets are even larger and more detailed. This has lead to much excitement in *big data* science

Reproducibility: Principles, Problems, Practices, and Prospects, First Edition. Edited by Harald Atmanspacher and Sabine Maasen.

(see also Stahel, this volume), where datasets are characterized by challenges in variety, velocity, volume, and some authors debate a fourth, veracity.

The most relevant aspect of data mining and knowledge discovery is its departure from established views of the scientific method. The conventional view is that scientific work starts by the formulation of a hypothesis about the behavior of some phenomena, followed by experimental design and the collection of data for analysis, repeating the experiment to check the reproducibility of its result, and eventually reaching a conclusion about a mechanism or a model that describes the data.

In contrast, the use of data mining methods to explore a larger population with datasets of uncontrolled sources and to obtain perhaps unreliable statistical inferences was called (in the 1960s) *data dredging, data fishing, and data snooping* to highlight them as bad practices for analyzing data without an *a priori* hypothesis. The practice was questioned because of concerns about the validity of any patterns discovered. That is the first distinctive characteristic: data mining is and remains an *exploratory* practice.

The data is collected without any preconceived hypothesis, or perhaps even without any particular phenomena in mind. There is no experimental design, and in many cases no experiment whatsoever. There is no repetition or reproduction, as there is nothing to repeat. The data is historical data. If anything, the data of one period is used to *validate* models constructed from earlier periods. Data sets usually consist of at least millions of measurements, with numerical, ordinal, or categorical attributes of, in some cases, several thousand dimensions.

The point is that *validity* (how can we trust the results?) is a central and fundamental theme of the inference processes in knowledge discovery and data mining. No wonder that there is debate in the big-data community about the exact meaning of "veracity." Some consider this as an issue of accuracy of the data itself, while others expect this to be an attribute of the accuracy of the models obtained.

Today, computers do much of the *scientific* inference work. Algorithms that discover patterns in huge datasets bring together advances from statistics, artificial intelligence, machine learning, and database technology. They have materialized into *business intelligence* systems, where large volumes of data (about transactions made by clients) are explored for many competitive advantages with commercial implications (identification of clusters, e.g., of customer behavior). Collecting data without a hypothesis, but with the goal of informed decision making, has been a long-standing practice – for instance, the census or other statistical surveys (cf. Keller, this volume). However, today *exploratory* algorithms are applied to many scientific endeavors that range from drug development to astronomy. In fact, some of the early success stories in the field were classifications of astronomical observations.

Nevertheless, *validity* remains an issue. With the risk of generalizing too much, most inference processes from data start by choosing a family of models based on some inductive principle. Then, we use the data to chose the model in the family that "best" explains the data (however "best" may be defined). The point is that most definitions of "best," when converted to a process that a computer will solve, imply the formulation of a corresponding optimization problem. In most cases, such optimization problems are computationally intractable, meaning that there is no efficient algorithm to solve it exactly. Thus, we apply approximation algorithms. Right here, we know that the model found by the computer, is an approximation to our desirable "best."

But, as the datasets get larger and larger, we have no choice but to trade the quality (accuracy) of the approximation (to the "best" model) against the time we expect the algorithm to provide a response. The speed of central processing units (CPUs), despite the advances in parallelism and other capacities to carry out computations faster, continues to be slower or improving at a lesser pace than the capacity to store and collect data. Thus, there is a tendency to accept an answer from the computer, without an understanding of the limitations or shortcomings of the underlying algorithms. More interestingly, there is a tendency to accept a result even if it is not reproducible.

8.1.2 Choices and Assumptions

Consider a classical approach to clustering. We assume that the data $\langle \vec{x}_i \rangle_{i=1,\ldots,n}$ is given as a sequence of n d-dimensional vectors $\vec{x}_i \in \mathbb{R}^d$, for $i = 1, \ldots, n$, also denoted as the set *set*(X) of elements of X.[1] The literature of parametric statistics, also referred to as statistical inference, will attempt to fit a probabilistic model to the data.[2] The most common idea in statistical inference is to fit a mixture model of normal distributions. In this case, one is inclined to believe that the data $\langle \vec{x}_i \rangle_{i=1,\ldots,n}$ is to be explained by a fixed number k of classes, each of which is distributed according to a multivariate normal distribution $\mathcal{N}_{\vec{\mu}_i,\Sigma_i}$ and that the classes are combined by a vector of non-negative proportions $\vec{\pi}^T = (\pi_1, \pi_2, \ldots, \pi_k)$ with $\sum_{i=1}^{k} \pi_i = 1$. Thus, the model defining the probability distribution is *Prob*$(\vec{x}) = \sum_{i=1}^{k} \pi_i \mathcal{N}_{\vec{\mu}_i,\Sigma_i}(\vec{x})$.

---

[1] Other forms of input for a clustering task are possible, like n object identifiers and a similarity function s that measures the separation $s(i, j)$ between each pair (i, j) of objects.

[2] Other communities, like the field of computational intelligence, may want to fit a Kohonen network, which is not a model defined by a formula and parameters.

Moreover, it is common to assume that each data item was independently drawn from that mixture and that future data will result from such a mixture. What is actually missing are the parameters $\vec{\pi}$, $\vec{\mu}_i$, and Σ_i (for $i = 1, \ldots, k$). Once the parameters are estimated, we can know the proportions of the clusters, their location, and scatter. More importantly, for each individual object, we know the probability that it belongs to the i-th cluster.

Statistical inference elects the "best" choice for the set $(\vec{\pi}, (\vec{\mu}_i, \Sigma_i)_{i=1,\ldots,k})$ guided by the inductive principle of maximum likelihood (*ML*): "chose the model that maximizes the probability of the data being generated by such a model" (Kalbfleisch 1985). For many models, the observed data has probability different from zero, but we asses the best model in proportion to its likelihood. This actually allows an explicit mathematical expression for the inductive principle in terms of an optimization problem (Papadimitriou *et al.* 1982):

$$\text{Maximize } ML(\vec{\pi}, (\vec{\mu}_i, \Sigma_i)_{i=1,\ldots,k}) = Prob(\langle \vec{x}_i \rangle_{i=1,\ldots,n} | \vec{\pi}, (\vec{\mu}_i, \Sigma_i)_{i=1,\ldots,k}) . \quad (8.1)$$

How do we construct an algorithm to solve this multi-variate optimization problem? In fact, what is the common algorithm used in this case? By equating the gradient of $\log ML$ to zero, sufficient (but not necessary) conditions for the optima that lead to an iterative algorithm are obtained (Tanner 1993, Titterington *et al.* 1985). This iterative algorithm converges to a *local optimum* and is known as the "expectation-maximization" method (Dempster *et al.* 1977).

Note that there have been assumptions and choices along the way: the choice of the model (sometimes referred to as the *model bias*),[3] assumptions about the independence of the samples, and also about the choice of method to find approximate solutions to the optimization problem (8.1). These assumptions may lead to misinterpretation, which we will discuss for the example of the algorithm called k-means.

The k-means method is a very fast algorithm for clustering; some believe it is capable of recuperating the parameters of a mixture of multi-variate normals as long as these normals have equal covariance matrices. In fact, this is incorrect, because k-means is statistically biased. That is, the result for k-means even in data drawn independently from $\sum_{i=1}^{k} \pi_i \mathcal{N}_{\vec{\mu}_i, \Sigma_i}(\vec{x})$ will not properly estimate $\vec{\mu}_i$.

But what is the induction principle that guides the k-means algorithm? What if we are convinced that each data item belongs to one class and one class

[3]In the machine-learning community any process that precludes some representation, structure, shape, or form for the possible models and focuses on a narrower family of models is called bias. This is so because models are as varied as support vector machines, Bayesian networks, and many others – beyond the mathematical models (defined by a formula and quantitative parameters) used by traditional statistical inference.

only, instead of having a degree of membership to each cluster assessed by a probability? Because we are to *partition* a set into more homogeneous clusters, we need to asses homogeneity.

The starting point for this purpose is the multivariate analysis of variance through the total scatter matrix T. A traditional measure of the size of this matrix is its *trace*. The trace of a square and symmetric matrix is just the sum of its diagonal elements, and thus minimizing $trace[T]$ is exactly the least sum of squares loss function known as L_2 (Rousseeuw and Leroy 1987):

$$\text{Minimize } L_2(C) = \sum_{i=1}^{n} \text{EUCLID}^2(\vec{x}_i, \mathit{Rep}[\vec{x}_i, C]), \tag{8.2}$$

where $\text{EUCLID}(\vec{x}, \vec{y}) = [(\vec{x} - \vec{y}) \cdot (\vec{x} - \vec{y})^T]^{1/2} = |\sum_{j=1}^{d}(x_j - y_j)^2|^{1/2}$ is the Euclidean metric, $C = \{\vec{c}_1, \ldots, \vec{c}_k\}$ is a set of k *centers*, or representative points in $\mathbb{R}^d$, and the point $\mathit{Rep}[\vec{x}_i, C]$ is the closest point in C to $\vec{x}_i$. The optimization problem (8.2) is then the search for a set C of k representatives. This is illustrative of representative-based clustering: the partition into clusters is defined by assigning each $\vec{x}_i$ to its representative $\mathit{Rep}[\vec{x}_i, C]$.

The k-means method produces an approximate solution to optimization problem (8.2) (rather than optimization problem (8.1)) by iteratively refining the partition encoded by the representatives. The inductive principle that is formalized in optimization problem (8.2) says "*pick the model (set of k representatives) that minimizes the total squared error.*" In this way, k-means returns as the representative of each class the mean of those data items it has labeled as members of such class (and by doing so, it returns a maximum-likelihood estimator per class).

More serious is the fact that the choice of k is crucial and that the mean is a very sensitive estimator of location (as opposed to the median, which is a robust estimator of location). Other critical issues with k-means derive from the initial choice of centers to start the iteration process, because the algorithm is performing a hill-climbing heuristic on optimization problem (8.2) (which is known to be computationally intractable). It is also well known that k-means is susceptible to noise or to the curse of dimensionality (i.e., when d is large, the Euclidean distance is hardly informative); and it is only applicable to "spherical" clusters.

Nevertheless, k-means remains one of the top 10 most used algorithms (Wu *et al.* 2008) and among the two in this category applied for clustering (the other one being "expectation–maximization"). Moreover, k-means is so much faster than many other clustering algorithms, that it remains a very popular choice. However, in experiments with over 20 datasets commonly used by the knowledge discovery community, the best answer found by 1,000 independently initialized k-means runs was inferior to the worst answer found by a five independently started runs of a genetic algorithm, obviously with respect to optimization problem (8.2) (Rahman and Islam 2014). What does this say about the suitability of k-means for realistic data, or about its facility to converge to inferior local optima?

There is a plethora of clustering algorithms as there are many inductive principles, i.e., many rational choices to impose or suggest structure in the data. This has been mostly explained as the fact that clustering is in the eye of the beholder, and that even alternative clusterings may be possible for the same observations (Estivill-Castro 2002). This diversity of inductive principles is multiplied as each one may lead to an alternative concept of what is the "best" model. And then it is multiplied even further by several computational methods to approximate such criteria – all for the same inductive principle. Most of this challenge is usually explained under the banner that clustering (or unsupervised learning) is an ill-defined problem (Siegelmann *et al.* 2001, Lange *et al.* 2004, Bae *et al.* 2006, Shamir and Tishby 2010).

Our first task here is to show that clustering can be analyzed on first principles. We will show that clustering can be provided with a more definite and precise goal somewhat independent of the induction principle and that it has more to do with stability or reproducibility. In this framework, the output of a clustering algorithm is to be regarded as valid if the supervised-learning problem that emerges from it (now the items are labels indicating the class they belong to) is stable. In other words, the accuracy of classifiers is reproducible under perturbations.

8.2 Supervised and Unsupervised Learning

Machine learning is concerned with automating induction procedures that obtain some general hypothesis from a finite set of examples (Thornton 1992, Mitchell 1997). This inductive procedure is, although natural in humans, simply unsound from the perspective of fundamental inference. Therefore, the issue of ensuring some confidence that the learned description is somehow valid became a focus of concern since the very start of the field (Weiss and Kulikowski 1991).

Traditionally, the scenarios of machine learning are both *supervised learning* and *unsupervised learning* (Thornton 1992). The supervised case has as input a dataset $\langle \vec{x}_i \rangle_{i=1,\ldots,n}$ and each case $\vec{x}_i$ has a label $c(\vec{x}_i, i) = c(\vec{x}_i) \in Y$.[4] Su-

[4]In supervised learning it is not unusual for datasets to have two (or more) incompatible

pervised learning aims to find a classifier F that "fits" the data and, thus, has connections with interpolation and regression.

However, the machine-learning approach also aims at generalizing for the future, so there are also connections with extrapolation. Hence, the *goal* of supervised learning is not so much to be correct on the data (i.e., $F(\vec{x}_i) = c(\vec{x}_i)$ is not so important) but to perform "best" in yet unseen examples. In fact, it became rapidly apparent that performance on the supplied data (the so-called training set) was not a good indicator of the quality of the classifier. The accepted notion of "best" performance in the classification problem is usually formalized by assuming (1) some model for the source of the cases, (2) that such model delivers the cases independently with some probability distribution P, and (3) that each case $\vec{x}$ belongs truly to one and only one class $c(\vec{x}) \in Y$. Then, any classifier $F : X \to Y$ that labels an unlabeled case $\vec{x}$ as belonging to $F(\vec{x})$ has a misclassification rate *MS* $= \mathbb{E}[F(\vec{x}) \neq c(\vec{x})]$ (expectation is with respect to P). The goal is to obtain a classifier that minimizes the misclassification rate.[5]

However, a model for the source of the data (and, thus, the probability distribution P) is known only in the simplest cases, and the misclassification rate *MS* can rarely be derived analytically. How, then, can one optimize a problem whose objective function cannot be evaluated? Or more simply, given two classifiers F_1 and F_2, how do we decide that one is better than the other? Since the misclassification rate is an expected value, the natural approach is to sample the distribution repeatedly and derive a simulated mean as the estimator for the distribution's mean.

However, the problem is again that one does not yet have the unseen cases. Assuming that the data $\langle \vec{x}_i \rangle_{i=1,\ldots,n}$ represents the distribution P closely, the notion of cross-validation appears naturally. Today, cross-validation is considered the standard for estimating the accuracy of machine-learning algorithms for supervised learning. In many applications, this estimated accuracy produces useful and very effective comparisons enabling decisions regarding implementations of the many techniques and methods to deploy classifiers (Witten and Frank 2000). Many computer systems (known as knowledge-based systems or recommender

---

examples $(\vec{x}_i, c)$ and $(\vec{x}_i, c')$ with $c \neq c'$. In what follows, we use the relation $c(\vec{x}_i)$ to mean the function $c(\vec{x}_i, i)$, because it is possible for $i \neq j$ to have $\vec{x}_i = \vec{x}_j$ but $c(\vec{x}_i) \neq c(\vec{x}_j)$.

[5]There are actually several sophistications of this formulation. First, the classes may not be crisp but incompatible examples may be present and there could be non-uniform costs associated with some types of classification errors. Second, the universe of models (or acceptable representations for F) may have restrictions. Also, we may take into account how complex the function F is among the permissible models.

systems) use classifiers in applications varying from assigning credit limits to loan applicants, detecting human shapes and gestures in video, scanning for tumors in medical imaging, or routing postage parcels.

On the other hand, the unsupervised case suffers from what many in the literature have characterized as an ill-defined problem (cf. Sections 8.1 and 8.2). Fundamental challenges ("what is a cluster?") remain as relevant now (Jain 2010) as they were two decades back (Jain and Dubes 1988). While the supervised case assumes a function $c(\vec{x})$ and attempts to fit a model $F(\vec{x})$, given the training sequence $\langle(\vec{x}_i, c(\vec{x}_i))\rangle$,[6] in the unsupervised setting we are only presented with the sequence of unlabeled cases $\langle\vec{x}_i\rangle$. Most likely we are performing such learning with no solid grounds for what is the actual (real-world) generator of the examples, and any assumption may actually be unjustifiably biased (e.g., the assumption that the examples are derived from a mixture model of k multivariate normal distributions). What in fact constitutes learning and what is the goal? In many cases, it has been argued that "in reality, a cluster is a subjective entity that is in the eye of the beholder" (Jain 2010).

From its very beginning, the field of knowledge discovery and data mining considered validity as a core property of any outcome (Fayyad and Uthurusamy 1996). How is it possible then to establish some confidence (or "credibility," as Witten and Frank 2000 say) concerning the result delivered by a clustering algorithm? This constitutes a fundamental question: "The statistical problem of testing cluster validity is essentially unsolved" (Duda *et al.* 2001).

The modern taxonomy (Halkidi and Vazirgiannis 2005, Xu and Wunsch 2009) of approaches to verify the validity of a clustering result considers *external criteria*, *internal criteria*, and *relative criteria*. Besides recalling the drawbacks of the existing approaches one could argue that they are imposing structure on the data, rather than letting the data speak for themselves (Openshaw 1999).

We propose to translate the issue of credibility in the output of unsupervised learning algorithms to the case of supervised learning. The fundamental idea is that, no matter what clustering algorithm is used, in the end one desires to obtain a model that can accurately answer the question: "are $\vec{x}_i$ and $\vec{x}_j$ in the same cluster?" When clusterings are partitions, this question has two disjoint answers (yes or no), so that the results of a clustering algorithm can be scrutinized by the tendency by which supervised learning algorithms can discover a suitable classifier.

Our approach refers to the true usability of the found clusters, and it does not fit within the three families of validity criteria mentioned above. Our ap-

[6]We outlined earlier how one can evaluate the quality of this fit, and many other solid alternatives exist (Mitchell 1997, Witten and Frank 2000, Halkidi and Vazirgiannis 2005).

proach has strong similarities with the notion of *clustering stability* (Levine and Domany 2001, Lange *et al.* 2004, Tibshirani and Walther 2005, Ben-David *et al.* 2006, Greene and Cunningham 2006, Ben-David and von Luxburg 2008, von Luxburg 2009, Shamir and Tishby 2008, 2010). We achieve this by introducing a formal notion of "instance easiness" to supervised learning and linking the validity of a clustering to how its output constitutes an easy instance for supervised learning. Our notion of instance easiness extends the notion of stability against perturbations, used to measure "clusterability" in the unsupervised setting (Ackerman and Ben-David 2009).

This relates clearly to the issue of reproducibility: A clustering is valid if it is reproducible in the presence of small perturbations. Our approach is practically applicable and leads us to the axiomatic and generic formulation for cluster-quality measures. As a result, we have an effective method to inform the trust we can place on a clustering result, which profits from the now standard validity methods for supervised learning (such as cross-validation). We then show that our proposal is not only suitable for partitions (or *crisp* clustering algorithms; cf. Atmanspacher, this volume) but applies to fuzzy clustering as well.

8.3 Cluster Validity as Easiness in Classification

8.3.1 Instance Easiness for Supervised Learning

We now present a notion of instance easiness for supervised learning. Intuitively, a supervised problem instance is easy, if classifiers that learn from it obtain high accuracy that is reproducible in another instance where the data has been slightly perturbed. We need some notation to make this precise.

The notion of instance easiness for supervised learning (Estivill-Castro 2011) follows the the notion of *clusterability*, or instance easiness for unsupervised learning. Let $\mathcal{X}$ be some domain set (Ackerman and Ben-David 2008, 2009) and consider an instance of the supervised learning problem given by

1. a sequence of n pairs $\langle(\vec{x}_i, c(\vec{x}_i))\rangle_{i=1}^{n}$, where $X = \langle\vec{x}_1, \ldots, \vec{x}_n\rangle$ is the training sequence of labeled examples ($set(X) \subset \mathcal{X}$), and Y is a finite [7] set of labels (i.e., $\bigcup_{i=1}^{n} c(\vec{x}_i) = Y$),
2. a family $\mathcal{F}$ of models, so that if a function $F \in \mathcal{F}$, then $F : \mathcal{X} \to Y$,[8]

[7] We consider small $|Y| \in \mathbb{N}$, thus focusing on classification, not interpolation or regression.

[8] A relation between sets A and B is any subset of the Cartesian product $A \times B$, but a mathematical function $f : A \to B$ is a relation between A and B with the additional property that $f(a) = b_1$ and $f(a) = b_2$ implies $b_1 = b_2$.

3. a real-valued loss function $\mathcal{L}$.[9]

The goal is to find $F_O \in \mathcal{F}$ that optimizes the loss function. We write $[X, Y]$ for an instance of the supervised learning problem. The loss function is then a function of the instance $[X, Y]$ and the classifier $F : X \to Y$, thus, we write $\mathcal{L}([X, Y], F)$ (Estivill-Castro 2011).

A function $d : \mathcal{X} \times \mathcal{X} \to \mathbb{R}$ is a *distance function* over $\mathcal{X}$ if

1. $d(\vec{x}_i, \vec{x}_i) \geq 0$ for all $\vec{x}_i \in \mathcal{X}$,
2. for any $\vec{x}_i, \vec{x}_j \in \mathcal{X}$, $d(\vec{x}_i, \vec{x}_j) > 0$ if and only if $\vec{x}_i \neq \vec{x}_j$,
3. for any $\vec{x}_i, \vec{x}_j \in \mathcal{X}$, $d(\vec{x}_i, \vec{x}_j) = d(\vec{x}_j, \vec{x}_i)$ (symmetry).

Note that a distance function is more general than a metric, because the triangle inequality is not required (Ackerman and Ben-David 2008).

Definition 8.1 *Two sample sequences X and X' are* ϵ-close *(with respect to a distance function d over $\mathcal{X}$) if there is a bijection $\pi : X \to X'$ so that $d(\vec{x}_i, \pi(\vec{x}_i)) \leq \epsilon$, for all $i = 1, \ldots, n$.*

Following alternative approaches to complexity (Ben-David 2006), we can consider an instance of supervised learning as easy if small perturbations of the sample sequence X result also in small perturbations of the loss.

Definition 8.2 *Let $[X, Y]$ and $[X', Y']$ be two instances of the supervised learning problem. We say they are ϵ-close if*

1. *$Y' \subseteq Y$ (no new classes are introduced),*
2. *there exists $\pi : X \to X'$ such that X and X' are ϵ-close,*
3. *$c(\vec{x}_i) = c(\pi(\vec{x}_i))$ (the label of the each ϵ-close pair is the same).*

That is, the training sets are ϵ-close with corresponding labels and there are no more class labels.

Definition 8.3 *Let $\mathrm{OPT}_{\mathcal{L}}(X, Y)$ be the optimum value of the loss function $\mathcal{L}$ for instance $[X, Y]$; i.e., $\mathrm{OPT}_{\mathcal{L}}(X, Y) = \min\{\mathcal{L}([X, Y], F) \mid F \in \mathcal{F}\} = \mathcal{L}(F_O)$. We say that the instance $[X, Y]$ is* (ϵ, δ)-easy *(with respect to $\mathcal{F}$ and $\mathcal{L}$) if for every classifier $F_0 : \mathcal{X} \to Y$ that optimizes the loss, and for all instances $[X', Y]$ that are ϵ-close to $[X, Y]$, we have*

$$\mathcal{L}([X', Y], F_0) \leq (1 + \delta) OPT_{\mathcal{L}}(X, Y).$$

[9]The loss function cannot be oblivious to the requirement that the classifier be a function in the following sense. Given an instance $[X, Y]$, at least for every classifier F_O that optimizes the loss function $\mathcal{L}([X, Y], F)$, the optimal value $\mathcal{L}([X, Y], F_O)$ cannot be the same to $\mathcal{L}([X', Y], F_O)$ when *set*$(X') \subset$ *set*(X) except that X' contains one or more additional incompatible examples.

The easiness of $[X, Y]$ is with respect to a family $\mathcal{F}$, to a loss $\mathcal{L}$, and to the distance function d in $\mathcal{X}$.

The loss $\mathcal{L}([X, Y], F)$ does not depend on any distance function on $\mathcal{X}$. It is based on the categorical/nominal nature of the set Y, and thus does not change if we rename the classes with any one-to-one function. Such loss functions are isomorphism-invariant. Nothing is assumed about the computability of the loss function expect its existence, like in the example of the misclassification rate *MS* with respect to the probability distribution P.

Using this notation, we rephrase what we mean for an instance $[X, Y]$ of supervised learning to be easy. If some other researcher R were to study data X' that is ϵ-close to X, and each x_i and $\pi(x_i)$ have the same labels, and researcher R could learn with similar accuracy for an inductive principle $\mathcal{F}$, then the instance is easy.

8.3.2 Clustering-Quality Measures Based on Supervised Learning

We now apply instance easiness in supervised learning to measure the quality of the result of a clustering algorithm. The idea is that a clustering algorithm should label the input into classes that should be easy to learn.

A *clustering-quality measure* (CQM) is a function $m[C, X, d]$ that returns a non-negative real number m when it is given a clustering C of the sample sequence X over $(\mathcal{X}, d)$. Many proposals of CQMs have been suggested for providing confidence (or ensuring validity) of the results of a clustering algorithm (Milligan 1981, Puzicha *et al.* 2000, Estivill-Castro and Yang 2003,Yang and Lee 2004).

Definition 8.4 *Given a sample sequence X, a (crisp) k-*clustering *of a sample $X = \langle \vec{x}_1, \ldots, \vec{x}_n \rangle$ is a k-partition, $C = \{C_1, C_2, \ldots, C_k\}$ such that*

- $\bigcup_{j=1}^{k} C_i = \text{set}(X)$, $C_j \neq \emptyset$ *(for all $j \in \{1, \ldots, k\}$) and*
- $C_i \cap C_j = \emptyset$ *(for all $i \neq j$).*

A clustering *of X is a k-clustering of X for some $k \geq 1$.*

A clustering is *trivial* if $|C_j| = 1$ for all $j \in \{1, \ldots, k\}$ or $k = 1$. For $\vec{x}_i, \vec{x}_j \in X$ and a clustering C of X, we write $\vec{x}_i \sim_C \vec{x}_j$ if $\vec{x}_i$ and $\vec{x}_j$ are in the same cluster of C, and we write $\vec{x}_i \not\sim_C \vec{x}_j$ if they are in different clusters.

A *clustering function* for some domain set $\mathcal{X}$ is a function A that takes as inputs a distance function d over $\mathcal{X}$ and produces, for every sample sequence X with *set*$(X) \subset \mathcal{X}$, a clustering $A(X)$ of X as output. As we discussed earlier, typically, such a clustering function is an algorithm that attempts to obtain the optimum of a loss function that formalizes some inductive principle (Cherkassky and Muller 1998, Estivill-Castro 2002). Now, we can specify a concrete goal for clustering (or unsupervised learning). The quality criterion is that the labels that the clustering suggests must represent a supervised-learning problem instance that is easy.

Definition 8.5 *Consider some family $\mathcal{F}$ and a loss function $\mathcal{L}$. Given a clustering $C = \{C_1, \ldots, C_k\}$ of the sample sequence X from $(\mathcal{X}, d)$, the CQM* by classification m_c *is the largest $\epsilon > 0$ so that instance $[X, C]$ is $(\epsilon, 0)$-easy with respect to the distance function d for a supervised-learning instance $[X, C]$ derived from the clustering by $Y = C$ and $c(\vec{x}_i) = C_j$ such that $\vec{x}_i \in C_j$.*

Definition 8.6 *Consider some family $\mathcal{F}$ and a loss function $\mathcal{L}$. Given a clustering $C = \{C_1, \ldots, C_k\}$ of the sample sequence X from $(\mathcal{X}, d)$, the CQM* by pairing m_p *is the largest $\epsilon > 0$ so that instance $[X \times X, \{1, 0\}]$ is $(\epsilon, 0)$-easy with respect to the distance function d for a supervised learning instance $[X \times X, \{1, 0\}]$ derived from the clustering C by $c(\vec{x}_i, \vec{x}_j) = \mathbb{I}[\vec{x}_i \sim_C \vec{x}_j,]$.*

The fact that $\mathcal{F}$ and $\mathcal{L}$ are parameters in these definitions reflects the fact that different researchers may apply different inductive principles and chose different loss functions. The diversity of families $\mathcal{F}$ reflects the freedom to chose diverse approaches to represent the structures that will encode the clusters. It is possible that different clusterings of the same dataset are (equally) valid.

Another virtue of Definitions (8.5) and (8.6) is that the two CQMs satisfy a first-principles approach to cluster validity (Ackerman and Ben-David 2008, Estivill-Castro 2011). Refining earlier work by Kleinberg (2002), measures of cluster quality must comply with four properties. The motivation for these properties goes back to studies of the fundamental properties of clustering algorithms due to Fisher and van Ness (1971).

1. *Scale invariance.* The desire for monotonic-admissible clustering algorithms suggests that scaling the distance between objects (or the similarity measure) should not change the results. Scale invariance means that the output is invariant to uniform scaling of the input.
2. *Invariance under isomorphism.* To complement the notion of monotonic admissibility, a notion of invariance with respect to the distance was defined by Ackerman and Ben-David (2008): Two clusters C and C' of the same domain $(\mathcal{X}, d)$ are *isomorphic* if there exists a distance-preserving isomorphism $\phi : \mathcal{X} \to \mathcal{X}$ such that $\forall \vec{x}_i, \vec{x}_j \in \mathcal{X}$, we have $\vec{x}_i \sim_C \vec{x}_j$ if and only if $\phi(\vec{x}_i) \sim_{C'} \phi(\vec{x}_j)$.
3. *Richness.* Naturally, no partition of the data should be outside the possible results of the clustering algorithm (Jain 2010).

4. *Consistency.* Most authors find it difficult to describe clustering without alluding to consistency: "the objects are clustered or grouped based on the principle of maximizing the inter-class similarity, and minimizing the intra-class similarity" (Han and Kamber 2000, p. 25). Thus, by compressing within-cluster distances and expanding between-cluster distances, the clustering results must remain stable (Jain 2010).

8.3.3 Using the Clustering-Quality Measures $\mathbf{m_p}$ and $\mathbf{m_c}$

The CQMs m_p and m_c can be used to evaluate and compare two clusterings C and C' over the same unsupervised dataset X. For each of the clusterings C and C', we can obtain a supervised problem (after clustering the data has labels). We then measure the instance easiness of each, i.e., we assess the supervised-learning problem resulting from the clusterings in terms of how "easy" they are. Here, easiness corresponds to how stable the supervised-learning instance is to perturbation or, equivalently, how reproducible prediction accuracy in classifiers is despite the data suffers from progressively larger perturbations. This means (Estivill-Castro 2011) to keep those classifiers learned from unperturbed C and C', and then use perturbed data as test data. The classifiers that derive from less-quality clustering degrade their accuracy more rapidly in the presence of the perturbation.

Data: Two clusterings C and C' for same dataset X.

Data: A learning bias, introduced by selecting a supervised-learning approach $(\mathcal{F}, \mathcal{L})$ and a distance d.

Result: Recommendation for which clustering is more valid.

Step 1: Use $(\mathcal{F}, \mathcal{L})$ for learning and cross-validation to obtain an estimate of the accuracy of classifiers for the supervised instances defined by $[C, X]$ and $[C', X]$.

Step 2: Choose a small ϵ and produce a random set X_ϵ that is ϵ-close to X by uniformly choosing $\vec{x}_{i,\epsilon} \in B_{d,\epsilon}(\vec{x}_i) = \{\vec{x} \in \mathcal{X} | d(\vec{x}, \vec{x}_i) \leq \epsilon\}$.

Step 3: Create two supervised instances $[C_\epsilon, X_\epsilon]$ and $[C'_\epsilon, X_\epsilon]$. In C_ϵ, the label of $\vec{x}_{i,\epsilon}$ is the index j of the cluster $C_j \in C$ so that $\vec{x}_i \in C_j$. In C'_ϵ, the label of $\vec{x}_{i,\epsilon}$ is the index j of the cluster $C'_j \in C'$ so that $\vec{x}_i \in C'_j$.

Step 4: Use $(\mathcal{F}, \mathcal{L})$ for learning and cross-validation to obtain an estimate of the accuracy of classifiers for the supervised instances defined by $[C_\epsilon, X_\epsilon]$ and $[C'_\epsilon, X_\epsilon]$.

Step 5. Choose a larger ϵ and repeat from step 2. Once $[C_\epsilon, X_\epsilon]$ or $[C'_\epsilon, X_\epsilon]$ deteriorates to a worse accuracy than in the unperturbed case, label it as the worse cluster.

The goal is to find for which clustering C or C', the value of $\epsilon > 0$ is largest under the condition that the corresponding supervised learning problem remains easy. This method empirically tests the density of the margin in the boundary of the classes.

8.4 Applying Clustering-Quality Measures to Data

CQMs that satisfy scale invariance, isomorphism invariance, richness, and consistency can be combined to produce new measures with these properties. This enriches the set of CQMs since m_c and m_p become generators to produce CQMs. Moreover, the methods to verify accuracy in supervised learning are now well established and many strong and solid implementations exist, such as WEKA (Hall *et al.* 2009).[10] Therefore, the issue of cluster quality can now be simplified. It is not surprising to find statements like: "Evaluation of clusterers is not as comprehensive as the evaluation of classifiers. Since clustering is unsupervised, it is a lot harder determining how *good* a model is" (Bouckaert *et al.* 2010).[11] The CQMs m_p and m_c enable the use of the machinery for evaluating supervised learning in order to tackle the issue of cluster validity without the need of already classified (supervised) instances.

In this chapter, it is impossible to address the differences and similarities with other clustering validity techniques in all detail. Therefore, we will leave aside approaches based on internal, external, and relative criteria; analyses of boundary sparsity; and consensus clustering and rather concentrate on a particular version of stability-based clustering that focuses on *prediction strength*.

8.4.1 Clustering Based on Prediction Strength

The measure *prediction strength* has been shown to be significantly superior to other earlier measures (Tibshirani *et al.* 2001). We contrast the CQMs m_p and m_c against cluster stability by first analyzing the theoretical results regarding the drop of *prediction strength* when the number k of clusters is too large. In fact, this elaborate result is only for a special case and the CQM approach handles this easily.

[10]WEKA is a free open-source software workbench for clustering algorithms and stands for *Waikato Environment for Knowledge Analysis*. It has been incorporated by the machine-learning group of the University of Waikato (New Zealand).

[11]Our intention is by no means demeaning to WEKA, as other recent methods and software for intelligent data analysis like KNIME and R (Berthold *et al.* 2010) display the same unbalance of assessing quality between supervised problems and unsupervised problems.

Data with *spherically separated clusters* (Tibshirani *et al.* 2001) consists of k_0 balls in d dimensions with $d > 1$ such that all centers are separated by at least 4 units (and the data in each component is uniform in the ball). Their Theorem 1 shows that, for k-means, the *prediction strength* $ps(k)$ has a strong drop; namely, $ps(k_0) = 1$ in probability, while for larger k, the *prediction strength* is less than $2/3$ in probability (i.e., $ps(k) \leq 3/2 + o_p(1)$, for $k > k_0$). We present the following result that matches the earlier "asymptotic properties of *prediction strength*".

Lemma 8.1 *With enough data, the CQMs m_c and m_p identify (in probability) the number k_0 of clusters for spherically separated clusters.*

Proof. Consider clustering using k-means with parameter $k = k_O$. Like many of the arguments on stability-based clustering, we also use that (with enough data) k-means converges in probability to centers on each of the balls when k is set to k_0 (Pollard 1982). That is, the algorithm k-means converges (in probability) to the global optimum of its objective function. But then, the instances of supervised learning we propose are easy for an ϵ that is large (an $\epsilon = 1$), since this ϵ still implies linear separability.

Now, for every $k > k_0$, one of the balls will receive more than one centroid by k-means. This ball will have cases from two classes and many values of $\epsilon < 1$ will shift points of this ball randomly across boundaries. This will affect the accuracy of classifiers and the accuracy will deteriorate. In fact, perturbations with $\epsilon = 1$ will not reach the high accuracies of clustering with k_0. □

8.4.2 Studies with Synthetic Data

Now we illustrate the performance of cluster validity by evaluating the CQMs m_p and m_c with synthetic datasets that have been used before to illustrate stability-based clustering techniques (Tibshirani *et al.* 2001, Lange *et al.* 2004). Such datasets were used to show that the number k of clusters was determined by the approach. Each scheme generates 50 datasets. The schemes are as follows.

Data Type I: *null (single cluster) data in 10 dimensions.* Generates 200 data points uniformly distributed over the unit square in 10 dimensions.

Data Type II: *three clusters in two dimensions.* Generates 100 points from a mixture of standard normal variables with proportions (1/4, 1/4, 1/2) and centered at (0,0), (0,5) and (5,-3). That is, the distribution P on $\mathcal{X} \subset \mathbb{R}^2$ is

$$\frac{1}{4}\mathcal{N}\left(\begin{bmatrix} 0 \\ 0 \end{bmatrix}, \vec{I}\right) + \frac{1}{4}\mathcal{N}\left(\begin{bmatrix} 0 \\ 5 \end{bmatrix}, \vec{I}\right) + \frac{1}{2}\mathcal{N}\left(\begin{bmatrix} 5 \\ -3 \end{bmatrix}, \vec{I}\right).$$

Data Type III: *four clusters in three dimensions.* First, four points $\vec{\mu}_i \in \mathbb{R}^3$ are chosen from $\mathcal{N}(\vec{0}, 5\vec{I})$. If they are pairwise separated by 1.0 or more, they are accepted as the centers of Gaussians $\mathcal{N}(\vec{\mu}_i, \vec{I})$ that are the components of a mixture. Then, four independent Bernoulli trials with probability $p = 1/2$ are performed (B_i, for $i = 1, \ldots, 4$). The number n of points in the data sample will be $\sum_{i=1}^{4} (25\mathbb{I}[B_i \text{ is } 0] + 50\mathbb{I}[B_i \text{ is } 1])$ and the proportion of the i-th component is $\frac{1+\mathbb{I}[B_i \text{ is } 1]}{4+\sum_{j=1}^{4}\mathbb{I}[B_j \text{ is } 1]}$.

Data Type IV: *four clusters in 10 dimensions.* Again, the structure of the mixture is chosen first by drawing randomly four points $\vec{\mu}_i \in \mathbb{R}^{10}$, now from $\mathcal{N}(\vec{0}, 1.9\vec{I})$. If they are pairwise separated by 1.0 or more, they are accepted as the centers of Gaussians $\mathcal{N}(\vec{\mu}_i, \vec{I})$ that are the components of a mixture. The size of the data sample and the proportions are chosen with four Bernoulli trials as above.

Data Type V: *two elongated clusters in three dimensions:* Each cluster has 100 points generated as follows. Set $x_i = y_i = z_i = t_i$ for $t_i = -0.5 + i/100$ for $i = 0, \ldots, 100$, then for each point, add white noise with standard deviation 0.1. That is, (x_i, y_i, z_i)+=$\mathcal{N}(\vec{0}, 0.1I)$, for $i = 0, \ldots, 100$. For the second cluster, each point is translated by (x_i, y_i, z_i)+=$(10, 10, 10)$.

While Types I–V represent convex clusters, the literature on stability-based clustering has also explored synthetic data consisting of concentric circles (Levine and Domany 2001, Lange *et al.* 2004, Shamir and Tishby 2008, Jain 2010).

Data Type VI: *three concentric rings in two dimensions:* This is a mixture of three components in polar coordinates, where the angle α is $U[0, 2\pi]$. The radius is distributed as $\mathcal{N}[R, \sigma]$ with (R=4.0, σ=0.2) for the outer ring with proportion 8/14; (R=2.0, σ=0.1) for the middle ring with proportion 4/14; and (R=1.0, σ=0.1) for the innermost ring with proportion 2/14. A dataset consists of 1,400 points obtained in this way.

We used two simple approaches from WEKA (Hall *et al.* 2009) for the supervised-learning algorithm and found no difference in the results. The results are positive for the proposed CQMs. First, by using WEKA's simple k-means on the 50 datasets from Data Type I and on the 50 datasets from Data Type VI, we observe that CQMs m_p and m_c do not recommend any value for k and suggests that there are no clusters (Estivill-Castro 2011). In Fig. 8.1 we see two plots[12] of randomly chosen datasets of each type (for the other 49 datasets, the plots are analogous).

[12]Plots in this section show 95% confidence intervals over 10 random perturbations by the corresponding ϵ producing 10 instances of supervised learning.

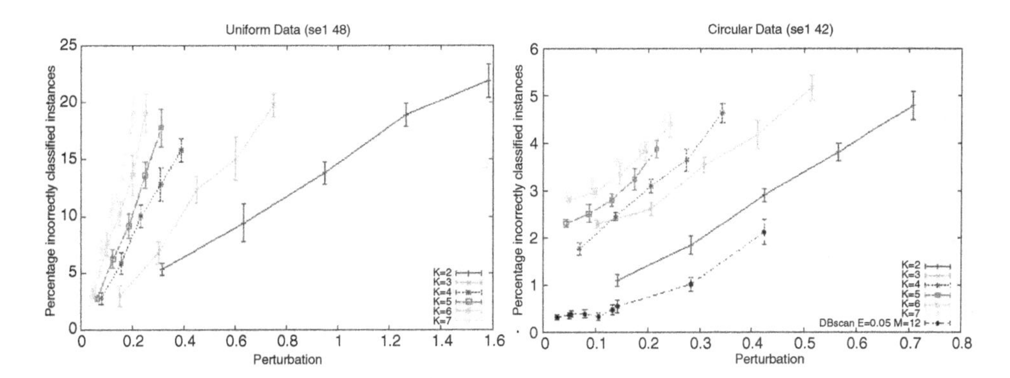

Figure 8.1: Percentage of incorrectly classified instances versus perturbation size for datasets of Type I (left) and of Type VI (right) for $2 < k < 7$. With WEKA's simple k-means there is no preferred value for k (the number of clusters) as the results show no stability of the error rate with respect to perturbation size.

The percentage of incorrectly classified instances grows for each value of k as the magnitude of the perturbation grows. Naturally, for smaller values of k, the rate of deterioration is slower than for larger values of k. Thus, the smallest slope is for $k = 2$, followed by $k = 3$. In general, k has a smaller slope than $k + 1$. This is to be expected because a larger k produces artificial partitions whose boundaries have a narrower margin, so that smaller perturbations will see more data swap sides on the supervised learning instance, causing a larger error rate for supervised learning.

For all 50 datasets of Data Type III, the evaluation of the CQMs m_p and m_c suggests the correct number $k = 4$ of components of the mixture that generated the dataset. This is very impressive in comparison with the performance reported in the literature (Tibshirani *et al.* 2001) of other earlier indexes (including *prediction strength*). Figure 8.2 displays illustrative results for one dataset of Type III on the left. Similarly remarkable outcomes are produced for Type IV. The CQMs method indicates $k = 2$ for 8 of the datasets, $k = 3$ for 13 times $k = 4$ for 24 datasets, and $k = 5$ for four times. Figure 8.2 displays illustrative results for one dataset of Type IV on the right.

For all 50 datasets of Type V the evaluation of CQMs immediately suggests that the clusters obtained by WEKA's simple k-means with $k = 2$ are the most appropriate results. Figure 8.3 presents samples of results for one randomly chosen dataset of each type. In each of the 50 datasets, the CQMs m_p and m_c identify the value of k for which the results by WEKA's simple k-means produces a most appropriate clustering. In fact, the results of Data Type V are very much practical illustrations of Lemma 8.1. The well-separated k clusters result in WEKA's simple k-means placing a centroid on each when k is the

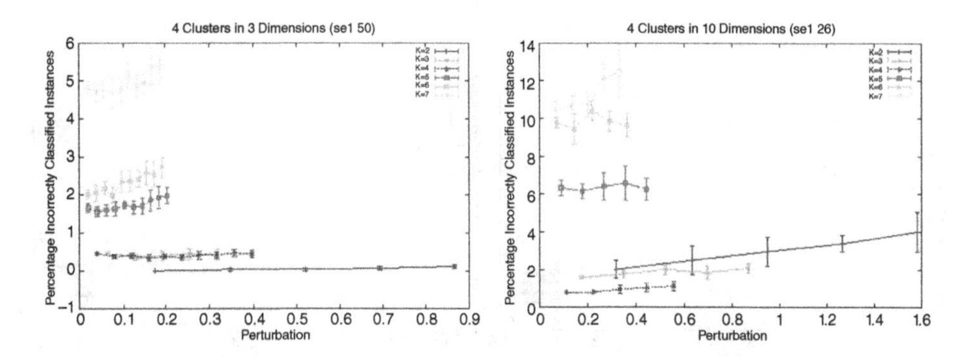

Figure 8.2: Analysis of one dataset of Type III on the left shows $k = 4$ as the largest value of k for which there is stability (in learned classifier accuracy) to perturbations when we consider the clustering results of WEKA's simple k-means as a supervised instance. On the right we have a similar (and typical) result of our analysis, but for dataset of Type IV. Again, $k = 4$ is the model-order identified for WEKA's simple k-means matching the number of components of the mixture generating the data.

number of clouds or fewer. When k is larger than the number of clusters, one of the clusters is artificially divided and this creates a boundary with a dense margin and perturbations result in larger misclassification rates in the resulting supervised instance.

For the 50 datasets of Type II the declaration of the number of clusters is clearly correct and much better than the results reported earlier in the literature.

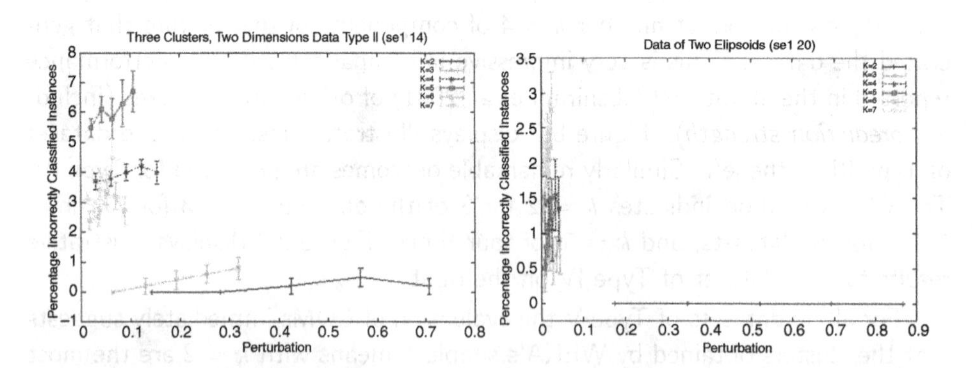

Figure 8.3: Reproducing previous studies, we determine the model order of WEKA's simple k-means for the datasets of Type II (left) and Type V (right). The analysis with the CQM method shows correctly that the largest values of k where there is stability in supervised learning to perturbation are $k = 3$ and $k = 2$, respectively. We illustrate the plot for one randomly chosen dataset out of the 50 of each data type.

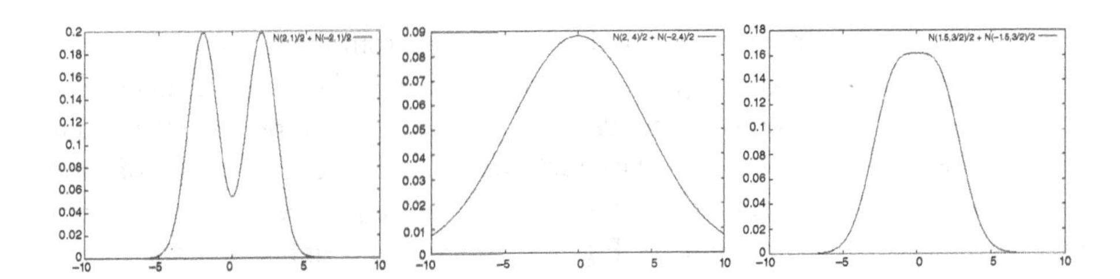

Figure 8.4: Distribution with mixture of two components with proportions $\pi_1 = \pi_2 = 1/2$ as per Eq. (8.3). Left: $\sigma < (\mu_1 - \mu_2)/2$, middle: $\sigma > (\mu_1 - \mu_2)/2$, right: $\sigma = (\mu_1 - \mu_2)/2$.

For Type II the CQMs indicate $k = 2$ for 20 times, $k = 3$ for 20 times, $k = 4$ for five times, and no clusters once.

Our results on these datasets were so impressive that we explored it with some possible limitations of the CQM method, namely, placing the centers of the components closer. If we replace the diagonal value of 5 for 1 in the selection of the four centers $\mu_i \in \mathbb{R}^3$ to generate the datasets, then the results for Type III also show less accurate determination of the number of components in the generating mixture. Among the 50 datasets, 9 times the CQM approach indicates that there are no clusters, while $k = 2$ is suggested for 19 times, $k = 3$ for 12 times, $k = 4$ for 5 times, and $k = 5$ for 5 times.

We explain these results by considering a mixture of two normal distributions in one dimension with centers $\mu_1 > \mu_2$ and equal variance. Then, the mixture is given by

$$\begin{aligned} P(x) &= \pi_1 \mathcal{N}(\mu_1, \sigma) + \pi_2 \mathcal{N}(\mu_2, \sigma) \\ &= \pi_1 \frac{1}{\sqrt{2\pi\sigma^2}} \exp^{(x-\mu_1^2)/2\sigma^2} + \pi_1 \frac{1}{\sqrt{2\pi\sigma^2}} \exp^{(x-\mu_2^2)/2\sigma^2} . \end{aligned} \tag{8.3}$$

For the proportions $\pi_1 = \pi_2 = 1/2$ one can show that the center of mass is at $(\mu_1 + \mu_2)/2$. But, more interestingly, this changes from a maximum or a local minimum of the distribution when $\sigma = (\mu_1 - \mu_2)/2$.

Fig. 8.4 (left) shows the plot of this distribution when $\sigma < (\mu_1 - \mu_2)/2$, while Fig. 8.4 (middle) illustrates the density when $\sigma = (\mu_1 - \mu_2)/2$, and Fig. 8.4 (right) displays an example when $\sigma > (\mu_1 - \mu_2)/2$. From the point of view of nonparametric analysis, it is clear that for the last two cases one could argue that there is only one cluster, while only in the first case one can suspect two clusters.

In fact, many algorithms used in the data-mining community, such as AUTOCLUST (Estivill-Castro and Lee 2002), DBSCAN (Ester *et al.* 1996), DENCLUE (Hinneburg and Keim 2003), and STING (Wang *et al.* 1997), that take a density-based approach will report only one cluster from any data generated by a distribution as per Fig. 8.4 (middle or right). Depending on the algorithm's parameters, these clustering methods would report $k = 1$ or $k = 2$ for data generated as per Fig. 8.4 (left). The fact that the statistical concept of *mode* (although formally more suited to discrete distributions) is relevant in many data analyses supports the idea that for Fig. 8.4 (middle and right) perhaps there is only one cluster.

For the modified ($\sigma = 1$) datasets of Type III there is significant overlap between the clusters. The partition suggested by k-means makes significant error with respect to the original distribution. For instance, if we take the centers of the components of the distribution as the centers of the clusters, then, on average, each cluster with 25 data points has 7 (or 28%) of its points wrongly labeled, and each cluster with 50 data points has 20 (or 40%) of its points wrongly labeled. For cylindrical components of the distribution with $\sigma = 1$, the analysis of the one-dimensional case results in two components merging and producing one mode.

This implies $\mu_1 - \mu_2 = 2$, or that, when the centers of the distributions are close to 2 units apart, the data appears as having one mode less. The value of 2 for this phenomenon decreases as the dimension of the data increases. In the modified dataset of Type III, the centers are approximately one unit apart, so in every dataset, there are at least two components that appear as one. As the dimension grows, the merging of components is much less frequent. For Data Type IV in 10 dimensions, the number of centers corresponds to the number of modes and the CQM method suggests clusterings that identify this. We found that, when the value 5 of the original generation process is replaced by a value as low as 2, datasets generated as described for Type III can still have four modes with the CQM approach.

8.4.3 Studies with Empirical Data

In addition to synthetic data, there have also been studies of stability-based validity on real data, e.g., with a focus on k-means and gene-expression (micro-array) data. We illustrate this with a commonly used dataset called *Leukemia*. Among the 7,130 genes, researchers chose about 50 to 100 as features for clustering (Golub *et al.* 1999, Dudoit and Fridlyand 2002, Lange *et al* 2004).

We analyzed the 50 original genes shown by Golub *et al.* (1999, Fig. 3B), since these are explicitly listed, and applied the same transformation of the input they used. All values were adjusted to the range [100,16000] with values outside the bound corrected to the nearest bound value. Each value was then replaced by $\log_{10}$, and then each of the 50 attributes was standardized (mean zero and

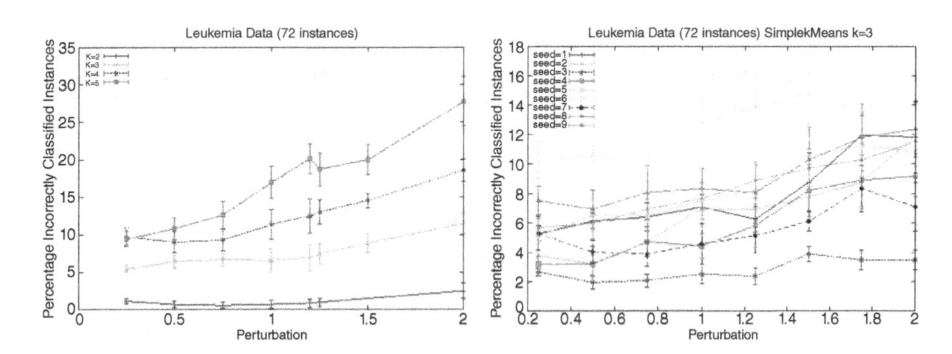

Figure 8.5: Plots for the analysis of the dataset *Leukemia* with WEKA's simple k-means. The classifier is WEKA's NaiveBayes. Left: Our analysis shows the preferred values of $k = 2$ and $k = 3$ since the supervised learning problem shows the same error rate for $k = 2$ until $\epsilon = 2$, while $k = 3$ does not start increasing until $\epsilon = 1.25$. On the other hand, $k = 4$ suffers from as early as $\epsilon = 0.75$. Right: Once $k = 3$ is determined, we analyze the different clustering obtained with different seeds.

variance one). We used WEKA's `weka.filters.unsupervised.attribute.Standardize` for this task.

The mentioned dataset has been used to evaluate methods for the task of determining the number k of clusters. The 72 instances of leukemia correspond to two major classes: acute lymphoblastic leukemia (ALL) or acute myeloid leukemia (AML) with 47 instances of the class ALL and 25 of the class AML. The class ALL has two subclasses: B-cell with 38 exemplars or T-cell with 9. Dudoit and Fridlyand (2002) and Lange *et al.* (2004) showed that the stability index is better or equal-best to methods like *Clest*, the *Gap Statistic*, Levine's FOM, and *prediction strength*. This stability index suggest two or three clusters. Several other indexes fail to suggest these values.

We applied the CQM approach which is not only much faster, but the availability of packages like WEKA (with methods for evaluating supervised learning) facilitates the task. We contrasted the values $k = 2, 3, 4, 5$ with WEKA's simple k-means approach. For each value of k, the algorithm simple k-means was executed with a different random seed. We used WEKA's method `getSquaredError()` to obtain the value of the objective function (the total squared representation error) and select a clustering among those resulting from 10 different seeds. We contrasted the resilience of the resulting supervised learning problems to perturbations (the same ϵ was applied 10 times). Figure 8.5 shows the results as ϵ is increased. Clearly the CQM approach suggests $k = 2$, and also $k = 3$ shows resilience at least until $\epsilon = 1.25$.

With this dataset, even for $k = 3$, we found at least 10 different clusterings emerge for different seeds (starting points) for WEKA's simple k-means. One surprising aspect in early research with *Leukemia* is that no one attempted to identify which of these different clusterings for $k = 3$ is more appropriate. Let us turn to this now.

Figure 8.5 (right) shows the result of applying our method with nine different seeds, where seed 1 is the clustering with smallest total squared error, while seed 9 results in a clustering that matches exactly the ground truth except one instance (this dataset contains its class labels). However, in the CQM method, the preferred clustering is the one with seed 3. This clustering has the worst performance with respect to k-means's objective function among the 10 different clusterings (i.e., has largest total squared error).

If one compares the clustering with the ground truth, clustering with seed 9 (which is equivalent to using the class labels, obtaining the mean of each class and classifying by nearest center) has 98.6% accuracy. But it is not as stable to perturbations as seed 3. The clustering with seed 1 produces three clusters and only achieves 60% accuracy when classifying between T-cell and B-cell (although it achieves 95.8% accuracy between ALL and AML). This is consistent with what was reported by Lange *et al.* (2004).

Why does the CQM method favor the clustering with seed 3? This clustering separates AML and ALL with 97.2% accuracy. In fact the cluster sizes are 45 instances of ALL, 25 instances of AML, one misclassification, and one instance by itself. This strongly indicates that the data actually reflects two different clusters rather than three. It also reflects that one of the instances is an outlier.

8.5 Other Clustering Models

8.5.1 Hierarchical Clustering

Stability-based clustering previously focused on approaches producing a clustering in one step. This focus on center-based clustering for stability-based validity perhaps prevented a different analysis that we now outline using *Leukemia* as an example. The data reflects well the anticipated partition into two clusters for AML and ALL. However, the low classification accuracy (around 60%–80%) when using $k = 3$ (Lange *et al.* 2004) suggests that k-means is not the most appropriate model. This is indicated by the poor performance in classification between T-cell and B-cell by the clustering with smallest total squared error.

Our analysis suggests that, for $k = 2$, we obtain resilience to perturbation in the corresponding supervised-learning problem (in fact, when $k = 2$, different seeds for simple k-means do not produce much different clusterings). Moreover,

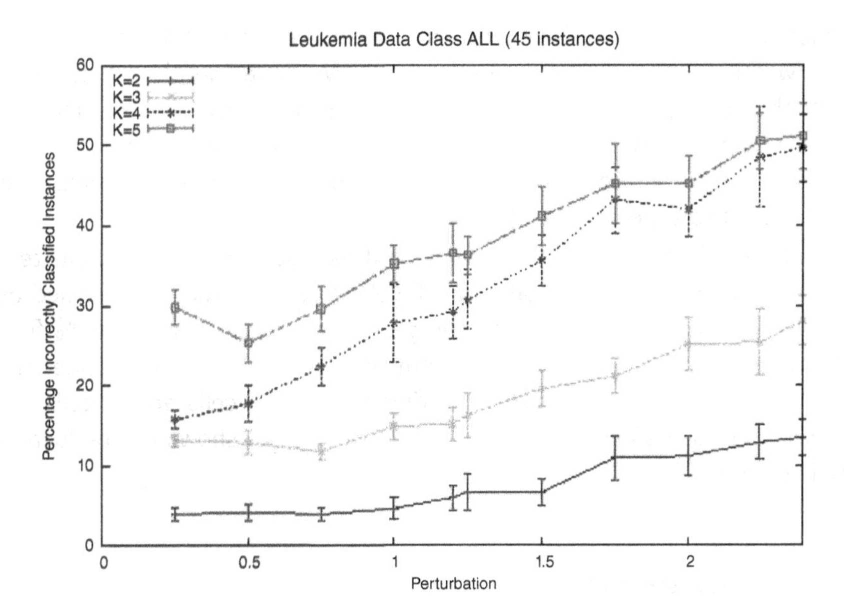

Figure 8.6: For the dataset *Leukemia* and the subclass ALL, obtained with an earlier clustering, we see that simple k-means with $k = 3$ preferrably (and maybe with $k = 2$) yields sub-clusters.

as indicated earlier, the most resilient supervised problem for $k = 3$ essentially coincides with $k = 2$. We propose to use top-down hierarchical clustering and apply $k = 2$ and simple k-means. We obtain two clusters AML and ALL. Each class is analyzed for potential subclusterings, and we continue recursively until the CQM method indicates it is unlikely that we have further clusters.

Figure 8.6 shows a sub-clustering of the class ALL preferably with $k = 3$ even using WEKA's simple k-means (this is the largest value of k that shows stability). The resulting supervised-learning problem is stable with respect to perturbations, and perhaps $k = 2$. Our analysis also reveals that there are no subclusters for the class AML.

We remark then that the structure we have found in the data is a top level split between AML and ALL, no further split for AML but a split into three subclasses for ALL. Moreover, when we test this against the ground truth we found T-cells with 78% accuracy (7 out of 9 instances correct), B-cells are split into two subclasses themselves, but together they are identified with 95% accuracy (36 out of 38 instance correct) and AML with 96% accuracy (24 out if 25 instances correct). This result is far better than any previous analysis of this dataset.

We attribute this to the fact that in previous studies the concern was to check that the numbers 2 or 3 would turn out for the order of k-means because the data was labeled with the ground truth. We have found with the CQM approach that the most prominent information in the data is the partition at the top level between AML and ALL. However, we also found that the data for ALL in fact reflects a sub-clustering of three subclasses, one matching T-cell accurately and two types of B-cells.

In fact, the data reflects this insofar as if we use $k = 2$ for sub-clustering ALL, then many B-cells are classified as T-cells and one T-cell is classified as a B-cell. The accuracy with respect to the ground truth decreases to 67% for T-cells (6 out of 9) and as low as 32% (12 out of 38) for B-cells. No wonder there were issues in classification in earlier studies between T-cells and B-cells, with several B-cells misclassified as T-cells. The data actually shows a sub-clustering of B-cells close to T-cells.

8.5.2 Fuzzy Clustering

So far we have discussed partitions (or so-called *crisp* clustering), but the literature also includes approaches to clustering where the clusters overlap. In this case the space of each cluster is not a partition cell, and thus exemplars may belong to different clusters. Despite the formal definitions of the measures m_p and m_c of cluster quality, this does not prevent the approach to be applied to probability-based clustering or fuzzy clustering.

The construction of a supervised instance from the resulting overlapping clustering can be achieved rather easily because all clustering algorithms represent membership of an instance to a cluster with a probability, or a degree of membership. It is then straightforward to construct a *crisp* supervised instance by assigning each case to the cluster of highest membership. For example, the algorithm EM in WEKA implements expectation–maximization, and the algorithm is in a subclass of `Clusterer`; therefore, the method `clusterInstance()` classifies a given instance by choosing the most likely cluster. For illustration, we can apply the CQM approach in this way to the mixture data (Estivill-Castro 2011, Fig. 1) with EM as the clustering algorithm with $k \in \{2, 3, 4, 5, 6\}$ to identify the number of clusters.

Figure 8.7 shows that most values of k offer smooth deterioration, but $k = 4$ is stable and achieves very low error rate. In fact, this dataset is suited for EM and the clusters for $k = 6$ are the same as for $k = 4$ (despite different starting seeds) except with two additional clusters each with fewer than 1% of the data points. Because of the high jump in incorrect classifications with $k = 5$, the CQM approach suggests correctly $k = 4$.

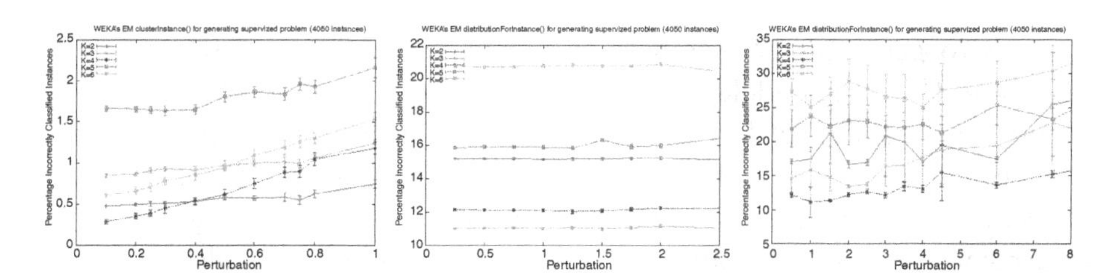

Figure 8.7: Identification of the number of clusters (components) in the mixture learned by EM on the simple data of Estivill-Castro (2011, Fig. 1). Left: Resilience to perturbation of NaiveBayes applied to a supervised-learning problem by forcing a crisp clustering of EM results selecting the most likely cluster. Middle: Resilience to perturbation of NaiveBayes applied to several supervised learning problems by choosing a class as per the distribution of EM results. Right: Resilience to perturbation of Voronoi partition classification applied to several supervised-learning problems by choosing a class as per the distribution of EM results.

However, the match between the CQM approach and assessing quality in probability-based clustering is more significant. For example, in WEKA's EM, although computationally costly, one can skip specifying the number of clusters (components). In that case, EM selects the number k of clusters by starting with $k = 1$ and iteratively evaluating the logarithm of the likelihood between k and $k+1$. As long as the value increases, the value of k increases. The logarithm of the likelihood is the sum over all instances of values proportional to membership probabilities. Membership is evaluated by 10-fold cross-validation classifying by the model.

That is, in this case, the objective function (loss function) optimized by the iterative local search of EM is the overall likelihood (Witten and Frank 2000). WEKA takes this objective function as the quality assessment of the clustering results. To evaluate it, the supervised-problem quality is estimated by the standard 10-fold cross-validation, where the classifier is the clustering model (Bouckaert *et al.* 2010, Witten and Frank, 2000).

This suggest that the CQM approach could follow analogous lines. That is, given a probabilistic-based clustering or a fuzzy clustering, we create supervised problems by assigning the class of each instance randomly but with a distribution as per the clustering result. We can create many of these supervised instances, and many random perturbations of each, also for each magnitude of perturbation ϵ. We then take the average accuracy we can reach of each ϵ, and we examine the resilience (lack of deterioration of classification accuracy) to larger ϵ. When this happens, we have a rather stable cluster that is also learnable, but we do not have to use the clustering result as the classifier.

Figure 8.7 (middle and right) show our results when we apply this sampling approach to generate the instances of supervised learning. If the classifier is NaiveBayes (Fig. 8.7, middle), then almost all values of k show little variation on a range of values of ϵ but $k = 3$ and $k = 4$ display clearly much less classification error. We notice here that the classifier's complexity (its size) grew. This opens another avenue to measure instance easiness of supervised learning, not only by the accuracy of classifiers, but also by the complexity of the model.[13]

In Fig. 8.7 (right), we used a Voronoi approach for nearest-neighbor classification limiting the number of data points that it can use to represent the classifier. By regulating, in this way, the complexity of the classifier, we see more variance but again, the CQM approach clearly shows $k = 4$ as the most resilient value to perturbation (relative to increased ϵ values) of the supervised-learning problems that result from the clustering.

Algorithms like fuzzy-c-means do not have an implementation in WEKA; thus we implemented this algorithm directly as described (Cherkassky and Muller 1998, p. 414). This algorithm, although each data item has a degree of membership to a cluster, is distance-based and computes as many centroids as clusters. Thus we create a supervised-learning instance by assigning each data to the cluster of its nearest centroid. Testing our approach for fuzzy-c-means (and another algorithm based on distances, like harmonic-k-means (Zhang *et al.* 2000) or those based on medians (Estivill-Castro and Houle 2001, Estivill-Castro and Yang 2004)), we found that they all produced results completely analogous to Fig. 8.7 (middle), determining $k = 4$ as the number of clusters for the mixture data in Estivill-Castro (2011).

8.6 Summary

Despite decades of research, the fundamental issue of clustering validity has remained unsolved – so much so that widely accepted software like WEKA (or KNIME or R) has minimal tools and resources for it. This contrasts with the large set of tools for validity for supervised learning. The present paper enables us to use the set of tools designed for the supervised case in the unsupervised case.

The intuition behind our work is a simple idea. When we have to discover groups (classes) in a dataset without corresponding information, whatever results we get must be assessed for validity. A clustering algorithm's output must be evaluated, and external validity approaches are mandatory since, if we had

[13] Complexity of the model, and not only accuracy, is also commonly used to contrast between two classifiers that have learned from a supervised problem. Preference is given to the less complex model as it is generalizing accurately better.

knowledge of the *true* clustering, "why would we be trying to find it?" (Jain 2010).

We expect that the classes obtained by the clustering function are in some way separable and constitute meaningful concepts. They should be robust to small perturbations. The classification rate must be reproducible. A classifier obtained from corresponding supervised-learning results should have a performance that degrades rather slowly when presented with data that is close. Such data can be obtained by perturbations and then the robustness of the classifier can be measured by the now standard approaches of supervised learning. We should not be discouraged by examples of failing cluster validity in some cases (see, e.g., the discussion by Ben-David and von Luxburg (2008, Fig. 1). This should not be surprising, learning from finite data is an unsound inference, and one could have easily attempted to fit the data to an inappropriate model.

Theoretical foundations have been provided (Estivill-Castro 2011) by formalizing a notion of instance easiness for supervised clustering and then deriving measures of cluster quality. Arguably, the m_p and m_c measures are instances of a Hamming distance between two clusterings or, alternatively, a detailed formulation of a 0–1 loss function (Duda *et al.* 2001).

These measures satisfy the generic properties of richness, scale invariance, isomorphism invariance, and consistency that are common to many measures. We showed that they do provide very competitive validity results at also very competitive algorithmic complexity. This approach assists the identification of model order (see our result on finding k for the algorithm k-means) as well as model selection. Thus, this approach enables a practically and theoretically useful mix.

References

Ackerman, M., and Ben-David, S. (2008): Measures of clustering quality: A working set of axioms for clustering. In *Advances in Neural Information Processing Systems 22 NIPS*, ed. by D. Koller, D. Schuurmans, Y. Bengio, and L. Bottou, MIT Press, Cambridge, pp. 121–128.

Ackerman, M., and Ben-David, S. (2009): Clusterability: A theoretical study. *Journal of Machine Learning Research – Proceedings Track* **5**, 1–8.

Bae, E., Bailey, J., and Dong, G. (2006): Clustering similarity comparison using density profiles. In *Advances in Artificial Intelligence*, ed. by A. Sattar and B.H. Kang, Springer, Berlin, pp. 342–351.

Ben-David, S. (2006): Alternative measures of computational complexity with applications to agnostic learning. In *Theory and Applications of Models of Computation*, ed. by J.-Y. Cai, S.B. Cooper, and A. Li, Springer, Berlin, pp. 231–235.

Ben-David, S., and von Luxburg, U. (2008): Relating clustering stability to properties of cluster boundaries. In *Proceedings of the 21st Annual Conference on Learning Theory*, ed. by R.A. Servedio and T. Zhang, Springer, Berlin, pp. 379–390.

Ben-David, S., von Luxburg, U., and Pál, D. (2006): A sober look at clustering stability. In *Proceedings of the 19th Annual Conference on Learning Theory*, ed. by G. Lugosi and H.-U. Simon, Springer, Berlin, pp. 22–25.

Berthold, M.R., Borgelt, C., Höppner, F., and Klawonn, F. (2010): *Guide to Intelligent Data Analysis: How to Intelligently Make Sense of Real Data*, Springer, Berlin.

Bouckaert, R.R., Frank, E., Hall, M., Kirkby, R., Reutemann, P., Seewald, A., and Scuse, D. (2012): *WEKA Manual for Version 3-6-2*, University of Waikato. Available at `www.nilc.icmc.usp.br/elc-ebralc2012/minicursos/WekaManual-3-6-8.pdf`.

Cherkassky, V., and Muller, F. (1998): *Learning from Data – Concept, Theory and Methods*, Wiley, New York.

Dempster, A.P., Laird, N.M., and Rubin, D.B. (1977): Maximum likehood from incomplete data via the EM algorithm. *Journal of the Royal Statistical Society B* **39**, 1–38.

Duda, R.O., Hart, P.E., and Stork, D.G. (2001): *Pattern Classification*, Wiley, New York.

Dudoit, S., and Fridlyand, J. (2002): A prediction-based resampling method for estimating the number of clusters in a dataset. *Genome Biology* **3**(7), 1–21.

Ester, M., Kriegel, H.P., Sander, S., and Xu, X. (1996): A density-based algorithm for discovering clusters in large spatial databases with noise. In *Proceedings of the 2nd International Conference on Knowledge Discovery and Data Mining*, ed. by E. Simoudis, J. Han, and U. Fayyad, AAAI Press, Menlo Park, pp. 226–231.

Estivill-Castro, V. (2002): Why so many clustering algorithms – A position paper. *SIGKDD Explorations* **4**, 65–75.

Estivill-Castro, V. (2011): The instance easiness of supervised learning for cluster validity. In *New Frontiers in Applied Data Mining*, ed. by L. Cao, J.Z. Huang, J. Bailey, Y.-S. Koh, and J. Luo, Springer, Berlin, pp. 197–208.

Estivill-Castro, V., and Houle, M.E. (2001): Robust distance-based clustering with applications to spatial data mining. *Algorithmica* **30**(2), 216–242.

Estivill-Castro, V., and Lee, I. (2002): Argument free clustering for large spatial point-datasets. *Computers, Environment and Urban Systems* **26**(4), 315–334.

Estivill-Castro, V., and Yang, J. (2003): Cluster validity using support vector machines. In *Data Warehousing and Knowledge Discovery*, ed. by Y. Kambayashi, M.K. Mohania, and W. Wöß, Springer, Berlin, pp. 244–256.

Estivill-Castro, V., and Yang, J. (2004): Fast and robust general purpose clustering algorithms. *Data Mining and Knowledge Discovery* **8**(2), 127–150.

Fayyad, U., and Uthurusamy, R. (1996): Data mining and knowledge discovery in databases. *Communications of the ACM* **39**(11), 24–26.

Fayyad, U., Piatetsky-Shapiro, G., and Smyth, P. (1996): The KDD process for extracting useful knowledge from volumes of data. *Communications of the ACM* **39**(11), 27–34.

Fisher, L., and van Ness, J.W. (1971): Admissible clustering procedures. *Biometrika* **58**(1), 91–104.

Golub, T.R., Slonim, D.K., Tamayo, P., Huard, C., Gaasenbeek, M., Mesirov, J.P., Coller, H., Loh, M., Downing, J.R., Caligiuri, M.A., Bloomfield, C.D., and Lander, E.S. (1999): Molecular classification of cancer: Class discovery and class prediction by gene expression monitoring. *Science* **286**, 531–537.

Greene, D., and Cunningham, P. (2006): Efficient prediction-based validation for document clustering. In *Machine Learning*, ed. by J. Fürnkranz, T. Scheffer, and M. Spiliopoulou, Springer, Berlin, pp. 663–670.

Halkidi, M., and Vazirgiannis, M. (2005): Quality assessment approaches in data mining. In *The Data Mining and Knowledge Discovery Handbook*, ed. by O. Maimon and L. Rokach, Springer, Berlin, pp. 661–696.

Hall, M., Frank, E., Holmes, G., Pfahringer, B., Reutemann, P., and Witten, I.H. (2009): The WEKA data mining software: An update. *SIGKDD Explorations* **11**, 10–18.

Han, J., and Kamber, M. (2000): *Data Mining: Concepts and Techniques*, Morgan Kaufmann, San Mateo.

Hinneburg, S., and Keim, D.A. (2003): A general approach to clustering in large databases with noise. *Knowledge and Information Systems* **5**, 387–415.

Jain, A.K. (2010): Data clustering: 50 years beyond k-means. *Pattern Recognition Letters* **31**, 651–666.

Jain, A.K., and Dubes, R.C. (1998): *Algorithms for Clustering Data*, Prentice-Hall, Englewood Cliffs.

Kalbfleisch, J.G. (1985): *Probability and Statistical Inference Vol. 2: Statistical Inference*, Springer, New York.

Kleinberg, J. (2002): An impossibility theorem for clustering. In *Advances in Neural Information Proesing Systems Vol. 15*, ed. by S. Becker, S. Thrun, and K. Obermayer, MIT Press, Cambridge, pp. 463–470.

Lange, T., Roth, V., Braun, M.L., and Buhmann, J.M. (2004): Stability-based validation of clustering solutions. *Neural Computation* **16**, 1299–1323.

Levine, E., and Domany, E. (2001): Resampling method for unsupervised estimation of cluster validity. *Neural Computation* **13**, 2573–2593.

Milligan, G.W. (1981): A Monte Carlo study of thirty internal criterion measures for cluster analysis. *Psychometrika* **46**, 187–199.

Mitchell, T.M. (1997): *Machine Learning*, McGraw-Hill, Boston.

Openshaw, S. (1999): Geographical data mining: Key design issues. In *Proceedings of the 4th International Conference on GeoComputation*. Available at `http://www.geocomputation.org/1999/051/gc_051.htm`.

Papadimitriou, C.M., and Steiglitz, K. (1982): *Combinatorial Optimization – Algorithms and Complexity*, Prentice-Hall, Englewood Cliffs.

Pollard, D. (1982): A central limit theorem for k-means clustering. *Annals of Probability* **10**, 919–926.

Puzicha, J., Hofmann, T., and Buhmann, J.M. (2000): A theory of proximity based clustering: Structure detection by optimization. *Pattern Recognition* **33**, 617–634.

Rahman, M.A., and Islam, M.Z. (2014): A hybrid clustering technique combining a novel genetic algorithm with k-means. *Knowledge-Based Systems* **71**, 345–365.

Rousseeuw, P.J., and Leroy, A.M. (1987): *Robust Regression and Outlier Detection*, Wiley, New York.

Shamir, O., and Tishby, N. (2008): Cluster stability for finite samples. In *Advances in Neural Information Processing Systems Vol. 20*, ed. by J.C. Platt, D. Koller, Y. Singer, and S.T. Roweis, MIT Press, Cambridge, pp. 1–8.

Shamir, O., and Tishby, N. (2010): Stability and model selection in k-means clustering. *Machine Learning* **80**, 213–243.

Siegelmann, H., Ben-Hur, A., Horn, D., and Vapnik, V. (2001): Support vector clustering. *Journal of Machine Learning Research* **2**, 125–137.

Tanner, M.A. (1993): *Tools for Statistical Inference*, Springer, New York.

Thornton, C.J. (1992): *Techniques in Computational Learning: An Introduction*, Chapman & Hall, London.

Tibshirani, R., and Walther, G. (2005): Cluster validation by prediction strength. *Journal of Computational & Graphical Statistics* **14**, 511–528.

Tibshirani, R., Walther, G., Botstein, D., and Brown, P. (2001): Cluster validation by prediction strength. Technical report, Stanford University, Department of Statistics.

Titterington, D.M., Smith, A.F.M., and Makov, U.E. (1985): *Statistical Analysis of Finite Mixture Distributions*, Wiley, New York.

von Luxburg, U. (2009): Clustering stability: An overview. *Foundations and Trends in Machine Learning* **2**, 235–274.

Wang, W., Yang, J., and Muntz, R. (1997): STING: A statistical information grid approach to spatial data mining. In *Proceedings of the 23rd International Conference on Very Large Data Bases*, ed. by M. Jarke, Morgan Kaufmann, San Mateo, pp. 186–195.

Weiss, S.M., and Kulikowski, C.A. (1991): *Computer Systems That Learn – Classification and Prediction Methods from Statistics, Neural Nets, Machine Learning and Expert Systems*, Morgan Kaufmann, San Mateo.

Witten, I., and Frank, E. (2000): *Data Mining – Practical Machine Learning Tools and Technologies with JAVA implementations*, Morgan Kaufmann, San Mateo.

Wu, X., Kumar, V., Quinlan, J.R., Ghosh, J., Yang, Q., Motoda, H., McLachlan, G., Ng, A., Liu, B., Yu, P., Zhou, Z.-H., Steinbach, M., Hand, D., and Steinberg, D. (2008): Top 10 algorithms in data mining. *Knowledge and Information Systems* **14**(1), 1–37.

Xu, R., and Wunsch II, D.C. (2009): *Clustering*, Addison-Wesley, Reading.

Yang, J., and Lee, I. (2004): Cluster validity through graph-based boundary analysis. In *Proceedings of the 2004 International Conference on Information and Knowledge Engineering*, ed. by H.R. Hamid and R. Arabnia, CSREA Press, Athens, pp. 204–210.

Zhang, B., Hsu, M., and Dayal, U. (2000): K-harmonic means – A spatial clustering algorithm with boosting. In *Proceedings of the International Workshop on Temporal, Spatial and Spatio-Temporal Data Mining*, ed. by J. Roddick and K. Hornsby, Springer, Berlin, pp. 31–42.

PART III
PHYSICAL SCIENCES

PART III: PHYSICAL SCIENCES
Introductory Remarks

Harald Atmanspacher

The regulative principle of reproducibility seems to be least difficult to manage in the so-called exact sciences, such as physics and parts of chemistry, where the objects of study are inanimate systems. However, this impression becomes less convincing upon closer inspection. Though many quasi-trivial examples, such as free falling bodies in classical mechanics, appear to be perfect instances for the reproducibility of specific values (within *measurement errors*) of particular properties of a system, things become subtler in less trivial scenarios.

For instance, large-scale physical systems as in geophysics and astrophysics offer the fundamental problem that a repetition of an "experiment" under laboratory conditions is obviously outside the range of possibilities. (Or the notion of an experiment is used differently, e.g., in the sense of observational instruments launched on a spacecraft.) Conditions in such large-scale systems are generically uncontrollable, and a successful reproduction of an observational result depends on how stable the observed system is, irrespective of lacking control.

A significant case in point is the relation between the variability of properties within an ecosystem and its stability, a central topic in the so-called "stability-diversity debate" (McCann 2000). Reducing the diversity within a system (as compared to its natural diversity) can destabilize it because an artificially narrow distribution of properties becomes more vulnerable to perturbations. This ramification of overly homogenized samples has been recognized as detrimental for the reproducibility of results in studies in which variability is narrowed in order to secure or even enhance reproducibility (see further examples in part IV).

Large-scale systems typically are many-body or many-particle systems. This raises the question of whether they should be treated in terms of the micro-properties of those many particles or whether one should look for global properties of the macro-system as a whole. Systems in or close to thermal equilibrium are well-understood examples for this situation, for which a whole new set of observables has been defined in thermodynamics: temperature, pressure, entropy, and so on (with further sophistication if thermal phases are coupled to electromagnetic fields, as in magnetohydrodynamics or other applications).

These thermodynamic observables can be related to the moments of distributions over the mechanical micro-observables, referring to properties of ensembles of particles. However, these distributions are not due to measurement errors but reflect the *actual variability* of system properties. In physics, this variability is often characterized by canonical distributions which can be derived and understood from theory.[1] (This is different in complex systems operating far from thermal equilibrium.) For instance, one can calculate the mean kinetic energy of particles from the canonical Maxwell–Boltzmann distribution of their momenta, and this mean kinetic energy is proportional to the temperature of the system.

One further issue in large-scale systems is their inherent tendency to exhibit rare and extreme events (records of which we have only on correspondingly long time scales). This area of research is covered in the contribution by *Holger Kantz* at the *Max-Planck Institute for the Physics of Complex Systems.* If events are rare and extreme, the usual statistical methodologies for data analysis based on limit theorems are in need of refinement – the number of events is *too small* for a law of large numbers to make sense, and the central limit theorem is *less informative* for "large deviations" within a distribution of "normal" events.[2]

This poses various problems to quantitative statements about extreme events (see Albeverio *et al.* 2005 for an overview). It is difficult to estimate their rate of occurrence, it is hard to make predictions about when the next event will happen, and it is problematic to model the time evolution of systems that are shaped by extreme events. All these difficulties are related to instabilities, which in turn derogate the reproducibility of their results.

Another example of a large-scale system is addressed in the article by *Georg Feulner* at the *Potsdam Institute for Climate Impact Research.* Climate change, its causes, and possible reactions to them are not only a most challenging scientific topic, but they are also hotly debated issues in environmental policy making and public discussions. The demands on reproducible results are, therefore, critical, and this is accentuated by the fact that there is only one specimen of our planet to be studied. Hence, climate science can either focus on the current climate system (with all the challenges of incomplete and noisy measurements) or use computer models (which can represent the system as a whole only in-

[1]Canonical distributions such as the classical Maxwell–Boltzmann distribution have their counterparts in quantum statistical physics, where different types of statistics (Fermi–Dirac, Bose–Einstein) pertain to systems with half-valued or integer-valued spin. An important conceptual difference between classical and quantum situations is that classical probabilities characterize our limited knowledge about the individual constituents of an ensemble, while probabilities for quantum systems indicate that chance is fundamental rather than merely ignorance-based.

[2]The field of "large deviations statistics" is concerned with fluctuations far from the center of a distribution, in its tails. See, e.g., Dembo and Zeitouni (1993) for a monograph, or Atmanspacher and Demmel (this volume) with respect to reproducibility in complex systems.

completely) or investigate the climate history (relying on approximate climate reconstructions from so-called proxy data).

A fundamental reason why reproducibility can be problematic in physics is that physical systems can be subject to *intrinsic dynamical instabilities*. It is well known that nonlinear dynamical systems, even with few degrees of freedom, may exhibit non-stationary, non-ergodic behavior, which cannot be exactly reproduced unless their initial conditions could be re-prepared exactly the same. The time interval after which such *chaotic systems* (cf., e.g., Strogatz 1994) become completely unpredictable depends on their stability properties, which can be derived from their equations of motion.

Even more sensitive to reproducibility are systems in which many constituents, each behaving nonlinearly, are coupled with one another (e.g., yielding turbulence and other forms of fullly developed spatiotemporal chaos). Such systems typically show *transient behavior* to an extent that makes it virtually impossible for them to ever relax onto a stable attractor state. Reproducibility becomes a sheer impossibility – not because of lacking external control but for intrinsic dynamical reasons. *Harald Atmanspacher (Collegium Helveticum Zurich) and Gerda Demmel (Nabios GmbH Munich)* discuss this severe complication.

On a more conceptual level, their contribution also outlines two additional basic methodological challenges of complex systems. One of them is the significant role of contexts for how to define complexity. Current literature contains many different definitions of complexity which all have their particular benefits for particular situations. Along those lines, a basic goal of traditional physics, the search for universality, needs to be reconsidered and complemented by the necessity to specify contexts. The other, related issue concerns the characterization of complexity by measures that go beyond purely syntactic information theory and lead us into semantics and pragmatics.

The article by *David Bailey, Jonathan Borwein, and Victoria Stodden* (a computer scientist, a mathematician, and a statistician) refers to problems of reproducibility in scientific computing. It covers the range from basic data analysis (e.g., statistical overfitting, big-data mining) to high-speed and high-precision numerical simulations (e.g., of nonlinear partial differential equations) and to mathematical physics and experimental mathematics. As the authors show, mathematical computing has evolved in ways that often do not meet the high standards required for reproducible results. Typical failures are: no records of the workflow that produced the published computational results, missing or significantly changed code descriptions during a study or after its completion, or computations that are subject to statistical errors or numerical variability that make it diffcult for others to reproduce results.

Finally, an important issue arises if an effect is looked for *somewhere* within a range of values, for instance within an energy spectrum. This is often the case in high-energy physics, if particle masses can only roughly be estimated from theory, or different theories predict different masses. If an effect, e.g., a resonance, is in fact found, its significance has to be corrected for the likelihood that it could have been *anywhere* within the observed spectrum. This is achieved by the so-called *trials factor*, a correction for *multiple testing* (see Collins, part I, and Stahel, part II) – the ratio between the probability of observing the resonance at a specific value and the probability of observing it anywhere (see, e.g., Gross and Vitells 2010). Without such corrections, the significance of a discovery can be considerably overestimated.

Summarizing: already the physical sciences, where the regulative principles of scientific methodology might be expected to be realized most stringently, present a number of key aspects of and problems with reproducibility:

- the strict reproduction of a measured value within measurement errors,
- the reproduction of a distribution of values with a variance that includes measurement error and variability of values,
- the events to be reproduced may be extreme and rare,
- reproducibility is restricted in intrinsically unstable dynamical systems,
- computations need to be carefully documented and may involve numerical variability and errors.

These aspects reappear in various ways in the life sciences and social sciences, where they will be joined by additional problems. One key issue in this respect (see Shiffrin and Chandramouli, part II) is that the variability of values and corresponding distributions are no longer determined by first principles.

References

Albeverio, S., Jentsch, V., and Kantz, H. (2005): *Extreme Events in Nature and Society*, Springer, Berlin.

Dembo, A., and Zeitouni, O. (1993): *Large Deviation Techniques and Applications*, Jones and Bartlett, Boston.

Gross, E., and Vitells, O. (2010): Trial factors for the look elsewhere effect in high energy physics. *European Physical Journal C* **70**, 525–530.

McCann, K.S. (2000): The diversity-stability debate. *Nature* **405**, 228–233.

Lyons, L. (2008): Open statistical issues in particle physics. *Annals of Applied Statistics* **2**, 887–915.

Strogatz, S.H. (1994): *Nonlinear Dynamics and Chaos*, Perseus, Cambridge.

9
Facilitating Reproducibility in Scientific Computing: Principles and Practice

David H. Bailey, Jonathan M. Borwein, and Victoria Stodden

Abstract. The foundation of scientific research is theory and experiment, carefully documented in open publications, in part so that other researchers can reproduce and validate the claimed findings. Unfortunately, the field of scientific and mathematical computing has evolved in ways that often do not meet these high standards. In published computational work, frequently, there is no record of the workflow process that produced the published computational results, and in some cases, even the code is missing or has been changed significantly since the study was completed. In other cases, the computation is subject to statistical errors or numerical variability that makes it difficult for other researchers to reconstruct results. Thus confusion often reigns.
That tide may be changing, though, in the wake of recent efforts that recognize both the need for explicit and widely implemented standards and also the opportunity to do computational research work more effectively. This chapter discusses the roots of the reproducibility problem in scientific computing and summarizes some possible solutions that have been suggested in the community.

9.1 Introduction

By many measures, the record of the field of modern high-performance scientific and mathematical computing is one of remarkable success. Accelerated by relentless advances of Moore's law, this technology has enabled researchers in many fields to perform computations that would have been unthinkable in earlier times. Indeed, computing is rapidly becoming central to both scientific theory and scientific experiment. Computation is already essential in data analysis, visualization, interpretation, and inference.

The progress in performance over the past few decades is truly remarkable, arguably without peer in the history of modern science and technology. For example, in the July 2014 edition of the top 500 list of the world's most powerful supercomputers (see Fig. 9.1), the best system performs at over 30 Pflop/s (i.e., 30 "petaflops" or 30 quadrillion floating-point operations per second), a level that exceeds the sum of the top 500 performance figures approximately 10 years earlier. Note also that a 2014-era Apple MacPro workstation, which features approximately 7 Tflop/s (i.e., 7 "teraflops" or 7 trillion floating-point operations

Reproducibility: Principles, Problems, Practices, and Prospects, First Edition. Edited by Harald Atmanspacher and Sabine Maasen.

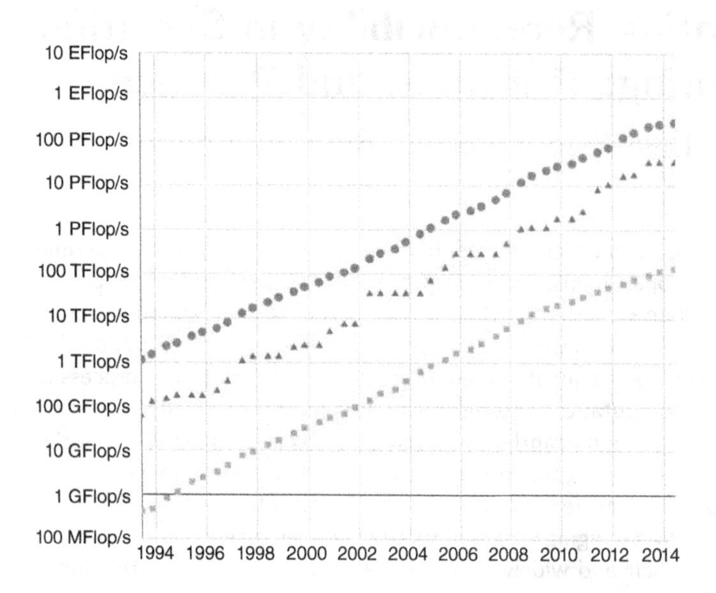

Figure 9.1: Performance development from 1994 to 2014 of the top 500 computers: medium curve = #1 system; lower curve = #500 system; upper curve = sum of #1 through #500. See `top500.org/statistics/perfdevel/`.

per second) peak performance, is roughly on a par with the #1 system of the top 500 list from 15 years earlier (assuming that the MacPro's Linpack performance is at least 15% of its peak performance).

Just as importantly, advances in algorithms and parallel implementation techniques have, in many cases, outstripped the advance from raw hardware advances alone. To mention but a single well-known example, the fast Fourier transform (FFT) algorithm reduces the number of operations required to evaluate the "discrete Fourier transform," a very important and very widely employed computation (used, e.g., to process signals in mobile phones), from $8n^2$ arithmetic operations to just $5n \log_2 n$, where n is the total size of the dataset. For large n, the savings are enormous. For example, when n is one billion, the FFT algorithm is more than six million times more efficient.

Yet there are problems and challenges that must be faced in computational science. One of them is how to deal with a growing problem of non-reproducible computational results. A December 2012 workshop on this topic, held at the Institute for Computational and Experimental Research in Mathematics (ICERM) at Brown University, USA, noted that (Stodden *et al.* 2013)

> science is built upon the foundations of theory and experiment validated and improved through open, transparent communication. With the increasingly central role of computation in scientific discovery, this means

> communicating all details of the computations needed for others to replicate the experiment. ... The "reproducible research" movement recognizes that traditional scientific research and publication practices now fall short of this ideal, and encourages all those involved in the production of computational science ... to facilitate and practice really reproducible research.

From the ICERM workshop and other recent discussions, four key issues have emerged:

1. the need to institute a culture of reproducibility.
2. the danger of statistical overfitting and other errors in data analysis.
3. the need for greater rigor and forthrightness in performance reporting.
4. the growing concern over numerical reproducibility.

In the remainder of this contribution, we shall discuss each of these items and will describe solutions that have been proposed within the community.

9.2 A Culture of Reproducibility

As mentioned above, the huge increases in performance due to hardware and algorithm improvement have permitted enormously more complex and higher fidelity simulations to be performed. But these same advances also carry an increased risk of generating results that either do not stand up to rigorous analysis, or cannot reliably be reproduced by independent researchers. As noted by Stodden *et al.* (2013), the culture that has arisen surrounding scientific computing has evolved in ways that often make it difficult to maintain reproducibility – to verify findings, to efficiently build on past research, or even to apply the basic tenets of the scientific method to computational procedures.[1]

Researchers in subjects like experimental biology and physics are taught to keep careful laboratory notebooks documenting all aspects of their procedures. In contrast, computational work is often done in a much more cavalier manner, with no clear record of the exact "workflow" leading to a published paper or result. As a result, it is not an exaggeration to say that in most published results in the area of scientific computing, reproducing the exact results by an outside team of researchers (or even, after the fact, by the same team) is often impossible. This raises the disquieting possibility that numerous published results may be unreliable or even downright invalid (Donoho *et al.* 2009, Ioannidis 2005).

[1]Franzen (this volume) expands this issue toward scholarly publishing in general, beyond the problems of scientific computation.

The present authors have personally witnessed these attitudes in the field. Colleagues have told us that they cannot provide full details of their computational techniques, since such details are their "competitive edge". Others have dismissed the importance of keeping track of the source code used for published results. "Source code? Our student wrote that, but he is gone now."

The ICERM report (Stodden *et al.* 2013) summarized this problem as follows:

> Unfortunately the scientific culture surrounding computational work has evolved in ways that often make it difficult to verify findings, efficiently build on past research, or even to apply the basic tenets of the scientific method to computational procedures. Bench scientists are taught to keep careful lab notebooks documenting all aspects of the materials and methods they use including their negative as well as positive results, but computational work is often done in a much less careful, transparent, or well-documented manner. Often there is no record of the workflow process or the code actually used to obtain the published results, let alone a record of the false starts. This ultimately has a detrimental effect on researchers' own productivity, their ability to build on past results or participate in community efforts, and the credibility of the research among other scientists and the public.

9.2.1. Documenting the Workflow

The first and foremost concern raised by Stodden *et al.* (2013) was the need to carefully document the experimental environment and workflow of a computation. Even if all of this information is not included in the published account of the work, a record should be kept while the study is underway, and the information should be preserved. Specific recommendations include the following:

1. a precise statement of assertions to be made in the paper,
2. the computational approach, and why it constitutes a rigorous test,
3. complete statements of, or references to, every algorithm employed,
4. auxiliary software (both research and commercial software),
5. test environment (hardware, software and number of processors),
6. data reduction and statistical analysis methods,
7. adequacy of precision level and grid resolution,
8. full statement or summary of experimental results,
9. verification and validation tests performed,
10. availability of computer code, input data, and output data,
11. curation: where are code and data available?

12. instructions for repeating computational experiments,
13. terms of use and licensing; ideally code and data "default to open,"
14. avenues explored and negative findings,
15. proper citation of all code and data used.

With regards to archiving data, McNutt (2015), editor of *Science*, recently announced that the journal is embarking on an effort to identify which data repositories are most reliable – well-managed, long-term support and responsive to community needs – and plans to connect papers submitted to the journal with these repositories. She envisions the day when a graph in a published paper, for instance, could include a digital tag that would lead the reader to the actual data underlying the graph. More effective scientific visualization tools, accessing this archived data, might lead to the discovery of new scientific phenomena.

9.2.2 Tools to Aid in Documenting Workflow and Managing Data

A number of the presentations at the ICERM workshop described emerging tools to assist researchers in documenting their workflow. Here is a brief summary of some of these tools, condensed and adapted from Stodden *et al.* (2013):

- *Literate programming, authoring, and publishing tools.* These tools enable users to write and publish documents that integrate the text and figures seen in reports with code and data used to generate both text and graphical results. This process is typically not interactive and requires a separate compilation step. Some examples here are WEB, Sweave, and knitr, as well as programming-language-independent tools such as Dexy, Lepton, and noweb. Other authoring environments include SHARE, Doxygen, Sphinx, CWEB, and the Collage Authoring Environment.
- *Tools that define and execute structured computation and track provenance.* Provenance refers to the tracking of chronology and origin of research objects, such as data, source code, figures, and results. Tools that record provenance of computations include VisTrails, Kepler, Taverna, Sumatra, Pegasus, Galaxy, Workflow4ever, and Madagascar.
- *Integrated tools for version control and collaboration.* Tools that track and manage work as it evolves facilitate reproducibility among a group of collaborators. With the advent of version control systems (e.g., Git, Mercurial, SVN, CVS), it has become easier to track the investigation of new ideas, and collaborative version control sites like Github, Google Code, BitBucket, and Sourceforge enable such ideas to be more easily shared. Furthermore, these web-based systems ease tasks like code review and feature integration, and encourage collaboration.

- *Tools that express computations as notebooks.* These tools represent sequences of commands and calculations as an interactive worksheet. Examples include both closed-source tools such as *MATLAB* (through the publish and app features), *Maple*, and *Mathematica*, as well as open-source tools such as IPython, Sage, RStudio (with knitr), and TeXmacs.
- *Tools that capture and preserve a software environment.* A major challenge in reproducing computations is installing the prerequisite software environment. New tools make it possible to exactly capture the computational environment. For instance, Docker, VirtualBox, VMWare, or Vagrant can be used to construct a virtual machine image containing the environment. Blueprint analyzes the configuration of a machine and outputs its text description. ReproZip captures all the dependencies, files, and binaries of the experiment and also creates a workflow specification for the VisTrails system in order to make the execution and exploration process easier. Application virtualization tools, such as CDE (Code, Data, and Environment), attach themselves to the computational process in order to find and capture software dependencies. Some computational environments can also be constructed and made available in the cloud, and others feature full workflow tracking. Examples include Synapse/clearScience and HUBzero including nanoHUB.
- *Interactive theorem proving systems for verifying mathematics and computation.* "Interactive theorem proving," a method of formal verification, uses computational proof assistants to construct formal axiomatic proofs of mathematical claims. Examples include coq, Mizar, HOL4, HOL Light, ProofPowerHOL, Isabelle, ACL2, Nuprl, Veritas, and PVS. Notable theorems such as the Four Color Theorem and the Prime Number Theorem have been verified in this way, and Thomas Hales's Flyspeck project, using HOL Light and Isabelle, aimed to obtain a formal proof of the Kepler conjecture. Each one of these projects produces machine-readable and exchangeable code that can be integrated in to other programs. For instance, each formula in the web version of NIST's authoritative Digital Library of Mathematical Functions may be downloaded in TeX or MathML (or indeed as a PNG image) and the fragment directly embedded in an article or other code. This dramatically reduces chances of transcription error and other infelicities being introduced.
- *Post-publication tools persistently connecting data, code, workflows, and articles.* Although digital scholarly objects may be made available, it is quite possible each object may be hosts by different services and may reside at different locations on the web (such as authors' websites, journal

supplemental materials documents, and various data and code repositories).[2]

The development of software tools enabling reproducible research is a rapidly growing area of research. We believe that the difficulty of working reproducibly will be significantly reduced as these and other tools continue to be adopted and improved – and their use simplified. But prodding the scientific computing community, including researchers, funding agencies, journal editorial boards, laboratory managers, and promotion committees, to broadly adopt and encourage such tools remains a challenge (LeVeque *et al.* 2012, Stodden *et al.* 2015).

9.2.3 Other Cultural Changes

While the above recommendations and tools will help, there is also a need to fundamentally rethink the "reward system" in the scientific computing field (and indeed, of all scientific research that employs computing). Journal editors need to acknowledge importance of computational details, and to encourage full documentation of the computational techniques, perhaps on an auxiliary persistent website if not in the journal itself.

Reviewers, research institutions, and, especially, funding agencies need to recognize the importance of computing and computing professionals, and to allocate funding for after-the-grant support and repositories. Also, researchers, journals, and research institutions need to encourage publication of negative results – other researchers can often learn from these experiences. Furthermore, as we have learned from other fields, only publishing "positive" results inadvertently introduces a bias into the field (cf. Ehm and Stahel, both in this volume).

9.3 Statistical Overfitting

The explosion of computing power and algorithms over the past few decades has placed computational tools of enormous power in the hands of practicing scientists. But one unfortunate side effect of this progress is that it has also greatly magnified the potential for serious errors, such as statistical overfitting.

Statistical overfitting, in this context, means either proposing a model for an input dataset that inherently possesses a higher level of complexity than that of the input dataset being used to generate or test it, or else trying many variations

[2]Services such as `ResearchCompendia.org`, `RunMyCode.org` and `ipol.im` attempt to colocate these objects on the web to enable both reproducibility and the persistent connection of these objects (Stodden *et al.* 2015).

of a model on an input dataset and then only presenting results from the one model variation that appears to best fit the data. In many such cases, the model fits the data well only by fluke, since it is really fitting only the idiosyncrasies of the specific dataset in question, and has little or no descriptive or predictive power beyond the particular dataset used in the analysis. Statistical overfitting can be thought of as an instance of "selection bias," wherein one presents the results of only those tests that support well one's hypothesis (Hand 2014).

The problem of statistical overfitting in computational science is perhaps best illustrated in the field of mathematical finance. A frequently encountered difficulty is that a proposed investment strategy or a fund based on this strategy looks great on paper, based on its "backtest" performance, but falls flat when actually fielded in practice.

In the field of finance, a *backtest* is the usage of a historical market dataset to test how a proposed investment strategy would have performed had it been in effect over the period in question. The trouble is that in our present era, when a computer program can easily analyze thousands, millions, or even billions of variations of a given investment strategy, it is almost certain that the optimal strategy, measured by its backtest performance, will be statistically overfit and thus of dubious predictive value.

Bailey *et al.* (2014a,b) derived formulas for (1) relating the number of trials to the minimum backtest length and (2) computing the probability of backtest overfitting. They also demonstrated that under the assumption of memory in markets, overfit strategies are actually somewhat prone to *lose* money.

How easy is it to overfit backtest data? Very easy! If only two years of daily backtest data are available, then no more than seven strategy variations should be tried. If only five years of daily backtest data are available, then no more than 45 strategy variations should be tried (see Fig. 9.2). In general, if a financial analyst or researcher does not report the number of trials N explored when developing an investment strategy, then it is impossible for a third party to properly "reconstruct" the results and correctly ascertain the true level of reward and risk. Indeed, Bailey *et al.* (2014a,b) showed that, given any desired performance goal, a financial analyst just needs to keep running his/her computer program, trying alternative parameters for his/her strategy, and eventually he/she will find a variation that achieves the desired performance goal, yet the resulting strategy will have no statistically valid predictive efficacy whatsoever.

Along this line, backtest overfitting can also be seen as an instance of the "filedrawer problem," i.e., trying many variations of a given algorithm but only reporting the "winner" (Ioannidis 2005, Romano and Wolf 2005, see also Ehm, this volume.

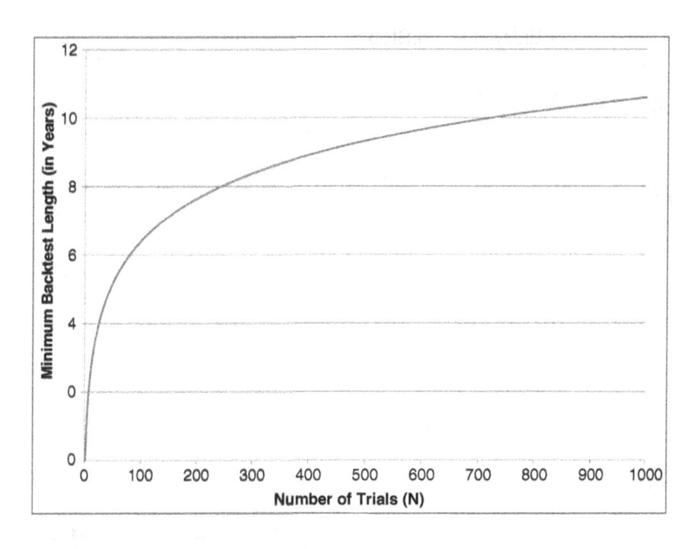

Figure 9.2: Minimum backtest length (in years) versus number of trials.

9.3.1 A Hands-on Demonstration of Backtest Overfitting

To better understand the phenomenon of backtest overfitting, two of the present authors, together with researchers from Boston College and the Lawrence Berkeley National Laboratory, developed an online simulator of backtest overfitting (see `datagrid.lbl.gov/backtest`). It first generates a pseudorandom time series representing, say, several years of daily market prices and then adjusts parameters of a simple investment strategy (monthly entry date, holding period, stop loss, etc.) to produce an "optimal" strategy. However, when this "optimal" strategy is then applied to a second dataset, it typically fails miserably, evidence of severe overfitting.

Figure 9.3 shows the performance of a typical "optimal" strategy, based on an input pseudorandom time series dataset, and Fig. 9.4 shows the performance of this "optimal" strategy on a new dataset. Note that the Sharpe ratio, which is a standard performance measures used in finance, is 1.32 for Fig. 9.3, indicating fairly good performance, while the same statistic for Fig. 9.4 is -0.03, indicating a completely ineffective strategy (see Bailey *et al.* 2014c for additional details).

In short, the field of mathematical/computational finance, like numerous other fields that employ "big data," must recognize the need to pay careful attention to the rules of statistical inference when drawing conclusions. This is not only the case for the special example of overfitting, but applies as well to strategies for pattern detection, such as cluster algorithms (see Estivill-Castro, this volume). Sloppy data analysis no longer suffices.

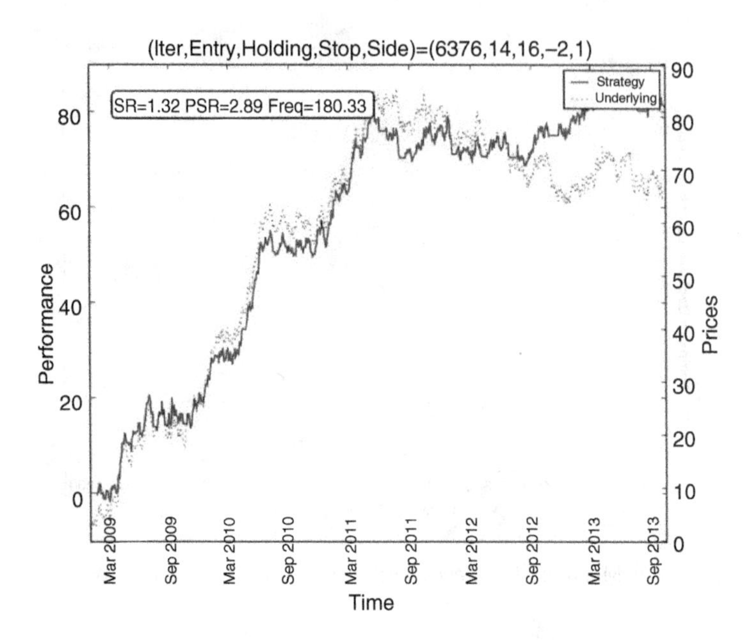

Figure 9.3: Final optimized strategy applied to the input dataset. Note that the Sharpe ratio is 1.32, indicating a fairly effective strategy on this dataset.

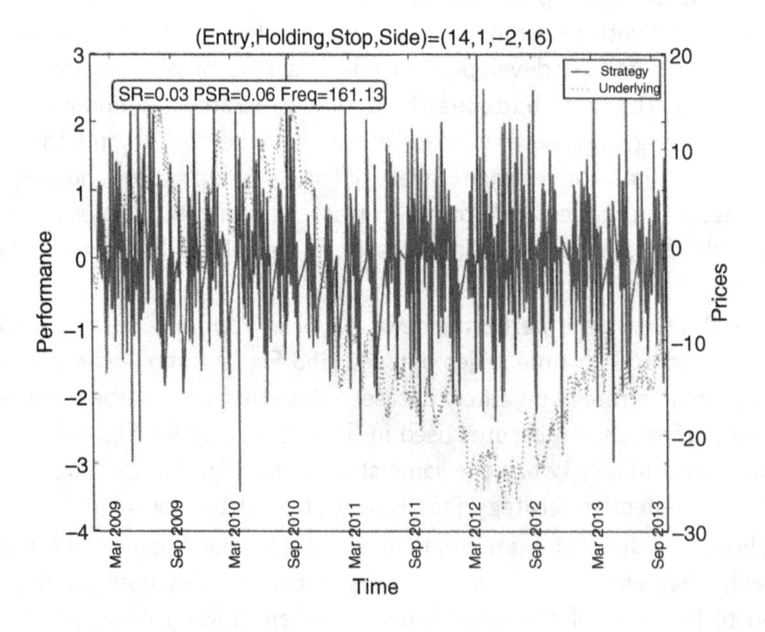

Figure 9.4: Final optimized strategy applied to the new dataset. Note that the Sharpe ratio is -0.03, indicating a completely ineffective strategy on this dataset.

9.3.2 Why the Silence?

The material highlighted above raises the question of why so many in computational science have remained largely silent about work in which, knowingly or not, questionable or non-reproducible practices are employed. In computational finance, this includes those who fail to disclose the number of models used to develop investment strategies, make vague market predictions that do not permit rigorous testing, misuse probability theory and statistics, or employ pseudoscientific charts, graphs or technical jargon (e.g., "Fibonacci ratios," "cycles," "waves," "pivot points," etc.). "Our silence is consent, making us accomplices in these abuses" (Bailey *et al.* 2014a). And ignorance is a poor excuse.

Similarly, in several instances when we have observed instances of reproducibility problems and sloppy data analysis within other arenas of scientific computing (see the next few sections), a surprisingly typical reaction is to acknowledge that this is a significant problem in the field, but that it is futile, unnecessary, or inappropriate to make much fuss about it. Given recent experiences, it is clear that this approach is no longer acceptable.

9.4 Performance Reporting in High-Performance Computing

One aspect of reproducibility in scientific computing that unfortunately is coming to the fore once again is the need for researchers to more carefully analyze and report data on the performance of their codes.

9.4.1 A 1992 Perspective

Some background is in order here. Back in the late 1980s and early 1990s, many new parallel systems had been introduced; each vendor claimed theirs was "best." What's more, many scientific researchers were almost as excited about the potential of highly parallel systems as were the computer vendors themselves. Few standard benchmarks and testing methodologies had been established, so it was hard to reproduce published performance results. Thus much confusion reigned. Overall, the level of rigor and peer review in the field was relatively low, and a number of bad-practice examples (without authors and titles of the respective papers) was listed by Bailey (1992):

- *Scaling performance results to full-sized systems.* In some published papers and conference presentations, performance results on small-sized parallel systems were linearly scaled to full-sized systems, without even clearly disclosing this fact. For example, in several cases 8,192-CPU performance results were linearly scaled to 65,536-CPU results, simply by multiplying by eight. Sometimes this fact came to light only in the question–answer period of a technical presentation. A typical rationale was "We can't afford a full-sized system."

- *Using inefficient algorithms on highly parallel systems.* In other cases, inefficient algorithms were employed for the highly parallel implementation, requiring many more operations, thus producing artificially high performance rates. For instance, some researchers cited partial differential equation simulation performance-based explicit schemes, for applications where implicit schemes were known to be much more efficient. Another paper cited performance for computing a 3-D discrete Fourier transform by direct evaluation of the defining formula, which requires $8n^2$ operations, rather than by using FFT, which requires $5n \log_2 n$ operations. Obviously, for sufficiently large problems, FFT-based computations can produce desired results with vastly fewer operations.
  This is not to say that alternate algorithms should not be employed, but only that when computing performance rates, one should base the operation count on the *best practical serial algorithm*, rather than on the actual number of operations performed (or cite both sets of figures).
- *Not actually performing a claimed computation.* One practice that was particularly troubling was that of not actually performing computations that are mentioned in the study. For example, in one paper the authors wrote in the Abstract that their implementation of a certain application on a CM-2 system (a parallel computer available in the 1992 time frame) runs at "300–800 Mflop/s on a full [64K] CM-2, or at the speed of a single processor of a Cray-2 on 1/4 of a CM-2." However, in the actual text of the paper, one reads that "this computation requires 568 iterations (taking 272 seconds) on a 16K Connection Machine."
  Note that it was actually run on 16,384 processors, not 65,536 processors – the claimed performance figures in the abstract were just the 16,384-CPU performance figures multiplied by four. One then reads that "in contrast, a Convex C210 requires 909 seconds to compute this example. Experience indicates that for a wide range of problems, a C210 is about 1/4 the speed of a single processor Cray-2." In other words, the authors did not actually perform the computation on a Cray-2, as clearly suggested in the abstract; instead they ran it on a Convex C210 and employed a rather questionable conversion factor to obtain the Cray-2 figure.
- *Questionable performance plots.* The graphical representation of performance data also left much to be desired in many cases. Figure 9.5 shows

a plot from one study that compares run times of various sizes of a certain computation on a parallel system, in the lower curve, versus the same computations on a vector system (a widely used scientific computer system at the time), in the upper curve. The raw data for this graph is given in Table 9.1.

In the text of the paper where this plot appears, one reads that in the last entry, the 3:11:50 figure is an "estimate" – this problem size was not actually run on the vector system. The authors also acknowledged that the code for the vector system was not optimized for that system. Note, however, by examining the raw data, that the parallel system is actually slower than the vector system for all cases, except for the last (estimated) entry. Also, except for the last entry, all real data in the graph is in the lower left corner. In other words, the design of the plot leaves much to be desired; a log–log plot should have been used for such data.

problem size (x axis)	parallel system run time	vector system run time
20	8:18	0:16
40	9:11	0:26
80	11:59	0:57
160	15:07	2:11
990	21:32	19:00
9600	31:36	3:11:50

Table 9.1: Run times on parallel and vector systems for different problem sizes (data for Fig. 9.5).

9.4.2 Fast Forward to 2014: New Ways of Bad Practice

Perhaps given that a full generation has passed since the earlier era mentioned above, present-day researchers are not as fully aware of the potential pitfalls of performance reporting. In any event, various high-performance computing researchers have noted a "resurrection" of some of these questionable practices. Even more importantly, the advent of many new computer architectures and designs has resulted in a number of "new" performance reporting practices uniquely suited to these systems. Here are some that are of particular concern:

- Citing performance rates for a run with only one processor core active in a shared-memory multicore node, producing artificially inflated performance (since there is no shared-memory interference) and wasting resources (since most cores are idle). For example, some studies have cited performance

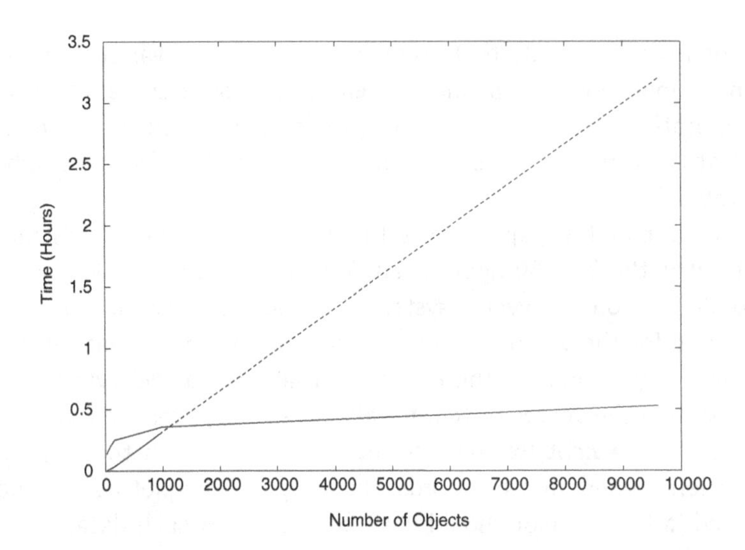

Figure 9.5: Performance plot of run time as a function of number of objects: parallel system (lower curve) vs. vector system (upper curve)]. Note that for all data points from the table except for the last entry, the vector system is *faster* than the parallel system.

on "1,024 cores" of a highly parallel computer system, even though the code was run on 1,024 multicore nodes (each node containing 16 cores), using only one core per node, and with 15 out of 16 cores idle on each node. Note that since this implementation wastes 15/16 of the parallel system being used, it would never be tolerated in day-to-day usage.

- Claiming that since one is using a graphics processing unit (GPU) system, that efficient parallel algorithms must be discarded in favor of "more appropriate" algorithms. Again, while some algorithms may indeed need to be changed for GPU systems, it is important that when reporting performance, one base the operation count on the *best practical serial algorithm*. See the second point in the previous subsection.
- Citing performance rates only for a core algorithm (such as FFT or linear system solution), even though full-scale applications have been run on the system (most likely because performance figures for the full-scale applications are not very good).
- Running a code numerous times, but only publishing the best performance figure in the paper (recall the problems of the pharmaceutical industry from only publishing the results of successful trials).
- Basing published performance results on runs using special hardware, operating system or compiler settings that are not appropriate for real-world production usage.

In each of these cases, note that in addition to issues of professional ethics and rigor, these practices make it essentially impossible for other researchers to reconstruct claimed performance results or to build on past research experiences to move the field forward. After all, if one set of researchers report that a given algorithm is the best-known approach for performing a given application on a certain computer system, other researchers may also employ this approach, but then wonder why their performance results seem significantly less impressive than expected. In other words, reproducible practices are more than just the "right" thing to do; they are essential for making real progress in the field.

9.5 Numerical Reproducibility

The ICERM report also emphasized rapidly growing challenges for numerical reproducibility (Stodden *et al.* (2013):

> Numerical round-off error and numerical differences are greatly magnified as computational simulations are scaled up to run on highly parallel systems. As a result, it is increasingly difficult to determine whether a code has been correctly ported to a new system, because computational results quickly diverge from standard benchmark cases. And it is doubly difficult for other researchers, using independently written codes and distinct computer systems, to reproduce published results.

9.5.1 Floating-Point Arithmetic

Difficulties with numerical reproducibility have their roots in the inescapable realities of floating-point arithmetic. "Floating-point arithmetic" means the common way of representing non-integer quantities on the computer, by means of a binary sign and exponent, followed by a binary mantissa – the binary equivalent of scientific notation, e.g., 3.14159×10^{13}. For most combinations of operands, computer arithmetic operations involving floating-point numbers produce approximate results. Thus roundoff error is unavoidable in floating-point computation.

For many scientific and engineering computations, particularly those involving empirical data, IEEE 32-bit floating-point arithmetic (roughly seven-digit accuracy) is sufficiently accurate to produce reliable, reproducible results, as required by the nature of the computation. For more demanding applications, IEEE 64-bit arithmetic (roughly 15-digit accuracy) is required. Both formats are supported on almost all scientific computer systems.

Unfortunately, with the greatly expanded scale of computation now possible on the enormously powerful, highly parallel computer systems now being deployed, even IEEE 64-bit arithmetic is sometimes insufficient. This is because numerical difficulties and sensitivities that are minor and inconsequential on a small-scale calculation may be major and highly consequential once the computation is scaled up to petascale or exascale size. In such computations, numerical roundoff error may accumulate to the point that the course of the computation is altered (i.e., by taking the wrong branch in a conditional branch instruction), or else the final results are no longer reproducible across different computer systems and implementations, thus rendering the computations either useless or misleading for the science under investigation.

9.5.2 Numerical Reproducibility Problems in Real Applications

As a single example, the ATLAS (acronym for "A Toroidal LHC Apparatus") experiment at the Large Hadron Collider (CERN) must track charged particles to within 10-micron accuracy over a 10-meter run and, with very high reliability, correctly identify them. The software portion of the experiment consists of over five million lines code (C^{++} and Python), which has been developed over many years by literally thousands of researchers.

Recently, some ATLAS researchers reported to us that in an attempt to improve performance of the code, they changed the underlying math library. When this was done, they found some collisions were missed and others were misidentified. That such a minor code alteration (which should only affect the lowest-order bits produced by transcendental function evaluations) had such a large high-level effect suggests that their code has significant numerical sensitivities, and results may even be invalid in certain cases. How can they possibly track down where these sensitivities occur, much less correct them?

As another example of this sort, researchers working with an atmospheric simulation code had been annoyed by the difficulty of reproducing benchmark results (see also Feulner, this volume). Even when their code was merely ported from one system to another, or run with a different number of processors, the computed data typically diverged from a benchmark run after just a few days of simulated time. As a result, it was very difficult even for the developers of this code to ensure that bugs were not introduced into the code when changes are made, and even more problematic for other researchers to reproduce their results. Some divergence is to be expected for computations of this sort, since atmospheric simulations are well known to be fundamentally "chaotic," but did it really need to be so bad?

After an in-depth analysis of this code, He and Ding (2001) found that merely by employing double–double arithmetic (roughly 31-digit arithmetic) in two key summation loops, almost all of this numerical variability disappeared.

With this minor change, benchmark results could be accurately reproduced for a significantly longer time, with very little increase in run time.

Researchers working with large applications in computational physics reported similar difficulties with reproducibility. As with the atmospheric model code, reproducibility of results can be dramatically increased by employing custom high-precision arithmetic in several critical summation loops (Robey *et al.* 2011).[3]

9.5.3 High-Precision Arithmetic and Numerical Reproducibility

As noted above, using high-precision arithmetic (i.e., higher than the standard IEEE 64-bit arithmetic) is often quite useful in ameliorating numerical difficulties and enhancing reproducibility.

By far the most common form of extra-precision arithmetic is known as "double–double" arithmetic (approximately 31-digit accuracy). This datatype consists of a pair of 64-bit IEEE floats (s, t), where s is the 64-bit floating-point value closest to the desired value, and t is the difference (positive or negative) between the true value and s. "Quad-double" arithmetic operates on strings of four IEEE 64-bit floats, providing roughly 62-digit accuracy. Software to support these datatypes is widely available, e.g., the QD package (Hida *et al.* 2001).

Extensions of such schemes can be used to achieve any desired level of precision. Software to perform such computations has been available for quite some time, e.g., in the commercial packages *Mathematica* and *Maple*. However, until 10 or 15 years ago, those with applications written in more conventional languages, such as C^{++} or Fortran-90, often found it necessary to rewrite their codes. Nowadays, there are several freely available high-precision software packages, together with accompanying high-level language interfaces, utilizing operator overloading, that make code conversions relatively painless. A current list of such software packages is given by Bailey and Borwein (2014).

Obviously there is an extra cost for performing high-precision arithmetic, but in many cases high-precision arithmetic is often only needed for a small portion of code, so that the total run time may increase only moderately.

Recently a team led by James Demmel at UC Berkeley have begun developing software facilities to find and ameliorate numerical anomalies in large-scale computations. These include facilities to: test the level of numerical accuracy required for an application; delimit the portions of code that are inaccurate; search

[3]For a discussion of other reproducibility problems in numerical studies of complex systems, see Atmanspacher and Demmel (this volume). See also Zimmerli (this volume) for some background from the philosophy of technology.

the space of possible code modifications; repair numerical difficulties, including usage of high-precision arithmetic; and even navigate through a hierarchy of precision levels (32-bit, 64-bit, or higher) as needed. The current version of this tool is known as "Precimonious". For details see Rubio-Gonzalez (2014); related research on numerical reproducibility is due to Nguyen and Demmel (2013a,b).

9.5.4 Computations Requiring Extra Precision

While it may come as a surprise to some, there is a growing body of important scientific computations that actually *requires* high-precision arithmetic to obtain numerically reliable, reproducible results. Typically such computations involve highly ill-conditioned linear systems, large summations, long-time simulations, highly parallel simulations, high-resolution computations, or experimental mathematics computations.

The following example, condensed from Bailey *et al.* (2012), shows how the need for high-precision arithmetic may arise even in very innocent-looking settings. Suppose one suspects that the data $(1, 32771, 262217, 885493, 2101313, 4111751, 7124761)$ are given by an integer polynomial for integer arguments $(0, 1, \ldots, 6)$. Most scientists and engineers will employ a familiar least-squares scheme to recover this polynomial (Press *et al.* 2007, p. 44).

Performing these computations with IEEE 64-bit arithmetic and the LINPACK routines,[4] with final results rounded to the nearest integer, finds the correct coefficients $(1, 0, 0, 32769, 0, 0, 1)$ (i.e., $f(x) = 1 + (2^{15} + 1)x^3 + x^6$). However, it fails when given the nine-long sequence generated by the function $f(x) = 1 + (2^{20} + 1)x^4 + x^8$. On a MacPro system, e.g., the resulting rounded coefficients are $(1, 6, -16, 14, 1048570, 2, 0, 0, 1)$, which differ badly from the correct coefficients $(1, 0, 0, 0, 1048577, 0, 0, 0, 1)$.

From a numerical analysis point of view, a better approach is to employ a Lagrange interpolation scheme or the more sophisticated Demmel–Koev algorithm (Demmel and Koev 2005). An implementation of either scheme with 64-bit IEEE-754 arithmetic finds the correct polynomial in the degree-8 case, but even these schemes both fail in the degree-12 case. By contrast, merely by modifying a simple LINPACK program to employ double–double arithmetic, using the QD software (Hida *et al.* 2005), all three problems (degrees 6, 8, and 12) are correctly solved without incident. See Bailey *et al.* (2012) for details.

There are numerous full-fledged scientific applications that also require higher precision to obtain numerically reliable, reproducible results. Here is a

[4] LINPACK is a collection of Fortran subroutines that analyze and solve linear equations and linear least-squares problems. It is available at `www.netlib.org/linpack/`.

very brief summary of some applications (see Bailey *et al.* (2012) and Bailey and Borwein (2014) for details):

1. a computer-assisted solution of Smale's 14th problem (18-digit arithmetic; Tucker 2002),
2. obtaining accurate eigenvalues for a class of anharmonic oscillators (18-digit arithmetic; Macfarlane 1999),
3. long-term simulations of the stability of the solar system (18-digit and 31-digit arithmetic; Lake *et al.* 1997, Farres *et al.* 2013),
4. studies of n-body Coulomb atomic systems (120-digit arithmetic; Frolov and Bailey 2003),
5. computing solutions to the Schrödinger equation (from quantum mechanics) for the lithium atom (Yan and Drake 2003) and, using some of the same machinery, to compute a more accurate numerical value of the fine structure constant of physics (approx. $7.2973525698 \times 10^{-3}$) (31-digit arithmetic; Zhang *et al.* 1994).
6. computing scattering amplitudes of collisions at the Large Hadron Collider (31-digit arithmetic; Ellis *et al.* 2009, Berger *et al.* 2008, Ossola *et al.* 2008, Czakon 2008),
7. studies of dynamical systems using the Taylor method (up to 500-digit arithmetic; Brrio 2005, 2006, Barrio *et al.* 2005),
8. studies of periodic orbits in dynamical systems using the Lindstedt–Poincaré technique of perturbation theory, together with Newton's method for solving nonlinear systems and Fourier interpolation (1000-digit arithmetic; Viswanath 2004, Viswanath and Sahutoglu 2010, Abad *et al.* 2011).

9.6 High-Precision Arithmetic in Experimental Mathematics and Mathematical Physics

Very high-precision floating-point arithmetic is essential to obtain reproducible results in experimental mathematics and in related mathematical physics applications (Borwein and Bailey 2008, Borwein *et al.* 2004). Indeed, high-precision computation is now considered one of two staples of the field, along with symbolic computing (see Section 9.7).

Many of these computations involve variants of Ferguson's PSLQ ("partial sum of squares, lower quadrature") integer relation detection algorithm (Bailey and Broadhurst 2000). Given an n-long vector (x_i) of floating-point numbers, the PSLQ algorithm finds the integer coefficients (a_i), not all zero, such that $a_1 x_1 + a_2 x_2 + \cdots + a_n x_n = 0$ (to available precision), or else determines that there is no such relation within a certain bound on the size of the coefficients. Integer relation detection typically requires very high precision, both in the input data and in the operation of the algorithm, to obtain numerically reproducible results.

9.6.1 The BBP Formula for π

One of the earliest applications of PSLQ and high-precision arithmetic was to numerically discover what is now known as the BBP formula for π (Bailey, Borwein and Plouffe 1997):

$$\pi = \sum_{k=0}^{\infty} \frac{1}{16^k}\left(\frac{4}{8k+1} - \frac{2}{8k+4} - \frac{1}{8k+5} - \frac{1}{8k+6}\right).$$

This formula had not been known previously and was discovered using computations with 200-digit arithmetic. What is particularly remarkable about this formula is that after a simple manipulation, it can be used to calculate a string of binary or base-16 digits of π beginning at the n-th digit, without needing to calculate any of the first $n-1$ digits (Borwein and Bailey 2008).

Since the discovery of the BBP formula in 1996, numerous other formulas of this type have been found for other well-known mathematical constants, again by employing the PSLQ algorithm and very high-precision arithmetic. For example, two formulas of this type were subsequently discovered for π^2. One permits arbitrary-position digits to be calculated in base 2 (binary), while the other permits arbitrary-position digits to be calculated in base 3 (ternary).

9.6.2 Ising Integrals

Very-high-precision computations, combined with variants of the PSLQ algorithm, have been remarkably effective in mathematical physics settings. Here is an example of this methodology in action. High-precision software was employed to study the following classes of integrals (Bailey *et al.* 2006). The C_n are connected to quantum field theory, the D_n integrals arise in the Ising theory of mathematical physics, while the E_n integrands are derived from D_n:

$$C_n = \frac{4}{n!}\int_0^{\infty}\cdots\int_0^{\infty}\frac{1}{\left(\sum_{j=1}^{n}(u_j+1/u_j)\right)^2}\frac{du_1}{u_1}\cdots\frac{du_n}{u_n}$$

$$D_n = \frac{4}{n!} \int_0^\infty \cdots \int_0^\infty \frac{\prod_{i<j} \left(\frac{u_i - u_j}{u_i + u_j}\right)^2}{\left(\sum_{j=1}^n (u_j + 1/u_j)\right)^2} \frac{du_1}{u_1} \cdots \frac{du_n}{u_n}$$

$$E_n = 2 \int_0^1 \cdots \int_0^1 \left(\prod_{1 \le j < k \le n} \frac{u_k - u_j}{u_k + u_j} \right)^2 dt_2 \, dt_3 \cdots dt_n.$$

In the last line $u_k = \prod_{i=1}^k t_i$.

In general, it is very difficult to compute high-precision numerical values of n-dimensional integrals such as these. But the C_n integrals can be written as one-dimensional integrals of the form

$$C_n = \frac{2^n}{n!} \int_0^\infty p K_0^n(p) \, dp,$$

where K_0 is the *modified Bessel function*,[5] and such integrals can be evaluated to high precision. For large n, these numerical values approach the limit

$$\lim_{n \to \infty} C_n = 0.6304735033743867961220401927108 \ldots.$$

This numerical value was quickly identified, using the "Inverse Symbolic Calculator 2.0,"[6] as

$$\lim_{n \to \infty} C_n = 2e^{-2\gamma},$$

where $\gamma = 0.5772156649\ldots$ is Euler's constant. This identity was then proven by Bailey *et al.* (2006). Some results were also obtained for the D_n and E_n integrals, although the numerical calculations involved there were much more expensive, requiring highly parallel computation (Bailey and Borwein 2012, Bailey *et al.* 2006). These computations required numerical values accurate to several hundred digits. Other studies in experimental mathematics have required as much as 50,000 digits (Bailey *et al.* 2013).

[5]Compare the digital library of mathematical functions of the National Institute of Standards and Technology of the United States, available at `http://dlmf.nist.gov`.

[6]The *Inverse Symbolic Calculator* is an online (floating point) number checker established in 1995 by P.B. Borwein, J.M. Borwein, and S. Plouffe of the Canadian Center for Experimental and Constructive Mathematics. It is available at `https://isc.carma.newcastle.edu.au`.

9.7 Reproducibility in Symbolic Computing

Closely related to the high-precision computations described in the past two sections is the usage of symbolic computing in experimental mathematics and mathematical physics. At the present time, the commercial packages *Maple* and *Mathematica* are most commonly used for symbolic computations, although other packages, such as *Sage* and *Magma*, are also used by some.

But like all other classes of computer software, symbolic manipulation software has bugs and limitations. In some cases, the software itself detects that it has encountered difficulties, and outputs messages acknowledging that its results might not be reliable. But not always. For example, consider the integral

$$W_2 = \int_0^1 \int_0^1 \left| e^{2\pi i x} + e^{2\pi i y} \right| \mathrm{d}x \, \mathrm{d}y, \tag{9.1}$$

which arose in research into the properties of n-step random walks on the plane (Borwein and Straub 2013). The latest editions of both *Maple* (version 18) and *Mathematica* (version 9) declare that $W_2 = 0$, in spite of the obvious fact that this integral is positive and nonzero. Indeed, $W_2 = 4/\pi$ is the expected distance traveled by a uniform random walk in two steps. It is worth pointing out that Eq. (9.1) can be rewritten as

$$W_2 = \int_0^1 |e^{2\pi i x}| \left(\int_0^1 |1 + e^{2\pi i (y-x)}| \mathrm{d}y \right) \mathrm{d}x \; = \; \int_0^1 |1 + e^{2\pi i y}| \, \mathrm{d}y, \tag{9.2}$$

which is in a form that both *Maple* and *Mathematica* can evaluate, but this "human" observation is not exploited by the commercial packages. Whether this behavior is a "bug" or a "feature" is matter of some dispute, although it is fair to expect the package to notify the user that the integrand function is problematic.

Symbolic computing applications often are quite expensive, with jobs running for many hours, days, or months. For example, it has recently been attempted to update an earlier effort to explore Giuga's 1950 conjecture that an integer $n \geq 2$ is prime if and only if

$$\sum_{k=1}^{n-1} k^{n-1} \equiv -1 \pmod{n}. \tag{9.3}$$

We verified this conjecture for all composite n with up to 4,771 prime factors, which means that any counterexample must have at least 19,907 digits. This multithreaded computation required 95 hours. Increasing this bound significantly would currently take decades or centuries (Borwein *et al.* 2013).

9.8 Why Should We Trust the Results of Computation?

These examples raise the question of why anyone should trust the results of any computation, numeric or symbolic. After all, there are numerous potential sources of error, including user programming bugs, symbolic software bugs, compiler bugs, hardware errors, and input/output errors. None of these can be categorically ruled out. However, substantial confidence can be gained in a result by reproducing it with different software systems, hardware systems, and programming systems or by repeating the computation with detailed checks of intermediate results.

For example, programmers of large computations of digits of π have for many years certified their results by repeating the computation using a completely different algorithm. If all but a few trailing digits agree, then this is impressive evidence that both computations are almost certainly correct. In a similar vein, programmers of large, highly parallel numerical simulations, after being warned by systems administrators that occasional system hardware or software errors are inevitable, are inserting internal checks into their codes. For example, in problems where the computation carries forward both a large matrix and its inverse, programmers have inserted periodic checks that the product of these two matrices is indeed the identity matrix, to within acceptable numerical error.

In other cases, researchers are resorting to formal methods, wherein results are verified by software that performs a computer verification of every logical inference in the proof, back to the fundamental axioms of mathematics. For example, Hales (2005) recently completed a computer-aided proof of the Kepler conjecture, namely, the assertion that the most efficient way to pack equal-sized spheres is the same scheme used by grocers to stack oranges. Some objected to this approach, so Hales (2014) re-did the proof using formal methods. This project is now complete (Hales 2014).

We conclude by briefly commenting on open-source versus commercial software. Some open-source software for mathematical computing, e.g., *GeoGebra*, which is based on *Cabri*, is now very popular in schools as replacement for *Sketchpad*. But the question is whether such software will be preserved when founders and developers lose interest or move on to other projects or employment. This is also an issue with large-scale commercial products, but it is more pronounced with open-source projects.

One advantage of *Maple* over *Mathematica* is that most of the *Maple* source code is accessible, while *Mathematica* is entirely sealed. As a result, it is often difficult to track down or rectify problems that arise. Similarly, *Cinderella* is very robust, unlike *GeoGebra*, mathematically sophisticated – using Riemann surfaces to ensure that complicated constructions do not crash – and only slightly commercial. In general, software vendors and open-source producers do not provide the support that is assumed on the textbook market.

9.9 Conclusions

The advent of high-performance computer systems, all the way from laptops and workstations to systems with thousands or even millions of processing elements, has opened new vistas to the field of scientific computing and computer modeling. It has revolutionized numerous fields of scientific research, from climate modeling and materials science to protein biology, nuclear physics, and even research in mathematics. In the coming years, this technology will be exploited in numerous industrial and engineering applications as well, including "virtual" product testing and market analysis.

However, like numerous other fields that are embracing "big data," scientific computing must confront issues of reliability and reproducibility, both in computed results and also in auxiliary statistical analysis, visualization, and performance. For one thing, these computations are becoming very expensive, both in terms of the acquisition and maintenance costs of the supercomputers being used, but even more so in terms of the human time to develop and update the requisite application programs.

Thus it is increasingly essential to design the workflow process right from the start to ensure reliability and reproducibility – carefully documenting each step, including code preparation, data preparation, execution, post-execution analysis, and visualization, through to the eventual publication of the results. A little effort spent early on will be rewarded with much less confusion and much more productivity at the end.

References

Abad, A., Barrio, R., and Dena, A. (2011): Computing periodic orbits with arbitrary precision. *Physical Review E* **84**, 016701.

Bailey, D.H. (1992): Misleading performance reporting in the supercomputing field. *Scientific Programming* **1**, 141–151.

Bailey, D.H., Barrio, R., and Borwein, J.M. (2012): High-precision computation: Mathematical physics and dynamics. *Applied Mathematics and Computation* **218**, 10106–10121.

Bailey, D.H., and Borwein, J.M. (2015): High-precision arithmetic: in mathematical physics, Mathematics, 3, 337–367, available at `http://www.mdpi.com/2227-7390/3/2/337/pdf`.

Bailey, D.H., Borwein, J.M., and Crandall, R.E. (2006): Integrals of the Ising class. *Journal of Physics A* **39**, 12271–12302.

Bailey, D.H., Borwein, J.M., Crandall, R.E., and Zucker, J. (2013): Lattice sums arising from the Poisson equation. *Journal of Physics A* **46**, 115201.

Bailey, D.H., Borwein, J.M., Lopez de Prado, M., and Zhu, Q.J. (2014a): Pseudo-mathematics and financial charlatanism: The effects of backtest overfitting on out-of-sample performance. *Notices of the American Mathematical Society* **61**(5), 458–471.

Bailey, D.H., Borwein, J.M., Lopez de Prado, M., and Zhu, Q.J. The probability of backtest overfitting. *Journal of Computational Finance*, to appear, 27 Feb 2015.

Bailey, D.H., Borwein, P.B., and Plouffe, S. (1997): On the rapid computation of various polylogarithmic constants. *Mathematics of Computation* **66**, 903–913.

Bailey, D.H. and Broadhurst, D. (2000): Parallel integer relation detection: Techniques and applications. *Mathematics of Computation* **70**, 1719–1736.

Bailey, D.H., Ger, S., Lopez de Prado, M., Sim, A., and Wu, K. (2014c): Statistical overfitting and backtest performance. Available at www.davidhbailey.com/dhbpapers/overfitting.pdf.

Barrio, R. (2005): Performance of the Taylor series method for ODEs/DAEs. *Applied Mathematics and Computation* **163**, 525–545.

Barrio, R. (2006): Sensitivity analysis of odes/daes using the Taylor series method. *SIAM Journal of Scientific Computing* **27**, 1929–1947.

Barrio, R., Blesa, F., and Lara, M. (2005): VSVO formulation of the Taylor method for the numerical solution of ODEs. *Computers & Mathematics with Applications* **50**, 93–111.

Berger, C.F., Bern, Z., Dixon, L.J., Cordero, F.F., Forde, D., Ita, H., Kosower, D.A., and Maitre, D. (2008): An automated implementation of on-shell methods for one-loop amplitudes. *Physical Review D* **78**, 036003.

Borwein, J.M., and Bailey, D.H. (2008): *Mathematics by Experiment: Plausible Reasoning in the 21st Century*, A.K. Peters, Natick.

Borwein, J.M., Bailey, D.H., and Girgensohn, R. (2004): *Experimentation in Mathematics: Computational Paths to Discovery*, A.K. Peters, Natick.

Borwein, J.M., Skerritt, M., and Maitland, C. (2013): Computation of a lower bound to Giuga's primality conjecture. *Integers* **13**. Available at www.carma.newcastle.edu.au/jon/giuga2013.pdf.

Borwein, J.M., and Straub, A. (2013): Mahler measures, short walks and log-sine integrals. *Theoretical Computer Science* **479**, 4–21.

Czakon, M. (2008): Tops from light quarks: Full mass dependence at two-loops in QCD. *Physics Letters B* **664**, 307–314.

Demmel J., and Koev, P. (2005). The accurate and efficient solution of a totally positive generalized Vandermonde linear system. *SIAM Journal of Matrix Analysis Applications* **27**, 145–152.

Donoho, D., Maleki, A., Shahram, M., Stodden, V., and Rahman, I.U. (2009): Reproducible research in computational harmonic analysis. *Computing in Science and Engineering* **11**(1), 8–18.

Ellis, R.K., Giele, W.T., Kunszt, Z., Melnikov, K., and Zanderighi, G. (2009): One-loop amplitudes for $W + 3$ jet production in hadron collisions. *Journal of High Energy Physics* 0901, 012.

Farres, A., Lsaskar, J., Blanes, S., Casas, F., Makazaga, J., and Murua, A. (2013): High precision symplectic integrators for the solar system. *Celestial Mechanics and Dynamical Astronomy* **116**, 141–174.

Frolov, A.M., and Bailey, D.H. (2003): Highly accurate evaluation of the few-body auxiliary functions and four-body integrals. *Journal of Physics B* **36**, 1857–1867.

Hales, T.C. (2005): A proof of the Kepler conjecture. *Annals of Mathematics* **162**, 1065–1185.

Hales, T.C. (2014): The flyspeck project. Available at `code.google.com/p/flyspeck/wiki/AnnouncingCompletion`.

Hand, D.J. (2014): *The Improbability Principle*, Macmillan, New York.

He, Y., and Ding, C. (2001): Using accurate arithmetics to improve numerical reproducibility and stability in parallel applications. *Journal of Supercomputing* **18**, 259–277.

Hida, Y., Li, X.S., and Bailey, D.H. (2001): Algorithms for quad-double precision floating point arithmetic. In *Proceedings of the 15th IEEE Symposium on Computer Arithmetic*, IEEE Computer Society Press, pp. 155–162.

Ioannidis, J. (2005): Why most published research findings are false. *PLoS Medicine* **2**(8), e124.

Lake, G., Quinn, T., and Richardson, D.C. (1997): From Sir Isaac to the Sloan survey: Calculating the structure and chaos due to gravity in the universe. In *Proceedings of the 8th ACM-SIAM Symposium on Discrete Algorithms*, pp. 1–10.

LeVeque, R., Mitchell, I., and Stodden V. (2012): Reproducible research for scientific computing: Tools and strategies for changing the culture. *Computing in Science and Engineering* **14**(4), 13–17.

Macfarlane, M.H. (1999). A high-precision study of anharmonic-oscillator spectra. *Annals of Physics* **271**, 159–202.

McNutt, M. (2015): Data, eternal. *Science* **347**, 7.

Nguyen, H.D., and Demmel, J. (2013a): Fast reproducible floating point summation. In *Proceedings of the 21st IEEE Symposium on Computer Arithmetic*, IEEE Computer Society Press, pp. 163–172.

Nguyen, H.D., and Demmel, J. (2013b): Numerical accuracy and reproducibility at exascale. In *Proceedings of the 21st IEEE Symposium on Computer Arithmetic*, IEEE Computer Society Press, pp. 235–237.

Ossola, G., Papadopoulos, C.G., and Pittau, R. (2008): Cuttools: A program implementing the OPP reduction method to compute one-loop amplitudes. *Journal of High Energy Physics* 0803, 04.

Press, W.H., Teukolsky, S.A., Vetterling, W.T., and Flannery, B.P. (2007): *Numerical Recipes: The Art of Scientific Computing*, Cambridge University Press, Cambridge.

Robey, R.W., Robey, J.M., and Aulwes, R. (2011): In search of numerical consistency incparallel programming. *Parallel Computing* **37**, 217–219.

Romano, J.P., and Wolf, M. (2005). Stepwise multiple testing as formalized data snooping. *Econometrica* **73**, 1237–1282.

Rubio-Gonzalez, C., Nguyen, C., Nguyen, H.D., Demmel, J., Kahan, W., Sen, K., Bailey, D.H., and Iancu, C. (2014): Precimonious: Tuning assistant for floating-point precision. In *Proceedings of the International Conference on High Performance Computing, Networking, Storage and Analysis*, available at `www.davidhbailey.com/dhbpapers/precimonious.pdf`.

Stodden, V., Bailey, D.H., Borwein, J.M., LeVeque, R.J., Rider, W., and Stein, W. (2013): Setting the default to reproducible: Reproducibility in computational and experimental mathematics. *SIAM News*, June 2013. Available at `www.davidhbailey.com/dhbpapers/icerm-report.pdf`.

Stodden, V., Miguez, S., and Seiler, J. (2015): Researchcompendia.org: Cyberinfrastructure for reproducibility and collaboration in computational science. *Computing in Science and Engineering* **17**, 12–19.

Tucker, W. (2002): Smale's 14th problem. *Foundations of Computational Mathematics* **2**, 53–117.

Viswanath, D. (2004): The fractal property of the Lorenz attractor. *Journal of Physics D* **190**, 115–128.

Viswanath, D., and Sahutoglu, S. (2010). Complex singularities and the Lorenz attractor. *SIAM Review* **52**, 294–314.

Yan, Z.-C., and Drake, G.W.F. (2003): Bethe logarithm and QED shift for Lithium. *Physical Review Letters* **81**, 774–777.

Zhang, T., Yan, Z.-C., and Drake, G.W.F. (1994). QED corrections of $O(mc^2\alpha^7 \ln \alpha)$ to the fine structure splittings of Helium and He-like ions. *Physical Review Letters* **77**, 1715–1718.

10
Methodological Issues in the Study of Complex Systems

Harald Atmanspacher and Gerda Demmel

Abstract. From an engineering perspective, it is well known that numerous problems hamper the control and prediction of complex systems that are essential for the reproducibility of their behavior. In addition, it is not obvious how the concept of complexity ought to be understood within the tradition of physics and its epistemology. Both issues have important ramifications for complex systems in other disciplines as well.
Three outstanding topics in this regard are discussed. (1) Many definitions of complexity stress its difference from randomness and are fundamentally context dependent rather than universal. (2) Complexity measures are often defined in information–theoretical terms, but extend the scope of pure syntax toward semantic and pragmatic dimensions. (3) Mathematical limit theorems, expressing the stability of a result, are often not straightforwardly applicable to complex systems.

10.1 Introduction

The concept of complexity and the study of complex systems represent an important focus of research in contemporary science. Although one might say that its formal core lies in mathematics and physics, complexity in a broad sense is certainly one of the most interdisciplinary issues scientists of many backgrounds have to face today. Beyond the traditional disciplines of the natural sciences, the concept of complexity has even crossed the border to areas like psychology, sociology, economics, and others. It is impossible to address all approaches and applications that are presently known comprehensively here; overviews from different eras and areas of complexity studies are due to Cowan *et al.* (1994), Cohen and Stewart (1994), Auyang (1998), Scott (2005), Shalizi (2006), Gershenson *et al.* (2007), Nicolis and Nicolis (2007), Mitchell (2009), and Hooker (2011).

The study of complex systems continues a whole series of interdisciplinary approaches, leading from system theory (Bertalanffy 1968) and cybernetics (Wiener 1948) to synergetics (Haken 1977) and self-organization (Foerster 1962), dissipative (Nicolis and Prigogine 1977) and autopoetic systems (Maturana and Varela 1980), automata theory (Hopcroft and Ullmann 1979), and others. In all these approaches, the concept of information plays a significant role in one

---

Reproducibility: Principles, Problems, Practices, and Prospects, First Edition. Edited by Harald Atmanspacher and Sabine Maasen.

or another way, first due to Shannon and Weaver (1949) and later also in other contexts (Zurek 1989, Atmanspacher and Scheingraber 1991, Kornwachs and Jacoby 1996, Marijuàn and Conrad 1996, Boffetta *et al.* 2002, Crutchfield and Machta 2011).

A most important predecessor of complexity theory is the theory of nonlinear dynamical systems, which originated from early work of Poincaré and was further developed by Lyapunov, Hopf, Krylov, Kolmogorov, Smale, and Ruelle – to mention just a few outstanding names. Prominent areas in the study of complex systems as far as it has evolved from nonlinear dynamics are fractals (Mandelbrot 1977), chaos (Stewart 1990), cellular automata (Wolfram 1986, 2002), coupled map lattices (Kaneko 1993, Kaneko and Tsuda 2000), symbolic dynamics (Lind and Marcus 1995), self-organized criticality (Bak 1996), computational mechanics (Shalizi and Crutchfield 2001), and network theory (Albert and Barabási 2002, Boccaletti *et al.* 2006, Newman *et al.* 2006).

This ample (and incomplete) list notwithstanding, it is fair to say that one important *open* question is the question for a *fundamental* theory with a *universal* range of applicability, e.g., in the sense of an axiomatic basis, of nonlinear dynamical systems. Although much progress has been achieved in understanding a large corpus of phenomenological features of dynamical systems, we do not have any compact set of basic equations (like Newton's, Maxwell's, or Schrödinger's equations), or postulates (like those of relativity theory) for a comprehensive, full-fledged, formal theory of nonlinear dynamical systems – and this applies to the concept of complexity as well.

Which criteria does a system have to satisfy in order to be complex? This question is not yet answered comprehensively, too, but quite a few essential issues can be indicated. From a physical point of view, complex behavior typically (but not always) arises in situations *far from thermal equilibrium*. This is to say that one usually does not speak of a complex system if its behavior can be described by the laws of linear thermodynamics. (In fact the entire framework of equilibrium thermodynamics may become inapplicable in such situations.) The thermodynamical branch of a system has to become unstable before complex behavior can emerge. In this manner the concept of *instability* becomes an indispensable element of any proper understanding of complex systems.[1]

In addition, complex systems are usually regarded as *open systems*, exchanging energy and/or matter (and/or information) with their environment. Other features that are most often found in complex systems are internal *self-reference* and external boundary conditions such as *control parameters*. Sometimes it is

[1] The contribution by Steinle in this volume addresses the significance of stability for the reproducibility of experimental results in detail.

argued that external boundary conditions gradually become internalized, i.e., become part of the internal dynamics of a system, if one is dealing with living organisms (see, e.g., Atlan 1991). In this context, Maturana and Varela (1980) established the concept of autopoiesis accounting for the fact that living system are able to develop (and modify) their own boundaries. Higher levels in the hierarchy, for instance, cognitive or even social systems, raise yet more sophisticated problems, above all the so-called "hard problem" (Chalmers 1995) of how consciousness can be understood in its relation to the physical world.

10.2 Definitions of Complexity

Problems with standard scientific methodology arise already for definitions of complexity. Subsequent to algorithmic complexity measures (Solomonoff 1964, Kolmogorov 1965, Chaitin 1966, Martin-Löf 1966), a remarkable number of different definitions of complexity have been suggested over the decades. Classic overviews are due to Lindgren and Nordahl (1988), Grassberger (1989, 2012), or Wackerbauer *et al.* (1994). Though some complexity measures seem to be more popular than others, there are no clear or rigorous criteria to select a "correct" definition and reject the rest.

It appears that for a proper characterization of complexity, one of the fundaments of scientific methodology, the search for *universality*, must be complemented by an unavoidable context dependence, or *contextuality*. An important example for such contexts are the role of the environment, including measuring instruments, in the measurement process of quantum theory (Atmanspacher 1997).[2] Another case in point is the model class an observer has in mind when modeling a complex system (Crutchfield 1994). For a more detailed account of some epistemological background for these topics compare Atmanspacher (1997).

A systematic orientation in the jungle of definitions of complexity is impossible unless a reasonable classification is at hand. Again, several approaches can be found in the literature: two of them are (1) the distinction of structural and dynamical measures (Wackerbauer *et al.* 1994) and (2) the distinction of deterministic and statistical measures (Crutchfield and Young 1989).[3] Another,

[2] A particularly relevant feature of quantum measurement for the discussion of reproducibility is that successive measurements do not commute. This is due to an uncontrollable backreation of the measurement process on the state of the system. This feature is addressed in more detail in the contributions by Collins and by Wang and Busemeyer in this volume.

[3] Note that deterministic measures are not free from statistical tools. The point of this distinction is that individual accounts are delineated from ensemble accounts.

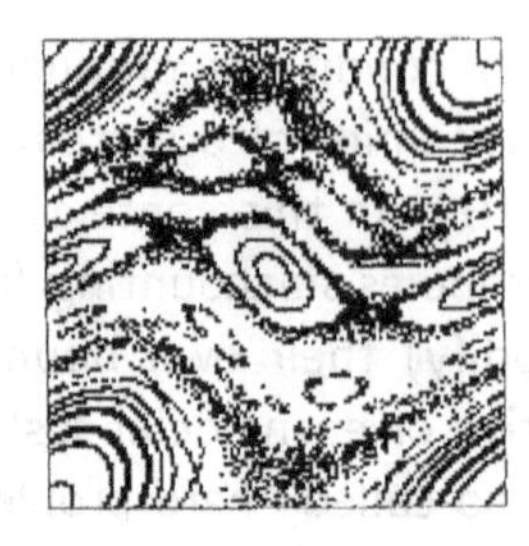
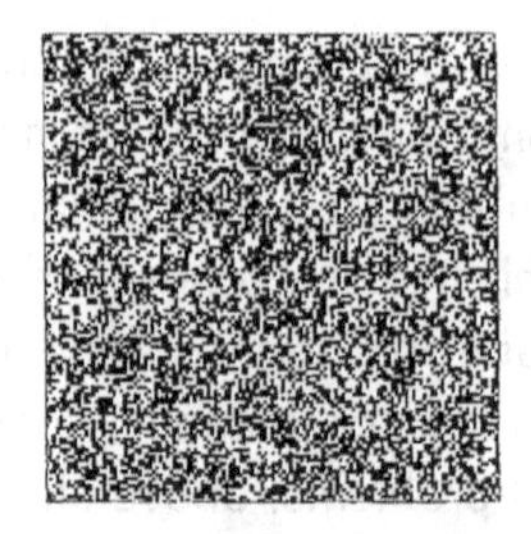

Figure 10.1: Three patterns used to demonstrate the notion of complexity. Typically, the pattern in the middle is intuitively judged as most complex. The left pattern is a periodic sequence of black and white pixels, whereas the pattern on the right is constructed as a random sequence of black and white pixels. (Reproduced from Grassberger (1986) with permission.)

epistemologically inspired (Scheibe 1973), scheme (3) assigns ontic and epistemic levels of description to deterministic and statistical measures, respectively (Atmanspacher 1994).

A phenomenological criterion for classification refers to the way in which a complexity measure is related to randomness, as illustrated in Fig. 10.1 (for an early reference in this regard, see Weaver 1968).[4] This perspective gives rise to two classes of complexity measures: (4) those for which complexity increases monotonically with randomness and those with a globally convex behavior as a function of randomness. Classifications according to (2) and (3) distinguish measures of complexity precisely in the same manner as (4) does: deterministic or ontic measures behave monotonically, while statistical or epistemic measures are convex. In other words, deterministic (ontic) measures are essentially measures of randomness, whereas statistical (epistemic) measures capture the idea of complexity in an intuitively appealing fashion.

Examples for monotonic measures are algorithmic complexity (Kolmogorov 1965) and various kinds of Rényi information (Balatoni and Rényi 1956), among them Shannon information (Shannon and Weaver 1949), multifractal scaling indices (Halsey *et al.* 1986), or dynamical entropies (Kolmogorov 1958). Examples for convex measures are effective measure complexity (Grassberger 1986), ϵ-machine complexity (Crutchfield and Young 1989), fluctuation complexity (Bates and Shepard 1993), neural complexity (Tononi *et al.* 1994), and variance complexity (Atmanspacher *et al.* 1997).

An intriguing difference (5) between monotonic and convex measures can be recognized if one focuses on the way statistics is implemented in each of them.

[4]It should be emphasized that randomness itself is a concept that is anything else than finally clarified. Here we use the notion of randomness in the broad sense of an entropy.

The crucial point is that convex measures, in contrast to monotonic measures, are formalized *meta*-statistically: They are effectively based on second-order statistics in the sense of "statistics over statistics."[5] Fluctuation complexity is the standard deviation (second order) of a net mean information flow (first order), effective measure complexity is the convergence rate (second order) of a difference of entropies (first order), ϵ-machine complexity is the Shannon information with respect to machine states (second order) that are constructed as a compressed description of a data stream (first order), and variance complexity is based on the global variance (second order) of local variances (first order) of a distribution of data. Monotonic complexity measures provide no such two-level statistical structures.

While monotonic complexity measures are essentially measures of randomness, intuitively appropriate measures of complexity are convex. Corresponding definitions of complexity are highly context dependent; hence, it is nonsensical to ascribe an amount of complexity to a system without specifying the precise context under which it is considered. As a consequence, already the assignment of some degree of complexity is reproducible only if all relevant contexts are explicitly known and taken into account. Since convex complexity measures do not obey a universal definition, strict reproducibility (as, e.g., in classical mechanics) cannot be taken for granted in complex systems.

10.3 Complexity and Meaning

Grassberger (1986) and Atlan (1987) were the first to emphasize a close relationship between complexity and the concept of meaning (semantic information). For instance, Grassberger (1989, his italics) wrote:

> complexity in a very broad sense is a *difficulty* of a *meaningful task*. More precisely, the complexity of a pattern, a machine, an algorithm etc. is the difficulty of the most important task related to it. ... As a consequence of our insistence on *meaningful* tasks, the concept of complexity becomes *subjective*. We really cannot speak of the complexity of a pattern without reference to the observer. ... A unique definition (of complexity) with a universal range of applications does not exist. Indeed, one of the most obvious properties of a complex object is that there is no *unique* most important task related to it.

[5] "Second-order statistics" does not mean that the second moment of a distribution has to be involved.

Although this remarkable statement by one of the pioneers of complexity research in physics dates almost 30 years back from now, it is still quite unclear how the relation between complexity and meaning looks in detail. Before we come to this, let us look at the notion of meaning in some more detail.

Traditionally, the concept of meaning has been of concern for philosophy, and later psychology and cognitive science. Early in the 19th century, Schleiermacher and Dilthey laid the foundations of what is today known as the "hermeneutic method", and late in the same century Brentano introduced the term "intentionality" as the reference relation that connects a mental representation with what it represents. Another approach, which will be of interest in the following, is due to Peirce and has been given an information theoretical framework by Morris (1955): the "semiotic approach."

Semiotics, the study of signs, is constituted by three different fields: syntactics, semantics, and pragmatics. While the syntactic level is relevant for the interrelations between signs (e.g., grammar), semantics deals with the relation between signs and what they designate (their meaning), and pragmatics focuses on the relation between signs, their meaning, and their users. Applying this to the realm of scientific models, one can distinguish between the syntactic level of the pure formalism of a model, the semantic level of its interpretation, and the pragmatic level of its application. Semantics addresses the meaning of the formalism, and pragmatics addresses its usage.

However, the apparent clarity of this distinction is somewhat artificial. As soon as one starts to consider aspects of constructing, testing, and working with a model concretely, any rigorous demarcation dissolves. Ultimately, syntactics, semantics, and pragmatics are no longer strictly separable (for a more detailed discussion, see Atmanspacher 1994, 2007). Nevertheless, their separation remains useful as a tool for conceptual analysis. Within the present context, this allows us to refer to meaning as the central notion of semantics without explicitly incorporating syntactics and pragmatics at the same time.

Weaver's contribution in the seminal work by Shannon and Weaver (1949) indicated early on that the purely syntactical component of information today known as Shannon information requires extension into semantic and pragmatic domains. For instance, imagine a "Babylonian library" of books most of which are meaningless to most readers because their texts merely satisfy some syntactic rules (if at all). Only a small fraction of the syntactic information contained in the library would amount to semantic information for a particular reader.[6]

---

[6]In this sense, a plausible condition for understanding meaning is that the structural organization of reader and text possesses commonalities. If this is the case, they are not independent, and the mutual information between them is greater than zero. This can be discussed in terms

How is it possible to check if a receiver understood the meaning of a message? Shortly after Shannon and Weaver's work, Bar Hillel and Carnap (1953) proposed a quantification of semantic information based on a receiver's ability to draw valid logical conclusions from a received message. Their approach tries to measure semantic information by its consequences. If meaning is understood, then it triggers action and changes the structure or behavior of the receiver. In this spirit, Weizsäcker (1974) has introduced a way to deal with the usage that is based on the meaning of a message in terms of pragmatic information.[7]

Pragmatic information is based on the notions of *novelty* ("Erstmaligkeit") and *confirmation* ("Bestätigung"). Weizsäcker argued that a message that does nothing but confirm the prior knowledge of a receiver will not change its structure or behavior. On the other hand, a message providing (novel) material completely unrelated to any prior knowledge of the receiver will not change its structure or behavior either, simply because it cannot be understood. In both cases, the pragmatic information of the message vanishes. A maximum of pragmatic information is assigned to a message that transfers an optimal combination of novelty and confirmation to its receiver. Purely syntactic Shannon information represents the limiting case of pragmatic information for complete confirmation. If novelty is added, Shannon information increases monotonically.

Pragmatic information can be made operationally accessible, as has been shown by Gernert (1985) and Kornwachs and Lucadou (1985) who applied pragmatic information to the study of cognitive systems. But also purely physical systems allow (though not require) a description in terms of pragmatic information. This has been shown by Atmanspacher and Scheingraber (1990) for physical systems far from thermal equilibrium. Instabilities in a laser system can be considered as meaningful in the sense of positive pragmatic information if they are accompanied by a change in the degree of complexity of the system.

One may object that this approach does not yield explanatory surplus over a purely physical model of such systems. However, if an explicit account of cognition becomes unavoidable, this objection dissolves. Atmanspacher and Filk (2006) demonstrated that the complexity of networks performing supervised learning behaves non-monotonically as learning proceeds. Their plausible suggestion is to interpret the maximum amount of complexity during the learning process as the point at which the learning task is represented in a way that becomes meaningful for the final solution of the task.

---

of monotonic complexity measures. For instance, Chaitin (1979) developed such a "common information" in the framework of algorithmic information theory.

[7]A collection of papers concerning this concept can be found in the journal *Mind and Matter* **4**(2) of 2006, including a more recent account by Weizsäcker and Weizsäcker (2006).

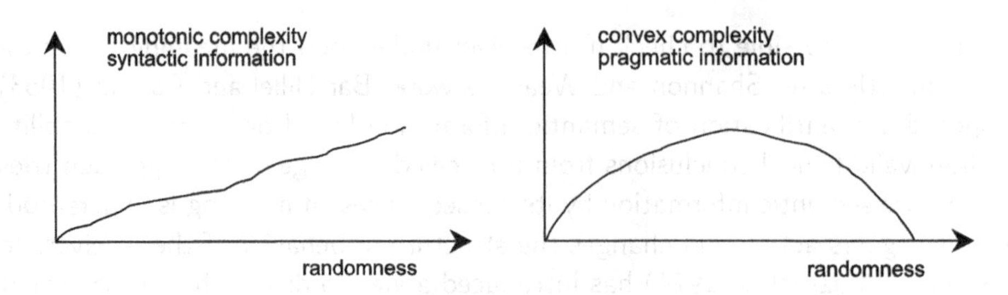

Figure 10.2: Two classes of complexity measures: Monotonic complexity measures essentially are measures of randomness, typically based on syntactic information. Convex complexity measures vanish for complete randomness and can be related to the concept of pragmatic information.

In a very timely and concrete sense, complexity and meaning are tightly related in a number of approaches in the rapidly developing field of semantic information retrieval as used in semantic search algorithms for databases (for an early reference see Amati and van Rijsbergen 1998, a recent collection of essays is due to de Virgilio *et al.* 2012). Many such approaches try to consider the context of lexical items by similarity relations, e.g., based on the topology of these items (see, e.g., Cilibrasi and Vitànyi 2007). This pragmatic aspect of meaning is also central in situation semantics on which some approaches of understanding meaning in computational linguistics are based (e.g., Rieger 2004).

Based on pragmatic information, an important connection between meaning and complexity can be established (cf. Atmanspacher 1994, 2007). Applying a proper algorithm in order to generate a regular pattern, e.g., a checker-board-like period-2 pattern, the corresponding generation process is obviously recurrent after the first two steps (compare Fig. 10.1 left). Considering the entire generation process as a process of information transmission, it is clear that any part of the process after its first two time steps just confirms these first steps. In this sense, a regular pattern of vanishing complexity corresponds to a process of information transmission with vanishing meaning as soon as an initial transient phase (the first two time steps) is completed. This argument holds for both monotonic (deterministic) and convex (statistical) definitions of complexity.

For a maximally random pattern the situation is more involved, since monotonic, deterministic complexity and convex, statistical complexity lead to different assessments. *Deterministically*, a random pattern is generated by an incompressible algorithm that contains as many steps as the pattern contains elements. The process of generating the pattern is not recurrent within the length of the algorithm. This means that it never ceases to produce elements that are unpredictable (except the entire algorithm were *a priori* known, such as the sequence of digits of the number π). Hence, the process generating a random pattern can be interpreted as a transmission of information completely lacking confirmation, and consequently with vanishing meaning.

If the *statistical* notion of complexity is focused on, the process of pattern generation is no longer related to the sequence of individually distinct elements. A statistical generation of the pattern is not uniquely specific with respect to its local properties; it is completely characterized by the global statistical distribution. This entails a significant shift in perspective. While monotonic complexity relates to syntactic information as a measure of randomness, the convexity of statistical complexity coincides with the convexity of pragmatic information as a measure of meaning (see Fig. 10.2). It is remarkable how the concepts of complexity and meaning are explicitly complementary in this respect.

Subtle combinations of regular and random behavior in complex systems correspond to subtle combinations of confirmation and novelty in terms of pragmatic information. There is a correspondence between convex complexity measures and pragmatic information as an operational measure of meaning. This correspondence can be understood phenomenologically (by the behavior of those measures as a function of randomness), formally (by their statistical structure), and conceptually (by epistemological arguments). The relation between complexity and meaning is important for learning theory, semantic information retrieval, and computational linguistics.

10.4 Beyond Stationarity and Ergodicity

Another important set of complications beyond the methodology of conventional science due to complex behavior refers to non-stationary, transient trajectories at instabilities between different modes of stable behavior. In addition, ergodicity cannot be generally presupposed in such cases. Roughly, this means that the ensemble average of a given system variable is no longer identical with its time average along an individual trajectory. For nonergodic systems, statistical measures from ergodic theory must be carefully scrutinized (Tanaka and Aizawa 1993).[8]

As a consequence, imprudent applications of *limit theorems* in statistical analyses of data from complex systems can lead to pitfalls and misinterpretations. The framework of *large deviations statistics* (LDS), originally applied to problems of statistical physics (Ellis 1985, Oono 1989), has been proposed as a useful tool

[8]Nonstationary and nonergodic behavior in this sense is expected to play a particularly significant role in living organisms and cognitive systems (cf. Freeman 1994, Nozawa 1994).

for the study of complex systems (Aizawa 1989, Seppäläinen 1995).[9] In cases where it is difficult to see how a system variable behaves, LDS allows us to estimate how long its convergence toward a limit will take.

If a variable is defined in the sense of an expectation value, then the relevant framework is that of a level-1 LDS description. The expectation value is defined in the limit of $N \to \infty$, e.g., in a "thermodynamic" limit, where N can be the number of particles, of degrees of freedom, of subensembles, and so on. For instance, the formalism of multifractal measures (Halsey *et al.* 1986, Paladin and Vulpiani 1987) is based on a thermodynamic limit; it is a level-1 theory and uses such first-order statistical measures.

The *law of large numbers* states that in the appropriate limit a distribution converges weakly to the unit point measure at the expectation value. LDS specifies how fast it converges in terms of an (exponential) *convergence rate*, the large deviation entropy (Ellis 1985). A more restrictive limit theorem that (other than a law of large numbers) presupposes the existence of the second moment of a distribution is the *central limit theorem.* It gives an estimate for the probability that the size of properly defined, i.e., normalized, fluctuations around the expectation value is of the order of $\sqrt{N}$.[10]

If the thermodynamic limit as a precondition for a law of large numbers in the sense of a level-1 description cannot be presupposed, one can consider a higher level at which the observed empirical distribution functions themselves (not single variables) are treated as stochastic objects. Measures on such a higher level are meta-statistical measures; they characterize the fluctuations of a distribution as a function of N. Distributions in a purely structural (nondynamical) sense then give rise to meta-statistical level-2 descriptions.

A good example is the behavior of histograms of scaling indices for finite N as a function of N, which become multifractal measures for $N \to \infty$ (cf. Halsey *et al.* 1986). For distributions covering structural as well as dynamical elements it can be reasonable to proceed to meta-statistical descriptions that are called level-3 descriptions in the terminology of LDS (Ellis 1985). The stochastic objects of these descriptions are trajectories or histories instead of level-2 distributions.

A level-$(n-1)$ theory can in general be obtained from the corresponding level-n theory ("contraction principle"; Ellis 1985). For instance, it is possible

[9] Relationships between LDS and ϵ-machine reconstruction in computational mechanics have been indicated by Young and Crutchfield (1994). A most up-to-date collection of LDS applications in physics is due to Vulpiani *et al.* (2014).

[10] Some background on standard limit theorems in statistics can be found in the contribution by Stahel, this volume. The extreme case of rare events, where the limit $N \to \infty$ is not achievable by definition, is treated by Kantz, this volume.

to infer the convergence rate toward an expectation value (assuming that it exists) from the convergence rate of its probability distribution. An analogous contraction principle does not in general apply to the *moments*. If the distribution function depends on time, ergodicity may break down so that ensemble averages cannot be expressed as time averages. If this is not explicitly taken care of, this can lead to severe problems with respect to the law of large numbers.

In a pertinent example, Pikovsky and Kurths (1994) clarified a corresponding misunderstanding in numerical studies of globally coupled maps (see also Griniasty and Hakim 1994). Coupled maps are networks of recursive nonlinear maps, such as the logistic map, which have been used to study the spatiotemporal properties of commplex systems. Coupled map lattices can be seen as discrete-time simulations of partial differential equations. Pioneering work in this direction is due to Cruthfield, Kaneko, and others in the early 1980s.

Pikovsky and Kurths (1994) showed that properly defined higher-order fluctuations are consistent with the law of large numbers at level 3 although they violate it at level 2 (Kaneko 1990). The reason is that stationarity and ergodicity break down in complex systems such as coupled map lattices or, more specifically, globally coupled maps (see also Aizawa 1989).

This is particularly interesting in view of the fact that such systems generically give rise to *long-living transients* – a field of research pioneered by Grebogi *et al.* (1985) and recently reviewed by Tel and Lai (2008). So-called *supertransients* are tightly connected to intrinsically unstable phases during which a system transits from previously stable behavior to a new stable solution. Their lifetimes scale exponentially or algebraically as a function of particular system parameters once these parameters exceed a critical value. Supertransients are frequent in spatially extended systems.

Coupled map lattices also offer an interesting playground for how intrinsically unstable behavior can be stabilized – a kind of "non-hierarchical control" due to the crucial dependence of each site in the lattice on its environment, e.g. its neighboring sites. Numerical studies by Atmanspacher *et al.* (2005) showed that this kind of feedback may stabilize the behavior of a system exactly in its *unstable* regime (e.g., at unstable fixed points). This counterintuitive result differs fundamentally from standard ways of controlling chaos by an ongoing external adjustment of parameters (Ott *et al.* 1990).

Basic assumptions in the statistical analysis of systems in traditional science are ergodicity, stationarity, stability, and so on. All these assumptions can be inappropriate in complex systems. Particular examples show that, as a consequence, statistical limit theorems need to be applied with great care, for instance, with the help of large deviations statistics. Since limit theorems are a basic statistical backbone of reproducible results, it is evident that the concept of reproducibility must not be uncritically assumed but carefully checked in each individual situation. Extremely long-living transients between stable behavior may be other indicators of intrinsically irreproducible behavior.

10.5 Conclusions

After several decades of research on complexity, there is now much evidence for an intimate relationship between the phenomenological classes of monotonic and convex complexity measures and the first-order and second-order statistical structure of these measures, respectively. Problems with the application of limit theorems (laws of large numbers, central limit theorem) can in a compact and general manner be addressed within the framework of LDS. First-order statistics (expectation values) may become irrelevant in systems requiring higher-order statistics (limit distributions, histories) as covered by LDS.

On the other hand, it is well known that first-order limit theorems (for the expectation value of an observable) often are not relevant for the statistical analysis of complex systems. For such systems it becomes mandatory to investigate limits of distributions or even limits of histories of distributions. The connection between convex complexity measures and higher-order LDS thus indicates promising novel directions with respect to a proper formal framework for the study of complex systems.

The significance of these ideas is further supported by a number of important epistemological issues (Atmanspacher 1994, 1997). Today it is a truism that the complexity of a system cannot be uniquely characterized unless an observer's model of this system is *explicitly* taken into account.[11] This implies that complexity is not a property of a system "out there" but rather a property of the relation between a system "out there" and the modeling framework by which it is assessed.

As a straightforward consequence, any model of this relationship itself has (at least) to be a meta-model (see, e.g., Casti 1992). A theory of complexity in this sense must, thus, be a second-order (meta-) theory. Its referents are not merely measured facts or data but *also* first-order models of these data *and* the relation between the two. This relation manifests itself most evidently in the process of model building, or learning.

---

[11]See Grassberger (1989) and also Ehm and Stahel, both in this volume. What Collins, also in this volume, addresses as "experimenters' regress" and its consequences can be seen under this aspect as well. This applies in particular to his recommendation for "meta-meta-analysis."

All this entails a profound change of perspective as compared to conventional methodological principles in science. One of the most remarkable points in this regard is the altered situation with respect to the issue of operationalization in a metatheoretical framework. It is obviously inadequate to confirm or reject a meta-model simply by a "naive" observation of a "pure" fact. Instead, proper "experiments" have to include the relationship between data and models.

In general, their analysis has to be a meta-analysis, and it has to be based on meta-statistics instead of conventional first-order statistics. The concepts of predictability and reproducibility, which needed critical reconsideration already for nonlinear, dynamical, particularly chaotic, systems (see the contributions by Zimmerli and Bailey *et al.*, this volume), will have to come under even more scrutiny in a theory of complex systems.

Acknowledgments

Thanks to Peter Grassberger for helpful comments and for the permission to reproduce Figure 10.1.

References

Aizawa, Y. (1989): Non-stationary chaos revisited from large deviation theory. *Progress in Theoretical Physics Supplement* **99**, 149–164.

Albert, R., and Barabási, A.-L. (2002): Statistical mechanics of complex networks. *Reviews of Modern Physics* **74**, 47–97.

Amati, G., and van Rijsbergen, K. (1998): Semantic information retrieval. In *Information Retrieval: Uncertainty and Logics* ed. by C.J. van Rijsbergen, F. Crestani, and M. Lalmas, Springer, Berlin, pp. 189–219.

Atlan, H. (1987): Self creation of meaning. *Physica Scripta* **36**, 563–576.

Atlan, H. (1991): Intentional self organization in nature and the origin of meaning: The role of interpretation. In *Ecological Physical Chemistry*, ed. by C. Rossi and E. Tiezzi, Elsevier, Amsterdam, pp. 311–331.

Atmanspacher, H. (1994): Complexity and meaning as a bridge across the Cartesian cut. *Journal of Consciousness Studies* **1**, 168–181.

Atmanspacher, H. (1997): Cartesian cut, Heisenberg cut, and the concept of complexity. *World Futures* **49**, 333–355.

Atmanspacher., H. (2007): A semiotic approach to complex systems. In *Aspects of Automatic Text Analysis*, ed. by A. Mehler and R. Köhler, Springer, Berlin, pp. 79–91.

Atmanspacher, H., and Filk, T. (2006): Complexity and non-commutativity of learning operations on graphs. *BioSystems* **85**, 84–93.

Atmanspacher, H., Räth, C., and Wiedenmann, G. (1997): Statistics and meta-statistics in the concept of complexity. *Physica A* **234**, 819–829.

Atmanspacher, H., and Scheingraber, H. (1990): Pragmatic information and dynamical instabilities in a multimode continuous-wave dye laser. *Can. J. Phys.* **68**, 728–737.

Atmanspacher, H., and Scheingraber, H., eds. (1991): *Information Dynamics*. Plenum, New York.

Atmanspacher, H., Filk, T., and Scheingraber., H. (2005): Stability analysis of coupled map lattices at locally unstable fixed points. *European Physical Journal B* **44**, 229–239.

Auyang, S.Y. (1998): *Foundations of Complex-System Theories*, Cambridge University Press, Cambridge.

Bak, P. (1996): *How Nature Works: The Science of Self-Organized Criticality*, Copernicus, New York.

Balatoni, J., and Rényi, A. (1956): Remarks on entropy. *Publications of the Mathematical Institute of the Hungarian Academy of Sciences* **9**, 9–40.

Bar Hillel, Y., and Carnap, R. (1953): Semantic information. *British Journal of the Philosophy of Science* **4**, 147–157.

Bates, J.E., and Shepard H. (1993): Measuring complexity using information fluctuations. *Physics Letters A* **172**, 416–425.

Bertalanffy, L. von (1968): *General System Theory*. Braziller, New York.

Boccaletti, S., Latora, V., Moreno, Y., Chavez, M., and Hwang, D.U. (2006): Complex networks: Structure and dynamics. *Physics Reports* **424**, 175–308.

Boffetta, G., Cencini, M., Falcioni, M., and Vulpiani, A. (2002): Predictability – A way to characterize complexity. *Physics Reports* **356**, 367–474.

Casti, J. (1992): The simply complex: Trendy buzzword or emerging new science? *Bulletin of the Santa Fe Institute* **7**, 10–13.

Chaitin, G.J. (1966): On the length of programs for computing finite binary sequences. *Journal of the ACM* **13**, 145–159.

Chaitin, G.J. (1979): Toward a mathematical definition of "life". In *The Maximum Entropy Formalism*, ed. by R.D. Levine and M. Tribus, MIT Press, Cambridge, pp. 447–498.

Chalmers, D.J. (1995): Facing up to the problem of consciousness. *Journal of Consciousness Studies* **2**(3), 200–219.

Cilibrasi, R.L., and Vitànyi, P.M.B. (2007): The Google similarity distance. *IEEE Transactions on Knowledge and Data Engineering* **19**, 370–383.

Cohen, J., and Stewart, I. (1994) : *The Collapse of Chaos*, Penguin, New York.

Cowan, G.A., Pines, D., and Meltzer, D., eds. (1994): *Complexity – Metaphors, Models, and Reality*, Addison-Wesley, Reading.

Crutchfield, J.P. (1994): Observing complexity and the complexity of observation. In *Inside Versus Outside*, ed. by H. Atmanspacher and G.J. Dalenoort, Springer Berlin, pp. 235–272.

Crutchfield, J.P., and Machta, J., eds. (2011): *Randomness, Structure, and Causality: Measures of Complexity from Theory to Applications*, Focus Issue of *Chaos* **21**.

Crutchfield, J.P., and Young K. (1989): Inferring statistical complexity. *Physical Review Letters* **63**, 105–108.

de Virgilio, R., Guerra, F., and Velegrakis, Y., eds. (2012): *Semantic Search over the Web*, Springer, Berlin.

Ellis, R.S. (1985): *Entropy, Large Deviations, and Statistical Mechanics*, Springer, New York.

Foerster, H. von (1962): *Principles of Self-Organization*, Pergamon, New York.

Freeman, W. (1994): Neural mechanisms underlying destabilization of cortex by sensory input. *Physica D* **75**, 151–164.

Gernert, D. (1985): Measurement of pragmatic information. *Cognitive Systems* **1**, 169–176.

Gershenson, C., Aerts, D., and Edmonds, B., eds. (2007): *Worldviews, Science, and Us: Philosophy and Complexity*, World Scientific, Singapore.

Grassberger, P. (1986): Toward a quantitative theory of self-generated complexity. *International Journal of Theoretical Physics* **25**, 907–938.

Grassberger, P. (1989): Problems in quantifying self-generated complexity. *Helvetica Physica Acta* **62**, 489–511.

Grassberger, P. (2012): Randomness, information, complexity. Available at `arxiv.org/abs/1208.3459`. Corrected and amended version of a publication in *Proceedings of the 5th Mexican School in Statistical Physics*, ed. by F. Ramos-Gómez, World Scientific, Singapore, pp. 59–99.

Grebogi, C., Ott, E., and Yorke, J.A. (1985): Super persistent chaotic transients. *Ergodic Theory of Dynamical Systems* **5**, 341–372.

Griniasty, M., and Hakim, V. (1994): Correlations and dynamics in ensembles of maps. *Phys. Rev. E* **49**, 2661–2667.

Haken, H. (1977): *Synergetics*, Springer, Berlin. Third, enlarged edition 1983.

Halsey, T.C., Jensen, M.H., Kadanoff, L.P., Procaccia, I., and Shraiman, B.I. (1986): Fractal measures and their singularities: The characterization of strange sets. *Phys. Rev. A* **33**, 1141–1151.

Hooker, C., ed. (2011): *Philosophy of Complex Systems*, Elsevier, Amsterdam.

Hopcroft, J.E. and Ullmann, J.D. (1979): *Introduction to Automata Theory, Languages, and Computation*, Addison-Wesley, Reading.

Kaneko, K. (1990): Globally coupled chaos violates the law of large numbers but not the central-limit theorem. *Physical Review Letters* **65**, 1391–1394.

Kaneko, K., ed. (1993): *Theory and Applications of Coupled Map Lattices*. Wiley, New York.

Kaneko, K., and Tsuda, I. (2000): *Complex Systems: Chaos and Beyond*, Springer, Berlin.

Kolmogorov, A.N. (1958): A new metric invariant of transitive dynamical systems and automorphisms in Lebesgue spaces. *Doklady Akademii Nauk SSSR* **119**, 861–864. See also Ya.G. Sinai: On the notion of entropy of a dynamical system. *Doklady Akademii Nauk SSSR* **124**, 768 (1959).

Kolmogorov, A.N. (1965): Three approaches to the quantitative definition of complexity. *Problems of Information Transmission* **1**, 3–11.

Kornwachs, K., and Jacoby, K., eds. (1996): *Information – New Questions to a Multidisciplinary Concept*, Akademie, Berlin.

Kornwachs, K., and Lucadou, W. von (1985): Pragmatic information as a nonclassical concept to describe cognitive processes. *Cognitive Systems* **1**, 79–94.

Lind, D., and Marcus, B. (1995): *Symbolic Dynamics and Coding*, Cambridge University Press, Cambridge.

Lindgren, K., and Nordahl, M. (1988): Complexity measures and cellular automata. *Complex Systems* **2**, 409–440.

Mandelbrot, B.B. (1977): *The Fractal Geometry of Nature*. Freeman, San Francisco. Third, updated edition 1983.

Marijuàn, P., and Conrad, M., eds. (1996): Proceedings of the conference on foundations of information science. *BioSystems* **38**, 87–266.

Martin-Löf, P. (1966): The definition of random sequences. *Information and Control* **9**, 602–619.

Maturana, H., and Varela, F. (1980): *Autopoiesis and Cognition*, Reidel, Boston.

Mitchell, M. (2009): *Complexity: A Guided Tour*, Oxford University Press, Oxford.

Morris, C.W. (1955): Foundations of the theory of signs. In *Intl. Encyclopedia of Unified Science, Vol. I/2*, ed. by O. Neurath (Chicago), pp. 77–137.

Newman, M., Barabási, A., and Watts, D. (2006): *The Structure and Dynamics of Networks*, Princeton University Press, Princeton.

Nicolis, G., and Nicolis, C. (2007): *Foundations of Complex Systems*, World Scientific, Singapore.

Nicolis, G. and Prigogine, I. (1977): *Self-Organization in Non-Equilibrium Systems*. Wiley, New York.

Nozawa, H. (1994): Solution of the optimization problem using the neural network model as a globally coupled map. *Physica D* **75**, 179–189.

Oono, Y. (1989): Large Deviations and Statistical Physics. *Progress in Theoretical Physics Supplement* **99**, 165–205.

Ott, E., Grebogi, C., and Yorke, J.A. (1990): Controlling chaos. *Physical Review Letters* **64**, 1196–1199.

Paladin, G., and Vulpiani, A. (1987): Anomalous scaling laws in multifractal objects. *Physics Reports* **156**, 147–225.

Pikovsky, A.S., and Kurths, J. (1994): Do globally coupled maps really violate the law of large numbers? *Physical Review Letters* **72**, 1644–1646. See also A.S. Pikovsky: Comment on: Noisy uncoupled chaotic map ensembles violate the law of large numbers. *Physical Review Letters* **71**, 653 (1993).

Rieger, B.B. (2004): On understanding understanding. Perception-based processing of NL texts in SCIP systems, or meaning constitution as visualized learning. *IEEE Transactions on System, Man, and Cybernetics C* **34**, 425–438.

Scheibe, E. (1973): *The Logical Analysis of Quantum Mechanics*, Pergamon, Oxford, pp. 82–88.

Scott, A., ed. (2005): *Encyclopedia of Nonlinear Science*, Routledge, London.

Seppäläinen, T. (1995): Entropy, limit theorems, and variational principles for disordered lattice systems. *Communications in Mathematical Physics* **171**, 233–277.

Shalizi, C.R. (2006): Methods and techniques of complex systems science: An overview. In *Complex System Science in Biomedicine*, ed. by T.S. Dreisboeck and J.Y. Kresh, Springer, Berlin, pp. 33–114.

Shalizi, C.R., and Crutchfield, J.P. (2002): Computational mechanics: Pattern and prediction, structure and simplicity. *Journal of Statistical Physics* **104**, 817–879.

Shannon, C.E., and Weaver, W. (1949): *The Mathematical Theory of Communication*, University of Illinois Press, Urbana.

Solomonoff, R.J. (1964): A formal theory of inductive inference. *Information and Control* **7**, 1–22, 224–254.

Stewart, I. (1990): *Does God Play Dice?* Penguin, New York.

Tanaka, K., and Aizawa, Y. (1993): Fine structures in stationary and non-stationary chaos. *Progress in Theoretical Physics* **90**, 547–567.

Tel, T., and Lai, Y.-C. (2008): Chaotic transients in spatially extended systems. *Physics Reports* **460**, 245–275.

Tononi, G., Sporns, O., and Edelman, G.M. (1994): A measure for brain complexity: Relating functional segregation and integration in the nervous system. *Proceedings of the National Academy of Sciences of the USA* **91**, 5033–5037.

Vulpiani, A., Cecconi, F., Cencini, M., Puglisi, A., and Vergni, D., eds. (2014): *Large Deviations in Physics: The Legacy of the Law of Large Numbers*, Springer, Berlin.

Wackerbauer, R., Witt, A., Atmanspacher, H., Kurths, J., and Scheingraber, H. (1994): A comparative classification of complexity measures. *Chaos, Solitons, & Fractals* **4**, 133–173.

Weaver, W. (1968): Science and complexity. *American Scientist* **36**, 536–544.

Weizsäcker, E. von (1974): Erstmaligkeit und Bestätigung als Komponenten der pragmatischen Information. In *Offene Systeme I*, ed. by E. von Weizsäcker, Klett-Cotta, Stuttgart, pp. 83–113.

Weizsäcker, E. von and Weizsäcker, C. von (2006): Information, evolution, and "error-friendliness". *Mind and Matter* **4**, 235–248.

Wiener, N. (1948): *Cybernetics*, MIT Press, Cambridge. Second, enlarged edition 1961.

Wolfram, S. (1986): *Theory and Applications of Cellular Automata*, World Scientific, Singapore.

Wolfram, S. (2002): *A New Kind of Science*, Wolfram Media Inc., Champaign.

Young, K., and Crutchfield, J.P. (1994): Fluctuation spectroscopy. *Chaos, Solitons, & Fractals* **4**, 5–39.

Zurek, W.H., ed. (1990): *Complexity, Entropy, and the Physics of Information.* Addison-Wesley, Reading.

11
Rare and Extreme Events

Holger Kantz

Abstract. Extreme events are defined to be events with large impact on a system exposed to it. Such events are rare. Their rareness poses various problems to quantitative statements about extremes. It is difficult to estimate the rate of occurrence of really large events, it is hard to make predictions about when the next event will happen, and there are difficulties in modeling the time evolution of systems that are shaped by extreme events. All these difficulties are related to instabilities in quantitative scientific approaches, which can be interpreted as lack of reproducibility.

11.1 Introduction

11.1.1 What Are Extreme Events?

While the mathematical theory of extreme value statistics has its roots in the late 1920s (Fisher and Tippett 1928, Gumbel 1935), the notion of *extreme events* as a field of scientific study is relatively recent. One of the first who repeatedly used this term was Didier Sornette (Sornette *et al.*. 1996). Meanwhile, it is an established keyword (Albeverio *et al.* 2005), there have been about a dozen of conferences, and two European research networks, "Extreme Event Ecology," and "Extreme Events – Causes and Consequences." Extreme events as a field of research spreads from the physics of critical phenomena, in particular self-organized criticality, to research on natural hazards and risk assessment (Ghil *et al.* 2011, Bunde *et al.* 2002).

Extreme events are, more precisely, large impact events. And as the name suggests, they are characterized by the extremity of some fluctuating quantity of short duration, so that the notion of an event is justified. Examples include earthquakes (short episode of strong seismic activity); extreme weather such as storms, extreme precipitation, and extreme temperatures (short episodes of large deviations from normal weather conditions); and breakdowns of power or communication networks (Kantz *et al.* 2005).

Extreme events are rare. In fact, being rare is a consequence of the requirement of having large impact. As a counterexample, assume the sea level at the shore of an idealized ocean: It will oscillate sinusoidally between low and

Reproducibility: Principles, Problems, Practices, and Prospects, First Edition. Edited by Harald Atmanspacher and Sabine Maasen.

high tide, so that the distribution of values of the sea level will be bounded between these two limits and, moreover, this distribution will have its maxima at these two extreme values. Hence, any living or nonliving object exposed to tidal change must be adapted in such a way that high and low tide will not have large impact, otherwise this object would not survive a single day. Hence, an event having large impact implies lack of adaptation of those being exposed, which in turn means that such events must be rare.

As the aspect of large impact implies, there is always some object that is exposed to an extreme event. Therefore, quite often, we have a couple of event-generating system and exposed system in mind. Events might not be considered extreme any more, if the system exposed to them is replaced by another system. Over land, an extreme precipitation event can cause flooding and thereby have large impact, whereas the same precipitation event over the ocean would not have any impact and would not be of much interest. Nonetheless, research on extreme events (and also this text) considers aspects that only depend on, and characterize, the event-generating system and its fluctuations, and other aspects related to the system exposed to extremes.

All systems that exhibit extreme events are driven or open systems in permanent motion. To be driven means that there is some flux through the system. This applies, e.g., to the Earth and in particular to its atmosphere, since there is a flux of energy: Solar energy enters the Earth mainly as light in the visible and ultraviolet range, and heat is emitted from the Earth in the form of infrared light. In the course of this transformation process, the Earth exhibits fluctuations and extremes in all participating components, which are in particular the atmosphere and the oceans. The sun is an open system, too: It is internally driven by nuclear fusion and emits energy, which is an outflux. Wild fluctuations are, e.g., expressed as solar flares and magnetic storms.

Typical measurements reflect such fluctuations by producing fluctuating signals. Rarely, the system's activity produces a huge deviation from its normal state, which in the measurement is visible as a particularly large fluctuation and which we call extreme event (usually under the additional requirement that it has large impact on another system).[1]

The fact that we discuss the general concept of extreme events instead of (or in addition to) the specific phenomena such as floods, earthquakes, or solar storms suggests that there might be some commonalities between the different phenomena that justify a more general point of view. In fact, our understanding

[1]We should refer to two books by Taleb (2005, 2010), who used the term "black swan" for phenomena which are, according to our definition, extreme events. As his focus is the lack of ability to foresee large crises, his thoughts are clearly related to this article.

of single phenomena might benefit from a more general and transdisciplinary approach. Two key issues are particularly relevant for the phenomena of interest.

The first one concerns the statistical properties of extremes: How frequently do events of a certain magnitude occur? Is there a largest possible event or is the event magnitude unbounded in principle? Are there temporal correlations between successive events? How can we predict the occurrence of individual extreme events?

The second issue concerns the underlying mechanism: How can a system undergo a fluctuation which is so large that we call it extreme? Are there feedback loops, amplifying mechanisms, or simply random fluctuations? Are extremes just very large and therefore rare normal events in the sense that they are created by the same dynamical mechanism as the normal fluctuations? Or are extremes created by switches into some other dynamical mode? This latter point has been addressed in particular by Sornette (2009).

11.1.2 Reproducibility of Extreme Events

This contribution focuses on the issue of reproducibility, which is relevant for extreme events due to their rareness. In physics, reproducibility first of all means that any scientifically sound statement about some phenomenon (e.g., about earthquakes) can be verified by any other physicist who has the same information as we do. This implies that our scientific results were correctly derived from the information which we use and that we have published all relevant steps of how we made use of this information.

In this *narrow* sense, reproducibility is a requirement for a statement to be scientifically sound. If a statement cannot be reproduced, then it is either wrong (due to some error in the derivation of the result) or important information about how the statement was obtained has not been disclosed (against good scientific practice).

Here, we use the term reproducibility in a *broader* sense. We mean that relevant scientific statements will not only be reproducible by someone else using the same input information and applying the same way of information processing. We also require that the results are robust against reasonable replacements of the used information and against replacing the scientific analysis, i.e., the way of information processing, by another, equally valid one.

In the case of earthquakes, changing input information could, e.g., mean to use a data set of recorded earthquakes covering a different time span, or recorded at a different location, or using a different measurement device to measure seismic activity. While it might indeed be possible that earthquakes in different regions of the world differ in their statistical properties, we would reasonably require that replacing one measurement instrument by another one will not affect our scientific conclusions.

An example for in-principle equivalent analysis techniques is the estimation of a probability density: It can be done by kernel estimators or by histograms, one can also obtain a cumulative distribution by rank ordering. Even though these different methods for the same purpose have their different strengths and flaws, their application to the very same data set should not lead to very different conclusions.

Reproducibility in a broader sense means that scientific results should not depend on details that have nothing to do with the phenomenon under study. This type of robustness is similar to what we encounter in the notion of causality: Causality in physics means that *exactly* the same cause has *exactly* the same consequence; hence, we must be able to reproduce an initial condition *exactly*. Typically we use a different version of this concept: We expect that *similar* causes have *similar* consequences, which is the basis of learning by experience. However, and this was an outstanding discovery 50 years ago, *chaotic* systems lack this robustness: Small changes in the initial conditions (i.e., causes) result in large changes in the effects (i.e., consequences).

We argue that a similar lack of robustness affects certain scientific results about extreme events in complex systems. This gives rise to irreproducibility in the broader sense. Its origin lies in the rareness of extremes, so that there is a certain lack of self-averaging.

11.2 Statistics of Extremes

Flood protection in Germany, as in many other countries, is designed to protect against the so-called 100-year flood. By this one means a water level which is, in statistical average, exceeded only once per 100 years, or, which is almost equivalent, has the probability 1/100 to occur per year. The evident difficulty in this requirement lies in the task to find out which water level at a given place on some river or ocean shore will on average be exceeded only once per 100 years. For a direct numerical estimation, one would take a data set of, say, 10,000 years and count how frequently every level was exceeded.

In technical terms, one would construct a robust cumulative distribution. The level with an exceedance probability of 1/100 per year estimated this way will have some statistical error, but most surely the result would be robust. Its robustness is ensured since the true level, which we do not know, would be exceeded about 100 times in this data sample. Unfortunately, we do not have such measurements for 10,000 years, and even if they existed, they would be useless, since the shore or the river bed has changed during this time span, so that old data cannot be expected to give useful information about the current situation. So we have two problems: the amount of available data and the issue of the stationarity of the data (or, respectively, of the underlying system). But if we have only, say, 50 years of reliable data, what would then be our estimate for the 100-year flood?

To deal with the problem of small data sets, one has to rely on assumptions which are the basis for extrapolation. One commonly used but not very sophisticated method is to assume a certain functional form of the distribution. Real-valued data are often Gaussian distributed, and the central limit theorem tells us why (cf. Stahel, this volume): When the observed quantity is the result of adding up many small increments, then there are good reasons to assume that this quantity is a Gaussian random number. A Gaussian is fully characterized by its mean value and its standard deviation. Hence, to know the frequency of events beyond a certain magnitude, one just tries to estimate, by means of statistical inference, the mean and the standard deviation and can then calculate the desired probability by integration using these two parameters.

There is, however, overwhelming evidence for quantities that are relevant for extreme events that in particular the tails of their distributions are often non-Gaussian. More precisely, they decay slower than the tails of a Gaussian. Such heavy-tailed distributions possess a much higher probability for extreme events than a similar Gaussian distribution. As a consequence, the assumption that tails are Gaussian would lead to an underestimation of the frequency of extremes. Distributions with power-law tails are one example, and indeed, one tries to fit power laws to observed data in many cases.

However, there is no guarantee that the distribution continues to behave like a power law beyond the range of observed data. Hence, the tails might show a different behavior. Despite the evidence for heavy tails of whatever nature, there is also the opposite effect: Finite-size effects (of, e.g., the area or volume of a system) introduce cut-offs in power laws. This is well known from, e.g., the numerical study of event magnitudes in systems exhibiting self-organized criticality (Bak *et al.* 1987), which should apply to many natural phenomena. Hence, we cannot assume that we know what functional form the tails of a distribution should follow, and we, therefore, cannot use fitted parameter values from the bulk of the distribution to estimate tail weights.

Another powerful method stems from the mathematical theory of "extreme value statistics." It has been developed to extrapolate from the probability to observe some maximal value in a short data set to the probability of observing even larger values in longer data sets. In principle, if the theory were applicable, one could infer from the largest observed events during 50 years of observation data the magnitude of an event which occurs on average once per 100 years, i.e., solve the above posed problem.

The basis of extreme value statistics is a certain stability assumption of the distribution of so-called block maxima: First, one needs to construct a sample of many data sets of identical length, normally by subdivision of the given time series of observations. In each data set, called block, one identifies the maximal value, called block maximum. The distribution of these block maxima, under suitable rescaling and shifting, should asymptotically remain unchanged when the block size is enlarged to infinity.[2]

This stability principle allows one to identify the shape of the asymptotic distribution, the generalized extreme value (GEV) distribution. It has three parameters, which need to be determined for the given data set in order to calculate the probability of any arbitrary event magnitude to happen. Hence, knowing these parameters for a given phenomenon, one can calculate, e.g., the magnitude of the 100-year event. In practice, one has to estimate the parameters from the block maxima of rather short block lengths, since one needs a sufficiently large sample of block maxima for statistically robust estimates.

There is an evident dilemma: One should estimate the parameters for as large as possible block lengths, in order to be close to the asymptotic limit of infinite block length. But if the finite time series should be cut into many blocks in order to have many block maxima for parameter estimation, the block length should be short. Every researcher who wants to make use of this approach will chose his/her own compromise between these two requirements.

In addition, using this procedure, one makes several assumptions. First, the data should have short-range correlations, i.e., extreme events do neither repel each other nor do they cluster. Second, the data is stationary. And third, the size of the data set should be large enough to be already in the asymptotic regime with the largest reachable block length.

In view of reproducibility, the results of this procedure can crucially depend on the input data and the block length used. What is the largest block length from which one estimates the three parameters? How is the data set cut into blocks? Also adding a few more years (i.e., repeating the analysis a few years later with some more data) might influence the estimated parameters quite severely.

In a recent model study, Garber and Kantz (2009) found that even the answer to the question whether the event magnitude is bounded from above,

---

[2]For more details, the interested reader is referred to the literature, e.g., Coles (2001).

i.e., whether there exists a largest possible event, can depend on the data set size. The reason is that the so-called "shape parameter" of the GEV distribution is found to have the wrong sign if the input time series is too short. Without entering technical details, we can summarize that the estimation of the event magnitude corresponding to a certain return period, using the theory of extreme value statistics, requires several choices and decisions, so that different researches might reach quite different conclusions even when using the same data set. This is a kind of intrinsic instability of the method, which arises as a consequence of the fact that one tries to extrapolate far beyond the observed time scale.

An interesting debate on this issue occurred in summer 2013 in Germany: In the year 2002, there was some unprecedented flooding of the river Elbe, which was qualified as a 100-year event. Only 11 years later, a flooding of similar extent occurred. Can one still consider this as a 100-year event, if it repeats itself after 11 years? Clearly, from the probabilistic point of view, there is no contradiction: If the chance of the event to happen is 1/100 per year, then probability theory tell us that the probability that two such events happen within a time span of 11 years is approximately 1/1,000, which is nonzero.[3]

The real issue is whether it is plausible that this event magnitude might occur more frequently than 1/100 in the future, because of the non-stationarity of the system. While it is popular to address climate change as responsible for modified precipitation patterns, hydrologists discuss anthropogenic modifications of river beds as more evident potential cause for changing flood frequencies. Regardless of whether climate change or modified river bed, we are back to the reproducibility issue: Can we assume the time series of water levels of the river Elbe to be sufficiently stationary for an application of extreme value statistics, or not? Hence, is it reasonable to assume that future floodings "obey rules" extracted from data of the past 100 years?

Two more examples of irreproducible statistical results derive from earthquake analysis. The first is the question for the average energy released by an earthquake. Naively, we would simply sum up the energies of a large number N of recorded earthquakes and divide this sum by N. This is what statisticians call the empirical sample mean. One can do so with a sequence of earthquakes from, e.g., the Southern California Earthquake Catalogue.[4] Computing this sam-

---

[3]We assume a Poisson process with a rate of $r = 0.01$ per year. The probability for two events to happen at time t and t_0 and no event in between is approximately r^2. If the second event should take place within 10 years after the first, we have 10 different choices for t_0, so that the total probability in this approximation is $10 \times r^2 = 10/10,000$.

[4]The Southern California Earthquake Catalogue is provided by the Southern California Earthquake Data Center, see `www.data.scec.org`.

ple mean for growing sample sizes N means to calculate this mean starting at the beginning of the catalog with all available data up to some time t. What would we observe for "well-behaved" data? We would observe that for ever increasing N, the sample means fluctuate less and less and settle down, for very large N, on the true mean value.

However, for real earthquake data, this value jumps up as soon as a new very strong earthquake happens (its energy increases the sample mean visibly), and it relaxes if then for a long time only smaller earthquakes contribute to the sample mean. The important flaw in the analysis arises due to the fact that the jumps are higher than the subsequent relaxations, so that the mean value continues to increase in the long run.

The mathematical explanation for this is as follows: The energy release of earthquakes is distributed according to a power law. That is, the relative number of earthquakes with energy release E decreases as a power of E so slowly, that the mean value of this distribution is infinite. Every finite sample of only N observations clearly has one event with a largest energy release, giving rise to a finite sample mean. But if the sample grows, then this largest observed value typically grows faster than linearly with N, hence the divergence of the sample mean. Someone who is not aware of this underlying power law and its consequences might try to estimate the mean energy release from some sample, and someone else using a different sample would find a very different value, since the true value is infinite.

The second example concerns the distinction of earthquakes into main shocks and sequences of aftershocks. There is currently no valid method to perform this distinction, but for purposes of understanding the temporal correlations of earthquakes, such a distinction is highly desirable. Hence, scientists use approximate distinctions, which may or may not lead to useful results. In any case, the number of main shocks thus identified is usually too small to perform a sound statistical analysis of the temporal relationship between successive main shocks.

11.3 Predictions of Extreme Events

The Italian region of L'Aquila was struck by a severe earthquake on April 6, 2009, in which more than 300 people died. Six days earlier, a national committee for risk assessment had appeased the population and said that the present sequence of smaller seismic shocks would not indicate that an earthquake would impend. *A posteriori*, this was evidently proven wrong. In the year 2012, seven members of this committee were sentenced to six years in prison each because of multiple involuntary manslaughter. The decision of the court was heavily criticized by a huge number of scientists, who pointed out that a precise prediction of when and where a major earthquake will take place is impossible even minutes before it happens.[5]

The call for precise predictions of impending natural disasters as a specific class of extreme events is nonetheless overwhelming. Whenever there is an extreme event threatening human life or welfare, not only those who are directly affected ask why this was not predicted beforehand. For instance, weather forecasts, if successful, should predict extreme events related to atmospheric fluctuations. But even there we have many examples of severe failure: The "Elbe flood" in Germany, August 2002, was a poorly predicted heavy precipitation event with 150–200 mm of rain within two days over the catchments of the tributaries of the upper Elbe river.

So we are faced with the situation that, due to the demand of forecasts, there are many (usually state-operated) institutions which try to produce forecasts which, however, are often unreliable. In particular, this means that the forecasts are hard to reproduce, if one uses different models, different underlying assumptions, or simply different input data.

An example for varying input data is the 1987 severe storm affecting the south of England:[6] Forecasts of the British Met Office several days before October 15, the day at whose evening the storm approached from southeast, indeed predicted that gale force winds would affect the English coast. Later runs of the same model, i.e., runs starting from initial conditions taking into account more recent atmospheric measurements, predicted a more southern storm track which would not strongly affect England. It is reported[6] that a French meteorological model made different, more accurate predictions indeed foreseeing the storm track to be along the Channel.

Current weather predictions are so-called *ensemble predictions.* Generally, a weather model, which is the computer implementation of some approximations to the physical, dynamical processes inside the atmosphere, is fed by initial conditions, which represent the current state of the atmosphere. Such initial conditions are generated from the most recent measurements such as temperatures, wind speeds, and air pressures.

While computer models represent the atmosphere on a three-dimensional grid of points in space, the measurements are taken from the available measure-

---

[5]For more details see the Wikipedia entry at `http://en.wikipedia.org/wiki/2009_L'Aquila_earthquake`.

[6]For more details see the Wikipedia entry at `http://en.wikipedia.org/wiki/Great_Storm_of_1987`.

ment stations. Not only are these not located exactly at grid points, but, more importantly, there are many more grid points in the model than measurement stations. Measurements provide only a fraction of the information needed to specify initial conditions. Therefore, the generation of the initial model state relies on interpolation techniques known as *data assimilation*.

In these techniques, the initial model state is constructed to be a good compromise between actual measurements and constraints according to the model. In particular, each new initial model state should be consistent with the outcome of the previous model run. Only with this additional information from the (potentially wrong) model state at the time of the new initialization, the missing information can be supplemented. In this sense, the new initial state is only an educated guess of what the current atmospheric state really is. This initial condition is then deterministically propagated into the future, i.e., the model equations allow one to follow the time evolution of the initial state, in principle arbitrarily far into the future.

However, atmospheric dynamics is unstable with respect to tiny perturbations, also known as chaos and turbulence (see also Feulner, this volume). This means that an infinitesimal change of the initial conditions will lead to a future time evolution, which exponentially fast deviates from the original one. Hence, in view of the fact that the model initial condition is only a good guess and nothing really precise, there is no good reason to believe that the far future of such a single simulation run provides a good forecast of the real weather system.

Ensemble forecasts cure this problem and, at the same, time yield an estimate of the uncertainty of the prediction. Instead of a single initial condition, an ensemble of many initial conditions (e.g., provided by the "European Center for Medium-Range Weather Forecasts" ensemble, consisting of 51 members) is propagated into the future. The different ensemble members are bred from the best guess initial condition by small but clever perturbations, and the idea is that each of the ensemble members represents an equally probable state of the atmosphere given the instrumental observations. Therefore, also the future weather represented by each single run is equally probable, given our lack of knowledge of the true current atmospheric state and all the model inaccuracies.

Due to sensitive dependence on initial conditions, ensembles indeed do diverge quite fast. Ensemble forecasts show that after about 10–14 days, their spread is so big that almost every weather which is compatible with the corresponding season at the location of the forecast can happen. A duration of 10–14 days is the current limit at which weather forecasts lose their predictive power.

In terms of reproducibility, it is interesting to see how much influence the choice of the initial conditions has for the forecast, even though this dependence is too complicated to be put into any simpler form than stating that the future is a deterministic consequence of the initial values, generated by the model equations. An ensemble forecast of weather in northern France / southern England of the French storm at December 26, 1999, is shown by Smith (2003). About 20% of all ensemble members predicted the storm about two days in advance, the majority did not.

This shows that without ensemble forecasts, running just single trajectories (as it is still done for many purposes), any two researchers using exactly the same model but slightly different sets of input data for initial conditions will predict different forecasts. And these can vary to an extent as remarkable as the difference between a severe winter storm and mild calm weather.[7]

What is true for weather forecasts is also true for all other attempts to forecast extreme events. The difference is that weather forecasting is a huge business with well-elaborated models and well-established schemes (including ensemble forecasts) operated by large organizations. In many other fields, predictions are individual endeavors of small groups of scientists.

In many cases there are not even well-established models for the dynamics of the system, e.g., for earthquakes, for economic processes, and for processes in human society (outbreak of a civil war). But also for freak waves (rogue waves), snow avalanches, and landslides, the modeling approaches are diverse. As soon as different forecasters use different models, it is plausible that they rely on different input data. Hence, it is not surprising if predictions of specific events are not reproducible unless one repeats exactly one forecasting scheme in all its details.

One might argue that if two different predictors for the same target both perform well, then an event predicted by one of the two should typically also be predicted by the other one. This is true if "to perform well" means that, e.g., about 90% of all events are correctly predicted, and that false alarms are rare. Extreme events, however, are rare. If a given weather service provides good forecasts, then this statement refers to "normal" weather, since most of the time we do have normal weather. But whether or not forecasts of average weather are good is almost irrelevant for extreme weather, since extreme weather is so rare.

Also, it is almost impossible to verify the performance of a predictor only in extreme situations. Extreme weather is too rare for the lifetime of a weather model to allow for a statistically significant performance test. There are about one to two severe windstorms per year in Europe, too few for the time span at

---

[7]For more details on this poorly predicted storm see also the entry in Wikipedia at `http://en.wikipedia.org/wiki/CycloneLotharandMartin`.

which weather models operate without updates that might modify their performance. But even ignoring these updates, one would wish to test the performance of any kind of predictor on at least 20–50 events for a statistically sound determination of the hit rate.

In addition, the prediction of extreme events is a binary (yes/no) prediction, so one should explore the false-alarm rate. Since this is practically impossible for really extreme events, predictors that perform forecasts routinely, such as weather forecasts, are optimized for normal weather conditions and might perform particularly poorly for extreme conditions. Hence, to repeat this point, two different weather models that both produce good predictions on average might both fail to predict extreme events.

11.4 Evolving Systems Exposed to Extreme Events

In the two previous sections we looked at extreme events from the perspective of the system that generates the extremes. Now we will consider the role of the impact of an event onto a second system which is exposed to it.

Very often extreme events of whatever nature have considerable influence on human life and can change an individual's life plan dramatically. Such events might be related to health, financial situation, education, or personal relationships. Following the concept of large-impact events, one could even define extreme events in the life of an individual as those that cause the most drastic change in his or her life.

The reasoning behind this nonphysical setting finds its counterpart, mostly on longer time scales, in inanimate systems. Many landscapes on Earth have been shaped by extreme geophysical events, such as landslides and floods, or by the effects of glaciation (which are "events" only on planetary time scales). The evolution of species has seen about five dramatic mass extinctions (such as the disappearance of the dinosaurs) each in relatively short time (compared to the total duration of the current agon) giving space for new species. So we see the same pattern: There are extreme events causing drastic changes, whose effect might be more visible and influential than the accumulated action of many years of "normal" fluctuations (see Pisarenko and Sornette 2003 for similar considerations).

Another, recent example for this is the legislation ruling German nuclear power plants. Shortly after the Fukushima nuclear disaster in March 2011, German politicians with strong support from the population decided, in June 2011, to quit using nuclear power for energy supply and to define rather short periods for the shutdown of nuclear power plants. In this case, a single extreme event caused more drastic changes than decades of activities of the German anti-nuclear movement, a political movement against nuclear power from which the green party emerged – and this even though the green party was part of the federal government from 1998 to 2005.

The issue of reproducibility here occurs in an ensemble view: Assume we had 20 copies of the planet Earth, starting some 200 million years ago with roughly identical initial conditions. We are interested in the question whether all 20 copies, if we let them evolve independently, would look identical today. It can be addressed using the *replica method*, a standard numerical tool in statistical physics to understand the spreading of local perturbations across a spatially extended system. In the physical setting, where perturbations are small by construction, the fate of our model-Earths will be strongly influenced by extreme events. Some of these events will be impacts from outside, others might be large fluctuations of the dynamics of the processes on Earth, which are irregular and irreproducible in the sense of weather fluctuations.

When will which copy be hit by a large meteorite? When and on which copy will the outbreak of some epidemics kill a large fraction of its population? How will continents drift and volcanoes erupt? The more extreme an event, i.e., the rarer an event, the less will its effects disappear on statistical averages. Every model Earth will experience its own sequence of rare extreme events, and every sequence will differ from all others in every statistical sense. In other words, we ask for the five most extreme events in every modeling run. If these five events have more impact on the system's evolution than all smaller events together, then it is evident that one cannot expect to see any self-averaging effect.

Hence, we expect that after 200 million years, the 20 Earths might look very different from each other. Perhaps there is one where the dinosaurs are still alive, and for sure there are many without humans. Indeed, already Bak, when introducing his concept of self-organized criticality, emphasized the idea of evolution as driven by extremes (Bak and Paczuski 1995). In his point of view, however, evolving systems drive themselves into a state where they are "critical" and exhibit extreme fluctuations without extreme inputs. In our context, this would mean that they are maximally vulnerable to extremes.

The consequence of the above considerations is that complex systems are "individuals": even if we have a large ensemble of similar subjects, each individual will be shaped by its individual history of extremes, and hence, their fate will be irreproducible by any model or in any experiment. It also implies that predictions can only be meaningful until the next extreme event happens. This is not surprising, e.g., predictions of the economic growth of states are made under certain reasonable assumptions about the future and cannot account for unexpected and extreme events such as the Fukushima nuclear disaster or the outbreak of a civil war.

A simple mathematical–physical model for such a scenario which can be studied quantitatively is a "continuous time random walk" (Montroll and Weiss 1965) with weak *ergodicity breaking* (Bouchaud 1992). In such systems, a random walker waits a randomly picked time at some position and then makes a random jump. If the waiting-time distribution has power-law tails so strong that the mean waiting time diverges, then, after every finite observation time, the largest waiting time dominates the whole process. As a consequence, the relevant statistical quantities by which one can characterize such random walks will be random variables themselves, even in the infinite time limit.

In less technical words, fluctuations of the waiting times will never average out, but there is always one largest waiting time which is of the same order of magnitude as the time over which the process had to evolve. And this is exactly the most extreme event that determines the precise value of observables. Hence, there is no need to study a "Whole Earth Simulator" in order to observe the lacking reproducibility of a system's evolution due to extreme events. This toy model highlights one of the potential physical mechanisms for this effect, but it does not explain why there might be power-law distributions of (in this case) waiting times.

Let us recall that we saw a similar effect in earthquakes: the power-law nature of the distribution for energy release had the consequence that its mean value is infinite, and hence, the mean energy release during every finite observation time is dominated by the most extreme event. Also there we can only speculate why there is a power-law distribution (often associated to self-organized criticality; Sornette and Sornette 1989). And there is even less theoretical understanding of the precise value of the observed power-law exponent.

Speaking about the evolution of real-world systems rather than toy models, it is but a hypothesis that the most severe extreme events have larger impact than the joint action of all smaller magnitude events. In most settings, this is not the case, and whether or not it is the case for a given setting depends not only on the statistics of the external perturbations but also on the vulnerability of the system itself.

If we look at the moon, we see a huge number of craters resulting from impacts of meteorites of all sizes. And even though there is definitely one largest size, we would not believe that this one had a particularly strong impact on the evolution of the moon. The moon is so much of a dead rock that it does not care about being hit by another object as long as this does not destroy the moon itself. So for the moon we would not say that after it assumed its stable orbit around the Earth, the five largest collisions with other objects had more impact on its surface structure than all the others. Instead, we see that there are craters of different size, but none of them is outstanding.

In contrast, the Earth with its atmosphere, its oceans, and its habitation by plants and animals, is vulnerable indeed. If dust in the upper atmosphere as a consequence of an impacting meteorite reduces solar radiation too much and for too long, vegetation and animals might die, one of the possible reasons for the extinction of the dinosaurs. A meteorite falling into the ocean might cause huge tsunamis and thereby eliminate a large fraction of the Earth's habitation.

We see a similar pattern on smaller scales: In deserted areas of the Earth, not much damage can happen, so that their vulnerability is low. Clearly, our anthropocentric point of view would first of all assign a large impact to events which endanger humans, their health, and their property. But, more generally, vulnerability with respect to perturbations is the larger, the more complex a system is. Not only does a complex system offer more possibilities to be perturbed, it can also be easily affected as a whole due to interdependencies among different system components – even if the perturbation originally only affects a small subsystem.[8]

Examples for this are manifold in complex technical assemblies. One of the most interesting ones was a planned and purposeful shutdown of two transportation power lines of the electric power grid in northwest Germany in November 2006. This event caused a failure of the power grid in western and central Europe with, eventually, 15 million consumers without power supply.[9]

The idea that the evolution of complex systems is nonaveraging and driven by details is also the basis of the counterfactual approach of historians to the study of extreme events in society. How would human civilization have evolved if some relevant politician had made a decision different from what he or she really did, or if some influential person had died before becoming influential? For example, Ferguson (1999) arrived at the conclusion that World War I was indeed a very improbable outcome of the July 1914 crisis and that it was caused since "decision makers made a series of blunders." Clearly, the chain of pre-war events was a sequence of small events that caused the extreme one. So this reasoning differs slightly from what we presented so far, but it also points out the instability in the evolution of complex systems, which might destroy reproducibility.

---

[8]For more discussion concerning problems of reproducibility in complex systems see Atmanspacher and Demmel, this volume.

[9]See the comprehensive report of the Union for the Coordination of Transmission of Electricity at `https://www.entsoe.eu/fileadmin/user_upload/_library/publications/ce/otherreports/IC-Interim-Report-20061130.pdf`.

11.5 Conclusions

Every scientifically sound statement should be reproducible, otherwise it does not accord with good scientific practice. However, in a strict sense, this only means that a precise repetition of an analysis should reveal the same results. In addition, we often assume (and experience usually supports this assumption) that scientific results are robust against small changes in the analysis, changes in input data, and so on. This is a much stronger version of reproducibility.

For extreme events, however, this strong version does not apply. The rareness of real extremes has the consequence that their analysis is always based on (too) small samples. And different small samples taken from the same underlying distribution differ from each other the more the smaller they are. Fluctuations are not averaged out.

We discussed three important aspects of lacking reproducibility with respect to highly desirable information about extremes: (1) the estimation of the rate of occurrence of events with a magnitude which is so large that only few or even none such events have been observed in the past; (2) the prediction when the next event of a given class will occur; (3) the evolution of systems over long time spans.

For the latter we argue that complex systems are often particularly vulnerable to extreme events, which, therefore, might have stronger impact on their evolution than the accumulated action of all smaller events. In such situations, every system is an individual with a different fate and an ensemble of such systems would represent a large variety of evolutions. Hence, modeling and prediction will suffer from severe limitations.

One field where we can see that this is indeed the case are the budgets of states: Every fiscal year, national parliaments try to balance state budgets and to limit public debts by careful planning. And almost every year huge unexpected extra expenses are required due to some extreme events in society, economy, nature, or foreign policy.

Acknowledgments

I am grateful for numerous and helpful discussions with the members of my research group at the Max-Planck-Institute for the Physics of Complex Systems. Particular thanks go to Nick Watkins, who contributed to this article by his tremendous knowledge of the literature and by invaluable suggestions and remarks.

References

Albeverio, S., Jentsch, V., and Kantz, H., eds. (2005): *Extreme Events in Nature and Society*, Springer, Berlin.

Bak, P., and Paczuski, M. (1995): Complexity, contingency, and criticality. *Proceedings of the Natonal Academy of Sciences of the USA* **92**, 6689–6696.

Bak, P., Tang, C., and Wiesenfeld, K. (1987): Self-organized criticality: An explanation of 1/f noise. *Physical Review Letters* **59**, 381–384.

Bunde, A., Kropp, J., and Schellnhuber, H.J., eds. (2002): *The Science of Disasters, Climate Disruptions, Heart Attacks, and Market Crashes*, Springer, Berlin.

Bouchaud, W.J.P. (1992): Weak ergodicity breaking and aging in disordered systems. *Journal de Physique* **2**, 1705–1713.

Coles, S. (2001): *An Introduction to Statistical Modeling of Extreme Values*, Springer, Berlin.

Ferguson, N. (1999): *The Pity of War*, Basic Books, New York.

Fisher, R.A., and Tippett, L.H.C. (1928): Limiting forms of the frequency distribution of the largest and smallest member of a sample. *Proceedings of the Cambridge Philosophical Society* **24**, 180–190.

Garber, A., and Kantz, H. (2009): Finite size effects on the statistics of extreme events in the BTW model. *European Physical Journal B* **67**, 437–443.

Ghil, M., Yiou, P., Hallegatte, S., Malamud, B.D., Naveau, P., Soloviev, A., Friederichs, P., Keilis-Borok, V., Kondrashov, D., Kossobokov, V., Mestre, O., Nicolis, C., *et al.* (2011): Extreme events: Dynamics, statistics and prediction. *Nonlinear Processes in Geophysics* **18**, 295–350.

Gumbel, E.J. (1935): Les valeurs extrêmes des distributions statistiques. *Annales de l'Insitute Henri Poincaré* **5**(2), 115–158.

Kantz, H., Altmann, E.G., Hallerberg, S., Holstein, D. and Riegert, A. (2005): Dynamical interpretation of extreme events: Predictions and predictability. In *Extreme Events in Nature and Society*, ed. by S. Albeverio, V. Jentsch, and H. Kantz, Springer, Berlin, pp. 69–93.

Montroll, E.W., and Weiss, G.H. (1965): Random walks on lattices II. *Journal of Mathematical Physics* **6**, 167–181.

Pisarenko, V.F., and Sornette, D. (2003): Characterization of the frequency of extreme events by the generalized Pareto distribution. *Pure and Applied Geophysics* **160**, 2343–2364.

Smith, L.A. (2003): Predictability and chaos. In *Encyclopedia of Atmospheric Sciences*, ed. by J.R. Holton, J. Pyle, and J.A. Curry, Academic, London, pp. 1777–1785.

Sornette, D. (2009): Dragon kings, black swans and the prediction of crises. *International Journal of Terraspace Science and Engineering* **2**(1), 1–18.

Sornette, D., Knopov, L., Kagan, Y.Y., and Vanneste, C. (1996): Rank-ordering statistics of extreme events: Application to the distribution of large earthquakes. *Journal of Geophysical Research* **101**, 13883–13893.

Sornette, A., and Sornette, D. (1989): Self-organized criticality and earthquakes. *Europhysics Letters* **9**, 197–202.

Taleb, N.N. (2005): *Fooled by Randomness: The Hidden Role of Chance in Life and in the Markets*, Random House, New York.

Taleb, N.N. (2010): *The Black Swan: The Impact of the Highly Improbable*, Random House, New York.

12
Science under Societal Scrutiny: Reproducibility in Climate Science

Georg Feulner

Abstract. Reproducibility should be a key concern for any scientific discipline. For climate science, however, the difficulties and demands appear even more pronounced than for other fields within science. The main difficulties arise from the fact that we have only one specimen of planet Earth which cannot be studied under laboratory conditions. This implies that climate scientists are left with either observing the present-day climate system (with all the challenges of incomplete and noisy measurements) or using computer models (which can provide only an incomplete representation of the complex Earth system) or investigating Earth's climate history (then being forced to rely on approximate climate reconstructions from so-called proxy data).
Despite these difficulties, the demands on reproducibility in climate science are particularly high because of the tremendous importance of the research results for our society facing the challenges of climate change. In practice, this societal scrutiny of climate science is exceedingly exacerbated by public assaults on particular results (and often also on the scientists responsible for providing these results) by individuals or organizations disagreeing with the notion of anthropogenic climate change or its adverse impacts. In this contribution I will discuss examples highlighting the challenges for reproducibility for observational climate science, climate modeling, and paleoclimate research. Furthermore, I will give specific recommendations how reproducibility in climate science can be further improved.

12.1 Reproducibility Challenges for Climate Science

Reproducibility is often considered one of the hallmarks of the modern scientific method. The concept of reproducible research demands that research results can be independently replicated by other scientists, irrespective of whether attempts of replication are actually undertaken or not (Thompson and Burnett 2012). Reproducibility thus ensures that the validity of published results can be tested in principle and should, therefore, be considered a key requirement for good research in each and every scientific discipline including, of course, climate science.

Climate science is concerned with understanding Earth's climate system and its changes over time. Main components of the climate system are the atmosphere, the hydrosphere (liquid water on Earth, including the oceans and fresh-

---

Reproducibility: Principles, Problems, Practices, and Prospects, First Edition. Edited by Harald Atmanspacher and Sabine Maasen.

water systems), the biosphere (all living organisms on the planet), the cryosphere (frozen water on Earth, including sea-ice, snow on land, glaciers, and continental ice sheets), and the lithosphere (the parts of the solid Earth body interacting with the other components of the climate system). A particular global climate state is characterized by variables determining the energy balance of the climate system (e.g., the amount of incoming solar radiation or the atmospheric concentrations of warming greenhouse gases and cooling aerosols) and by interactions and feedback between the components of the climate system. Global climate changes are, therefore, the result of changes in variables, which can affect the planetary energy balance.

"Climate" is often casually defined as "average weather," indicating that there is a fundamental difference between short-lived weather conditions and the climate which is characterized by longer-term (typically 30-year) averages of meteorological variables like temperature or precipitation. As an example, the highest near-surface air temperature since the beginning of regular weather measurements at Potsdam (Germany) was established to be 39.1°C on August 9, 1992[1]. This extreme value is a result of a particular weather situation on that day and does not tell us much about the typical conditions at Potsdam.

On the other hand, one can, e.g., compute the 30-year average of the monthly mean maximum temperatures for August (see Fig. 12.1). For the time period from 1984 to 2013, this average value is 24.17°C which gives a feeling for the typical maximum temperatures for August at Potsdam during this time period. The corresponding value for the years 1893 to 1922 at the beginning of the Potsdam climate series is 22.47°C, by the way, which would suggest that climate has indeed changed over the course of the 20th century, although the statistical variations are about as large as the temperature difference in this case of measurements at one single location. For global temperatures, the warming can be much more robustly established; see, e.g., Hartmann *et al.* (2013).

The distinction between climate and weather is important because the chaotic nature of weather prohibits longer-term predictions, whereas climate science investigates the long-term means, which are generally stable. For example, different complex models or different simulations with the same model exhibit different trajectories of future warming due to the chaotic nature of the system, but they will all more or less agree on how the climate will warm on average (see Fig. 12.3). Finally, not acknowledging the difference between short-term weather events and longer-term climatic trends can be a means to cast doubt on modern climate change; see, e.g., Schneider *et al.* (2014).

---

[1]Data source: Long-term meteorological station at Potsdam Telegrafenberg at `www.klima-potsdam.de`, accessed 1 August 2014.

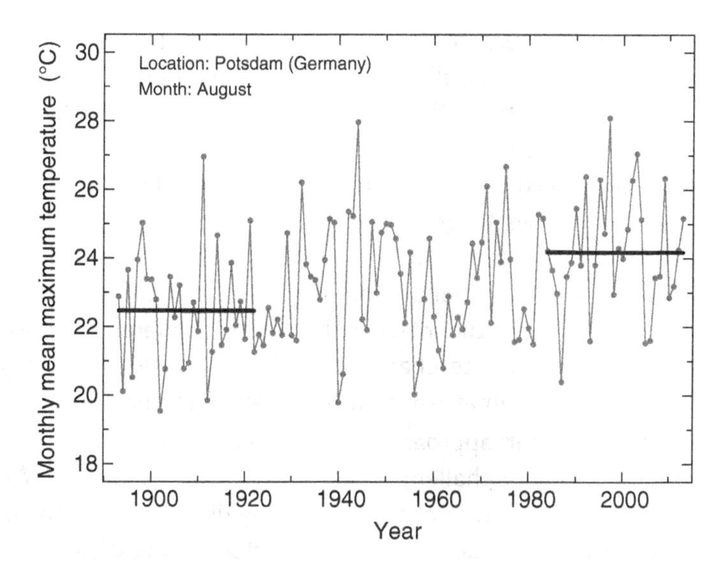

Figure 12.1: Monthly average maximum temperatures during August at Potsdam (gray circles and connecting lines) from 1893 to 2013 exhibiting large interannual variations due to weather phenomena and short-term climatic variability. For investigations of longer-term trends, averages over extended periods of time are more meaningful, as indicated by the 30-year climatological averages at the beginning and at the end of the time series (horizontal black lines).

The most frequent definitions of climate (and the example above) focus on meteorological variables measured in the atmosphere, in particular temperature, precipitation, or wind. Because the climate system consists not only of the atmosphere, however, it is important to note that the description of a climate state is not limited to atmospheric variables, but includes characteristics like ocean currents or vegetation distributions as well.

Given the nature of the physical, chemical, and biological processes within individual components of the climate systems and the multiple, often nonlinear interactions both within and between the different compartments, the Earth's climate system is a highly complex entity. It is also characterized by processes on spatial and temporal scales spanning many orders of magnitude. Spatial scales cover 11 orders of magnitude, ranging from 10^{-7} meters (less than a millionth of a meter) for the size of aerosols (solid or liquid air pollution particles) to 10^4 meters (10,000 kilometers) for the scale of our planet. Temporal scales extend over at least 14 orders of magnitude, from 10^2 seconds (minutes) for fast atmospheric processes to 10^{16} seconds (billions of years) for slow changes in solar luminosity. Understanding this complex system is the central challenge of climate science.

The special physical characteristics of the climate system also lead to particular challenges for this scientific discipline as far as reproducibility is concerned. The main difficulties arise from the complexity of the system described above and from the fact that there is only one specimen of planet Earth which cannot be studied under controlled conditions in the laboratory. This leaves climate scientists with only three ways to go. First, they can use detailed observations of the present-day climate system to understand the processes and feedbacks within the system. Second, they can build computer models of the climate system to improve our understanding of the interplay between its different components and of past as well as future climate changes. And third, they can turn to Earth's past to work out how the climate system has reacted to changes in its history.

Each of these different approaches is associated with particular scientific problems and reproducibility challenges discussed in the remainder of this chapter. In principle, these issues are not fundamentally different from the ones faced by other scientific disciplines dealing with data, historical investigations, or results based on computer models. In addition to these scientific challenges, however, the field of climate science also happens to be of particular importance to our society because of the observed warming of the planet due to anthropogenic emissions (Stocker *et al.* 2013) and the expected impacts (Field *et al.* 2014, Barros *et al.* 2014) of current and future climate change.

Avoiding these potentially dangerous effects of global warming requires decisive actions ranging from behavioral changes of individuals to nontrivial transformations of how humanity organizes power generation, industrial production, transport and agriculture. Society rightfully demands that the findings in climate science on which these far-reaching decisions are based adhere to the highest level of scientific rigor, including the principle of reproducibility. In that sense climate science is indeed under closer societal scrutiny than other scientific disciplines.

This scrutiny, however, is unduly exacerbated by organizations and individuals which are skeptical about the notion of anthropogenic climate change, its impacts, or the actions required to mitigate future warming and to adapt to climate change impacts. This was most obvious in the aftermath of the so-called "climategate" affair beginning in late 2009 when more than 1,000 e-mails were hacked from the Climatic Research Unit in the United Kingdom (Schiermeier 2009). In public debates, climate skeptics often express suspicions about data manipulation or raise issues of public access to climate science data and climate model source code. Thus, in a sense these attacks amplify public scrutiny and reiterate the strong demand for high standards of reproducibility within climate science. Given the difficulties described above, meeting these standards can be a challenge for the climate research community.

The reproducibility challenges for climate science will be discussed in the following. In Section 12.2 we describe some of the problems related to climate data and discuss reproducibility challenges for global temperature datasets. Section 12.3 discusses the problems for reproducibility in climate modeling. Finally, Section 12.4 is concerned with reproducibility issues in paleoclimate research, before conclusions and recommendations are presented in Section 12.5.

12.2 Reproducibility in Observational Climate Science

Much of our knowledge about the climate system is derived from the analysis of measurements and observational data. These include, e.g., meteorological data taken at weather stations around the globe as well as satellite observations. The reproducibility challenge for data-based climate science is very similar to any other scientific discipline working with data, and the standard solutions to these problems apply for climate science as well. Reproducibility relies on public databases where long-term storage of data, metadata, and data-handling software in non-proprietary formats is guaranteed (Kattge *et al.* 2014).

To illustrate some of the challenges specific to climate science, the debates surrounding global temperature datasets will be briefly discussed. Global surface air temperature is a key quantity defining the energy balance of our planet. Global temperatures over land are constructed from thermometer data gathered at several thousand weather stations around the world. To construct global temperatures, measurement errors and inhomogeneities have to be corrected (e.g., by comparison with data from adjacent stations) and the data have to be interpolated to fill areas where no measurements were taken. There are four different global land temperature datasets constructed by independent institutions which all yield very similar results (Hartmann *et al.* 2013). Land-surface temperature data can also be combined with data on sea-surface temperature (from ship or buoy measurements) to obtain a global temperature dataset covering both land and ocean areas.

Because global temperature datasets directly demonstrate the warming which Earth has experienced since the late 19th century, they have attracted considerable criticism by global-warming skeptics. In particular the land-temperature data (Jones *et al.* 2012b) from the Climatic Research Unit (CRU) have been under attack, including a surge of freedom-of-information-act requests to obtain the raw station data driven by suspicions of possible scientific misconduct (Heffernan 2009). Dealing with these requests proved rather difficult because of a lack of personnel (Russell *et al.* 2010) and because some of the raw data were obtained from national weather services that do not allow for public redistribution due to commercial interests (Thorne *et al.* 2011). In any case, a thorough review cleared the CRU scientists of any allegations of misconduct, but has criticized them to some degree for a lack of openness (Russell *et al.* 2010).

The often personal and unjustified attacks on CRU scientists cannot be excused. Nevertheless, this unpleasant episode has certainly further increased the awareness about reproducibility issues and the need for better data management within the climate science community (Stott and Thorne 2010, Santer *et al.* 2011, Thorne *et al.* 2011, Maibach *et al.* 2012, Osborn and Jones 2014).

The story of the CRU temperature data also raises the issue that one has to be very clear about what is meant by "data." It has been argued, e.g., that the concept of data in the philosophy of science may be conceived of as too simplistic (McAllister 2012). In the case of the CRU example, the patchy, unsystematic raw station data are not as crucial as critics think, while the processed, filtered data (which in a sense were already made public by the CRU scientists) are more important.

In any case, in terms of reproducibility, it is important to document both raw and processed data as well as all processing steps undertaken to translate the former into the latter. I think it is fair to say that climate science is certainly aware of these issues and is on track of providing public access to ever more data. It is also clear (see Fig. 12.2) that the warming signal in the global temperature data sets is robust as it has been reproduced by four independent research groups using different sets of station data and different methodologies (Hartmann *et al.* 2013).[2]

12.3 Reproducibility in Climate Modeling

Because the complex climate system of Earth cannot be studied under laboratory conditions to understand the interactions between system components or, say, the impact of future increases in greenhouse gas concentrations, computer models of the climate system are an important tool for climate scientists. Of course, these climate models cannot reflect the full complexity of the real Earth system and are not able to represent the processes at all temporal and spatial scales discussed in Section 12.1. Climate models prove invaluable, however, if

[2]The data sources are as follows. Berkeley: `http://berkeleyearth.lbl.gov/auto/Global/Land_and_Ocean_summary.txt`, accessed 31 October 2014; HadCRUT4: `http://www.cru.uea.ac.uk/cru/data/temperature/HadCRUT4-gl.dat`, accessed 6 October 2014 NASA GISS: `http://berkeleyearth.lbl.gov/auto/Global/Land_and_Ocean_complete.txt`, accessed 6 October 2014; NOAA: `http://www.ncdc.noaa.gov/cag/time-series/global/globe/land_ocean/ytd/12/1880-2014.csv`, accessed 31 October 2014.

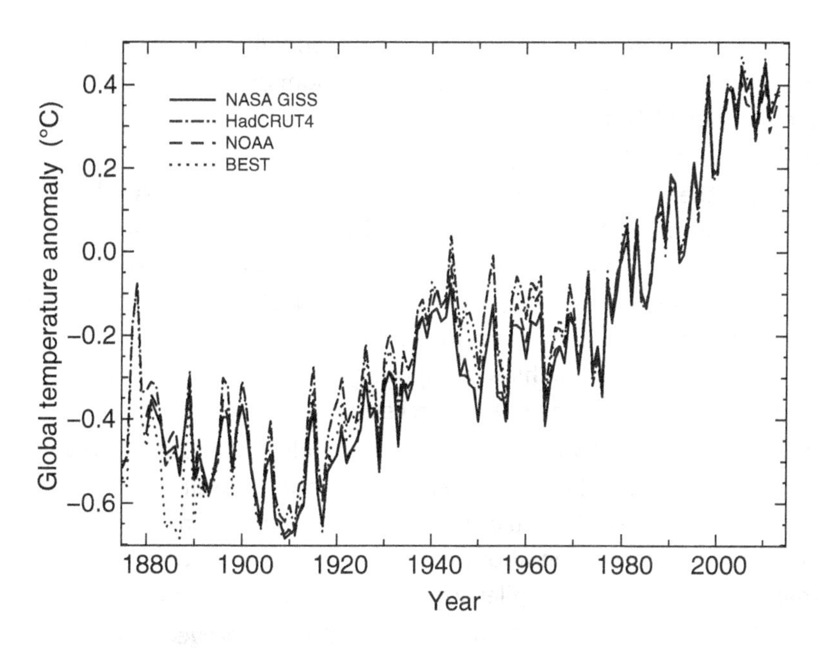

Figure 12.2: Global surface air temperature anomalies (relative to the average over the period 1971–2000) for the four different datasets and analysis methods described in the text.

their specific range of applicability and their limitations are properly taken into account. As George Box has put it in an often-cited quote (Box and Draper 1987): "Essentially, all models are wrong, but some are useful."

The limitations of climate models are often in the focus of climate skeptics. It should be pointed out, however, that the evidence for anthropogenic global warming is mostly based on empirical data. Models are just tools to better understand the processes and to project future warming under different assumptions of future greenhouse gas emissions. Nevertheless, the wide-spread use of computer models in climate science results in significant problems as far as reproducibility is concerned.

Reproducibility of climate models is deeply connected to model reliability and quality control. Standard procedures ensuring model reliability (Oreskes *et al.* 1994) include code verification in the narrow sense (ensuring that the model code correctly represents the mathematical model on which it is based), model validation (ensuring that the model results are consistent with measurements or observations), and independent confirmation (e.g., by comparison with other models). It should be pointed out that one of the most critical steps in model development, the calibration or "tuning" of free parameters in the model has to be performed independently of the model validation exercise. Note that overfitting (see Bailey *et al.*, this volume) is generally not an issue in climate modeling because the number of degrees of freedom of the system is much larger than the number of tunable parameters.

The reproducibility challenges for climate modeling are not fundamentally different from the challenges for other fields where simulation-based methods are used (Schwab *et al.* 2000, Peng 2011, see also Bailey *et al.*, this volume). Difficulties reproducing climate model-based results can arise from the unavailability of the original source code of the climate model or of input data required to run the simulations, from the platform dependence of climate model computations or from the requirement of high computing power (in particular for complex climate models) which might not be accessible to other groups.

Standard solutions to these challenges include the proper archiving of the relevant version of the climate model source code as well as of all input and output data of the model simulations performed for the investigation in question. In addition, information about the platform on which the experiments were performed and about the compiler version and settings are important.

Arguably the most important element of an improved reproducibility of computer-based science is to ensure open access to the models' source code (Peng 2011, Ince *et al.* 2012, Easterbrook 2014). Providing open-source code can be a challenge because it requires human resources and a technological platform which guarantees long-term availability in nonproprietary formats. Sometimes it can also involve legal difficulties related to copyright issues for commercial parts of the code. Nevertheless, climate science has taken up this challenge, and an increasing number of climate model source codes are publicly available.

In addition to the recommendations discussed above which aim at a replication of results with the same code (although potentially on a different platform), replication in a wider sense can also be attempted with a different methodology, in particular with a different climate model or a model of a different type. There are two common concepts in climate science which can be regarded as efforts related to the reproducibility of results: *coordinated model intercomparison* projects and the use of a *hierarchy of models* of different complexity. These efforts will be briefly discussed in the following.

Model intercomparison projects are a standard tool in climate science to compare results from different models and thus assess the uncertainty due to poorly constrained model parameters and differences in model design. They can also serve as a first step towards understanding differences between different models. Since different models are used to investigate the same question, the primary aim is not perfect replication, of course, but the more general reproduction of important research results. Note that although model intercomparison

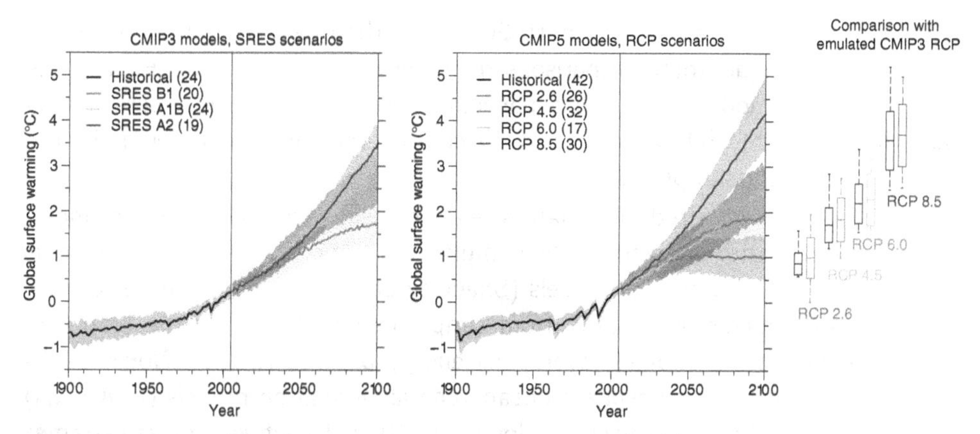

Figure 12.3: Climate model intercomparison of projections for future global temperatures under different emission scenarios for the fourth (left) and fifth (right) assessment report of the Intergovernmental Panel on Climate Change. (Reproduced from Knutti and Sedláček (2013) with permission by Nature Publishing Group.)

projects might appear to share some characteristics with meta-analyses (Ehm, this volume) they differ in the sense that the models used in intercomparison projects are typically driven by identical input data.

As an example for the application of model intercomparisons in climate science, the fifth phase of the Coupled Model Intercomparison Project (CMIP5, Taylor *et al.* 2012) and intercomparison projects of simpler models (Eby *et al.* 2013, Zickfeld *et al.* 2013) used a combination of idealized input data, historical climate forcing data, and scenarios for future emissions to assess the sensitivity of the climate system to carbon-dioxide increase, to understand historical climate change and to project future warming for different emission scenarios. These results from a range of models form the basis for the Fifth Assessment Report of Working Group I of the Intergovernmental Panel on Climate Change published by Stocker *et al.* (2013).

As an example, Fig. 12.3 shows projections of future global temperatures for different scenarios of future greenhouse-gas emissions from this report. The graph shows both the ensemble mean of the simulations with different climate models and the range of temperatures from the individual ensemble runs. The general agreement of the models increases the credibility of the projections, while the spread indicates the uncertainty resulting from model differences, parameter uncertainties, and dependence on initial conditions.

It should be kept in mind, however, that many models share model components or have a common evolutionary history which means that they are not fully independent (Masson and Knutti 2011). Furthermore, from a philosophical point of view, model intercomparisons might promote conformity. Finally, the fact that many models agree on a particular result does not necessarily imply that they are right. They might just all contain the same errors or make the same simplifying assumptions.

Another strategy used in climate science to deal with model-related uncertainties and to better understand the complex interactions in the Earth system is the use of different classes of models (Shackley *et al.* 1998). In climate science, models form a hierarchy ranging from simple models based on the planetary or latitude-dependent (zonally-averaged) energy balance (EBMs, North 1975) to the highly complex atmosphere-ocean general circulation models (AOGCMs) used, e.g., in the CMIP5 project (Taylor *et al.* 2012). In between these extremes are Earth-system models of intermediate complexity (EMICs, Claussen it et al. 2002), which comprise a very broad and diverse range of models, so in terms of definition this category may not be entirely helpful.

In any case, using a variety of models of different complexity to investigate the same problem is important in terms of reproducibility because it allows to test for shortcomings of models based on their design (e.g., because of a lack of certain interactions or feedbacks). Again, the focus is not on perfect replication but on a general confirmation of other studies. For example, the expected impact from a potential future grand solar minimum was first estimated using an EMIC (Feulner and Rahmstorf 2010) and later confirmed by studies using an EBM (Jones *et al.* 2012a) and AOGCMs (Anet *et al.* 2013, Meehl *et al.* 2013), firmly establishing this result. In addition, the use of faster models allows for a more systematic quantification of model uncertainties than would be possible with computationally expensive complex models, helping to assess the credibility of climate model results.

12.4 Reproducibility in Paleoclimatology

The third and final methodological approach to better understand Earth's climate system is the study of past climates and climate changes, a subdiscipline within climate science called paleoclimatology. Typical problems include the question of what kept early Earth warm enough despite a faint young Sun (Feulner 2012), the question of what triggered the global "Snowball Earth" glaciations around 700 million years ago (Pierrehumbert *et al.* 2011), the question of what can be learned for our future from a sudden warming event 55 million years ago (McInerney and Wing 2011), the question of how the periodic patterns of glacial and interglacial periods during the past 2.5 million years can be explained (Paillard 2001) or the question of what caused climate variations during the last millennium (Feulner 2011).

From a philosophical point of view, paleoclimate research is restricted by the fact that it can study only one planet with one history which is partly shaped by purely accidental events like impacts from space or major evolutionary changes (see also Kantz, this volume). The fact that such limitations can bias our scientific understanding is evident from the problems faced by theories of planetary-system formation modelled after our solar system when confronted with newly discovered systems of planets around other stars which did not at all look like the solar system (Howard 2013).

In scientific practice, however, these philosophical considerations are marginalized by more mundane challenges of paleoclimate research. In contrast to the instrumental period, no more-or-less precise measurements are available, so scientists have often to live with indirect measures (so-called proxies) for important climate variables like temperatures. The climate reconstructions from proxy data are usually patchy and rather noisy, with the quantity and quality of available data getting worse the further back one goes in time. Climate models can then be important tools to connect and make sense of these proxy data, provided that sufficient information about boundary conditions is available to actually run the models. In turn, paleoclimate information can also serve as validation information for climate models, as exemplified by the comparison of simulated and observed Northern-hemisphere temperatures over the last millennium shown in Figure 12.4.

As far as paleoclimatology is based on empirical measurements and climate model experiments, the practical challenges for reproducibility are similar to the ones discussed in Secrions 12.2 and 12.3 for observational and model-based climate science, respectively. The same is true for solutions improving reproducibility of results. In terms of data, there are extensive repositories for paleoclimate data, e.g., the World Data Center for Paleoclimatology operated by the National Climatic Data Center of the National Oceanic and Atmospheric Administration (USA), guaranteeing public access to a vast number of datasets. In terms of paleoclimate modeling, coordinated programmes like the third phase of the Paleoclimate Modelling Intercomparison Project (Braconnot *et al.* 2012) help to understand the reliability of individual climate models applied to paleoclimate problems by comparing them with other models.

Paleoclimate research is not purely of academic interest and not free from critical debates in the public. Provided that sufficient information is available and that uncertainties are carefully considered, paleoclimate studies can help to better understand the workings of the climate system and to better quantify

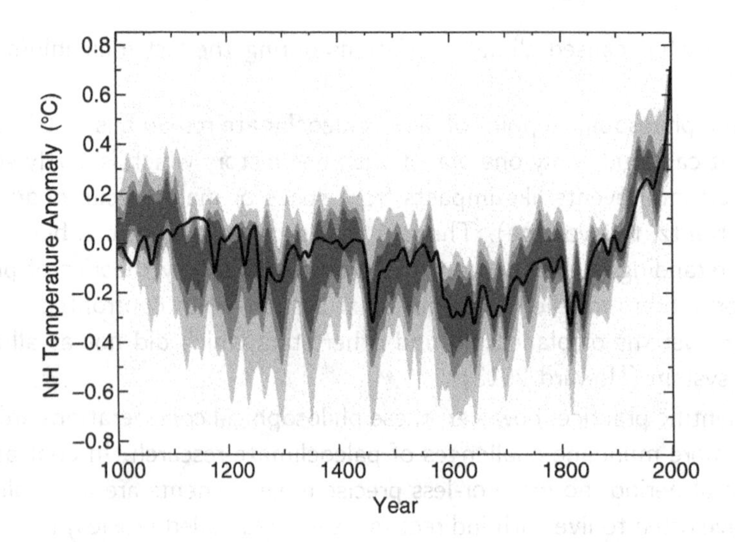

Figure 12.4: Ensemble reconstructions of Northern-hemisphere surface air temperature anomalies (with respect to the reference period 1370–1420) over the last millennium (Frank *et al.* 2010). The shaded regions (from light to dark gray) indicate the ranges covered by 90%, 70%, and 50% of the ensemble reconstructions. The solid black line shows a climate model simulation (Feulner 2011) which compares very well with the reconstructed temperatures.

ill-constrained parameters in climate models. Furthermore, it can serve as an important reference point providing the background on which global warming is unfolding. This is best illustrated by the controversies surrounding temperature constructions over the last millennium which demonstrate that temperatures in the recent past are higher than any time during the past 600 years (Mann *et al.* 1998).

The famous "hockey stick" graph (a more recent version is shown in Fig. 12.4) visualizing this has been prominently featured in the third report of the Intergovernmental Panel on Climate Change and has become one of the prime targets for global warming skeptics. Again, the researchers were cleared of any allegations of scientific misconduct by a thorough independent review commissioned by the National Research Council (USA) which broadly confirmed their findings, but highlighted the need for improved efforts ensuring the reproducibility of the results (National Research Council 2006).

12.5 Conclusions and Recommendations

This chapter has highlighted examples for the reproducibility challenges in climate science for observational studies, research based on climate modeling and investigations of Earth's past climate. It has also illuminated the societal context in which climate science is performed, where the importance of climate science for political decisions and attacks by global warming-skeptical individuals and organizations increase the pressure to adhere to strict principles of reproducible research. Of course, the fundamental motivation to increase the reproducibility of results originates within the scientific community itself. However, the critical questions from skeptical persons or organizations have undoubtedly played a role in the intensification of these efforts, as has the increasing demand for evidence-based policy (e.g., Head 2010).

The general recommendations for climate science are obvious. For observational work, data, metadata, and processing scripts should be thoroughly documented, placed in long-term storage and made available in public repositories. For modeling work, the same should be done with the model source code as well as all model input and output data. These recommendations do not imply that climate science is not doing this already. On the contrary, the field is increasingly aware of these issues; many initiatives and a wealth of data repositories are constantly improving the reproducibility of results in climate science. One can always do more, however, and the community should further intensify these efforts, because only in this way can climate science ensure its credibility with the public.

The methods for increasing reproducibility in climate science outlined here might appear to be an additional burden increasing the workload of scientists. Apart from the fact that they should be considered a key element of good scientific practice, however, they also have the potential to enhance the productivity both within research groups and the wider scientific community. Anyone who has wasted precious time trying to retrace badly documented research results will certainly agree.

Concluding on a personal note, I had worked in astrophysics before becoming a climate scientist, so I am in the position to compare these fields. Astrophysics is – in my humble opinion – of utmost cultural importance, of course, but has much less direct relevance for our society than climate science. Fortunately, astrophysics enjoys a lot of public fascination and attention, but in contrast to climate science, there is much less public scrutiny of its findings. No one (except for Johannes Kepler) cared much that Tycho Brahe was not exactly an enthusiastic proponent of open access to his data (to put it mildly). And the same holds true today – although there are open data and code initiatives in modern astrophysics today.

I must admit that the change into a field with so many heated public debates took some getting used to. But in the end, climate science has no choice. It cannot discard the societal implications of its findings, it cannot hide in the proverbial ivory tower to escape public scrutiny. The only thing it can do is to take up the challenge and to strive for the highest standards of reproducibility to enhance its public credibility and to prepare society for what lies ahead.

References

Anet, J.G., Rozanov, E.V., Muthers, S., Peter, T., Brönnimann, S., Arfeuille, F., Beer, J., Shapiro, A.I., *et al.* (2013): Impact of a potential 21st century "grand solar minimum" on surface temperatures and stratospheric ozone. *Geophysical Research Letters* **40**, 4420–4425.

Barros, V.R., Field, C.B., Dokken, D.J. Mastrandrea, M.D., Mach, K.J., Bilir, T.E., Chatterjee, M., Ebi, K.L., *et al.*, eds. (2014): *Climate Change 2014: Impacts, Adaptation, and Vulnerability. Part B: Regional Aspects*, Cambridge University Press, Cambridge.

Box, G.E.P., and Draper, N.R. (1987): *Empirical Model Building and Response Surfaces*, Wiley, New York.

Braconnot, P., Harrison, S.P., Kageyama, M., Bartlein, P.J., Masson-Delmotte, V., Abe-Ouchi, A., Otto-Bliesner, B., and Zhao, Y. (2012): Evaluation of climate models using paleoclimatic data. *Nature Climate Change* **2**, 417–424.

Claussen, M., Mysak, L.A., Weaver, A.J., Crucifix, M., Fichefet, T., Loutre, M.-F., Weber, S.L., Alcamo, J., *et al.* (2002): Earth system models of intermediate complexity: Closing the gap in the spectrum of climate system models, *Climate Dynamics* **18**, 579–586.

Easterbrook, S.M. (2014): Open code for open science? *Nature Geoscience* **7**, 779–781.

Eby, M., Weaver, A.J., Alexander, K., Zickfeld, K., Abe-Ouchi, A., Cimatoribus, A.A., Crespin, E., Drijfhout, S.S., *et al.* (2013): Historical and idealized climate model experiments: An intercomparison of Earth system models of intermediate complexity. *Climate of the Past* **9**, 1111–1140.

Feulner, G. (2011): Are the most recent estimates for Maunder Minimum solar irradiance in agreement with temperature reconstructions? *Geophysical Research Letters* **38**, L16706.

Feulner, G. (2012): The faint young sun problem. *Reviews of Geophysics* **50**, RG2006 (1–29).

Feulner, G., and Rahmstorf, S. (2010): On the effect of a new grand minimum of solar activity on the future climate on Earth. *Geophysical Research Letters* **37**, L05707.

Field, C.B., Barros, V.R., Dokken, D.J., Mach, K.J., Mastrandrea, M.D., Bilir, T.E., Chatterjee, M., Ebi, K.L., *et al.*, eds. (2014): *Climate Change 2014: Impacts, Adaptation, and Vulnerability. Part A: Global and Sectoral Aspects*, Cambridge University Press, Cambridge.

Frank, D.C., Esper, J., Raible, C.C., Büntgen, U., Trouet, V., Stocker, B., and Joos, F. (2010): Ensemble reconstruction constraints on the global carbon cycle sensitivity to climate. *Nature* **463**, 527–530.

Hartmann, D.L., Klein Tank, A.M.G., Rusticucci, M., Alexander, L.V., Brönnimann, S., Charabi, Y., Dentener, F.J., Dlugokencky, E.J., *et al.* (2013): Observations: Atmosphere and Surface. In: *Climate Change 2013: The Physical Science Basis*, ed. by T.F. Stocker, D. Qin, G.-K. Plattner, *et al.*, Cambridge University Press, Cambridge.

Head, B.W. (2010): Reconsidering evidence-based policy: Key issues and challenges. *Policy and Society* **29**, 77–94.

Heffernan, O. (2009): Climate data spat intensifies. *Nature* **460**, 787.

Howard, A.W. (2013): Observed properties of extrasolar planets. *Science* **340**, 572–576.

Ince, D.C., Hatton, L., and Graham-Cumming, J. (2012): The case for open computer programs. *Nature* **482**, 485–488.

Jones, G.S., Lockwood, M., and Stott, P.A. (2012a): What influence will future solar activity changes over the 21st century have on projected global near-surface temperature changes? *Journal of Geophysical Research (Atmospheres)* **117**, D05103 (1–13).

Jones, P.D., Lister, D.H., Osborn, T.J., Harpham, C., Salmon, M., and Morice, C.P. (2012b): Hemispheric and large-scale land-surface air temperature variations: An extensive revision and an update to 2010. *Journal of Geophysical Research (Atmospheres)* **117**, D05127 (1–29).

Kattge, J., Diaz, S., and Wirth, C. (2014): Of carrots and sticks. *Nature Geoscience* **7**, 778–779.

Knutti, R., and Sedláček, J. (2013): Robustness and uncertainties in the new CMIP5 climate model projections. *Nature Climate Change* **3**, 369–373.

Maibach, E., Leiserowitz, A., Cobb, S., Shank, M., Cobb, K.M., and Gulledge, J. (2012): The legacy of climategate: Undermining or revitalizing climate science and policy? *Wiley Interdisciplinary Reviews: Climate Change* **3**, 289–295.

Mann, M.E., Bradley, R.S., and Hughes, M.K. (1998): Global-scale temperature patterns and climate forcing over the past six centuries. *Nature* **392**, 779–787.

Masson, D., and Knutti, R. (2011): Climate model genealogy. *Geophysical Research Letters* **38**, L08703 (1–4).

McAllister, J.W. (2012): Climate science controversies and the demand for access to empirical data. *Philosophy of Science* **79**, 871–880.

McInerney, F.A., and Wing, S.L. (2011): The Paleocene-Eocene thermal maximum: A perturbation of carbon cycle, climate, and biosphere with implications for the future. *Annual Review of Earth and Planetary Sciences* **39**, 489–516.

Meehl, G.A., Arblaster, J.M., and Marsh, D.R. (2013): Could a future "Grand Solar Minimum" like the Maunder Minimum stop global warming? *Geophysical Research Letters* **40**, 1789–1793.

National Research Council (2006): *Surface Temperature Reconstructions for the Last 2000 Years*, Committee on Surface Temperature Reconstructions for the Last 2000 Years, National Academies Press.

North, G.R. (1975): Theory of energy-balance climate models. *Journal of Atmospheric Sciences* **32**, 2033–2043.

Oreskes, N., Shrader-Frechette, K., and Belitz, K. (1994): Verification, validation, and confirmation of numerical models in the earth sciences. *Science* **263**, 641–646.

Osborn, T.J., and Jones, P.D. (2014): The CRUTEM4 land-surface air temperature data set: Construction, previous versions and dissemination via Google Earth. *Earth System Science Data* **6**, 61–68.

Paillard, D. (2001): Glacial cycles: Toward a new paradigm. *Reviews of Geophysics* **39**, 325–346.

Peng, R.D. (2011): Reproducible research in computational science. *Science* **334**, 1226–1227.

Pierrehumbert, R.T., Abbot, D.S., Voigt, A., and Koll, D. (2011): Climate of the Neoproterozoic. *Annual Reviews of Earth and Planetary Sciences* **39**, 417–460.

Russell, M., Boulton, G., Clarke, P., Eyton, D., and Norton, J. (2010): *The Independent Climate Change E-mails Review*, accessible at `http://www.cce-review.org/pdf/FINAL%20REPORT.pdf`.

Santer, B.D., Wigley, T.M.L., and Taylor, K.E. (2011): The reproducibility of observational estimates of surface and atmospheric temperature change. *Science* **334**, 1232–1233.

Schiermeier, Q. (2009): Storm clouds gather over leaked climate e-mails. *Nature* **462**, 397.

Schneider, B., Nocke, T., and Feulner, G. (2014): Twist and shout: Images and graphs in skeptical climate media. In *Image Politics of Climate Change*, ed. by B. Schneider and T. Nocke, Transcript, Bielefeld, pp. 153–186.

Schwab, M., Karrenbach, M., and Claerbout, J. (2000): Making scientific computations reproducible. *Computing in Science & Engineering* **2**, 61–67.

Shackley, S., Young, P., Parkinson, S., and Wynne, B. (1998): Uncertainty, complexity and concepts of good science in climate change modeling: Are GCMs the best tools? *Climatic Change* **38**, 159–205.

Stocker, T.F., Qin, D., Plattner, G.-K., Tignor, M., Allen, S.K., Boschung, J., Nauels, A., Xia, Y., *et al.*, eds. (2013): *Climate Change 2013: The Physical Science Basis*, Cambridge University Press, Cambridge.

Stott, P.A., and Thorne, P.W. (2010): How best to log local temperatures? *Nature* **465**, 158–159.

Taylor, K.E., Stouffer, R.J., and Meehl, G.A. (2012): An overview of CMIP5 and the experiment design. *Bulletin of the American Meteorological Society* **93**, 485–498.

Thompson, P.A., and Burnett, A. (2012): Reproducible research. *CORE Issues in Professional and Research Ethics* **1**, paper 6.

Thorne, P.W., Willett, K.M., Allan, R.J., Bojinski, S., Christy, J.R., Fox, N., Gilbert, S., Jolliffe, I., *et al.* (2011): Guiding the creation of a comprehensive surface temperature resource for twenty-first-century climate science. *Bulletin of the American Meteorological Society* **92**, ES40–ES47.

Zickfeld, K., Eby, M., Weaver, A.J., Alexander, K., Crespin, E., Edwards, N.R., Eliseev, A.V., Feulner, G., *et al.* (2013): Long-term climate change commitment and reversibility: An EMIC intercomparison. *Journal of Climate* **26**, 5782–5809.

PART IV
LIFE SCIENCES

PART IV: LIFE SCIENCES
Introductory Remarks

Harald Atmanspacher

In the life sciences, issues with reproducibility are subtler than in the so-called exact sciences, as their sources are more diverse. One point is that ensembles of living systems are almost always characterized by distributions with natural variability, not only measurement errors. And, different from physics, this variability is not canonical, and can hardly be derived from first principles. This requires that "best-fit" distributions must be estimated from data, which is a statistical task (cf. Stahel as well as Shiffrin and Chandramouli, part II) not easy to accomplish. They are often substituted by the simplifying (and often wrong) assumption of a normal distribution and can entail questionable conclusions.

Moreover, particular areas of the life sciences explore the behavior of living systems in terms of their (allegedly) most basic constituents – molecular biology tries to reduce biosystems to the molecular level, or parts of neuroscience try to understand brain activity in terms of neurons (the so-called "neuron doctrine"). These attempts were guided by the idea that complex systems may be easiest to comprehend if they are reducible to smaller subsystems.

This reductive program has generated many important insights indeed, but it also became increasingly clear (e.g., in systems biology) that much of the complexity of living systems is only properly addressed at the level of systems as a whole rather than their constituents. The notion of the "mereological fallacy" points to this problem. It does not only express the fact that reductive explanations are typically not exhaustive, it also entails the challenge of how to identify proper macro-system properties and their relation to the micro-properties by appropriate coarse grainings.

Which properties should be reproduced under what conditions becomes a highly contextual question. And it is not restricted to the micro–macro distinction – it also includes how living systems are embedded in their physical environment (ecology); how political and economic factors shape the decisions and actions of funding agencies, public health institutions, pharmaceutical companies, and other stakeholders; and how epistemological mainstream positions change. All these issues play a role and are (ideally) considered in what has been called *translational research* (Woolf 2008, Tageja 2011).

The article by *Marianne Martic-Kehl and August Schubiger (Collegium Helveticum Zurich)* addresses such translational research "from bench to bedside," i.e., from pre-clinical animal experiments with medical drugs to their clinical applications. A threatening result of literature studies and meta-analyses of drug interventions in various areas is the extreme failure rate in clinical trials (highest in oncology) versus considerably higher success rates in animal studies. The authors report that a most conspicuous set of reasons for the low translational success is the low quality of preclinical experiments.

One particularly interesting point in this context relates to the misplaced belief that a *homogenization* of experimental conditions facilitates reproducible results in animal experiments. Mounting evidence indicates that even subtle differences in test conditions can lead to conflicting test outcomes, and experiments conducted under highly homogenized conditions may reveal local "truths" with little overall validity. Richter *et al.* (2009) reviewed this hypothesis based on data from a multilaboratory study on behavioral differences between inbred mouse strains. They showed that environmental and genetic homogenization are causes of, rather than cures for, poor reproducibility of experimental outcomes.

This conclusion calls for research into practicable and effective ways of implementing systematic *heterogenization* to attenuate scientific, economic, and ethical costs. From a systematic point of view, a major issue in the "homogenization fallacy" is the erroneous assumption that reproducibility is easier to achieve for artificially narrow distributions. As mentioned in the introduction to part III, the opposite is often the case: A distribution reflecting natural variability (or diversity) can stabilize reproducibility, an issue also addressed by Shiffrin and Chandramouli, part II. – For more current research on animal experiments in preclinical studies see the upcoming volume by Martic-Kehl and Schubiger (2015), as well as Freedman *et al.* (2015).

Gerd Folkers and Sabine Baier (Collegium Helveticum Zurich) address the problem area of drug development by considering the traditional strategy of the so-called *magic bullet* from an "ecological" point of view. Usually, a single target (typically a protein) is selected for a certain disease, and biochemists try to develop substances that block or stimulate it as selectively as possible. This works well for relatively "simple" diseases, but it turns out to be less efficacious for complex diseases, such as cancer.

Ecologically, complex diseases can be regarded as highly challenging environments for drugs. In this case, so-called "promiscuous" multi-target drugs have become interesting alternatives to single-target drugs. Obviously, their impact cannot be reproducible and efficient at the same level of specificity as it is for magic bullets. But considering multi-target designs in terms of broad target distributions, similar arguments apply as for the discussion of homogenization versus heterogenization above. Overly specific single-target drugs become insufficient to react to a "disease environment" that requires broad-band "adaptation."

Of course, lacking reproducibility can also be due to flawed designs and data analyses, especially in big-data sciences like genomics, proteomics, or other "-omics" research. For a new interesting direction in this context the notion of *forensic bioinformatics* has been coined. Two pioneers of this field, Baggerly and Coombes, described their approach as the attempt to reconstruct what might have gone wrong in particular experiments and their data analysis if the reported results are inconsistent with what their authors claimed to have done. In the spectacular "Potti case," more than 1,500 hours of tedious work (Baggerly and Coombes 2009) proved that a series of trivial errors and mistakes had led to serious ramifications, including clinical studies under false preconditions and the termination of their authors' employment.

But there are also constructive examples, where difficulties with reproducibility provide prospects for novel insight into situations from which reproducibility problems arise. One pertinent example is described in the contribution by *Johannes Lengler and Angelika Steger, computer scientists at ETH Zurich* working in the area of computational neuroscience. They investigate how irreproducible random behavior of individual neurons can lead to enhanced reproducibility at the level of neuronal ensembles. More aspects of relations between descriptive levels in neuroscience have been reviewed by Atmanspacher and Rotter (2008).

The message of this lesson is that randomness, often thought to be harmful for designing reliable systems, can – somewhat counterintuitively – become a stabilizing factor to improve control and predictability. The essential trick is to change the level of description at which the system is considered from individual neurons to neural networks. Focusing on inappropriate details hampers reproducibility, and a sound theoretical understanding becomes more difficult.

Michael Anderson at Franklin & Marshall College, Lancaster, chairs a program for scientific and philosophical studies of mind and is interested in neural correlates of mental states. His chapter builds upon his earlier meta-analysis of 1469 experiments correlating 9 different mental tasks with activated cortical regions. His basic result was that correlations are many-to-many – far from the naive hope that the "decades of the brain" might reveal a new phrenology based on new imaging technology.

This result is a slap in the face for reproducible mind–brain links. But Anderson does not stop short at this point. He tries to understand the sources of lacking reproducibility, and this leads him to propose an evolutionary-developmental hypothesis to explain it. It is based on the observation that adaptive dynamical systems often favor reusing existing components for new tasks over developing new circuits *de novo*. This idea of *neural reuse* has a number of substantial implications and opens up novel directions of research.

Kai Vogeley at the University Clinic Cologne leads us back to the problem of the mereological fallacy in its largest scope as far as biomedicine is concerned: the tension between *persons and things*. Roughly speaking, this tension boils down to the difference between how we ascribe causes to events in (folk) physics and in (folk) psychology. While physical causes derive from "objective" laws of nature (e.g., Newtonian mechanics) with high reproducibility, causes of psychological events are usually related to "subjective" motivations or intentions, typically with limited reproducibility, and hard to measure anyway.

Social neuroscience tries to reveal how much reproducibility there is still possible in social interactions, correlated with the two basic neural mechanisms of understanding other persons: the *social neural network* and the *mirror neuron system*. As is often the case, intriguing steps toward progress are based on pathologies, and Vogeley's prime examples of psychopathological deviations of experiencing others come from the area of autism spectrum disorders. Subjects with autism recruit the social neural network significantly less than control subjects do. And they sometimes judge their lack of empathy as an advantage for reacting (objectively) properly to challenging situations rather than being impeded by (subjective) distractions.

References

Atmanspacher, H., and Rotter, S. (2008): Interpreting neurodynamics: Concepts and facts. *Cognitive Neurodynamics* **2**, 297–318.

Baggerly, K.A., and Coombes, K.R. (2009): Deriving chemosensitivity from cell lines: Forensic bioinformatics and reproducible research in high-througput biology. *Annals of Applied Statistics* **3**, 1309-1334.

Freedman, L.P., Cockburn, I.M., and Simcoe, T.S. (2015) The economics of reproducibility in preclinical research. *PLOS Biology* **13**(6): e1002165.

Martic-Kehl, M., and Schubiger, P.A., eds. (2015): *Animal Models for Cancer: Discovery and Development of Novel Therapeutics*, Wiley, New York.

Richter, S.H., Garner, J.P., and Würbel, H. (2009): Environmental standardization: Cure or cause of poor reproducibility in animal experiments? *Nature Methods* **6**(4), 257–261.

Tageja, N. (2011): Bridging the translation gap – New hopes, new challenges. *Fundamental & Clinical Pharmacology* **25**(2), 163–171.

Woolf, S.H. (2008): The meaning of translational research and why it matters. *Journal of the American Medical Association* **299**, 211–213.

13
From Mice to Men: Translation from Bench to Bedside

Marianne I. Martic-Kehl and P. August Schubiger

Abstract. Drug development is time consuming and highly prone to attrition. One critical step within the process is the translation of results from preclinical animal studies to human patients. This concerns animal efficacy studies as well as safety testing in animals. In this chapter, we focus on efficacy studies, particularly on the translation of data from animal models to phase-II clinical trials, where small groups of human patients are treated with a new intervention.
A literature study by Thomson-Reuters revealed that the success rate of new development projects in phase-II clinical trials dropped from 28% to below 18% between 2009 and 2010 (Arrowsmith 2011), with insufficient efficacy being the reason for failure in more than half of the cases. A meta-analysis of results in the field of acute stroke research revealed that almost 500 interventions showed positive outcome in animal models, whereas only three are proven to be successful in humans suffering from acute stroke (Perel *et al.* 2007). Success rates differ greatly depending on the field of research, with oncology showing the highest clinical failure rates (Begley and Ellis 2012).
We will discuss potential reasons for the low translational success from bench to bedside with a focus on the quality of preclinical animal experiments. Examples from the literature will be examined, particularly from stroke research, as well as from our own research on the quality of animal experiments in the development of anti-angiogenic cancer drugs.

13.1 The Drug Development Process

The process of drug development, from first ideas to the product's launch on the market, takes 15 years on an average and costs millions of Euros. One out of thousands of potential drug candidates might be admitted to the market after successfully passing through the various sub-phases of the development process. These can be divided into research and development phases. Approximately two-thirds of the costs are invested in the latter, which include preclinical and clinical phases. The preclinical phase covers *in vitro* tests of action mechanisms and mutagenicity, *in vivo* efficacy studies, and *in vivo* toxicity tests (acute and chronic toxicity, mutagenicity, carcinogenicity, local tolerability, potential interactions with essential systems like lungs, kidneys, the nervous and cardiovascular systems

Reproducibility: Principles, Problems, Practices, and Prospects, First Edition. Edited by Harald Atmanspacher and Sabine Maasen.

and reproduction toxicology). *In vivo* efficacy tests are mainly performed in rodents like mice or rats. For acute and chronic toxicity testing the investigation of two species is required, one rodent and one non-rodent, with dogs typically being used as the non-rodent species.

The clinical part consists of three phases:

- clinical phase I: the substance is tested in healthy volunteers (10–50 subjects, mainly young men);
- clinical phase II: efficacy testing in a small population of patients (100–500 subjects);
- clinical phase III: application in a larger patient population (1,000–5,000 subjects).

Drug candidates that pass this process successfully are given market permission by the authorities of individual countries (e.g., the Food and Drug Administration, FDA, in the Umited States). After release onto the market, the product is still carefully monitored and evaluated in the entire patient population.[1]

The development process as it is known today has its origin in the 1960s. Since then approximately 150 drugs available on prescription have been withdrawn from the market due to safety problems. The withdrawal rate in Western countries has reached 3–4% (Zhang *et al.* 2012). Within the development process, the attrition rate of potential drug candidates is naturally far higher. One particularly critical step is the translation of results from preclinical animal models to the clinic (Folkers and Baier, this volume). A recent study by Thomson-Reuters showed that the success rates for new development projects in phase-II trials dropped from 28% to 18% between 2009 and 2010. The reason for failure was identified as insufficient efficacy in 51% of cases (Arrowsmith 2011).

Here is a brief overview of the historical background that finally led to the introduction of the regulatory drug development process as practiced today, in particular the requirement that animals be used as surrogate organisms.[2] Back in the 19th century, healthy animals were investigated for chemistry-based understanding of drug effects on physiological function. In 1906, the Pure Food and Drugs Act in the United States prohibited the interstate commerce of unsafe drugs. Official standards for drugs were identified and proper labeling became required.

---

[1]For details see www.bio-pro.de/magazin/thema/0018/index.html?lang=de and www.interpharma.ch.

[2]Compare the website of the FDA on the history of drug regulation in the United States at www.fda.gov/AboutFDA/WhatWeDo/History/FOrgsHistory/CDER/CenterforDrugEvaluationandResearchBrochureandChronology/ucm114465.htm. See also Hildebrandt (2004).

This Act was extended to the Food, Drug, and Cosmetic Act in 1938, with a requirement for proof of safety and authorized inspections emerging in the course of the sulfanilamide disaster of 1937, which killed over 100 people in the United States. The antibiotic sulfanilamide was administered as a cough syrup. It was easy to administer and particularly popular among children due to its sweet taste. The reason for the syrup's toxic effect was the sweet-tasting solvent diethylene glycol.

Streptomycin, which became available in 1945, was the first privately financed, nationally supervised and coordinated drug evaluation in history. The introduction of regulatory authorities and the movement toward supervised drug development was restricted to the Anglo-Saxon world. In continental Europe, diseases were regarded as non-comparable processes and, therefore, not suitable for statistical evaluation "since treatment never concerns populations but only individual patients" (Virchow 1847).

Approximately 10 years later, in 1954, the German company Grünenthal patented the sedative thalidomide, which was launched as Contergan® on the West German market in 1957.[3] Contergan® was considered particularly safe and, therefore, was recommended as a sleeping pill for pregnant women, as well as against morning sickness in early pregnancy. In 1959, first reports suggesting nerve damage caused by Contergan® appeared, and two years later Grünenthal was forced to withdraw the drug from the market. In the course of this tragedy, commonly referred to as the thalidomide disaster, thousands of babies were born with extremity abnormalities and Germany passed its first drug law in 1961. A year later, the United States called for revision of their drug law to include proof of efficacy and sufficient pharmacological and toxicological research in animals before a license for market authorization could be granted by independently acting authorities.

Experiments with animals need ethical justification. Ethical boards decide whether permission for animal experimentation is warranted, after balancing the benefits to humans against the suffering of animals. "Virtually every medical achievement of the last century has depended directly or indirectly on research with animals" is a prominent general justification found in official documents published by the US Department of Public Health (1994), the Royal Society (2002) in the United Kingdom or the UK Department of Health (2001).

This particular statement was not formally validated until 2008 when Robert

---

[3]See Grünenthal's webpage at www.contergan.grunenthal.info/grt-ctg/GRT-CTG/Die_Fakten/Chronologie/152700079.jsp; in addition see www.contergan.grunenthal.info/grt-ctg/GRT-CTG/Die_Fakten/Das_deutsche_Arzneimittelrecht_nach/152700071.jsp;jsessionid=B82E77391EBCF93D7DCD34660297CBF2.drp1.

Matthews published an article carefully investigating it (Matthews 2008). He came to the conclusion that the statement is anecdotal and does not generally hold true. Nevertheless, he is convinced that "animal models can and have provided many crucial insights that have led to major advances in medicine and surgery" (Matthews 2008).

13.2 Contributions of Animals to Medical Progress

Animal models have indeed provided crucial insights to major advances in medicinal history, some of which will be described in some detail in the following subsections.

13.2.1 Louis Pasteur and Vaccine Development against Anthrax and Rabies

In the second half of the 19th century, the French chemist and microbiologist Louis Pasteur discovered that chicken did not die from the bacteria causing fowl cholera, if they had been inoculated with an old culture of the pathogen prior to infection. Having discovered this vaccination principle, Pasteur translated it to other infective diseases, one of them being anthrax, the cause of millions of productive livestock deaths in the France of his time.

In a dramatic public experiment, Pasteur demonstrated the efficacy of attenuated anthrax as a vaccine for sheep and cows. The Agriculture Society of Melun, France, provided 60 sheep, 2 goats, and 10 cows for Pasteur's experiment. Pasteur demonstrated the efficacy of anthrax vaccination before an audience of several hundred people. All vaccinated animals survived infection with the virulent anthrax bacillus and appeared healthy, in contrast to the unvaccinated animals, which all died within two days (sheep and goats) or developed severe edema and body temperature increase (cows).

In the following years, Pasteur successfully translated the principle of attenuated microbe vaccination from animals to humans and succeeded in developing a human vaccination against rabies. Being unable to visualize the rabies virus, he inoculated the agent by intra-cerebral inoculation in the rabbit. To attenuate the infective agent, he dissected spinal cords from infected animals. Unaware of the fact that the viruses were mostly killed by this procedure, he created the basis for the production of subunit vaccines (Schwartz 2001, Pasteur 2002).

13.2.2 Paul Ehrlich and the Magic Bullet against Syphilis

The medical student Paul Ehrlich, who investigated histological staining in his doctoral thesis, proposed the magic bullet concept for toxins specifically targeting pathogens without affecting the host's cells.

In his quest for the magic bullet, Ehrlich investigated the effect of dye on parasites. In 1902, he systematically evaluated several hundred compounds in animals infected with *Trypanosoma equinum* and succeeded in treating infected mice with trypan red stain. However, trypanosomes developed resistance to the treatment, so his first attempts involving animal models were unsuccessful. But the story changed once Ehrlich continued his research to find a chemotherapeutic treatment for syphilis, which is caused by the bacterium *Treponema pallidum*, discovered in 1905 by Fritz Schaudinn and Erich Hoffmann.

Having turned his focus to derivatives of aminophenyl arsenic acid (Atoxyl®), synthesized in 1859 by a French biologist, he reevaluated all synthetic arsenic acid derivatives in rabbit models of syphilis developed by his Japanese co-worker. This search culminated in the identification of arsphenamine as the magic bullet against syphilis. Due to precautions associated with arsenic therapy, salvarsan – as Ehrlich named the substance in 1909 – was not tested in patients before animal experiments had proven safety and efficacy.

Second round clinical trials with larger patient populations revealed unpleasant side effects arising from the substance's insolubility in water and the fact that it degraded to a toxic product when dissolved without basification. After chemical modification, a water-soluble analogue was identified and marketed as Neosalvarsan®.

Paul Ehrlich won the Nobel Prize together with Ilya Ilyich Mechnikov in 1908 "in recognition of their work on immunity." Beside demonstrating pharmacological resistance and developing the first man-made antibiotic as a chemotherapeutic against syphilis, Ehrlich "introduced a modern research system based on the synthesis of multiple chemical structures for pharmacological screening in animal models of disease states" (Bosch and Rosich 2008, Kaufmann 2008).

13.2.3 Christiaan Eijkman and Frederick Gowland Hopkins and the Discovery of Vitamins

In the 19th century it was believed that good nutrition should contain proteins, carbohydrates, fats, inorganic salts, and water. It was at the end of the 19th century (1881) that Nikolai Lunin performed nutrition experiments in mice showing that young animals fed with purified components of milk did not grow. However, when small amounts of milk were added to their diet, the mice thrived.

Lunin concluded from this observation that milk contains "unknown substances" crucial for life.

Lunin's results were not recognized or followed up, and it was only at the turn of century (1897) that Christiaan Eijkman hypothesized a factor present in raw meat and unpolished rice that prevented chicken from developing beriberi, a disease responsible for the deaths of approximately 50% of children in the Dutch East Indies (modern day Indonesia) as well as Dutch soldiers in Sumatra. Eijkman misinterpreted his results and concluded that polished rice contained toxins that are neutralized by a factor in the rice cover and raw meat. He referred to it as the anti-beriberi factor.

Eijkman's successor in the Dutch East Indies, Gerrit Grijns, hypothesized in 1901 that instead of polished rice containing a toxic factor, it lacks a factor that is essential for life. This factor was named vitamine in 1912 by Casimir Funk, based on his assumption that the factor was an essential amine. In the 1920s the "e" in vitamine was dropped, as there was no supporting evidence that these essential nutrition compounds were amines.

Finally, it was due to the recognition of Eijkman's results by Frederick Gowland Hopkins, who emphasized the strong evidence that scurvy and rickets are caused by malnutrition connected with factors belonging to the same group as the anti-beriberi factor in 1912. He suggested that the area required further investigation. Presumably as a consequence, Elmer Verner McCollum and Marguerite Davis discovered the fat-soluble vitamin A, and Edward Mellanby found vitamin D and its crucial role to prevent rickets. Eijkman and Hopkins were awarded the Nobel Prize in 1929 for their crucial work in the discovery of vitamins (Rosenfeld 1997, Zetterström 2006).

13.2.4 Alexander Fleming, Howard Walter Florey, and Ernst Boris Chain and the Discovery and Development of Penicillin

At the beginning of the 20th century, hospital patients often suffered from severe and frequently fatal bacterial infection. The discovery of penicillin by Alexander Fleming through pure serendipity, therefore, was of extraordinary importance for medical treatment and opened the path for the development of antibiotics.

In 1929, the bacteriologist Fleming discovered mold in a petri dish containing staphylococci colonies after returning to his messy laboratory after a month of vacation. He observed that the bacteria did not grow closer to the mold than a distance of two centimeters. Fleming immediately recognized the great potential of his discovery for medical use. He named the active substance penicillin, as the mold producing it turned out to be a *Penicillium*. He performed several experiments on agar plates with different bacteria, as well as with dilutions of the mold in liquid media, showing that even at a dilution of 1:1000 its antibacterial activity was still present. Intravenous and intraperitoneal injections of penicillin solution into mice and rabbits revealed no harmful effects for the test animals.

Fleming's findings did not have a major impact at the time, as it was not possible for Fleming, or for his collaborating chemist, to extract and purify penicillin. It took a decade before the findings were picked up by the pathologist Howard Walter Florey and his co-worker and biochemist Ernst Boris Chain. Chain successfully purified penicillin and the two scientists proved the potency of the agent in mice injected with a virulent strain of streptococcus in 1939. First clinical trials were performed in 1941. The commercial success of penicillin came with the successful production of large amounts via a deep-tank fermentation process in the United States in 1941. Fleming, Florey, and Chain received the Nobel Prize in 1945 "for the discovery of penicillin and its curative effect in various infectious diseases" (Ligon 2004a,b, Wong 2003).

13.3 Translation Challenges in Different Fields of Research

The outstanding scientific discoveries and developments described in Section 13.2 provide inspiring examples of the successful translation of animal experimental results to human patients. Nonetheless, criticism of the predictive power of animals as human surrogates has become more and more pronounced in recent decades (Langley 2009, Knight 2007, Macleod 2011, Matthews 2008, Sena *et al.* 2007). Translation of animal experiments to the clinic is challenging and often fails completely.

An internal group at Bayer Healthcare investigated the reproducibility rate of published data within their company. In-house data from 67 projects in different fields of research were collected and it was shown that only 20–25% of the relevant published data was fully reproducible (Prinz *et al.* 2011).The biotechnology company Amgen in California performed a similar study, which revealed a mere 11% successful reproducibility fraction (Begley and Ellis 2012). Reasons for these rather poor results are diverse, extending from poor study design, incorrect statistical analyses, and insufficient study power to an over-representation of positive results in the literature[4] and high publication pressure.

The following subsections describe some of the problems in different fields of research when it comes to the translation of animal data to human patients.

---

[4]This is known as publication bias; see Ehm, this volume, for more details.

13.3.1 Vaccines against Human Immunodeficiency Virus (HIV)

Since the 1980s, human immunodeficiency virus (HIV) infection has spread steadily, and the WHO estimates that about 34 million people were affected by HIV in 2012. Although prevention and progress with antiretroviral drugs led to a stagnation of new infections in Western countries, this is not the case worldwide. A cure for HIV infection is still not available and AIDS (acquired immunodeficiency syndrome) is among the 10 most frequent causes of death.

Beside humans only non-human primates have a similar virus, the so-called simian immunodeficiency virus (SIV). The assumption is that SIV from chimpanzees (SIVcpz) was transferred to humans many times and thus HIV developed.

It is obvious that in this context the most frequently used animal models like rodents, cats, or dogs would not be suitable for the development of anti-HIV drugs. Since chimpanzees have a 98–99% genetic similarity to humans, this species has been considered to be the optimal animal model for efficacy trials of new anti-HIV vaccines. But it has also been shown that successful trials in chimpanzees did not translate to humans. This may indicate that high genetic similarity does not correspond to a sufficiently high similarity of chimpanzee and human physiology (Bailey 2005, 2008).

At the same time it was found that several species of macaques are susceptible to SIV infection, which leads to a disease very similar to human AIDS. Therefore, macaque models with SIV are being investigated for the development of AIDS treatments (Ambrose *et al.* 2007). The aim of these investigations is to find an efficient use of this model to increase the likelihood of finding a successful vaccine against HIV.

13.3.2 Acute Stroke Research

Acute stroke research is a field with very well-investigated translation rates and problems therewith. A systematic search of clinical and preclinical stroke treatments revealed that 494 interventions out of 1026 had a positive effect in animal models, whereas only three of them showed treatment success in patients suffering from acute stroke (O'Collins *et al.* 2006, Sena *et al.* 2007).

This strikingly high discrepancy between preclinical and clinical outcome was investigated on the preclinical side by meta-analyses in order to identify shortcomings in animal research that might explain the low translation rate. The Collaborative Approach to Meta-Analysis and Review of Animal Data from Experimental Studies (CAMARADES) identified 10 quality criteria for animal studies in acute stroke research (Sena *et al.* 2007):

1. publication in a peer-reviewed journal,
2. statement of control of temperature,
3. randomization of treatment or control,
4. allocation concealment,
5. blinded assessment of outcome,
6. avoidance of anesthetics with marked intrinsic properties,
7. use of animals with hypertension or diabetes,
8. sample size calculation,
9. statement of compliance with regulatory requirements,
10. statement regarding possible conflict of interest.

Most of the criteria listed are not specific to stroke research but apply to any disease field. Typical study design features like randomization, allocation concealment, blinded assessment of outcome, and sample size calculation are standard for clinical trials and would prohibit the official administration of a trial upon non-fulfillment.

Sena *et al.* (2007) assigned a quality score to each published study they identified, reflecting the total number of quality points fulfilled. The highest quality score reached was 7 (out of 10). By pooling effect sizes of all studies with the same quality score, they found a trend toward overestimation of efficacy with lower study quality.

The quality criteria that were mainly neglected included randomization, allocation concealment, blinded assessment of outcome, use of animals with hypertension or diabetes, and sample size calculation, i.e., the standard study design criteria for clinical trials and appropriately reflecting patient states with animal models (typical co-morbidities; Sena *et al.* 2007).

A meta-analysis on the effect of tirilazad for stroke treatment found poor standards in the performance of animal experiments as well (Perel *et al.* 2007). It was stated that treatment onset in mice took place substantially sooner (median 10 minutes) than in patients (median 5 hours).

In the field of stroke research, detailed investigations revealed shortcomings in animal experimentation regarding general study design features (which are standard for clinical trials) but also difficulties with the appropriate reflection of the disease in the model (lack of comorbidities) and discordances with the treatment schedule. These issues might well be crucially responsible for translational problems from bench to bedside.

13.3.3 Anti-Angiogenic Drugs in Cancer Research

The field of oncology has been reported to have the highest failure rates compared with other disease fields when it comes to translation of results from bench to bedside. An editorial in *Nature Reviews Clinical Oncology* stated that the licensing success rate for anticancer drugs is only 5% compared to 20% for cardiovascular disease (Hutchinson and Kirk 2011).

In an attempt to investigate the field of preclinical cancer research analogously to that of acute stroke research, we performed a systematic review of the treatment success of anti-angiogenic drugs in animal models of cancer (Martic-Kehl *et al.* 2015). In the 1980s, Judah Folkman postulated the anti-angiogenic treatment concept of starving cancer by cutting off its blood supply. It took more than a decade until the first preclinical studies were published on that topic, and we, therefore, identified research papers from the late 1990s up to 2011 for review. Analogous to the stroke researchers, we established a quality checklist with criteria we consider relevant for preclinical cancer research:

1. regulatory requirements fulfilled,
2. conflict of interest statement,
3. sample size calculation,
4. allocation concealment,
5. randomized allocation to test groups,
6. blinded assessment of outcome,
7. genetic variety (use of inbred, outbred strains),
8. inoculation type (sub-cutaneous, orthotopic inoculation of tumor cells, genetic modification, spontaneous tumor development),
9. tumor type (primary, metastatic),
10. immunodeficiency / comorbidity.

The 232 articles fulfilling the inclusion criteria for this study described a total of 299 different drugs and drug candidates in 1,538 individual experiments and outcomes. Similar to the investigators of stroke studies, we found rather low reporting of study design features like randomization, sample size calculation, or blinded assessment of outcome:

- fulfillment of regulatory requirements: 47%,
- statement of potential conflict of interest: 12%,
- sample size calculation: 0.5%,
- allocation concealment: 0%,
- randomization: 41%,
- blinded assessment of outcome: 2%.

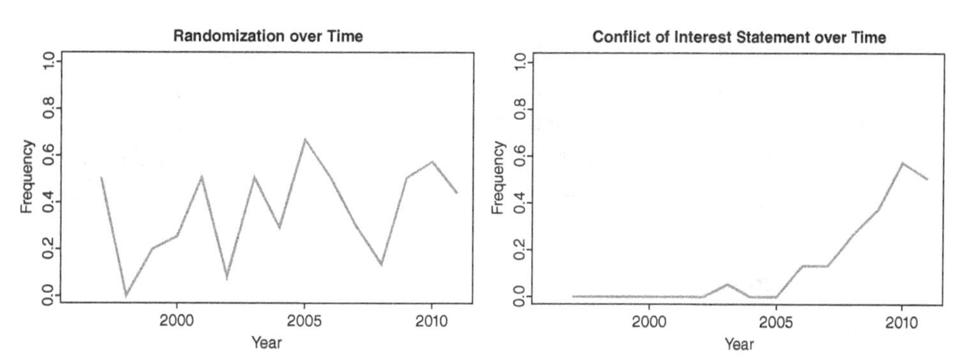

Figure 13.1: Study design features for 232 publications for the period from 1997 to 2011. Left panel: randomization; right panel: conflict of interest statement.

We hypothesized that over time, study design features such as randomization, or at least a statement mentioning it, might have increased over time, as the awareness of the bias that occurs without it has been raised in recent years (Bebarta *et al.* 2003, Sena *et al.* 2007). Nevertheless, time-line analyses of the general study design quality criteria revealed rather sobering results. With the exception of a "conflict of interest statement," no criteria showed any increase in reporting (Fig. 13.1). The increase of conflict-of-interest reporting in the middle of the first decade of the 21st century is not surprising, as numerous scientific journals began to demand such statements in their author guidelines.

The highest quality score identified was 6 (out of 10). Analyzing a potential connection between study quality and effect size, we found a trend toward a higher proportion of the undesirable outcome "cancer progression" with increasing study quality (Fig. 13.2), with the exception of the highest quality score of 6, possibly due to the very small number of studies within that category (3 studies describing 7 experiments versus 229 studies describing 1,531 experiments).

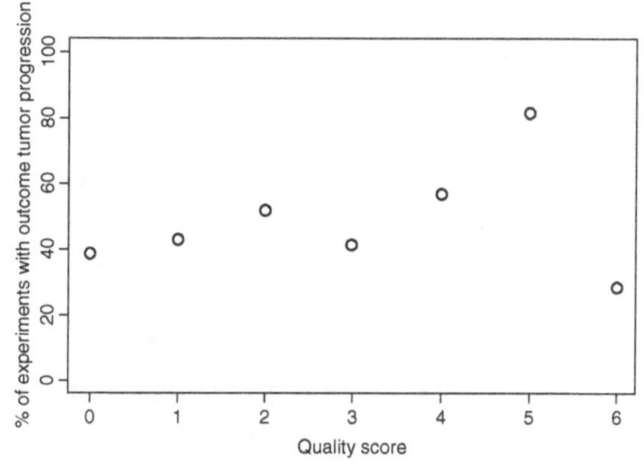

Figure 13.2: Incidence of experiments with outcome "tumor progression" versus quality score for 1,538 experiments described in 232 publications.

A rather unsurprising result of this analysis still seems worth mentioning. For every experiment reported, we determined whether the drug investigated was the main focus of the published study or whether it just served as a control substance. Within these two categories, we assessed the outcome distribution for the following outcome classes:

(1) cure,
(2) tumor regression,
(3) stable disease,
(4a) moderate tumor progression,
(4b) tumor progression.

The categorization of outcome was adapted from the Response Criteria Solid Tumors (RECIST) defined for human solid tumors (Eisenhauer *et al.* 2009). In addition to the RECIST categories, we introduced a refinement of "tumor progression," since this category was the one most frequently reported. The cut-off line for moderate progression was a minimum of 50% reduced tumor volume compared to control tumors.

outcome	main focus	control
1 and 2	50	1
3	88	4
4a	266	29
4b	233	43
N_{tot}	637	77

Table 13.1: Outcome distribution for drugs in the "main focus" category and in the "comparison" category, within category "drug type."

Table 13.1 shows that the chance of reporting a positive outcome (classes 1–3) was approximately 20% when a drug was the main focus of a study, compared to 7% for control drugs. Matching to this result, the chance of reporting outcome class 4b was 37% for drugs in the main focus of the study, compared to 56% for control drugs. This result can be assumed to arise from the high rate of failure to perform blinded assessment of outcome, as it shows a clear bias of investigators toward confirming their hypothesis.

Taken together, these results clearly agree with the findings of stroke researchers described in Section 13.3.2. It is probably fair to assume that the shortcomings of animal experimental studies in anti-angiogenic cancer research described above account for a large proportion of translation difficulties from the preclinical laboratory to the clinical bedside. Furthermore, it can be presumed that experiments in other fields of anti-cancer drug development suffer from similar problems.

13.3.4 Amyotrophic Lateral Sclerosis (ALS)

ALS is a degenerative disorder of the motoric nervous system. Even though it is estimated that more than 1 in 500 people will die of this disease, it is considered an orphan disease with a rare condition (Ludolph *et al.* 2010). ALS research is one of the first disease areas where, although therapeutic advances are still in the early stage, reliable and reproducible animal models exist (Ludolph *et al.* 2010). Nevertheless, the translation of numerous successful drug candidates in animal models of ALS has failed to date.

Benatar (2007) performed a meta-analysis of treatment trials in the SOD1 mouse model in an attempt to identify reasons for these failures. This model is a genetically modified model with mutations in the SOD1 gene. The analysis revealed that there was a strong publication bias toward studies using small animal numbers with positive outcome.

Furthermore, the majority of the 86 investigated publications suffered from methodological shortcomings. In only 31% of all cases, randomized allocation of animals to test groups was reported, and in even fewer (20%), investigators were blinded toward treatment allocation. Both lack of blinding and randomization are known to introduce avoidable bias to study outcome (Bebarta *et al.* 2003). Among all scientific publications included, only one discussed the power of the study performed.

Other potential problems with the translation of results were directly connected to the animal models used and the treatment protocol applied. Mutations in the SOD1 gene are known to be an important cause of familial ALS. The treatment of mice was initiated pre-symptomatically in 82% of the investigated studies. These two facts cast substantial doubt on the predictability of study outcome for patients with the sporadic form of ALS, as the pathophysiology of the disease is not fully understood. Thus, there is some likelihood of crucial biological differences between the two disease forms. Pre-symptomatic treatment might have some potential for the prevention of familially predisposed individuals, but it has no clinical value for people suffering from the sporadic form (Benatar 2007).

Finally, Benatar pointed out that rather than the statistical significance of the outcome, clinical relevance should be the driving force for translation decisions. Using high enough numbers of animals per group, a life span prolongation of one day might be statistically significant, but its clinical relevance is highly questionable.

In a consensus meeting in 2010, guidelines for preclinical animal research in ALS were defined (Ludolph *et al.* 2010). A first crucial point was the call for a clear statement of the purpose of a study. The study design of proof-of-concept studies can differ from preclinical studies, and it is important to make that distinction. Second, they require that a statement regarding possible conflict of interest be included, minimally reporting on financial sources and whether or not an applied intervention was provided by pharmaceutical industries. Other checklist points are as follows (Ludolph *et al.* 2010):

1. genetic background of animals, including breeding protocol, animal source, and gender,
2. potential source of variability (gender, litter, etc.),
3. animal numbers per group,
4. select key read-outs from phenotype,
5. proof of mechanism, including target tissue concentrations,
6. experimental design, including blinding, onset of clinical signs, disease progression, survival, histology, and statistical analysis,
7. external validity of data: data of control animals should be comparable with other published results; interlaboratory comparisons should be considered,
8. claims about therapeutic value,
9. replication of effect, preferably in an independent laboratory or in an additional mammalian animal model,
10. post-symptomatic efficacy,
11. efficacy linked to exposure or biomarker,
12. additional outcome measure, e.g. MRI, physiology, comprehensive histopathology, gene profiling, and proteomics.

The authors are confident that better standardization of preclinical ALS studies can be achieved using the above guidelines, facilitating nonhypothesis-driven and unbiased research, improving sensitivity of results as well as interlaboratory comparison of outcome (Ludolph *et al.* 2010). These efforts will hopefully translate to the development of successful clinical treatment for ALS.

13.3.5 Microglia

Microglia play a major role in immunological and anti-inflammatory processes of the central nervous system and brain aging. Probably due to 65 million years of evolutionary divergence between rodent and human immunological and neurological processes, it is challenging to translate results from mouse microglia research to humans (Smith and Dragunow 2014a). Another reason for the inadequate modeling potential of mouse models for neuroinflammatory diseases in humans is the absence of the heterogeneity inevitably found in human patients; this is due to inbreeding and the sterile laboratory conditions that test animals are kept in (Smith and Dragunow 2014a). In addition, the life span of rodents and humans differs crucially, and laboratory mice are rarely tested in old age, which would best reflect the state of neurological diseases in humans.

It is well established that there are crucial biochemical and pharmacological differences between mouse and human microglia behavior. Growth of microglia in culture dishes is one example, as rodent microglia grow on top of each other, whereas human microglia strongly adhere to the culture surface (Smith and Dragunow 2014a).

The frequently assessed neuroinflammatory endpoint measure of nitric oxide (NO) production is also worth mentioning. For the investigation of inflammatory processes mouse microglia are easily stimulated to produce NO; in contrast, human microglia are not so readily induced (Smith and Dragunow 2014a). It was also found that, when exposed to the drug valproic acid, human and rodent microglia show opposite responses. Another example is the efficacious performance of the drug propentofylline in rodent models of pain despite its showing no beneficial effect in human patients (Smith and Dragunow 2014a).

Smith and Dragunow suggested the use of human brain tissue samples from different individuals in order to bridge rodent research with the clinic, also with regard to appropriately reflecting the genetic variability in the human population. In a critical response to the opinion article by Smith and Dragunow in *Trends of Neurosciences*, Watkins and Hutchinson (2014) argued that commercially available samples of microglia are not suitable for rodent comparison, as they are mainly collected from first-trimester abortions where fetal brain tissue is damaged. Furthermore, fetal microglia are not fully developed and nothing is known about the environmental and genetic predisposition regarding the life style and genome of the mother.

In their response to Watkins and Hutchinson, Smith and Dragunow (2014b) pointed out the importance of studying adult microglia from post-mortem and biopsy tissues. In their opinion, the "best hope for understanding and treating human brain disorders is highest if animal and human approaches are working in unison."

13.4 Increasing Translational Success: Summary and Conclusions

There is a common understanding that efficacy testings of a newly developed drug almost always have to rely on animal models as surrogates for the human situation. However, clinical reality reveals that the majority of animal efficacy results could not be translated to the clinic (Langley 2009). The reasons for this deplorable fact are manifold and may be summarized in three groups: (1) suitable animal models and experimental protocol, (2) study design and statistical data evaluation, and (3) proper reporting. In order to increase translational success, we have to improve on all three groups.

We did already refer to the difficulties involved in finding suitable animal models. Consider the case of Contergan®, which was tested in a mouse model showing no teratogenic side effects (Shanks *et al.* 2009). However, later when human application revealed severe deformations of the extremities in newborns, it was discovered that the teratogenic side effect occasionally occurred in a variety of species such as cats, armadillos, guinea pigs, swine, ferrets, 10 strains of rats, 15 strains of mice, 11 breeds of rabbits, 2 breeds of dogs, 3 strains of hamster, and 8 species of primates (Shanks *et al.* 2009).

The results of antitumor drug translation are also highly dissatisfying. It has been shown that anti-angiogenic cancer drugs can lead to increased metastasis formation in patients, despite very promising preclinical test results. Retrospectively, it was shown by Ebos *et al.* (2009) that this complication could have been foreseen by using an appropriate metastatic mouse model in addition to the traditional primary tumor model. The lesson to learn from this example is the careful consideration of the appropriate animal tumor type, e.g., human cells, metastatic, orthotopic (Ebos *et al.* 2009).

Another very promising approach toward more reliable preclinical cancer drug testing could be the investigation of human companion animals like cats and dogs with spontaneous tumors. Companion animals usually share their environment very closely with humans, unlike laboratory rodents which live in rather artificial surroundings. Furthermore, it can be assumed that spontaneously growing tumors react to anticancer drugs more comparatively than induced grafts originating from a largely different species (i.e., humans). Pet dogs and cats are outbred species like humans, unlike laboratory animals which are mostly inbred strains (Vail and MacEwen 2000). The use of inbred animals might be one reason for lacking external validity of results, since sometimes different animal strains can react variably to specific treatments – as the thalidomide example demonstrated (Shanks *et al.* 2009).

In the case of breast cancer drug development, it has been shown that pet cats are highly suitable surrogates. There is considerable similarity between human and feline mamma carcinoma progression regarding age of disease onset, incidence, histopathology, biological behavior, and metastasis pattern (Weijer *et al.* 1972). The feline Her2 kinase domain shows 92% homology to the human Her2. This fact is particularly important, as Her2 positive human females generally develop more aggressive carcinomas and have a worse prognosis of disease progression (de Maria *et al.* 2005).

Apart from very disease-specific animal model questions, it has also been shown that the housing and husbandry of laboratory animals can have a substantial influence on study outcome and the external validity of results (Richter *et al.* 2009, Crabbe *et al.* 1999, Wahlsten *et al.* 2003). An illustrative example of the influence of animal husbandry on experimental outcome is the stress response in adult laboratory rodents depending on maternal care as neonates. It was shown that the adaptation of rodent pups to their environment in later months and years begins with the quality and duration of contact with their mothers and associated levels of stress hormones.

Experiments have demonstrated that laboratory rodents react with the least stress upon experimental handling as adults when they experienced mild stressors as neonates, such as short-term separation from their mothers (e.g., for 15 minutes). In contrast, stress proneness was elevated for animals that experienced either high stress or no stress at all as pups (Macri *et al.* 2011). That young animals experiencing no challenges during their early life period react with higher stress levels upon experimental handling later was referred to as phenotypic mismatch (Bateson *et al.* 2004). Such animals are simply not sufficiently prepared for the potential challenges they might experience in their laboratory environment later on. It is, therefore, important to be aware of the early life experiences of experimental animals, as they might have a certain impact on outcomes and, even more, on the external validity of the results.

Another important aspect is experimental design in general. For example, lack of appropriate randomization is known to bias results toward higher efficacy (Sena *et al.* 2007, Bebarta *et al.* 2003). Unfortunately, the simple reporting of randomization performance in a scientific paper does not necessarily mean that randomization was carried out appropriately. A health policy expert at the University of San Francisco put it as follows (Couzin-Frankel 2013):

> I was trained as an animal researcher. Their idea of randomization is, you stick your hand in the cage and whichever one comes up to you, you grab.

This procedure is strongly prone to bias. Animals that can be grabbed easily in a group of mice are often not as dominant and strong as the ones that avoid capture. Full randomization should be performed by using a random number generator; however, full randomization is not always the best method for animal experimental design. Complete randomization requires substantial numbers of animals in order to fully randomly balance all possible confounding factors (animal strain, age, gender, housing, etc.). A better approach would be the use of randomized block designs, where different sources of variability are distributed in a controlled manner to the individual block entities to which individual animals are assigned randomly. Other options include Latin square, crossover design, repeated measure design, or sequential design (Festing and Altman 2002, Festing *et al.* 1998).

The randomization dilemma describes one major problem of the meta-analytical approach in research. Authors of scientific papers may or may not report the performance of a specific experimental detail. And even if they report it, there is no guarantee that a specific aspect of randomization has been performed in an appropriate way. Proper reporting is another often-neglected point in scientific papers; it will be discussed later in this section.

Nonperformance of blinded assessment of outcome, like randomization, has been shown to bias outcome toward positive results or too high effect sizes (Sena *et al.* 2007, Bebarta *et al.* 2003). The blinding concept carries some potential for misunderstanding as well. The stroke researcher Dirnagl in his literature analysis came across studies in which double-blinded animal experiments had been reported (Germano *et al.* 1987, Xu *et al.* 2005). This finding made him suspicious regarding the peer-reviewing process, which acts as a general quality check for scientific publications (Dirnagl 2006).

Finally, one very important study design feature is the determination of optimal sample sizes prior to study performance. In animal research, sample sizes are often too small, which hampers the power of the study dramatically (see Stahel, this volume). Low study power reduces the chances of finding true effects, if they are present, but – which might weigh even heavier – it also increases the chance of detecting false positives (Button *et al.* 2013). False positives, again, are measurement artifacts with no external validity whatsoever.

The nonreporting or inappropriate reporting of data is yet another topic to mention. It has been convincingly revealed that investigators sometimes present only the experiment that supported their hypothesis, even if several other experiments they also performed did not confirm those results (Begley and Ellis 2012). Generally, it is estimated that only about half of all preclinical and clinical studies are reported at all (Chan *et al.* 2014, ter Riet *et al.* 2012).

Furthermore, positive results are by far more likely to be published than negative or neutral ones. This phenomenon is referred to as publication bias and has a number of consequences. First, effect sizes determined via meta-analyses are prone to overestimation, which leads to the misestimation of potential drug effects (Ehm, this volume). Sena and colleagues analyzed 16 systematic reviews in the field of acute stroke research and estimated that approximately one third of the effect sizes determined were due to publication bias overestimation (Sena *et al.* 2010). It is, therefore, fair to assume that many potential drug candidates are moving on in the developmental process, based on overestimated effect sizes in the preclinical phase. This can lead to high financial losses and, even worse, might put patients at unnecessary risk.

Another aspect of the unbalanced publication of scientific results is an unnecessary repetition of experiments. If negative and neutral results remain unpublished, researchers are simply unaware that the same or similar tests had been performed before. Again, this is highly cost-intensive and inefficient. The Dutch epidemiologist Gerben ter Riet tried to assess reasons for higher rates of positive versus negative or neutral results being published by surveying animal researchers in the Netherlands (ter Riet *et al.* 2012). Reasons for not publishing results listed by Dutch animal researchers were lack of statistical significance, technical problems, supervisors, and peer reviewers. As a potential solution to this dilemma, researchers believe that mandatory publication of study protocols and results might help (ter Riet *et al.* 2012).

The problems of low translation rates from bench to bedside, insufficient study design, inappropriate reporting, and publication bias have been investigated and discussed increasingly in recent years. The good news is that there have been initiatives toward improving the situation: e.g., the suggestion of a registry of preclinical experiments similar to clinical trials or the suggestion that journals should decide on the publication of scientific papers based on background information, hypothesis, and the material and methods section, but regardless of outcome (ter Riet *et al.* 2012).

Apart from such suggestions, there have also been efforts to actively improve the landscape of preclinical (animal) research.The first such initiative to mention here are the ARRIVE guidelines (Animal Research: Reporting of In Vivo Experiments), developed by the National Centre for the Replacement, Refinement and Reduction of Animals in Research (NC3Rs) in the United Kingdom.[5] These guidelines have been published in *PLOS ONE* and come with a checklist containing all the important aspects of study design and reporting for a scientific publication.

Moreover, *PLOS ONE* launched a reproducibility initiative in 2006 in order to provide a platform for scientists to publish their negative data or replication

[5]See `www.nc3rs.org.uk/page.asp?id=1357`.

results of existing studies in order to counteract publication bias and increase the availability of replication studies.[6] In May 2013, *Nature* adapted their authors' guidelines for life science articles, giving clear instructions on study design description, statistical analysis, and study performance.[7] The editor-in-chief of *Science* published an editorial entitled "reproducibility" in March 2014, in which she presented a new initiative toward more reliable and reproducible experiments, with special emphasis on preclinical studies as "one of the targets of recent concern" (McNutt 2014). Like the new guidelines of *Nature*, the initiative's aim is to obtain detailed information about pre-experimental plans of data handling, as well as study design and performance.

All the above-mentioned findings, suggestions, and initiatives raise hopes that a better estimation of drug effects will be possible through meta-analyses and systematic reviews and that data reproducibility will increase with better reported and appropriately designed studies. However, even with the very best evaluation and reporting of data, a successful translation of efficacy from animal models to the clinic cannot be guaranteed (cf. the HIV-model chimpanzee).

For each disease, specific and profound evaluations should be performed (genetically, physiologically, pharmacologically, and epidemiologically) to find an animal model that comes as close as possible to the human situation of disease (Ebos *et al.* 2009) in order to ensure the best achievable translation of efficacy results. Only if the most suitable human surrogate is investigated, along with careful and sound experimental planning and performance, will results reach the highest possible translation rates from bench to bedside.

References

Ambrose, Z., KewalRamani, V.N., Bieniasz, P.D., and Hatziioannou T. (2007): HIV/AIDS: In search of an animal model. *Trends in Biotechnology* **25**, 333–337.

Arrowsmith, J. (2011): Trial watch: Phase II failures: 2008-2010. *Nature Reviews Drug Discovery* **10**, 1.

Bailey, J. (2008): An assessment of the role of chimpanzees in AIDS vaccine research. *Alternatives to Laboratory Animals* **36**,381–428.

Bailey, J. (2005): Non-human primates in medical research and drug development: A critical review. *Alternatives to Laboratory Animals* **39**, 527–540.

Bateson, P., Barker, D., Clutton-Brock, T., Deb, D., D'Udine, B., Foley, RA., Gluckman, P., Godfrey, K., Kirkwood, T., Lahr, M.M., McNamara, J., Metcalfe, N.B.,

[6]See blogs.plos.org/everyone/2012/08/14/plos-one-launches-reproducibility-initiative/.

[7]See www.nature.com/authors/policies/repor-ting.pdf.

Monaghan, P., Spence, H.G., and Sultan, S.E. (2004): Developmental plasticity and human health. *Nature* **430**, 419–421.

Bebarta, V., Luyten, D., and Heard, K. (2003): Emergency medicine animal research: Does use of randomization and blinding affect the results? *Academic Emergency Medicine* **10**, 684–687.

Begley, C.G., and Ellis, L.M. (2012): Raise standards for preclinical cancer research. *Nature* **483**, 531.

Benatar, M. (2007): Lost in translation: Treatment trials in the SOD1 mouse and in human ALS. *Neurobiology of Disease* **26**, 1–13.

Bosch, F., and Rosich, L. (2008): The contributions of Paul Ehrlich to pharmacology: A tribute on the wccasion of the centenary of his nobel prize pharmacology. *Pharmacology* **82**, 171–179.

Button, K.S., Ioannidis, J.P.A., Mokrysz, C., Nosek, J.F., Robinson, E.S.J., and Munafo, M.R. (2013): Power failure: Why small sample size undermines the reliability of neuroscience. *Nature Reviews Neuroscience* **14**, 365–376.

Chan, A.W., Song, F., Vickers, A., Jefferson, T., Dickersin, K., Getzsche, P.C., Krumholz, H., Ghersi, D., and van der Worp, H.B. (2014): Increasing value and reducing waste: Addressing inaccessible research. *Lancet* **383**, 257–266.

Couzin-Frankel, J. (2013): When mice mislead. *Science* **342**, 922–925.

Crabbe, J.C., Wahlsten, D., and Dudek, B.C. (1999): Genetics of mouse behavior: Interactions with laboratory environment. *Science* **284**, 1670–1672.

De Maria, R., Olivero, M., Iussich, S., Nakaichi, M., Murata, T., Biolatti, B., and Di Renzo, M.F. (2005): Spontaneous feline mammary carcinoma is a model of HER2 overexpressing poor prognosis human breast cancer. *Cancer Research* **65**, 907.

Dirnagl, U. (2006): Bench to bedside: the quest for quality in experimental stroke research. *Journal of Cerebral Blood Flow & Metabolism* 26,1465-1478.

Ebos, J.M.L., Lee, C.R., Cruz-Munoz, W., Bjarnason, G.A., Christensen, J.G., and Kerbel, R.S. (2009): Accelerated metastasis after short-term treatment with a potent inhibitor of tumor angiogenesis. *Cancer Cell* **15**, 232–239.

Eisenhauer, E.A., Therasse, P., Bogaerts, J., Schwartz, L.H., Sargent, D., Ford, R., Dancey, J., Arbuck, S., Gwyther, S., Mooney, M., Rubinstein, L., Shankar, L., Dodd, L., Kaplan, R., Lacombe, D., and Verweij, J. (2009): New response evaluation criteria in solid tumours: Revised RECIST guideline (version 1.1). *European Journal of Cancer* **45**, 228–247.

Festing, F.M.W., Baumans, V., Combes, R.D., Halder, M., Hendriksen, C.F.M., Howard, B.R., Lovell, D.P., Moore, G.J., Overend, P., and Wilson, M.S. (1998): Reducing the use of laboratory animals in biomedical research: Problems and possible solutions. *Alternatives to Laboratory Animals* **26**, 283–301.

Festing, F.M.W., and Altman, D.G. (2002): Guidelines for the design and statistical analysis of experiments using laboratory animals. *Institute for Laboratory Animal Research Journal* **43**, 244–258.

Germano, I.M., Bartkowski, H.M., Cassel, M.E., and Pitts, L.H. (1987): The therapeutic value of nimodipine in experimental focal cerebral ischemia. Neurological outcome and histopathological findings. *Journal of Neurosurgery* **67**, 81–87.

Hildebrandt, A.G. (2004): Pharmacology, drug efficacy, and the individual. *Drug Metabolism Reviews* **36**, 845–852.

Hutchinson, L., and Kirk, R. (2011): High drug attrition rates – Where are we going wrong? *Nature Reviews Clinical Oncology* **8**, 189–190.

Kaufmann, S.H.E. (2008): Paul Ehrlich: Founder of chemotherapy. *Nature Reviews Drug Discovery* **7**, 373.

Knight, A. (2007): Systematic reviews of animal experiments demonstrate poor human utility. *Alternatives to Animal Testing and EXperimentation* **14**, 125–130.

Langley, G. (2009): The validity of animal experiments in medical research. *Revue Semestrielle de Droit Animalier* **1**, 161–168.

Ligon, L. (2004a): Penicillin: Its discovery and early development. *Seminars in Pediatric Infectious Diseases* **15**, 52–57.

Ligon, L. (2004b): Sir Howard Walter Florey – The force behind the development of penicillin. *Seminars in Pediatric Infectious Diseases* **15**, 109–114.

Ludolph, A.C., Bendotti, C., Blaugrund, E., Chio, A., Greensmith, L., Loeffler, J.-P., Mead, R., Niessen, H.G., Petri, S., Pradat, P.-F., robberecht, W., Ruegg, M., Schwalenstcker, B., Stiller, D., Van den Berg, L., Vieira, F., and von Horsten, S. (2010): Guidelines for preclinical animal research in ALS/MND: A consensus meeting. *Amyotrophic Lateral Sclerosis* **11**, 38–45.

MacLeod, M. (2011): Why animal research needs to improve. *Nature* **477**, 511.

Macri, S., Zoratto, F., and Laviola G. (2011): Early-stress regulates resilience, vulnerability and experimental validity in laboratory rodents through mother-offspring hormonal transfer. *Neurosciences and Biobehavioral Reviews* **35**, 1534–1543.

Martic-Kehl, M.I., Wernery, J., Folkers, G., and Schubiger, P.A. (2015): Quality of Animal Experiments in Anti-Angiogenic Cancer Drug Development – A Systematic Review. *Plos One* **10**, e0137235.

Matthews, R.A.J. (2008): Medical progress depends on animal models – doesn't it? *Journal of the Royal Society of Medicine* **101**, 95–98.

McNutt, M. (2014): Reproducibility. *Science* **343**, 229.

O'Collins, V.E., Macleod, M.R., Donnan, G.A., Horky, L.L., van der Worp, B.H., and Howells, D.W. (2006): 1026 Experimental treatments in acute stroke. *Annals of Neurology* **59**, 467–477.

Pasteur, L. (2002): Summary report of the experiments conducted at Pouilly-le-Fort, near Melun, on the anthrax vaccination. *Yale Journal of Biology and Medicine* **75**, 59–62.

Perel, P., Roberts, I., Sena, E., Wheble, P., Briscoe, C., Sandercock, P., Macleod, M., Mignini, L.E., Jayaram, P., and Khan, K.S. (2007): Comparison of treatment effects between animal experiments and clinical trials: Systematic review. *British Medical Journal* **334**, 197–202.

Prinz, F., Schlange, T., and Asadullah K. (2011): Believe it or not: How much can we rely on published data on potential drug targets? *Nature Reviews Drug Discovery* **10**, 712–713.

Richter, S.H., Garner, J.P., Zipser, B., Lewejohann, L., Sachser, N., Touma, C., Schindler, B., Chourbaji, S., Brandwein, C., Gass, P., van Stipdonk, N., van der Harst J., Spruijt, B., Voikar, V., Wolfer, D.P., and Würbel, H. (2009): Effect of population heterogenization on the reproducibility of mouse behavior: A multi-laboratory study. *Nature Methods* **6**, 257–261.

Rosenfeld, L. (1997): Vitamine – vitamin. The early years of discovery. *Clinical Chemistry* **43**, 4.

Royal Society (2002): *Statement of the Royal Society's Position on the Use of Animals in Research*, Royal Society, London.

Schwartz, M. (2001): The life and works of Louis Pasteur. *Journal of Applied Microbiology* **91**, 597–601.

Sena, E.S., van der Worp, B., Howells, D., and Macleod, M. (2007): How can we improve the pre-clinical development of drugs for stroke? *Trends in Neuroscience* **9**, 433–439.

Sena, E.S., van der Worp, H.B., Bath, P.M.W., Howells, D.W., and Macleod, M.R. (2010): Publication bias in reports of animals stroke studies leads to major overstatement of efficacy. *PLOS Biology* **8**, e1000344.

Shanks, N., Greek, R., and Greek, J. (2009): Are animal models predictive for humans? *Philosophy, Ethics and Humanities in Medicine* **4**, 2.

Smith, A.M., and Dragunow, M. (2014a): The human side of microglia. *Trends in Neurosciences* **37**, 125–134.

Smith, A.M., and Dragunow, M. (2014b): Response to Watkins and Hutchinson. *Trends in Neurosciences* **37**, 190.

ter Riet, G., Korevaar, D.A., Leenaars, M., Sterk, P.J., Van Noorden, C.J.F., Bouter, L.M., Lutter, R., Oude Elferink, R.P., and Hooft, L. (2012): Publication bias in laboratory animal research: A survey on magnitude, drivers, consequences and potential solutions. *PLOS ONE* **7**, e43404.

UK Department of Health (2001): *Memorandum for House of Lords Select Committee on Animals In Scientific Procedures*, UK Department of Health, London.

US Department of Public Health (1994): The importance of animals in biomedical and behavioral research. *The Physiologist* **37**, 107.

Vail, D.M., and MacEwen, E.G. (2000): Sponaneously occurring tumors of companion animals as models for human cancer. *Cancer Investigation* **18**, 781–792.

Virchow, R. (1847): Über die Standpunkte in der wissenschaftlichen Medizin. *Virchows Archiv* **1**, 3–19.

Wahlsten, D., Metten, P., Phillips, T.J., Boehm II, S.L., Burkhart-Kash, S., Dorow, J., Doerksen, S., Downing, C., Fogarty, J., Rodd-Henricks, K., Hen, R., McKinnon, C.S., Merrill, C.M., Nolte, C., Schalomon, M., Schlumbohm, J.P., Sibert, J.R., Wenger, C.D., Dudek, B.C., and Crabbe, J.C. (2003): Different data from different labs: Lessons from studies of gene-environment interaction. *Journal of Neurobiology* **54**, 283–311.

Watkins, L.R., and Hutchinson, M.R. (2014): A concern on comparing "apples" and "oranges" when differences between microglia used in human and rodent studies go far, far beyond simply species: Comment on Smith and Dragunow. *Trends in Neurosciences* **37**, 189–188.

Weijer, K., and Hart, A.A.M. (1972): Prognostic factors in feline mammary carcinoma. *Journal of the National Cancer Institute* **49**, 1697–1704.

Wong, J. (2003): Dr. Alexancer Fleming and the discovery of penicillin. *Primary Care Update for OB/GYNS* **10**, 124–126.

Xu, Z., Ford, G.D., Croslan, D.R., Jiang, J., Gates, A., Allen, R., and Ford, B.D. (2005): Neuroprotection by neuregulin-1 following focal stroke is associated with the attenuation of ischemia-induced pro-inflammatory and stress gene expression. *Neurobiology of Disease* **19**, 461–470.

Zhang, W., Roederer, M.W., Chen, W.-Q., Fan, L., and Zhou, H.-H. (2012): Pharmacogenetics of drugs withdrawn from the market. *Pharmacogenomics* **13**, 223–231.

Zetterström, R. (2006): C. Eijkman (1858-1930) and Sir F.G. Hopkins (1861-1947): The dawn of vitamins and other essential nutritional growth factors. *Acta Paediatrica* **95**, 1331–1333.

14
A Continuum of Reproducible Research in Drug Development

Gerd Folkers and Sabine Baier

Abstract. The increasing lack of efficacy in recent times is one of the major challenges of modern drug development that goes hand in hand with impressive low reproducibility rates of the associated experimental settings and results. A range of efforts has been made to identify and improve potential reasons. Commonly, the reasons are seen in technical failures like bad documentation and inappropriate animal models.
While these efforts are undeniable important and necessary, we go one step further and scrutinize the underlying assumptions of modern drug development. We reflect the process of drug development from an ecological perspective and claim that not all drugs need the same requirements concerning reproducibility. We argue that adapting the process of development to individual drugs, or at least groups of similar drugs, according to their specific "continuity requirements," a more appropriate assessment of reproducibility and efficacy could be achieved.

14.1 Introduction

In this paper we propose a possible *continuum of reproducible research* in biomedical science, particularly in the field of drug development, where the question of reproducibility has been especially pressing. Various recent studies showed that the success rate of a newly developed drug after first-in-human-testing is only about 10% (Hornberg 2012). If we consider cancer, the number goes down even more to a striking 5%.

The main reasons for these remarkably low success rates are, on the one hand, poor safety leading to dangerous toxic responses after the translation from animal models to humans. The most famous (besides thalidomide) and recent example of poor safety is probably the drug named TGN1412, a supposed immuno-modulatory substance that caused severe inflammatory reactions and multiple organ dysfunctions in healthy humans in clinical phase-I testing (Schraven and Kalinke 2008). Pure lack of efficacy of new drugs seems to be another reason of poor success rates. This means that many of the carefully designed drugs seem to do nothing else than cost lots of money spent during the elaborate process of their development (Arrowsmith 2011).

Reproducibility: Principles, Problems, Practices, and Prospects, First Edition. Edited by Harald Atmanspacher and Sabine Maasen.

But what are the reasons for these difficulties, and how does the question of reproducibility relate to them? To understand the connections between the success rate of a new drug and the reproducibility of their associated experimental settings and their specific results, we will first take a quick look at today's predominant drug development practices.

14.2 The Strategy of the Magic Bullet

Usually, biochemists start to develop a new drug by identifying an appropriate *target*. This can be a cellular or molecular structure that plays an important role for the disease to be cured. Today, most of the targets are proteins. At first glance, this may sound easy – in the manner of a simple key-and-lock system – but in fact, it can be severely difficult to identify the appropriate target. Once a possible target is identified, biochemists try to develop a corresponding key-substance to that target that either blocks it – as an antagonist – or stimulates it – as an agonist. Since only one biochemical target is attacked, this whole process from the identified target to the new drug in clinical use is called *single-target drug development*. A popular nickname for these single-target drugs is the "magic bullet."

At first glance, this kind of drug development may be tempting: Designing a highly specialized drug that does all the work in one shot – like a real magic bullet. And indeed, the single-target approach was able to produce many and very effective drugs for the treatment of a lot of diseases. These achievements cannot be denied. Nonetheless, as mentioned above, this wave of success of single-target drugs seems to subside due to an increasing lack of efficacy.

If we take a look at the experimental settings that accompany the process of drug development, we find that these settings not only seem no longer capable of producing effective drugs but also that they are hard to reproduce. In 2012, a survey found that 47 out of 53 medical research papers on the subject of cancer were not reproducible (Begley and Ellis 2012). Only 6 studies could be successfully reproduced. But, even trying to reproduce the results in exactly the same perfect setting – that is to say: same subjects, same equipment, same laboratory, same tools and assays, etc. – yields only substantially varying reproducibility rates between 32% and 99% in the field of drug development (Prinz *et al.* 2011).

The reasons for low reproducibility rates have been discussed for many years and, generally, are seen in bad documentation of the experimental settings (Gibb 2014), inappropriate statistical analysis or insufficient sample sizes, unsuitable animal models and the problem of implicit knowledge. Also bad target choices have been brought under suspicion. Besides these concrete faults, the high competition between laboratories may play a role, as well as the publishing practice of the so-called "good journals" preferring positive results and rejecting negative results (Martic-Kehl and Schubiger, this volume).

14.3 Specialists and Generalists

In order to meet this problematic situation, we have been trying to address the lack of both efficacy and reproducibility by looking for potential *continua* in the process of drug development. One important hint is that the low reproducibility rates of complex diseases like cancer correlate with the high lack of efficacy rates in cancer. Thus, if it were possible to identify a continuously distributed entity responsible for both problems this might indicate a starting point for their solution. But, to make a long story short, there is no such entity. In fact, trying to act as if there were one or trying to construct one retrospectively, the problem becomes even worse.

Now, this is the bad news. The good news is that if we switch our perspective from one "hidden" cause for lacking reproducibility and efficacy to a more *ecological and evolutionary perspective*, this can shed new light on these problems. In this sense, we propose to consider appropriate *continuity requirements for specific drugs*. But what does this mean in detail?

To answer this question, a quote from Edward Drinker Cope's book *The Primary Factors of Organic Evolution* may be helpful to begin with. In the chapter entitled "The Law of the Unspecialized," Cope distinguished the "specialists" and the "generalists" and compared them to each other (Cope 1896, p. 174; our italics):

> Degeneracy is a fact of evolution ... and its character is that of an extreme specialization, which has been, like an *overperfection of structure*, unfavorable to survival.

The specialist, so Cope, is extremely depending on stable environmental circumstances. Within its preferred environment, a specialist can be highly effective. But if circumstances change, maybe even just a little, the specialist is no longer able to maintain its existence and, therefore, cannot guarantee its future reproduction. The "overperfection of structure," as Cope depicted it, can be boon and bane at the same time.

The generalist, by contrast, may neither be the most effective nor the most perfect form of existence, but it has the ability to adapt itself to changing environments and reproduce itself even under inferior conditions. To put it briefly, the robustness, or resilience, of a generalist is higher as compared to that of a specialist. In fact, under changing circumstances any specialization will be an evolutionary dead end road in the truest sense of the word. On the long run, Cope wrote, the generalist will be in a better position for survival – hence, the "law of the unspecialized."

If we leave Cope's theory at this point and return to the problem of the poor efficacy and reproducibility rates, we can now explain why the exploration of *appropriate continuity requirements* is crucial for future drug development processes. Every (over-)perfectly designed single-target drug can be regarded as a specialist in Cope's sense. As a consequence, its continuity requirements are very high because it depends on maximal stable environmental conditions.

The evolutionary and ecological perspective on drug development not only explains the lacking efficacy of "magic bullets" but also the inferior reproducibility rates of clinical studies. Because the more specialized a drug is designed to act, the more likely it is to fail in curing highly complex diseases like cancer, whose prominent characteristics are massive inner robustness combined with an elevated versatility and even mobility over time. To put it in ecological words: Diseases can be considered as highly challenging, robust, and constantly changing environments for drugs.[1] Each new drug design has to meet these specific continuity requirements or else it will fail or depend on mere luck.

Following recent publications, clinical and pharmaceutical researchers seem to realize the limitations of an exclusive single-target approach in drug development. Increasing attention becomes focused on the design of multi-target drugs that act, in the sense of a generalist, like a "magic shotgun" rather than a "magic bullet" (Csermely *et al.* 2005, Hopkins 2008, Bolognesi *et al.* 2007, Lu *et al.* 2012, Zheng *et al.* 2014, Li *et al.* 2014).

Multi-target drugs attack more than one target at a time and can combat complex diseases much better than a single-target drug. Furthermore, as studies have shown, complex diseases can easily cope with the loss of single features because of their inner robustness and systemic nature so that a single-target approach is destined to fail either way. An effective drug simply has to eliminate more than one target to combat a complex disease as a whole.

[1]Such high variability is actually typical for living organisms and cannot be properly modeled by overly homogeneous distributions with spuriously minimized dispersion. It is tempting to speculate that an artificially small variability may have a destabilizing effect. This is indeed known in ecosystems: reducing the biodiversity of such systems can increase their vulnerability against small perturbations dramatically (McCann 2000, Pereira *et al.* 2012).

14.4 From Single-Target to Multi-Target Drugs

Regarding the lack of efficacy, pharmaceutical research is beginning to adapt to the different continuity requirements of different drugs and diseases. The more complex a disease is, the more should its potential curing drug be a generalist in the sense that it is effective even under challenging and changing circumstances. The less complex a disease is, the more likely will a single-target drug be able to combat it successfully.

Let us now turn to the continuity requirements regarding the problem of reproducibility. As opposed to natural species, drugs are not just given to us by nature and then simply deal with our diseases. We control and manipulate their development. That is to say, if we know the continuity requirements of a specific disease in humans, we are able to tailor the drug development process according to them.

If the disease requires a single-target drug solution, the continuity requirements are high. As a consequence, we will have to focus on and force the realization and fulfilment of maximal continuity requirements in the process of development and, thus, increase its reproducibility rate to a maximum. In other words, if a disease is identified to be fairly "simple," the continuity requirements are high and maximizing reproducibility is useful and practicable.

However, in case of complex diseases like cancer, it becomes disputable whether or not strict reproducibility is useful (Daston 2002). If the poor reproducibility rates of cancer studies as mentioned above can be ascribed to technical issues (bad experimental documentations, different tools, persons and assays, insufficient sample sizes, etc.), we just need to be more careful in the process of drug development. But if we think of it from an ecological perspective, the poor rates might have their origin in different reasons.

So-called "dirty" and "promiscuous" multi-target drugs do not exhibit any high specialization to act at complex diseases. In fact, they are built as generalists, able to deal with diseases providing low stability environments. This implies that they must be designed in a way that we can expect them to react with incompletely known, maybe even unknown, biochemical targets in unexpected ways. But it is exactly this unexpected, and perhaps even uncontrollable, behavior which makes them efficacious to an extent that single-target drugs will never be able to achieve.

To go one step further, dirty drugs seem to entail a lot of unknown curing abilities that are just waiting to be explored. For example, we may think of various anti-depressants or even the good old aspirin, which could be successfully re-positioned on the market in a new way with new curing effects. These success stories indicate that the way dirty drugs work may often not be addressed appropriately if conventional methodological principles such as strict reproducibility are applied.

A most recent and spectacular case is Sunitinib. The drug is marketed as Sutent® by Pfizer and is one of the most successful small-molecule kinase inhibitors in cancer therapy. Its anti-tumor activity is related to its ability to block tumor cell and tumor vasculature cell signaling via several tyrosine kinase inhibitor receptors (Finke *et al.* 2011). Although it emerged from a single-target drug design program, Sunitinib is now known to target a variety of kinases and inhibit cellular signaling at their related receptors such as PDGF, VEGF, and others.

In 1999, the potential of blocking the receptors of "platelet-derived growth factors", the "vascular endothelial growth factor" and the "fibroblast growth factor" were evaluated. Substituted indolinones as Sunitinib were in the focus of the investigation, which for the first time was directed at a substance with significant binding properties of the receptors VEGF-R2 and FGF-R1, in addition to high specificity for the PDGF receptor.

The following year the start-up company SUGEN (Redwood City, California) published a corresponding study and presented the newly found substance as SU6668. It was shown that SU6668 had good anti-angiogenetic properties and was effective against a plurality of tumors at least in the animal model. These results brought SU6668 into phase-I clinical trials.

Furthermore, SU6668 and its related compounds had been shown by crystallography to bind in the ATP-binding cleft at the PDGF receptor (Sun *et al.* 2000). Since ATP acts as an energy carrier in the cells throughout the body and is involved in numerous biochemical pathways, it is surprising that nevertheless such specificity for a handpicked number of receptors is possible. Considering such a mechanism was almost like breaking a taboo in the history of drug development.

Despite the high number of promising preclinical and clinical studies of the effect of a structurally related specific VEGF-R2 inhibitor, the compound Semaxanib (SU5416) yielded poor results in phase-III clinical trials for colorectal cancer. This caused SUGEN to withdraw it prematurely in 2002 (Expert Panel 2002). The next structural generation was characterized by improved pharmaceutical properties: water solubility, good receptor binding, and good bioavailability – and it immediately entered phase-I clinical trials (Sun *et al.* 2003). In addition to the breakthrough in the pharmaceutical properties, those findings, central for the process of drug development, also break with conventional methodological approaches.

14.5 Conclusions

Although characterized by an improved bio-availability and, hence, improved solubility, the two inhibitor molecules mentioned above actually were hardly soluble in water. This is quite surprising, since water solubility of a drug is one of the fundamental factors ensuring that the ligand can reach its site of action at all. One can only speculate about the reasons for this quite drastic miscalculation. The fragmentation of preclinical and clinical work into different scientific traditions and methods of chemistry, biology, and medicine (Fleck 1994) may have played an important role.

It is interesting to note that SUGEN scientists, when choosing a suitable inhibitor, apparently refrained from following a key doctrine of rational drug design. This doctrine states that the specificity of an inhibitor should be as high as possible, to make sure that only one target structure is modulated and unwanted side effects can be excluded.

The development exemplified by SUGEN reflects an emerging change from the principle of the magic bullet, which provides the modulation of one target only and perceives any multiple effects pejoratively as "dirty," toward the idea of a "drug with rich pharmacology" (sometimes called "poly-pharmacology"). On the one hand, this may be due to the fact that the current approach led and keeps leading to a "poor harvest." On the other hand, more sophisticated analyses show that the desired effect in a variety of approved drugs is elicited by binding to far more places than intended (Frantz 2005).

Considering this, it seems to be disputable whether or not the value of a maximized reproducibility rate in experimental settings and clinical studies at all costs is able *to reflect* low continuity requirements for drugs in complex diseases like cancer. Viewed from a less technical and more epistemological perspective, it appears quite ambiguous whether we just have not pushed the implementation of reproducibility far enough or whether the value of reproducibility itself has to be eased in the case of specific drugs. Maybe we have to reconsider some of the historical path dependencies and their values in standardized processes of drug development and adjust them to the specific continuity requirements of specific drugs and their corresponding diseases.

The idea of an ecologically induced "continuum of reproducibilities" in the process of drug development does not necessarily imply that reproducibility has to be given up completely for multi-target drugs. But it implies for sure that we must reconsider the specific continuity requirements of each specific drug anew by taking into account the expected characteristics of each future environment, respectively, vessel, of the drug – be it human beings or early experimental arrays.

Even back in early modern time, every alchemist knew that substances have their own and individual requirements, and that it is not up to the adept to force these requirements into standardized procedures and vessels. In fact, the art of alchemy was not so much about changing and transmuting substances but about making the right choice of the right vessel and the right procedure for every substance and every use of it in each individual instance (Baier 2015).

With this meaning and importance of the more general concept "vessel" in mind, this would mean to move from investigating the detailed micro-mechanisms by which a drug acts to its "emergent" impact on all possible scales – in other words, to move from precision to reliable performance.

References

Arrowsmith, J. (2011): Trial watch: Phase II failures: 2008–2010. *Nature Reviews Drug Discovery* **10**, 328–329.

Baier, S. (2015): *Feuerphilosophen. Alchemie als performative Metaphysik des Neuen*, Dissertation ETH Zurich.

Begley, C.G., and Ellis, L.M. (2012): Drug development: Raise standards for preclinical cancer research. *Nature* **483**, 531–533.

Bolognesi, M.L., Cavalli, A., Valgimigli, L., Bartolini, M., Rosini, M., Andrisano, V., Recanatini, M., and Melchiorre, C. (2007): Multi-target-directed drug design strategy: From a dual binding site acetylcholinesterase inhibitor to a trifunctional compound against Alzheimer's disease. *Journal of Medicinal Chemistry* **50**, 6446–6449.

Cope, E.D. (1896): *The Primary Factors of Organic Evolution*, Open Court, Chicago.

Csermely, P., Ágoston, V., and Pongor, S. (2005): The efficiency of multi-target drugs: The network approach might help drug design. *Trends in Pharmacological Sciences* **26**, 178–182.

Daston, L. (2002): Knowledge and science: The new history of science. In *Las Ciencias Sociales y la Modernización, La Función de las Academias*, ed. by M.H.R. de Miñón and J.-M. Scholz, Real Academia de Ciencias Morales y Politicas, Madrid, pp. 33–52.

Expert Panel (2002): Pharmacia's SU5416 not effective. *Expert Review of Anticancer Therapy* **2**, 5.

Finke, J., Ko, J., Rini, B., Rayman, P., Ireland, J., and Cohen P. (2011): MDSC as a mechanism of tumor escape from sunitinib mediated anti-angiogenic therapy. *International Immunopharmacology* **11**, 856–861.

Fleck, L. (1994): *Entstehung und Entwicklung einer wissenschaftlichen Tatsache*, Suhrkamp, Frankfurt. Originally published in 1935.

Frantz, S. (2005): Drug discovery: Playing dirty. *Nature* **437**, 942–943.

Gibb, B.C. (2014): Reproducibility. *Nature Chemistry* **6**, 635–636.

Hopkins, A.L. (2008): Network pharmacology: The next paradigm in drug discovery. *Nature Chemical Biology* **4**, 682–690.

Hornberg, J.J. (2012): Simple drugs do not cure complex diseases: The need for multi-targeted drugs. In *Designing Multi-Target Drugs*, ed. by J.R. Morphy and C.J. Harris, The Royal Society of Chemistry, Cambridge, pp. 1–13.

McCann, K.S. (2000): The diversity-stability debate. *Nature* **405**, 228–233.

Li, K., Schurig-Briccio, L.A., Feng, X., Upadhyay, A., Pujari, V., Lechartier, B., Fontes, F.L., Yang, H., Rao, G., Zhu, W., Gulati, A., *et al.* (2014): Multitarget drug discovery for tuberculosis and other infectious diseases. *Journal of Medicinal Chemistry* **57**, 3126–3139.

Lu, J.-J., Pan, W., Hu, Y.-J., and Wang, Y.-T. (2012): Multi-target drugs: The trend of drug research and development. *PLOS ONE* **7**(6): e40262.

Pereira, H.M., Navarro, L.M., Martins, I.S.S. (2012): Global biodiversity change: The bad, the good, and the unknown. *Annual Review of Environment and Resources* **37**, 25–50.

Prinz, F., Schlange, T., and Asadullah, K. (2011): Believe it or not: How much can we rely on published data on potential drug targets? *Nature Reviews Drug Discovery* **10**, 712.

Schraven, B., and Kalinke, U. (2008): CD28 superagonists: What makes the difference in humans? *Immunity* **28**, 591–595.

Sun, L., Liang, C., Shirazian, S., Zhou, Y., Miller, T., Cui, J., Fukuda, J.Y., Chu, J.Y., Nematalla, A., Wang, X., Chen, H., Sistla, A., Luu, T.C., Tang, F., Wei, J., and Tang C. (2003): Discovery of 5-[5-fluoro-2-oxo-1,2- dihydroindol-(3Z)-ylidenemethyl]-2,4- dimethyl-1H-pyrrole-3-carboxylic acid (2-diethylaminoethyl)-amide, a novel tyrosine kinase inhibitor targeting vascular endothelial and platelet-derived growth factor receptor tyrosine kinase. *Journal of Medicinal Chemistry* **46**, 1116–1119.

Sun, L., Tran, N., Liang, C., Hubbard, S., Tang, F., Lipson, K., Schreck, R., Zhou, Y., McMahon, G., and Tang, C. (2000): Identification of substituted 3-[(4,5,6, 7-tetrahydro-1H-indol-2-yl)methylene]-1,3-dihydroindol-2-ones as growth factor receptor inhibitors for VEGF-R2 (Flk-1/KDR), FGF-R1, and PDGF-Rbeta tyrosine kinases. *Journal of Medicinal Chemistry* **43**, 2655–2663.

Zheng, H., Fridkin, M., and Youdim, M. (2014): From single target to multitarget/network therapeutics in Alzheimer's therapy. *Pharmaceuticals* **7**, 113–135.

15
Randomness as a Building Block for Reproducibility in Local Cortical Networks

Johannes Lengler and Angelika Steger

Abstract. Neurons in the brain show highly irreproducible behavior. Attempts to repeat controlled experiments lead to very different responses of the very same neuron. Nevertheless, on the system level the behavior of a neuronal ensemble is pretty much reliable. In this chapter we discuss possible explanations of why such irreproducible behavior can occur, thereby giving a case study for building a mathematical model to describe a biological observation.
Then we discuss the effects that random, irreproducible behavior has on ensembles of neurons. Surprisingly, the randomness at the individual level leads to more (not less, as one might intuitively expect) predictable behavior of the system. We also show some of the benefits that emerge from random individuals. Thus, neuronal ensembles provide an example where reproducibility fails if the object of study is badly chosen (single neurons level), but reproducibility is achieved if we shift our focus to the statistical properties on a system level.

15.1 Introduction

In the brain, a neuron conveys information to its neighbors in form of spike trains, i.e., by a sequence of electric pulses. Pulses *per se* do not differ from each other, so the information must lie either in the number or the timing of the pulses. Both options do not exclude each other, and they are known as rate coding and temporal coding, respectively. The question of what exactly is the encoding of the brain has been studied intensively for many decades, but understanding is still embarrassingly poor.

More than 20 years ago, Softky and Koch (1993) analyzed the variance of interspike intervals (ISIs), i.e., the duration between two successive spikes. Surprisingly, they found that the lengths of these intervals have an extremely high variance: The sequence of ISIs has a coefficient of variation (CV) of roughly 1, an indication that they are governed by random processes (cf. Section 15.3.1 for technical background and definitions).

Sequences with a CV of 1 are not uncommon in nature. For example, a sample of radionuclides that undergo radioactive decay also show such behavior. But

Reproducibility: Principles, Problems, Practices, and Prospects, First Edition. Edited by Harald Atmanspacher and Sabine Maasen.

then, for radioactive decay this is very plausible, as it is assumed to be stochastic in nature. However, that the behavior of individual neurons is governed to such a large extend by randomness came as a surprise. The surprise turned into a puzzle when computer simulations could not reproduce this phenomenon: in simulations spike trains had usually been much more regular, and in subsequent years, people realized that it is actually not so easy to build systems achieving a CV close to one.

As it seems from today's perspective, one of the reasons is that we, as persons and researchers, intuitively believe that randomness is harmful for designing a reliable system. Using a high level of randomness clearly contradicts the standard principles of engineering. In a computer chip every transistor works highly accurately, and great efforts are undertaken to make the output of the chip perfectly predictable. Nevertheless, it seems that nature is using randomness as a design principle that allows to achieve predictable results.

The aim of this chapter is to discuss such phenomena in the context of neuronal spiking, both for single neurons and small networks of neurons. Note that this chapter discusses reproducibility on a microscopic level (single neurons, or clusters of only a few thousands of neurons). The macroscopic level of large-scale brain activity is discussed by Anderson (this volume). Although methodology and focus are very different, both chapters come essentially to the same conclusion: If we focus our attention on unsuitable specific details in the brain, then reproducibility is low, and it seems rather hopeless to gain a thorough theoretical understanding. On the other hand, even immensely complex systems as the brain rely on reproducible effects – and by shifting the perspective just a little (considering CV instead of single spikes, or considering networks instead of single neurons) these effects can indeed be measured.

15.2 Spike Trains and Reproducibility

Intuitively, one expects that repeated trials with identical stimuli yield identical responses. On a neuronal level this has been tested by observing a fixed neuron's response to several presentations of exactly the same stimulus, and investigating how much the different spike trains of the same neuron deviate from each other. It turns out that reproducibility is pretty high in sensory areas like the retina, and in the first areas of processing like the thalamus (Liu *et al.* 2001, Keat *et al.* 2001), but that reproducibility ceases as the signal is further processed.

Already in the primary visual cortex variability is high. Figure 15.1 shows the firing rate of a single cell in response to a cycle of drifting grating: (a) shows the result of three trials, while (b) shows the firing rate averaged over seven

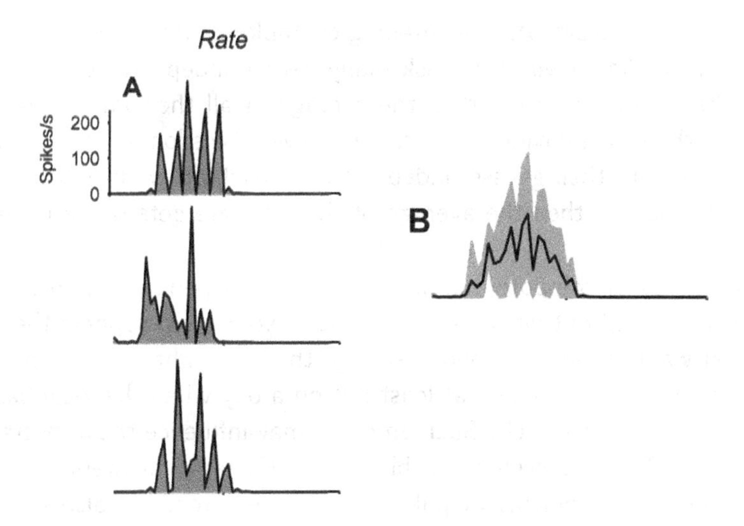

Figure 15.1 (from Carandini 2004): Response of a single cell to an identical stimulus. (a) three different trials, and (b) average over seven trials.

trials. Clearly, the reproducibility of a single neuron is not very high. This may be less surprising than it seems on first sight. There is strong evidence that information is not encoded by single neurons but by populations of hundreds of neurons (population coding). So the behavior of individual neurons need not be reproducible. It suffices that the population behaves in some reproducible way.

Unfortunately, until today the exact mechanism for such an encoding is still unknown. In particular, we do not yet know what kind of reproducibility we should look for. There are strong indications, like the one mentioned above, that not reproducibility in a deterministic sense might be decisive, but a reproducibility that is hidden in properties of probability distributions.

To gain some intuition for this, consider a collection of 100 coin tosses of different coins that each have a different (and unknown) bias for turning up head. What can we learn about these biases from tossing each coin just once? Essentially nothing. Even if the probability for head is very small, it is not impossible that head comes up, so we cannot draw any solid conclusions from single tosses. However, we do gain quite a lot of information about the total head count. Even if all biases are different, the law of large numbers tells us that the total head count will be very similar in all subsequent trials, so it is a reproducible quantity.

The problem in neuroscience is that we are not dealing with collections of coins that are independent of each other, but with neurons that are complicated and heavily interacting objects. Interaction may have positive or negative effects, depending on the context. An enlightening example comes from a phenomenon called *wisdom of the crowd*: If we ask many people independently to give estimates for the weight of an ox, then the average of all their guesses is a much better match than the individual guesses. However, "wisdom of the crowd" only works if people make their guesses independent of each other. If they are allowed to discuss the matter, then the average of the estimate gets worse (Surowiecki 2005).

To get an idea of the potential advantages of independent answers imagine that you want to find out whether the people in your town support the government. If you want to do a telephone survey, then it might be a good idea not to ask all people the same day – at least not on a day when the local basketball team won an important match. Such an event may influence the answers (Healy *et al.* 2010), and it is a *systematic* bias that affects all answers in the same direction. For some problems of public opinion research, see also Keller (this volume).

Some techniques how to recognize and overcome such biases are discussed by Ehm and by Stahel (this volume). Similarly, if you observe a neural population for a short time (very short, perhaps a few milliseconds), then it is rather easy to estimate the average spike rate if they all spike independently – the law of large numbers applies in the same way as it applies to independent coin tosses. However, if the spikes do *not* behave independently of each other, but rather behave like a townsfolk sharing a common short-lived trend, the task becomes more difficult. In the presence of correlations or systematic biases the law of large numbers does no longer apply and an observation over a short time period only cannot be used for reliable statements. The effects and origins of biases in various other areas of research are discussed by Martic-Kehl and Schubiger, Porter, Ehm (all in this volume), and many other publications. Methods to deal with them statistically are described by Stahel (this volume).

It should be emphasized that the effects of correlations in neural populations are poorly understood. It is not even clear how dependent or independent neuronal spike trains are. There has been some support for the hypothesis that different neurons act rather independently (Ranganathan and Koester 2011), but direct measurements face substantial technical difficulties, and the debate is far from being settled.

As mentioned in Section 15.1, the spike train of a *single* neuron has a CV close to one, i.e., it looks like a random (Poisson) sequence. If we move from a single neuron to a *population* of neurons, then we could face two situations: (1) every single neuron follows a Poisson spike train, but all neurons fire simultaneously, or (2) every single neuron follows a Poisson spike train *and* the spike train of the union of all neurons also behaves like a Poisson spike train.

As we will see in Section 15.4, the second case has many favorable properties, and we will exhibit a potential candidate that leads to such a behavior: randomness. We will compare two network architectures. The first one involves little randomness, the neurons in the network have strongly correlated spike trains, and the spike trains are far from Poisson. The second one involves much more randomness, but not more than seems biologically plausible. In this case, the neurons exhibit a behavior similar to when they act independently and the CVs of both the individual neurons and the population are close to one.

In Section 15.3, we will study how large CVs can arise, and discuss several biological mechanisms that may influence the CV of neural spike trains. Thereby we provide a case study for building a mathematical model of a biological property. We should mention that there have been successful attempts to reproduce a CV of one by simulations (Shadlen and Newsome 1998, Troyer and Miller 1997). However, as Bailey *et al.* (this volume) point out there are some severe problems when other laboratories want to reproduce simulations, even if reports on the simulations are faithful. These problems can be avoided by mathematical models, since a mathematical proof contains (by definition) all information that is needed to reproduce the results. In particular, for mathematical models it is possible and customary to reproduce the results (follow the proof) in the peer-reviewing process.

An important difference between mathematical models and statistical ones is discussed by Stahel (this volume): In a statistical model one has to take great care to separate the process of hypothesis forming from the process of verifying this hypothesis. One of the origins of systematic biases in medical literature identified by Martic-Kehl and Schubiger (this volume) is that some studies lack a clear formulation of their hypotheses before they start their research.

In contrast, for mathematical models the resulting properties are *provably* correct, i.e., a mathematical model is a (quite exceptional) example of *perfect* reproducibility. Of course, this does not mean that a mathematical model is a perfect model of what it is supposed to model. But it provides a theory that explains a particular property. How useful the theory is depends, on the one hand, on how realistic its assumptions are and, on the other hand, on its predictive power for additional properties.

15.3 Spike Trains

The molecular biology of single nerve cells is very well understood, see, e.g., Levitan and Kaczmarek (2001). For our purposes it is sufficient to know that a neuron receives input from other neurons via synapses and conveys it into output signals. The input opens some ion channels, which in turn changes the membrane potential of the neuron. Once this potential reaches a certain threshold the neuron fires, i.e., it sends an action potential along its axon. For a single neuron one can track the moments in time, say, $t_1, t_2, \ldots$, at which such an action potential, a "spike," is generated. This sequence is called the spike train of a neuron.

Usually one is not much interested in the actual moments in time, but in properties of the sequence, like number of spikes per second (the firing rate) or properties of the interspike intervals (ISIs) defined as $X_i := t_{i+1} - t_i$. In this paper we are mainly concerned with the coefficient of variation (CV, to be defined formally below) of the ISIs. As mentioned, it has been observed experimentally by Softky and Koch (1993) that neurons have a CV that is approximately one. However, it turned out to be a challenging task to come up with mathematical models of neurons that match this observation.

Before we come to this, we first try to give some intuition for why we need a mathematical model. A neuron is a very complicated object, and textbooks on neurons (such as Levitan and Kaczmarek 2001, Fain *et al.* 2014) easily fill hundreds of pages. However, a single neuron tells us very little about how the brain works. For that we need to study the interaction between neurons by simulations. For the simulations we need to compress hundreds of pages of knowledge into a few facts that we can implement. The simpler the facts (the model), the easier and faster the implementation. On the other hand, simplifying the model too much bears the risk that it will not reflect key properties of the "true" network that we would like to understand.

In the remainder of this section we illustrate this quest for an appropriate model. In our exposition we partly follow Lengler and Steger (2015). We start with a very simple model, check its CV (and observe that it is close to zero instead of one) and subsequently add more and more additional features so that we eventually end up with a neuron that exhibits a CV of approximately one. Along the way, we gain some understanding of models that have a CV of one and those which don't.

15.3.1 Some Technical Background: Poisson Spike Trains and Coefficient of Variation

A *Poisson spike train* is a sequence of spikes that, intuitively speaking, is as random as possible. This can be quantified as follows: a Poisson spike train has the property that a nonoccurrence of a spike for some time interval Δ has no influence on the likelihood that a spike will occur in, say, the next second. In mathematical terms this is formalized as follows: all ISIs (i.e., all time intervals between subsequent spikes) are stochastically independent and exponentially distributed: The probability $P(t)$ to see an interval of length t is given by

$$P(t) = e^{-\lambda t} ,$$

where λ is the rate (spikes per second) of the spike train. The exponential distribution is *memoryless*, which implements our above assumption that the likelihood of a spike in the next second does not change, regardless of how long we already waited. Note that for most distributions this is not the case. For example, for the uniform distribution in the interval $[0, 1]$, the information that we have not seen a spike between 0 and 0.5 changes the likelihood for a spike between 0.5 and 1 from $1/2$ to 1.

The discrete counterpart of the exponential distribution is the so-called geometric distribution defined as follows. Consider flipping a (biased) coin with probability p of seeing head and probability $1-p$ of seeing tail. Then the number of coin tosses until we see head for the first time is memoryless: regardless of how many consecutive tails we have already seen, the probability of seeing a head in the next toss of the coin is still exactly p.

The exponential distribution and the geometric distribution are actually related: if we repeatedly flip a coin with success probability $1/n$, and after each coin flip wait for time $1/\lambda n$, then the time until we see a success for the first time is (in the limit of $n \to \infty$) exponentially distributed with parameter $1/\lambda$. Note that this implies that a Poisson spike train can essentially be obtained by a sequence of very unlikely events. We will later use this characterization.

We close this section by defining the so-called *coefficient of variation* (CV). For this we recall that the *expectation* of a random variable is a formalization of the long-run average value of repetitions of the experiment it represents. For example, if we consider a fair coin, then we expect that we have to toss the coin twice until we see a head. Formally, a geometrically distributed random variable with parameter $p = 1/2$ has expectation 2: If the probability for head is p, then the probability that we need k tosses until we see head for the first time is $(1-p)^{k-1} \cdot p$. The expectation $\mathbb{E}[X]$ for a geometrically distributed random variable X with parameter p is thus given by

$$\mathbb{E}[X] = \sum_{k \geq 1} k \cdot \Pr[X = k] = \sum_{k \geq 1} k \cdot (1-p)^{k-1} \cdot p = 1/p.$$

The variance of a random variable captures the (squared) average deviation of the variable from its expectation: $\mathrm{Var}[X] = \mathbb{E}[(X - \mathbb{E}[X])^2]$. For a geometrically distributed random variable with parameter p some elementary calculations show that in this case $\mathrm{Var}[X] = 1/p^2$. Similarly, an exponentially distributed random variable with parameter λ has expectation $1/\lambda$ and variance $1/\lambda^2$.

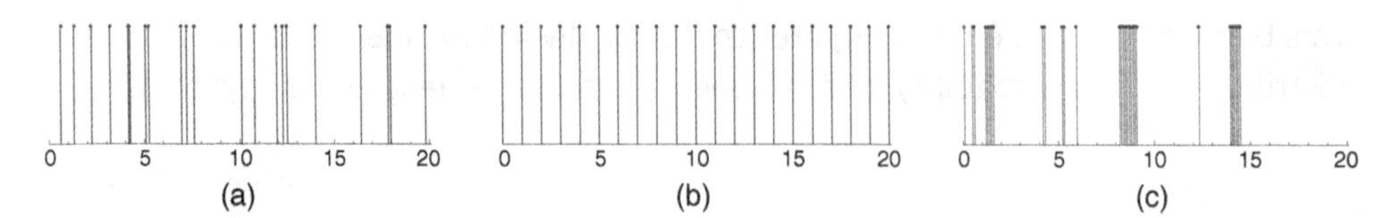

Figure 15.2: Different types of spike trains. The x-axis shows time (arbitrary units) and each peak corresponds to a spike. (a) Poisson spiking (CV $= 1$), (b) regular spiking (CV $= 0$), and (c) random spike bursts (CV > 1).

The *coefficient of variation* (CV) of a random variable X can now be defined by the expectation and variance of its distribution:

$$\text{CV} = \frac{\sqrt{\text{Var}[X]}}{\mathbb{E}[X]}.$$

For a Poisson spike train the ISIs are exponentially distributed and thus, independent of the parameter λ, a Poisson spike train always has a CV of 1. Theoretically it is possible to design other types of spike trains that also have a CV of 1. However, because such spike trains are rather strange and artificial, one usually assumes that a spike train with a CV of 1 *is* a Poisson spike train. Figure 15.2 shows (a) a Poisson spike train (CV $= 1$), (b) regular spiking (CV $= 0$), and (c) an example for a spike train with CV > 1.

15.3.2 A Simple Model: A Counter

A neuron essentially sums up incoming spikes and if this sum reaches a certain threshold then (and only then) the neuron spikes. In a very simple abstraction, we can thus model a neuron as a counter: It counts incoming spikes and if this count reaches some threshold value, say k, then the neuron spikes. For such a counter we can analyze its spike time variability (measured by the CV) precisely. To do so we need a few facts: Merging the spike trains from the various synapses of a neuron, we get a single (input) spike train. Now if all the spike trains of the synapses are Poisson spike trains, then the combined input spike train will also be a Poisson spike train – and thus have a CV of 1.

Let (X_i) be the ISIs of the combined input spike train. Then the ISIs of the spike train of the neuron (modeled as a counter that needs k inputs in order to spike) is given by the sums of k consecutive X_i's. Some elementary calculations then yield the simple relation:

$$\text{CV (output spike train)} = \frac{1}{\sqrt{k}} \cdot \text{CV (input spike train)}.$$

In other words, the CV of the output spike train is by a factor of $\sqrt{k}$ smaller than the CV of the input spike train. Feeding this output into another neuron would again lower the CV by a factor of $\sqrt{k}$, and so on. So even if we start with a perfectly Poissonian spike train, we obtain CVs close to 0 after only a few iterations. This shows that a simple counter is not an adequate model of a neuron.

15.3.3 Low Rates: Membrane Leakage

By modeling a neuron as a counter we indirectly assumed that the membrane potential remains unchanged in the absence of spikes. However, in reality the membrane potential decays over time. In a reasonably good approximation, the leakage is proportional to the accumulated voltage, so in the absence of input spikes a voltage of $U(t)$ decays exponentially to $U(t+\Delta t) = U(t)e^{-\Delta t/\tau}$, where τ is a constant that determines the speed of leakage. In the brain, $\tau \approx 20$ms is realistic. Assume that we receive spikes at times $t_1, t_2, \ldots, t_N$. Then the membrane potential at time t_i is

$$U(t_i) = \sum_{j=1}^{i} e^{-\frac{1}{\tau}(t_i - t_j)} .$$

To get a feeling for the effect of the decay of the membrane potential, we assume for a moment that the ISIs of the input spike train are constant with value $1/\lambda$. Some calculation shows that then the number of spikes N that we need is given by

$$N = -\tau\lambda \cdot \log(ke^{-\frac{1}{\tau\lambda}} - (k-1)) .$$

If λ is too small (as a function of k and τ), the above equation becomes undefined because the neuron will not spike if the input comes in completely regular intervals that are very large. However, if the ISIs are not all equal, then it may spike due to variations in the input. This happens in particular if a consecutive sequence of input spikes comes with smaller ISIs than expected.

Naturally, the probability for this to happen is small. So the neuron has to wait for an unlikely event, similarly as flipping a coin and waiting for a head when the probability for heads is very small. Thus, the next output spike can (qualitatively) be described as the first success in a long sequence of trials with low success probability, which matches our description of a Poisson process in Section 15.3.1. Indeed, as $\lambda \to 0$, the CV of the output spike train approaches 1. For larger λ a variation of the input ISIs will still result in a varying number of spikes N that the neuron needs to receive in order to spike. This varying number of input spikes does contribute to the CV of the output spike train. However, in general, it does not entail that CV comes close to 1.

15.3.4 Stochastic Synapses

So far we tacitly assumed that an output spike from one neuron contributes as input to the next neuron. In reality, however, synapses in the brain are not perfect input–output machines. Rather the opposite: a synaptic site transmits the incoming spike only with a rather small probability of 20–40% (Branco *et al.* 2008). In other words, the transmission fails in 60–80% of the cases! On the other hand, two (connected) neurons are typically connected by more than just one site, so there is still some decent probability that at least one of the sites will transmit the signal.

For the model we will assume that a connection between two neurons consists of s sites, each of which releases a vesicle with probability $1/s$. Biologically plausible are values for s in the range of 5–10 (Markram *et al.* 1997, Feldmeyer and Sakmann 2000). If i vesicles are released, then the effect on the postsynaptic neuron is simply i-fold stronger than the effect of a single vesicle. Thus, every spike has an expected effect of one, which is the same as in our simple model above. Under these assumptions, we get:

$$\text{CV (output spike train)} \approx \frac{1}{\sqrt{k}} \cdot \sqrt{1 + (\text{CV (input spike train)})^2 - \tfrac{1}{s}}\,.$$

If the input is a Poisson spike train with CV $= 1$, then the CV of the output train will be roughly $\sqrt{2/k}$, i.e., by a factor $\sqrt{2}$ higher than in our simple model.

We also get the following important fact: if the input ISIs have a CV of $\sqrt{(1-1/s)/(k-1)}$, then the output ISIs have approximately the same CV. That is, even if a signal is piped through a sequence of neurons, then the CV will not approach 0. Rather it will stabilize at an intermediate value which is roughly $1/\sqrt{k-1}$. This does not give us a Poisson spike train, but it already prevents the ISIs from becoming perfectly regular.

15.3.5 Balanced Excitation and Inhibition

Besides the so-called excitatory neurons that we considered so far, there exists also a second type of neurons, the so-called inhibitory neurons. When a neuron receives input from an inhibitory neuron, its effect is the *opposite* of that from an excitatory neuron: It *decreases* the memory potential instead of increasing it.[1] There is evidence that inhibition and excitation are balanced (Wehr and Zador 2003), i.e., that mostly a neuron receives about the same amount of excitatory

[1] Biologically speaking, all our models revert plus and minus, as all involved voltages are negative. We chose this convention because it is more intuitive that excitation corresponds to "plus" and inhibition to "minus."

and inihibitory drive. Such a balanced regime supports spike train variability in a remarkable way, as we outline next.

Unbounded voltage. Let us consider a neuron with two input spike trains, one excitatory and one inhibitory and let us assume that both spike trains are Poisson distributed, say with parameters λ^+ and λ^- of the underlying exponential distribution. The memoryless property of the exponential distribution implies that the probability for the next spike to be excitatory is $\lambda^+/(\lambda^+ + \lambda^-)$. If, under these conditions, we reconsider our simple counting neuron in Section 15.3.2, then we can model its behavior as a so-called random walk. It moves up by one unit if it receives an excitatory input and it moves down by one unit if it receives an inhibitory input.

The theory of random walks allows us to compute the CV of the output spike trains (under the assumption that $\lambda^+ > \lambda^-$):

$$\mathsf{CV}(X) = \frac{1}{\sqrt{k}} \cdot \left(\frac{\lambda^+ + \lambda^-}{\lambda^+ - \lambda^-} \right)^{1/2}.$$

A closer inspection of this formula tells us that we can achieve a wide range of CVs (including one, but also larger values) by adjusting the rates λ^+ and λ^- appropriately.

We conclude that for an appropriate ratio between excitation and inhibition, the CV of the output spike train can be 1. However, this result is not completely satisfactory yet. Firstly, it relies on the fine tuning between excitatory and inhibitory input rate. A deviation in either direction leads to different CVs. Secondly, as the possibility of CVs larger than 1 indicates, even for a CV of 1 the output spike train will be variable, but still no Poisson spike train. Thirdly, the output rate is fixed if we insist on the CV to be 1 (for a given input rate λ^+). Fortunately, it turns out that these problems arise only because our model is still oversimplified.

Bounded voltage. So far we studied the CV in the presence of inhibition. The mindful reader will recall that we could only analyze the case that there is more excitatory than inhibitory input. This is due to the fact that otherwise it would we possible (and not even unlikely) that we see a large burst of inhibition which causes the neuron potential to become so negative that it does not recover again. Clearly, this is only a problem of our model. In reality the membrane potential will never move below a certain threshold. We thus adapt our model by introducing an absolute lower bound k^- on the voltage: as soon as the voltage reaches k^-, inhibitory spikes are simply ignored. Biologically the exact value of k^- depends on many things, but values of 5–10 are realistic.

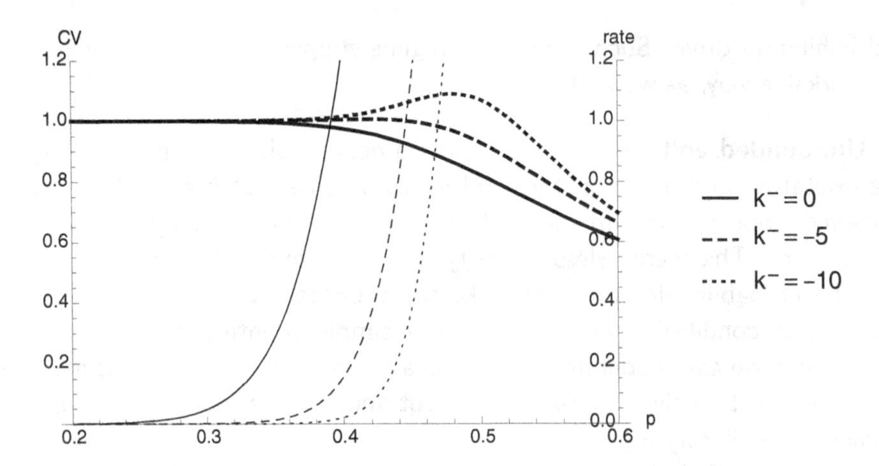

Figure 15.3: Inhibition with leakage for lower voltage bounds of $k^- = 0$ (solid), $k^- = -5$ (dashed), or $k^- = -10$ (dotted). *Bold lines, left axis:* CV of output spike train; *thin lines, right axis:* output rate for an input rate of 100Hz. On the x-axis, we plot the probability $p = \lambda^+/(\lambda^+ + \lambda^-)$ that the next spike is excitatory. So $p < 0.5$ corresponds to excess inhibition, while $p > 0.5$ corresponds to excess excitation. The y-axis is the CV of the resulting output spike train. For excess inhibition the output CV is close to 1, regardless of the value of k^-. In contrast to the model without lower voltage bound, the rate is not fixed by insisting on a CV of 1.

This model can still be analyzed rigorously. However, the resulting formulas are rather technical, so we omit them and show only plots of the results in Figure 15.3. In contrast to the model without leakage, the CV is robustly at 1 for excess inhibition. Moreover, the output rate can vary over several orders of magnitude while a CV of 1 is maintained.

15.4 Neuronal Populations

A standard approach in mathematical modeling in many disciplines is to first understand how a system works in a "pure" setting and then generalize it step by step in order to transfer it to a more noisy "real-world" scenario. In this section we show, based on results by Lengler *et al.* (2013), that in the case of neuroscience such an approach may lead to misconceptions of fundamental principles of information processing in the brain.

While in the previous section we studied single neurons, we now consider a simulation of a population of neurons that responds to some input. For such a simulation we need to specify properties of the neurons and their connectivity.

Lengler *et al.* (2013) studied the effect of two ways to do this: (1) homogeneous networks in which all neurons are identical copies of each other, and synapses are 100% reliable, and (2) inhomogeneous networks in which the properties of each neuron are drawn randomly from a distribution with mean identical to the neurons in the homogeneous model, with variance closely matched to biological data, and with unreliable synapses as described in Section 15.3.4. For more specific details we refer the interested reader to the original article.

Intuitively, it is plausible that both networks should behave similarly: by the law of large numbers, the effect of single neurons should average out in a large population of neurons. This is indeed true for the average spiking rate of the neurons. However, a careful analysis of the intrinsic dynamics of the network exhibits large differences. One of them is illustrated in Fig. 15.4a and 15.4b: by changing the input slightly, we always get the same reaction from a heterogeneous network. The homogeneous network, on the other hand, shows very different responses from trial to trial. In this sense, the reaction of the heterogeneous network is reproducible, while the reaction of the homogeneous one is not.

We should mention that the y-axis of Fig. 15.4a gives the *population CV*, i.e., we take the sequence of all spikes in the whole population and compute the CV of this sequence. This is not exactly the same as computing the CV of each neuron individually (and possibly averaging), but the results for single neuron CVs are similar (not shown).

We have argued before that it is desirable that the spike trains of different neurons are not strongly correlated with each other. The *cross-correlation (CC)* is a measure for how similar the responses of different neurons are. A CC of 1 indicates almost identical spike trains, while independent Poisson spike trains lead to a CC of 0. Figure 15.4c shows the CC for three chained populations, i.e., the output of the first population is the input to the second, and so on. It turns out that the homogeneous network has a high CC, while the heterogeneous one is decorrelating.

The reason is that the homogeneous network shows very strong and precisely timed oscillations. This reduces dramatically the information contained in the network, as each single neuron carries basically the same information in its spike train. Thus the whole network behaves merely like a single "superneuron." Since each neuron behaves the same, population codes become meaningless. On the other hand, for the heterogeneous network each neuron shows its own individual response, and it is possible to use them for efficient population coding. It turns out that the different sources of randomness are to some extent interchangeable. Both the individual cells and the unreliable synapses reduce the CV. However, the network performs best if both aspects are combined with each other.

Another positive effect of heterogeneous networks is that they are more sensitive to input. Figure 15.4d shows how the networks respond to very weak inputs of only a few spikes. The heterogeneous network reacts faster and needs less input to show a reaction.

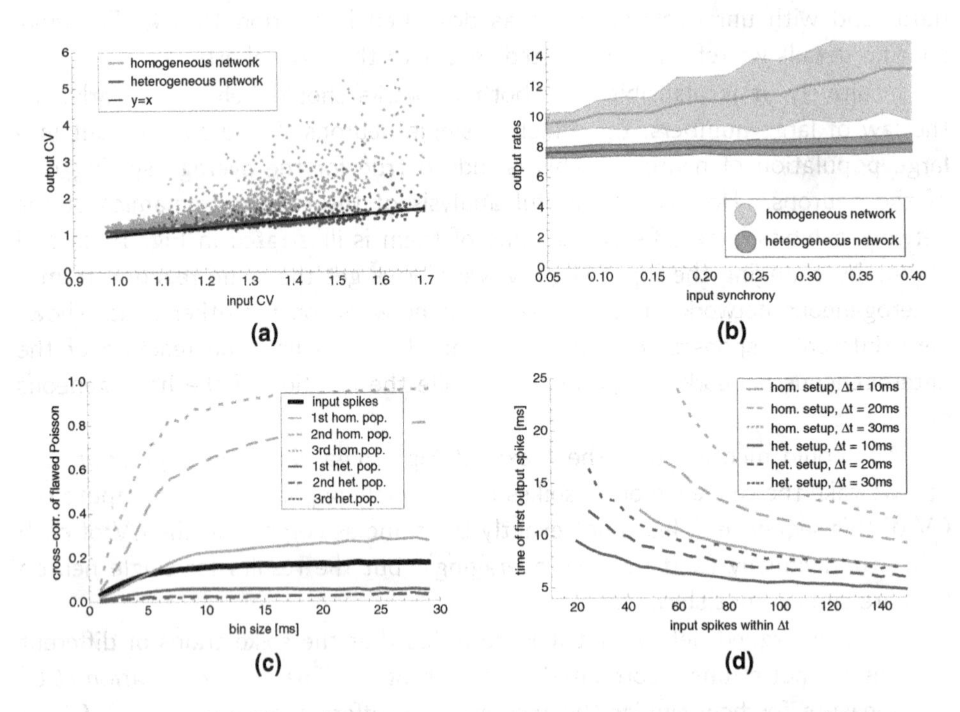

Figure 15.4 (from Lengler *et al.* 2013): Response of a heterogeneous (dark) and a homogeneous (light) network with balanced excitation and inhibition for Poisson input of varying rates. (a) Reaction to input which is not perfectly Poisson. The plot shows that the heterogeneous network has strictly smaller CVs. Moreover, the homogeneous network shows strongly fluctuating responses to similar inputs. (b) Reaction to Poisson input that is disturbed by spontaneous bursts of input synchronization. The x-axis gives the fraction of input spikes that belong to bursts. The input rate is constant. For the heterogeneous network the output rate is robust and concentrated, while the homogeneous network shows much less predictable responses. The shaded areas show the standard deviations. (c) Behavior if the output of the population is fed in as input to an additional population; populations 1, 2, and 3 are shown (from solid to dotted). For the heterogeneous network the spike trains of different neurons are decorrelated as the signal propagates, while the homogeneous network increases cross-correlations. (d) Reaction of the network to a small number of neurons (x-axis) each of which feeds one input spike into the network within a time interval of 10 ms (solid lines), 20 ms (dashed lines), or 30 ms (dotted lines). The y-axis shows the time of the first spike in the population. The curves start at the input size where all 100 trials produced at least one spike. The heterogeneous network can be activated by fewer input spikes and reacts faster.

15.5 Summary

Spike train variability is observed in the brain, but is usually not observed in simulations unless they incorporate explicit mechanisms that increase the variance in the system. Early research tended to disqualify the observed variability as noise that needs to be overcome, but more recently focus has shifted to investigate benefits of this variability.

In this chapter we have seen how randomness in the model assumptions actually result in a higher reproducibility of the properties of the underlying system. In terms of the convex complexity measures discussed by Atmanspacher and Demmel (this volume), the complexity of the system is *increased* by *decreasing* the complexity of its components.

References

Branco, T., Staras, K., Darcy, K.J., and Goda, Y. (2008): Local dendritic activity sets release probability at hippocampal synapses. *Neuron* **59**, 475–485.

Carandini, M. (2004): Amplification of trial-to-trial response variability by neurons in visual cortex. *PLoS Biology* **2**(9), e264.

Fain, G.L., O'Dell, T., and Fain, M.J. (2014): *Molecular and Cellular Physiology of Neurons*, Harvard University Press, Cambridge.

Feldmeyer, D., and Sakmann, B. (2000): Synaptic efficacy and reliability of excitatory connections between the principal neurones of the input (layer 4) and output layer (layer 5) of the neocortex. *Journal of Physiology* **525**, 31–39.

Healy, A.J., Malhotra, N., and Mo, C.H. (2010): Irrelevant events affect voters' evaluations of government performance. *Proceedings of the National Academy of Sciences of the USA* **107**, 12804–12809.

Keat, J., Reinagel, P., Reid, R.C., and Meister, M. (2001): Predicting every spike: A model for the responses of visual neurons. *Neuron* **30**, 803–817.

Lengler, J., Jug, F., and Steger, A. (2013): Reliable neuronal systems: The importance of heterogeneity. *PLoS ONE* **8**(12), e80694.

Lengler, J., and Steger, A. (2015): Note on the coefficient of variations of neuronal spike trains. Submitted for publication.

Levitan, I.B., and Kaczmarek, L.K. (2001): *The Neuron: Cell and Molecular Biology*, Oxford University Press, Oxford.

Liu, R.C., Tzonev, S., Rebrik, S., and Miller, K.D. (2001): Variability and information in a neural code of the cat lateral geniculate nucleus. *Journal of Neurophysiology* **86**, 2789–2806.

Markram, H., Lübke, J., Frotscher, M., Roth, A., and Sakmann, B. (1997): Physiology and anatomy of synaptic connections between thick tufted pyramidal neurones in the developing rat neocortex. *Journal of Physiology (London)* **500**, 409–440.

Ranganathan, G.N., and Koester, H.J. (2011): Correlations decrease with propagation of spiking activity in the mouse barrel cortex. *Frontiers in Neural Circuits* **5**, 8.

Softky, W.R., and Koch, C. (1993): The highly irregular firing of cortical cells is inconsistent with temporal integration of random EPSPs. *Journal of Neuroscience* **13**, 334–350.

Shadlen, M.N., and Newsome, W.T. (1998): The variable discharge of cortical neurons: Implications for connectivity, computation, and information coding. *Journal of Neuroscience* **18**, 3870–3896.

Surowiecki, J. (2005): *The Wisdom of Crowds*, Anchor, New York.

Troyer, T.W., and Miller, K.D. (1997): Physiological gain leads to high ISI variability in a simple model of a cortical regular spiking cell. *Neural Computation* **9**, 971–983.

Wehr, M., and Zador, A.M. (2003): Balanced inhibition underlies tuning and sharpens spike timing in auditory cortex. *Nature* **426**, 442–446.

16
Neural Reuse and In-Principle Limitations on Reproducibility in Cognitive Neuroscience

Michael L. Anderson

Abstract. For much of the past century, it was assumed that individual regions of the brain were selective and functionally dedicated – that they responded to a restricted class of inputs and supported singular well-defined cognitive processes. But over 50 years of electrophysiology, and especially the past 20 years of neuroimaging, have revealed instead a brain with a far more complex and dynamic functional architecture. Most individual regions of the brain are multi-modal and active across a wide range of cognitive tasks. Moreover, two kinds of developmental plasticity ensure that this architecture is continually being remodeled: Hebbian learning that changes the strength of synaptic connections to tune local function, and a neural "search" process that acts to establish the functional partnerships between regions that will support newly acquired abilities. These facts call for a reconsideration of both the mathematical and conceptual tools we bring to bear in our understanding of the brain. This paper discusses some of these tools – including functional fingerprinting, machine learning, and matrix decomposition – and show how they can be used to capture the brain's complexity. Moreover, I will offer an analysis of what kind and degree of reproducibility we should expect these tools to reveal. To what degree can we talk about "the" architecture of the brain? I will argue that there are both evolutionary and developmental reasons to expect a great deal of inter-individual functional similarity; but there will also be differences. The tools I advocate for can be used to capture and quantify these individual differences.

16.1 Introduction

Cognitive neuroscience has long been dominated by the concepts of modularity (Barrett and Kurzban 2006, Fodor 1983) and localization (Posner *et al.* 1988). According to this way of understanding brain architecture, individual mental operations – whatever their nature, and there is naturally some debate about this question – are strictly localized to circumscribed neural territories. Thus are many portions of the occipital lobe (from cells to small regions) thought to be specialized for visual tasks, including contrast, motion and edge detection (Geisler and Albrecht 1997, Hubel and Wiesel 1962), visual texture processing (Beason-Held *et al.* 1998), and the like. Similarly, Broca's area has been strongly associated with syntactic processing (Embick *et al.* 2000), and it has been argued

Reproducibility: Principles, Problems, Practices, and Prospects, First Edition. Edited by Harald Atmanspacher and Sabine Maasen.

that parts of fusiform gyrus are specialized for face processing (Kanwisher *et al.* 1997).

A frequent ally of this localizationist view has been a form of nativism according to which achieving species-typical adult neurofunctional architecture is largely a matter of letting pre-established genetically encoded functional specializations express themselves in the course of normal development (Atkinson 1984, Kanwisher 2010). Thus, Nancy Kanwisher (among the more prominent advocates of genetically encoded modularity) wrote that "genes may be largely responsible for wiring up the face system, with little or no role [for] experience with faces" (Kanwisher 2010, p. 5).

There are significant theoretical considerations in support of this framework for understanding neurofunctional architecture. I'll note two that are especially relevant to the current discussion. First, the neural segregation of mental operations is presumed to minimize interference and maximize the potential for information throughput via parallel processing (Fodor 1983). Second, this arrangement would apparently allow for the neural modules to be independently modified (Carruthers 2006) and thereby offer a target for selection pressure (Barret and Kurzban 2006), allowing for mosaic evolution of function (Barton 1998).

Over the past few decades, however, this modular, localizationist view has been slowly eroded on three fronts (whether it nevertheless remains the dominant view, and if so for how much longer that will remain true, I leave for others to decide). First has been an increasing appreciation of the role of individual experience in shaping local neural organization. Second has been the growing recognition that local circuits are not specialized, at least not in the way that has long been assumed. And third has been an emphasis not just on local organization, but on the establishment of distributed networks of local nodes, which mutually shape one another.

Before moving on to describe these lines of research and the evidence that supports them, it is worth characterizing at the outset the challenge these developments pose to reproducibility in the cognitive neurosciences. If it were the case that neurofunctional architecture was largely determined by genetic predisposition, one could reasonably hope that variation in this architecture would be little greater than the variation in the underlying genetics and would moreover be constrained by the same evolutionary pressures that shaped the genetically specified modules in the first place. In this case one might expect variation around a central, species-typical tendency, and characterizing that tendency would constitute a valid scientific project likely leading to highly reproducible results.

If, however, individual experience – which must of necessity be idiosyncratic to some degree – plays an important role in shaping local organization, this introduces additional degrees of freedom in the eventual functional characteristics of that local circuit. Moreover, insofar as functional development involves the creation of networks of nodes, then differences in the functional characteristics of individual nodes will lead to differences in which ones get incorporated into networks, and these differences are likely to be magnified by differences in individual learning strategies, and self-reinforcing, as the networks shape the nodes which shape the networks, and so on.

Finally, these additional layers and degrees of freedom in the development of the overall neural system seem likely to interact in a super-additive fashion, with each difference potentially magnifying the effects of the others.

If this picture is closer to the truth than the nativist, modular localization account sketched above, then one might wonder about the possibility of severe and in-principle limitations on reproducibility in the cognitive neurosciences. It would not, for instance, necessarily be the case that there is a species-typical central tendency in neurofunctional architecture; that is, although mathematically there is of course a central tendency for any population, in a dynamic developmental system with multiple degrees of freedom, there is no guarantee that the central tendencies of multiple samples will resemble one another. Moreover, intra-individual differences in architecture over developmental time could be significant, leading to very poor test–retest reliability in measures of functional organization.

We will return to this issue of reproducibility, as well as to the theoretical considerations regarding the evolution of function noted above, once we have a fuller picture of the emerging evidence pertaining to neurofunctional architecture and its development. We will try to characterize what we know about the degree of reproducibility, given current scientific practices, and describe some of the tools that might be used to better capture and quantify individual differences. Finally, in light of evidence that there is only moderate reproducibility at the individual level, but robust reproducibility at the group level, we will discuss some implications for the evolution and development of the brain.

16.2 The Erosion of Modular Thinking

As mentioned in the last section, the nativist modular viewpoint is being eroded on three primary fronts. First has been an increasing appreciation of the role of individual experience in shaping local neural organization. For instance, it has been observed that the location of peak brain activity shifts from frontal to parietal areas as proficiency improves in a visuomotor sequence learning task (Sakai *et al.* 1998). In addition, Gauthier *et al.* (1999) have shown that the ability to recognize individual and categorize families of novel objects ("greebles") involves activation in fusiform gyrus – the same region supposed by Kanwisher and others to be genetically specialized for face processing. Both of these findings appear to fit more closely with the notion that a region acquires its functional role as a result of experiential shaping, and not solely as the result of genetic prespecification.

A more radical example of the impact of experience on neural development comes from a series of experiments involving the redirection of afferent inputs from one part of the brain to another (Sur *et al.* 1988, 1990). In these experiments, it was observed in neonatal ferrets that rerouting to deafferented primary auditory cortex connections that would normally innervate primary visual cortex resulted in the induction of neurofunctional structures such as color pinwheels that would normally be found in visual cortex. Moreover, these animals remained behaviorally similar to controls, able, e.g., to orient normally to visual stimuli (von Melchner *et al.* 2000).

Such effects are neither uncommon nor restricted to surgical manipulations. For example, according to Barton, the observed increase in the Parvocellular:Magnocellular neuron ratio in the lateral geniculate nucleus should be seen as a special adaptation for frugivory in diurnal primates (Barton 1998, 2004). However, Finlay *et al.* (2013) showed that the cause was likely to be the developmental cell death of the Magnocellular neurons representing the visual periphery. Over developmental time, these cells compete unsuccessfully for synaptic space in primary visual cortex, which instead becomes dominated by foveal inputs. The cause of the changed ratio, therefore, is not a genetic change directly producing more Parvocellular neurons, but rather a side effect of the greater amount of experiential input from the fovea.

Similarly, it has been shown that the visual cortex of early-blind subjects is activated by Braille reading (Sadato *et al.* 1996) and that the application of repetitive transcranial magnetic stimulation (rTMS) to disrupt processing in the activated visual areas disrupts Braille character recognition (Cohen *et al.* 1997, Kupers *et al.* 2007), thus establishing the functional relevance of the observed activation. Moreover, these findings also hold true of sighted participants taught Braille character recognition while blindfolded (Merabet *et al.* 2008). This finding both aligns local function with individual experience, and indicates that recruitment of regions for multiple purposes is an early and ongoing developmental process, an observation we will examine in more detail, below.

This brings us to the second front along which modular localization is being eroded: the growing recognition that local circuits are not functionally specialized, at least not in the way that has long been assumed. In a series of studies over the past several years (Anderson 2007, 2008, 2010, Anderson and Pessoa 2011, Anderson *et al.* 2013, Anderson and Penner-Wilger 2013), my colleagues and I have demonstrated that each individual region of the brain is typically active during a wide range of tasks across the domains of language, emotion, mathematics, reasoning, visual perception, audition, and more. This observation holds true for a number of different ways of grouping experiments into domains, and regardless of whether the region is defined by anatomical area using Brodmann's maps; or the FreeSurfer or Harvard-Oxford atlases; or by random seeding; or by systematically measuring the whole brain voxel by voxel. Each region of the brain is *differentiated* from the others – none may have exactly the same functional profile as any other – but they do not generally appear to specialize in a single kind of task, nor a single sort of input, nor a single sort of sensory modality.

These results are consistent with the conclusions emerging from the intensive study of individual regions of interest. For instance, although Broca's area has been strongly associated with language processing, it turns out to also be active more frequently in nonlanguage tasks (Poldrack 2006), including general action-related tasks such as movement preparation (Thoenissen *et al.* 2002), action sequencing (Nishitani *et al.* 2005), action recognition (Decety *et al.* 1997, Hamzei *et al.* 2003, Nishitani *et al.* 2005), imagery of human motion (Binkofski *et al.* 2000), and action imitation (Nishitani *et al.* 2005; for reviews, see Hagoort 2005, Tettamanti and Weniger 2006, Grodzinsky and Santi 2008).

Similarly, visual and motor areas – long presumed to be among the most highly specialized in the brain – appear to also be involved in various sorts of language processing and other "higher" cognitive tasks (Damasio and Tranel 1993, Damasio *et al.* 1996, Glenberg *et al.* 2007, 2008a,b, Glenberg and Kaschak 2002, Hanakawa *et al.* 2002, Martin *et al.* 1995, 1996, 2000, Pulvermüller 2005, Pulvermüller and Fadiga 2010; see Schiller 1995 for a related discussion). The fusiform gyrus earned the label "Fusiform Face Area" because of its apparent selectivity for faces (Kanwisher *et al.* 1997), but we have since learned that the area also responds to cars, animals, sculptures, and other stimuli (Gauthier and Nelson 2000, Gauthier *et al.* 1999, 2000, Grill-Spector *et al.* 2006, Rhodes *et al.* 2004, Hanson and Schmidt 2011).

The above findings are based on neuroimaging data, and thus reflect the functional properties of very large collections of neurons, but the observations are echoed in electrophysiological studies even at the level of single neurons. For instance, during tasks where different movements must be made depending on the particular content of tactile or visual stimuli (e.g., whether successive stimuli are the same or not), both the sensory encoding and the decision-making processes appear to rely on the very same neurons in premotor areas. As Cisek and Kalaska (2010, p. 274) concluded in a review of such findings:

> In all of these cases, the same neurons appear to first reflect decision-related variables such as the quality of evidence in favor of a given choice and then later encode the metrics of the action used to report the decision (Cisek and Kalaska 2005, Kim and Basso 2008, Roitman and Shadlen 2002, Schall and Bichot 1998, Yang and Shadlen 2007).

Similarly, single neurons in posterior parietal cortex (PPC) appear to be involved in processing information about object location and action intentions, among other things, and their activity is known to be modulated by attention, behavioral context, expected utility, and other decision-related variables. Reviewing evidence for the functional complexity of neurons in PPC, Cisek and Kalaska (2010, p. 274) write:

> In short, the PPC does not appear to fit neatly into any of the categories of perception, cognition, or action; or alternatively, the PPC reflects all categories at once without respecting these theoretical distinctions. Indeed, it is difficult to see how neural activity in the PPC can be interpreted using any of the concepts of classical cognitive psychology (Culham and Kanwisher 2001).

All this evidence also fits quite comfortably with emerging evidence for the ubiquity of "mixed selectivity" in individual neurons: those under-studied cells whose response profiles don't appear to reflect single sensory features, decision variables, motor parameters, and the like, but any number of these at a time or successively (Rigotti *et al.* 2013).

Finally, moving to the third front on which the localizationist view is being challenged, recent years have seen an increased emphasis on the importance of the establishment of distributed, overlapping networks of local nodes, which mutually shape one another's functional properties (Johnson 2001, 2011, Sporns 2011). Learning involves not just Hebbian plasticity, the up- and down-regulation of local synaptic connections, but also the search for and establishment of new local and long-distance neural partnerships. Some observations of the neural changes induced by learning to control an artificial limb via a brain–machine interface (BMI) illustrate the basic idea quite well (Lebedev and Nicolelis 2006, p. 542):

> Continuous BMI operations in primates lead to physiological changes in neuronal tuning, which include changes in preferred direction and direction tuning strength of neurons (Taylor *et al.* 2002, Carmena *et al.* 2003, Lebedev *et al.* 2005). In addition, broad changes in pair-wise neuronal

> correlation can be detected after BMIs are switched to operate fully under brain-control mode (Carmena *et al.* 2003, Lebedev *et al.* 2005).

At least in so far as oscillatory coherence between cells is a sign of functional cooperation, the learning process here involved not just local changes, but also the establishment of new sets of functional partnerships. There is evidence that this process occurs not just between non-adjacent neurons in a local region, but also across the brain as a whole. For instance, Fair *et al.* (2007, 2009) performed a series of coherence (functional connectivity) analyses of resting-state fMRI data in children and adults. They reported that development entails both "segregation," decreased short-range connectivity indicating differentiation of the response properties of neighboring local circuits, and "integration," increased long-range connectivity between regions implementing functional networks.

These findings were largely consistent with those reported in a similar study by Supekar *et al.* (2009). Learning is a matter not just of tuning local receptivity and modulating local synaptic connections but also of finding the right set of neural partnerships that can be brought to bear to solve or master the task in question (an outcome already illustrated by the incorporation of "visual" areas of the brain into the network responsible for implementing Braille reading). The end result of such a process would be an architecture consisting of multiple, overlapping functional networks, in which the active partnerships change as the tasks demand. Is there evidence that this is what typical adult functional architecture looks like?

Indeed there is. Evidence from the primate (including human) literature demonstrates the importance of large-scale modulation of neural partnerships in support of cognitive function. For instance, there is evidence relating changes in the oscillatory coherence between brain regions (local and long-distance) to sensory binding, modulation of attention, and other cognitive functions (Varela *et al.* 2001, Steinmetz *et al.* 2000). Friston (1997) demonstrated that whether a given region of inferotemporal cortex was face selective depended on the level of activity in posterior parietal cortex. And McIntosh *et al.* (1994) investigated a region of inferotemporal cortex and a region of prefrontal cortex that both support face identification and spatial attention. They showed that during the face-processing task, the inferotemporal region cooperated strongly with a region of superior parietal cortex; while during the attention task, that same region of parietal cortex cooperated more strongly with the prefrontal area. More recently, Cole *et al.* (2013) reported evidence for "flexible hubs" in the brain, "regions that flexibly and rapidly shift their brain-wide functional connectivity patterns depending on the task."

These experimental findings are consistent with both the expectation of large-scale modulation of neural partnerships and with analyses of large collections of neuroimaging studies (Anderson 2008, Anderson and Penner-Wilger 2013), showing that while the same brain regions supported various tasks across disparate cognitive domains – e.g., semantics, motor control emotion, and visual perception – the regions cooperated with different neural partners under different circumstances. That is to say, differences in the neural supports for different cognitive domains came down to differences in overall patterns of neural cooperation, and not to differences in which specific regions were differentially supporting tasks in the different domains.

Learning, then, as noted above, involves in part the search for the right set of neural partnerships to achieve the desired aim. Which partnerships become consolidated in this process will depend on factors such as how the functional dispositions of individual regions have been shaped, and how available they are to support new task demands, and these factors seem likely to vary from individual to individual.

But how important a factor this is has only recently come under significant scrutiny, putting aside the vast literature investigating the neural underpinnings of various disorders and pathologies. In the next section I will offer a brief and limited overview of some attempts to quantify the degree of inter- and intra-individual variation in the "normal" population and discuss these implications of these findings for the larger issue of reproducibility.

16.3 Intrinsic Limits on Reproducibility

To illustrate the issue in a simple way, consider Fig. 16.1, reprinted from Scheiber (2001), representing the motor map in a single individual squirrel monkey before, during, and after learning a simple manual task. As can be easily seen in these motor maps, there is evidence for both reproducibility – the *same* neurons being dedicated to the behavior both times it is learned – and also for flexibility – *different* neurons being used for identical behaviors, with no detectable difference in performance.

Similarly, Fig. 16.2, reproduced from Barch *et al.* (2013), showcase the contrast between the apparently robust findings obtained from group averages (left panel) and the rather small proportion of individual participants who show activation at each individual voxel. In light of the results discussed in the preceding section, this should hardly be surprising. It appears that neurofunctional architecture is the result of a number of different experience-dependent, learning-driven processes, given which one might reasonably expect significant inter-individual differences between those learning a task and significant intra-individual differ-

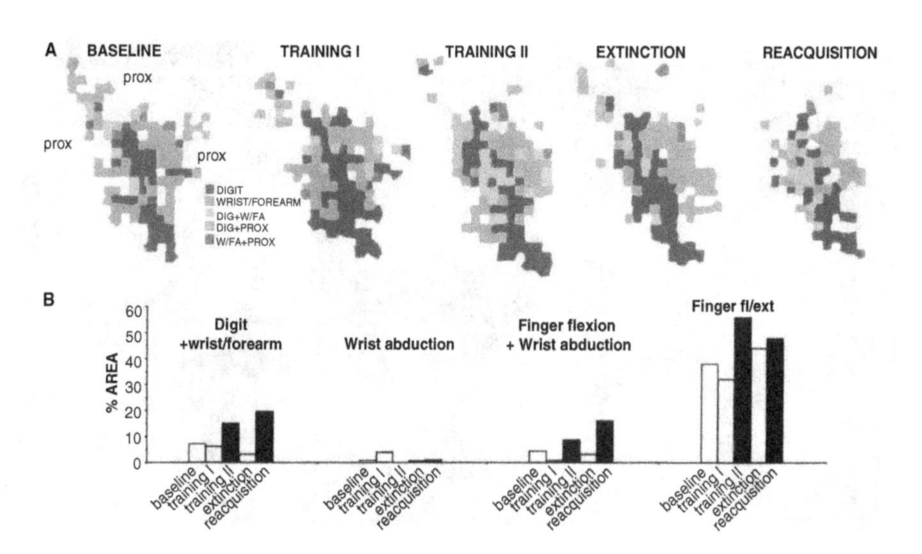

Figure 16.1: Panel A offers a spatial map of motor cortex, showing which local regions exhibited neural activity during movements of the digit, wrist, etc. Panel B summarizes the total amount of the mapped area showing neural activity during the specified movements. Note both the similarities and differences in the spatial map during the two training phases and reacquisition. The same behavior can be supported by somewhat different neural arrangements. (Reprinted from Scheiber (2001) with permission license # 3577790361692.)

ences over developmental time. Moreover, one might reasonably expect these differences to relate to performance variability, at least in some cases.

In fact, there are some initial indications that this is the case. For instance, Kosslyn *et al.* (1996) observed that participants who performed well on a certain visual imagery task appeared to rely in part on Brodmann area 17, whereas those who performed poorly seemed to rely instead on a region in the parietal lobes. This would seem to indicate that these two groups differed in which of these two regions was incorporated into the overall functional network responsible for this task – a difference that could reflect different strategies for the task, different degrees of functional burden from other tasks on Brodmann area 17, or both.

Similarly, Baldassarre *et al.* (2012) showed that differences in functional connectivity between regions predict differences in task performance, and Bassett *et al.* (2011) demonstrated that "network flexibility" – the average number of times a region of the brain changes its functional connectivity profile during learning – predicts the degree of learning gain. In other words, differences in brain networks are associated with differences in learning outcomes, and the degree of change to individual brain network is correlated with the degree of learning. Interestingly enough, Merabet *et al.* (2008) also found that there

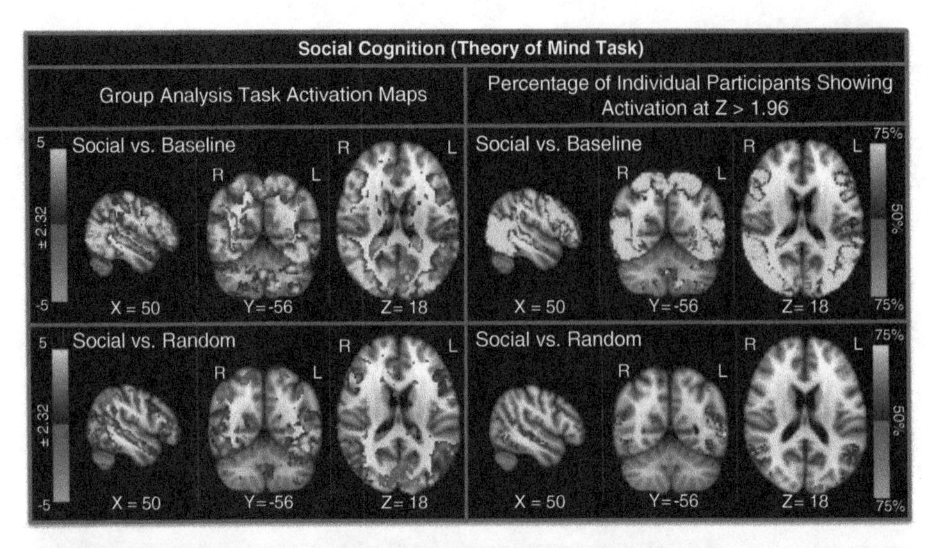

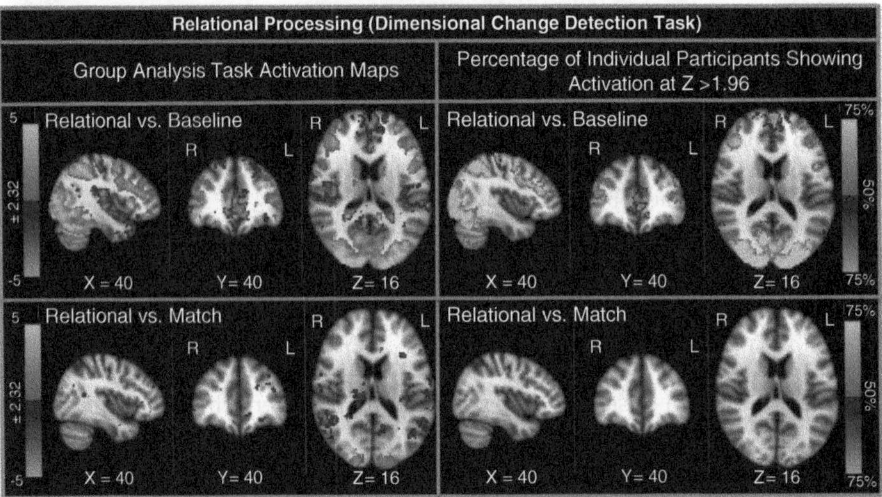

Figure 16.2: Top panel shows neural activation during a theory of mind task, and bottom panel during a change detection task. The left side of each panel displays the group average activation maps, and the right side the percentage of individuals who had activation at each location in the group average map. Note that for the stricter comparisons (social vs. random, and relational vs. match, which show differences in activity in the experimental task vs. a control task, rather than vs. baseline activity, and thus represent the regions of the brain specific to the experimental task) very little of the group average activation was displayed by more than 50% of individual participants. (Reprinted from Barch *et al.* 2013 with permission license # 3577790735856.)

could be differences in which brain networks were used to support a task within individuals, depending on circumstances, and that this difference *need not* result in performance differences. All of this points once again to the possibility of significant individual and contextual differences.

That being said, there has been little consistent, concerted effort to *quantify* these differences across a range of tasks and populations; nevertheless, a review of the literature can offer some initial clues. To keep the discussion manageable, here I will stick to human participants, and only to reproducibility and flexibility in measurements of functional and physical connectivity.

In what follows I will make a sharp conceptual distinction (it is not always possible to make a sharp *empirical* distinction) between intrinsic and methodological limits on reproducibility. Methodological limits on reproducibility are those imposed by the accuracy and precision of the instruments with which we measure. An excellent review of those limitations is offered by Fornito *et al.* (2013) and Yan *et al.* (2013). They will not be further discussed here.

What I have in view instead are the limits imposed by the sort of dynamical system the brain is, governed by the developmental principles canvassed above. In a system like that, there will be irreducible inter- and intra-individual differences that will persist even when using instruments with an arbitrarily high degree of accuracy and precision. It is of course not possible to get a quantitative estimate of intrinsic reproducibility without taking into account the methodological contributions to variability – and I will thus be in no position to offer precise estimates here. But we can get some sense for the likely *bounds* on intrinsic reproducibility.

In one of the earliest attempts to map the network structure of the human cortex, Hagmann *et al.* (2008) performed diffusion-weighted tractographic imaging (DTI) with five participants. One participant was scanned twice during sessions two days apart. They estimated the strength of the structural connectivity between 998 regions of interest, and found a fairly high between-participant correlation for the connection matrix: $R^2 = 0.65$. The within participant correlation was $R^2 = 0.78$.

In a more recent set of studies, Zuo *et al.* (2010) estimated the test–retest reliability in the relative amplitude of low-frequency neural oscillations across different regions of the brain, an indirect measure of stability in network architecture. They performed three scans: an initial scan, followed 5–16 months later by two more scans taken 45 minutes apart. They found that the intra-individual rank correlation between the same-day scans ranged from 0.7 to 0.95 (R^2-values between 0.49 and 0.90), and between the first scan and the average of scans two and three ranged from 0.6 to just under 0.9 (R^2-values between 0.36 and

0.81; all results here read off the figures; individual results were not reported by the authors).

This gives some sense of both intra- and inter-individual differences, which appear to be in line with those reported by Hagmann *et al.* (2008), although between-participant variation was not directly assessed in this study. Interestingly, despite the only moderate-to-high rank inter-individual correlations across scans, the group average ranks were *extraordinarily* reproducible, with an R^2 for same-day reliability of 0.97 and for long-term reliability of 0.90.

Similarly, Liao *et al.* (2013) estimated the test–retest reliability of the weighted degree of each voxel in the brain. Degree is a measure of the total strength of the functional connection of each voxel to the others. In this case, it is an estimate of its average degree of synchrony with the other voxels in the brain during a resting-state scan. As with the study above, degree stability serves as an indirect (and somewhat imperfect) measure of the stability of the overall functional network. Here the authors performed two scans, approximately one week apart, and calculated an intraclass correlation coefficient (ICC) comparing the between-subject mean square degree (BMS) with the within-subject mean square degree (WMS) for each voxel:

$$ICC = (BMS - WMS) / (BMS + WMS)$$

ICC values ranged from −0.5 to 0.9; mean 0.31; standard deviation 0.28 for weighted overall degree, with similar results for other degree estimates. Within- and between-subjects comparisons were not separately reported.

The authors also estimated the test–retest reliability of group averages using normalized (z-scored) degree measures. As in the case of Zuo *et al.* (2010), the group averages were extraordinarily reliable; Pearson's $r = 0.96$ ($R^2 = 0.92$), despite the low-to-moderate ICC.

In contrast to these findings, Markov *et al.* (2011) claimed, based on an extensive study of cortical connectivity in Macaques, that "quantitative connectivity profiles do not differ significantly" between individuals. Therefore, they suggested that there is a species-specific "robust signature (connectivity profile) ... for each [brain] area" (Markov *et al.* 2011, p. 1263). In fact, however, it is not clear that this conclusion is warranted by their analysis, as it is based on finding no main effect of a BRAIN factor[1] in their analysis of variance. If I understand their design, this would indicate only that differences in signatures were

[1]That is, they treated each individual brain as one level in a multilevel independent variable (IV) in their analysis. The dependent variable (DV) was the connectivity pattern of each individual region. They did not find a significant difference between their DV at various levels of the IV.

not sufficiently consistent across areas within individual brains to be explained by the BRAIN factor. This is a weaker conclusion than there being no significant differences between individuals in the connectivity profile for a given region. It remains clear that more concerted effort on this question is called for.

The overall pattern of results appears to suggest that while there is indeed significant inter- and intra-individual variability in the physical and functional architecture of the brain, there is also a robust species-typical prototype around which the variation clusters. Note the claim is not that there is a population average – that is a trivial mathematical necessity. Rather, the facts that (1) there is a fairly robust but far from perfect correlation between individual's structural connectivity patterns; (2) there is significant inter-individual change in functional connectivity over time; and (3) the group averages are, nevertheless, nearly *identical* over time, strongly suggest that the observed variation is occurring around a temporally stable central tendency.[2] This did not have to be the case; there could easily have been drift in the sample averages, leading to low test–retest reliability in those averages. Thus, this is an observation in need of some explanation, as it appears to suggest the presence of some species-typical constraints on the variability of brain architecture.

Joint work with Barb Finlay (Anderson and Finlay 2013) might illuminate the matter. How is it that the developmental principles outlined in the introduction, which might have been expected to lead to rampant inter- and intra-individual variability are in fact constrained to operate within species-typical norms? The key lies in seeing how evolutionary and developmental processes interact to give rise to adult neurofunctional architecture.

In light of the findings discussed in the previous section, I have been urging the field to speak not of the functional *specialization* of various regions of the brain, but rather of their functional *differentiation* – for it is abundantly clear that the parts of the brain differ from one another, even as they each continue to support multiple tasks across distinct task categories. For similar reasons, I have been led to speak not of *the function* of a particular part of the brain, but rather of its *functional biases*, which might be thought of as its underlying set of causal dispositions that could be useful under a variety of circumstances (Anderson 2015). Expressed using these terms, brain development involves experience-driven local plasticity that sculpts the functional biases of the brain, leading to functional differentiation, along with the process of neural reuse by which regions with the right functional biases are woven together as we acquire new skills.

---

[2] Lengler and Steger, this volume, investigate and in fact exploit a related situation between precision and reliability across levels of individual (neural) and ensemble (network) descriptions. Although their focus is clearly different, there may be structural similarities.

What apparently constrains these processes to remain close to species-typical norms is the existence of early-developing, highly stereotyped, presumably genetically specified sensory afferents, which project to highly conserved locations in the brain. This ensures – assuming of course that the environment is largely inherited, such that the statistical properties of the sensor inputs are relevantly similar across generations – that early experience-driven local plasticity will reproduce the same early-developing functional biases in the same regions of the brain. Insofar as individuals tend to begin development with the same or very similar set of regional functional biases, and to the extent to which skills are acquired according to a similar developmental timetable, individuals are highly likely to construct similar skill-supporting networks.

That there will be variation in these processes is evident, and as noted above in fact observed. But the fact that the processes are initially working with the same set of species-typical functional biases will tend to center development along species-typical trajectories. Although this is clearly a sketch in need of a great deal of supporting detail (Anderson and Finlay 2013), it does appear that it offers an evolutionary-developmental framework capable of explaining the twin observations of significant between- and within-individual variation, and robust, stable sample means.

A very recent finding that appears to support this overall framework is described by Takerkart *et al.* (2014). The authors develop a version of multivoxel pattern analysis (Haxby *et al.* 2001) and apply it to the problem of classifying (sorting) functional brain images according to the experimental task, using only information about the observed patterns of neural activity. In other words, the machine learner was trained to analyze patterns of neural activity in order to identify the experimental condition under which these data were observed.

The interesting result of this study was that including information about the *relative* spatial location of the activations significantly improved performance. That is to say, while the *absolute* location of the most informative activity varied from person to person (presumably due to individual variation in cortical folding, vascularization, and other factors), the *relative* location of that activity was much more consistent. The overall spatial structure of the functional network is more conserved than the absolute locations of the nodes.

16.4 Going Forward

The results summarized in the preceding sections offer two different lessons. First, because of the dynamic nature of the brain and the principles guiding its development, we should expect to see significant, irreducible individual differences in neurofunctional architecture. Individuals will differ from one another, and differ from themselves from moment to moment, with respect to the exact location of the neural tissue they dedicate to particular functions, the strength of the physical and functional connections between regions, and the precise functional biases for each of the nodes in their functional networks. Second, these differences are sufficiently constrained that we can have confidence in the validity and reproducibility of species-typical means.

This being said, we remain largely in the dark about the amount by which and the precise manner in which individuals differ from one another. For instance, functional biases will evidently differ from individual to individual, but by how much, and in what way? Will such differences be themselves systematic and meaningful, relating to other individual variables, such as proficiency? We do not currently know the answers to these questions, and they are arguably crucial to the future success of the cognitive neurosciences.

In other work (Anderson *et al.* 2013) I have urged the adoption of functional fingerprinting as a way of capturing and quantifying the functional biases of brain regions and networks. A functional fingerprint is a vector in a multi-dimensional functional space that represents the activity of a region or network in terms of its likelihood of being active across a range of tasks or conditions. Figure 16.3 represents the functional fingerprints for some regions within left and right auditory cortex, left intraparietal sulcus and right anterior cingulate cortex, across 20 cognitive and behavioral task domains.

As is obvious, the functional fingerprints (and therefore, we argue, the underlying functional biases) differ from region to region. Moreover, because the fingerprint captures that bias not just visually, but quantitatively, we can give a precise estimate of just how *much* the biases differ. Although we have so far used these techniques only on group-averaged data, there is no reason in principle they cannot be applied to individual data and used to begin to precisely quantify both inter- and intra-individual differences, as well as to plot developmental trajectories and quantify the similarities and differences in those.

Similarly, with brain activity data in this format, it is possible to apply various factor analysis and matrix decomposition techniques to uncover any consistent underlying structure.[3] In fact, I have argued at length (Anderson 2015) that one fruitful avenue for the field to pursue would involve specifying a set of underlying neuroscientifically relevant psychological factors and characterizing brain regions in terms of their different loadings on this set of shared factors.

---

[3]These and other machine-learning tools for big-data analysis come with their own specific challenges as far as reproducibility is concerned. See Stahel and, in particular, Estivill-Castro, this volume.

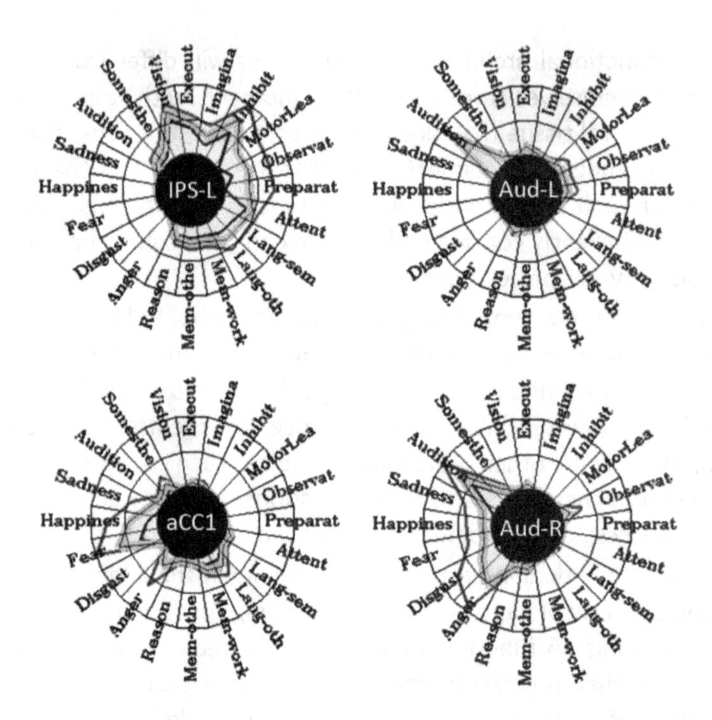

Figure 16.3: Functional fingerprints for selected regions of the brain. Clockwise from top left: left intraparietal sulcus, left auditory cortex, right auditory cortex, and right anterior cingulate. The lines represents the relative amount of the overall activity (with confidene intervals) for each region that was observed in each of 20 task domains.

Here again, although I have applied the techniques on group-averaged data to begin to quantify the differences between brain regions in terms of these "neural personalities," the technique can certainly be used also to quantify the differences between individuals, and between the same individual at different times.

In sum, and in light of the general, increasing concern over the issue of reproducibility in science, the time may have come to make a more consistent, concerted effort to actually quantify and specify the precise degree, amount, and sort of reproducibility we can expect in the cognitive neurosciences.

References

Anderson, M.L. (2015): *After Phrenology: Neural Reuse and the Interactive Brain*, MIT Press, Cambridge.

Anderson, M.L. (2007): Evolution of cognitive function via redeployment of brain areas. *The Neuroscientist* **13**(1), 13–21.

Anderson, M.L. (2008): Circuit sharing and the implementation of intelligent systems. *Connection Science* **20**(4), 239–251.

Anderson, M.L. (2010): Neural reuse: A fundamental organizational principle of the brain. *Behavioral and Brain Sciences* **33**(4), 245–266.

Anderson, M.L., and Finlay, B.L. (2013): Allocating structure to function: the strong links between neuroplasticity and natural selection. *Frontiers in Human Neuroscience* **7**, 918.

Anderson, M.L., Kinnison, J., and Pessoa, L. (2013): Describing functional diversity of brain regions and brain networks. *NeuroImage* **73**, 50–58.

Anderson, M.L., and Penner-Wilger, M. (2013): Neural reuse in the evolution and development of the brain: Evidence for developmental homology? *Developmental Psychobiology* **55**(1), 42–51.

Anderson, M.L., and Pessoa, L. (2011): Quantifying the diversity of neural activations in individual brain regions. In *Proceedings of the 33rd Annual Conference of the Cognitive Science Society*, ed. by L. Carlson, C. Hölscher and T. Shipley, Cognitive Science Society, Austin, pp. 2421–2426.

Atkinson, J. (1984): Human visual development over the first six years of life: A review and a hypothesis. *Human Neurobiology* **3**, 61–74.

Baldassarre, A., Lewis, C.M., Committeri, G., Snyder, A.Z., Romani, G.L., and Corbetta, M. (2012): Individual variability in functional connectivity predicts performance of a perceptual task. *Proceedings of the National Academy of Sciences of the USA* **109**, 3516–3521.

Barch, D.M., Burgess, G.C., Harms, M.P., Petersen, S.E., Schlaggar, B.L., Corbetta, M., *et al.* (2013): Function in the human connectome: Task-fMRI and individual differences in behavior. *NeuroImage* **80**, 169–189.

Barrett, H.C., and Kurzban, R. (2006): Modularity in cognition: Framing the debate. *Psychological Review* **113**, 628–647.

Barton, R.A. (1998): Visual specialization and brain evolution in primates. *Proceedings of the Royal Society of London B* **265**, 1933–1937.

Barton, R.A. (2004): Binocularity and brain evolution in primates. *Proceedings of the National Academy of Sciences of the USA* **101**, 10113–10115.

Bassett, D.S., Wymbs, N.F., Porter, M.A., Mucha, P.J., Carlson, J.M., and Grafton, S.T. (2011): Dynamic reconfiguration of human brain networks during learning. *Proceedings of the National Academy of Sciences of the USA* **108**, 7641–7646.

Beason-Held, L.L., Purpura, K.P., Krasuski, J.S., Maisog, J.M., Daly, E.M., Mangot, D.J., *et al.* (1998): Cortical regions involved in visual texture perception: A fMRI study. *Cognitive Brain Research* **7**(2), 111–118.

Binkofski, F., Amunts, K., Stephan, K.M., Posse, S., Schormann, T., Freund, H.-J., *et al.* (2000): Broca's region subserves imagery of motion: A combined cytoarchitectonic and fMRI study. *Human Brain Mapping* **11**, 273–285.

Carmena, J.M., Lebedev, M.A., Crist, R.E., O'Doherty, J.E., Santucci, D.M., Dimitrov, D.F., *et al.* (2003): Learning to control a brain-machine interface for reaching and grasping by primates. *PLOS Biology* **1**(2), e42.

Carruthers, P. (2006): *The Architecture of the Mind: Massive Modularity and the Flexibility of Thought*, Clarendon, Gloucestershire.

Cohen, L.G., Celnik, P., Pascual-Leone, A., Corwell, B., Faiz, L., Dambrosia, J., *et al.* (2007): The cognitive control network: Integrated cortical regions with dissociable functions. *NeuroImage* **37**(1), 343–360.

Cole, M.W., Reynolds, J.R., Power, J.D., Repovs, G., Anticevic, A., and Braver, T.S. (2013): Multi-task connectivity reveals flexible hubs for adaptive task control. *Nature Neuroscience* **16**, 1348–1355.

Cisek, P., and Kalaska, J.F. (2010): Neural mechanisms for interacting with a world full of action choices. *Annual Review of Neuroscience* **33**, 269–298.

Cisek, P., and Kalaska, J.F. (2005): Neural correlates of reaching decisions in dorsal premotor cortex: Specification of multiple direction choices and final selection of action. *Neuron* **45**, 801–814.

Culham, J.C., and Kanwisher, N.G. (2001): Neuroimaging of cognitive functions in human parietal cortex. *Current Opinion in Neurobiology* **11**(2), 157–163.

Damasio, A., and Tranel, D. (1993): Nouns and verbs are retrieved with differently distributed neural systems. *Proceedings of the National Academy of Sciences of the USA* **90**, 4957–4960.

Damasio, H., Grabowski, T.J., Tranel, D., Hichwa, R.D., and Damasio, A.R. (1996): A neural basis for lexical retrieval. *Nature* **380**, 499–505.

Decety, J., and Grèzes, J. (1999): Neural mechanisms subserving the perception of human actions. *Trends in Cognitive Sciences* **3**, 172–178.

Decety, J., Grèzes, J., Costes, N., Perani, D., Jeannerod, M., Procyk, E., *et al.* (1997):. Brain activity during observation of actions. Influence of action content and subject's strategy. *Brain* **120**, 1763–1777.

Embick, D., Marantz, A., Miyashita, Y., O'Neil, W., and Sakai, K.L. (2000): A syntactic specialization for Broca's area. *Proceedings of the National Academy of Sciences of the USA* **97**, 6150–6154.

Fair, D.A., Dosenbach, N.U.F., Church, J.A., Cohen, A.L., Brahmbhatt, S., Miezin, F., *et al.* (2007): Development of distinct control networks through segregation and integration. *Proceedings of the National Academy of Sciences of the USA* **104**, 13507–13512.

Fair, D.A., Cohen, A.L., Power, J.D., Dosenbach, N.U., Church, J.A., Miezin, F.M., *et al.* (2009): Functional brain networks develop from a "local to distributed" organization. *PLOS Computational Biology* **5**, e1000381.

Finlay, B.L., Charvet, C.J., Bastille, I., Cheung, D.T., Muniz, J.A.P.C., and de Lima Silveira, L.C. (2013). Scaling the primate lateral geniculate nucleus: Niche and neurodevelopment in the regulation of magnocellular and parvocellular cell number and nucleus volume. *Journal of Computational Neuroscience* **522**, 1839–1857.

Fodor, J.A. (1983): *The Modularity of Mind: An Essay on Faculty Psychology*, MIT Press, Cambridge.

Fornito, A., Zalesky, A., and Breakspear, M. (2013): Graph analysis of the human connectome: Promise, progress, and pitfalls. *NeuroImage* **80**, 426–444.

Friston, K.J. (1997): Imaging cognitive anatomy. *Trends in Cognitive Sciences* **1**, 21–27.

Gauthier, I., and Nelson, C. (2001): The development of face expertise. *Current Opinion in Neurobiology* **11**, 219–224.

Gauthier, I., Skudlarski, P., Gore, J.C., and Anderson, A.W. (2000): Expertise for cars and birds recruits brain areas involved in face recognition. *Nature Neuroscience* **3**(2), 191–197.

Gauthier, I., Tarr, M.J., Anderson A.W., Skudlarski, P., and Gore, J.C. (1999): Activation of the middle fusiform "face area" increases with expertise in recognizing novel objects. *Nature Neuroscience* **2**, 568–573.

Geisler, W.S., and Albrecht, D.G. (1997): Visual cortex neurons in monkeys and cats: Detection, discrimination, and identification. *Visual Neuroscience* **14**, 897–919.

Glenberg, A.M., Brown, M., and Levin, J.R. (2007): Enhancing comprehension in small reading groups using a manipulation strategy. *Contemporary Educational Psychology* **32**, 389–399.

Glenberg, A.M., and Kaschak, M.P. (2002): Grounding language in action. ıPsychonomic Bulletin and Review **9**, 558–565.

Glenberg, A.M., Sato, M., and Cattaneo, L. (2008a): Use-induced motor plasticity affects the processing of abstract and concrete language. *Current Biology* **18**, R290–R291.

Glenberg, A.M., Sato, M., Cattaneo, L., Riggio, L., Palumbo, D., and Buccino, G. (2008b): Processing abstract language modulates motor system activity. *Quarterly Journal of Experimental Psychology* **61**, 905–919.

Grill-Spector, K., Sayres, R., and Ress, D. (2006): High-resolution imaging reveals highly selective nonface clusters in the fusiform face area. *Nature Neuroscience* **9**, 1177–1185.

Grodzinsky, Y., and Santi, A. (2008): The battle for Broca's region. *Trends in Cognitive Sciences* **12**, 474–480.

Hagoort, P. (2005): On Broca, brain and binding. *Trends in Cognitive Sciences* **9**, 416–423.

Hamzei, F., Rijntjes, M., Dettmers, C., Glauche, V., Weiller, C., and Büchel, C. (2003): The human action recognition system and its relationship to Broca's area: An fMRI study. *NeuroImage* **19**, 637–644.

Hanakawa, T., Honda, M., Sawamoto, N., Okada, T., Yonekura, Y., Fukuyama, H., and Shibasaki, H. (2002): The role of rostral Brodmann area 6 in mental-operation tasks: An integrative neuroimaging approach. *Cerebral Cortex* **12**, 1157–1170.

Hagmann, P., Cammoun, L., Gigandet, X., Meuli, R., Honey, C.J., Wedeen, V. J., and Sporns, O. (2008). Mapping the structural core of human cerebral cortex. *PLOS Biology* **6**(7), e159.

Hanson, S.J., and Schmidt, A. (2011): High-resolution imaging of the fusiform face area (FFA) using multivariate non-linear classifiers shows diagnosticity for non-face categories. *NeuroImage* **54**, 1715–1734.

Haxby, J.V., Gobbini, M.I., Furey, M.L., Ishai, A., Schouten, J.L., and Pietrini, P. (2001): Distributed and overlapping representations of faces and objects in ventral temporal cortex. *Science* **293**, 2425–2430.

Hubel, D.H., and Wiesel, T.N. (1962): Receptive fields, binocular interaction and functional architecture in the cat's visual cortex. *Journal of Physiology* **160**, 106–154.

Johnson, M.H. (2011): Interactive specialization: A domain-general framework for human functional brain development? *Developmental Cognitive Neuroscience* **1**(1), 7–21.

Johnson, M.H. (2001): Functional brain development in humans. *Nature Reviews Neuroscience* **2**, 475–483.

Kanwisher, N. (2010). Functional specificity in the human brain: A window into the functional architecture of the mind. *Proceedings of the National Academy of Sciences of the USA* **107**, 11163–11170.

Kanwisher, N., McDermott, J., and Chun, M.M. (1997): The fusiform face area: A module in human extrastriate cortex specialized for face perception. *Journal of Neuroscience* **17**, 4302–4311.

Kim, B., and Basso, M.A. (2008): Saccade target selection in the superior colliculus: A signal detection theory approach. *Journal of Neuroscience* **28**, 2991–3007.

Kosslyn, S.M., Thompson, W.L., Kin, L.J., Rauch, S.L., and Alpert, N.M. (1996): Individual differences in cerebral blood flow in area 17 predict the time to evaluate visualized letters. *Journal of Cognitive Neuroscience* **8**, 78–82.

Kupers, R., Papperns, M., de Noordhout, A.M., Schoenen, J., Ptito, M., and Fumal, A. (2007): rTMS of the occipital cortex abolishes Braille reading and repetition priming in blind subjects. *Neurology* **69**, 691–693.

Lebedev, M.A., Carmena, J.M., O'Doherty, J.E, Zacksenhouse, M., Henriquez, C.S., Principe, J.C., and Nicolelis, M.A. (2005): Cortical ensemble adaptation to represent velocity of an artificial actuator controlled by a brain-machine interface. *Journal of Neuroscience* **25**, 4681–4693.

Lebedev, M.A., and Nicolelis, M.A. (2006): Brain machine interfaces: Past, present and future. *Trends in Neurosciences* **29**, 536–546.

Liao, X.H., Xia, M.R., Xu, T., Dai, Z.J., Cao, X.Y., Niu, H.J., *et al.* (2013): Functional brain hubs and their test-retest reliability: A multiband resting-state functional MRI study. *NeuroImage* **83**, 969–982.

Martin, A., Haxby, J.V., Lalonde, F.M., Wiggs, C.L., and Ungerleider, L.G. (1995): Discrete cortical regions associated with knowledge of color and knowledge of action. *Science* **270**, 102–105.

Martin, A., Ungerleider, L.G., and Haxby, J.V. (2000): Category-specificity and the brain: The sensorymotor model of semantic representations of objects. In *The New Cognitive Neurosciences*, ed. by M.S. Gazzaniga, MIT Press, Cambridge, pp. 1023–1036.

Martin, A., Wiggs, C.L., Ungerleider, L.G., and Haxby, J.V. (1996): Neural correlates of category-specific knowledge. *Nature* **379**, 649–652.

McIntosh, A.R., Grady, C.L., Ungerleider, L.G., Haxby, J.V., Rapoport, S.I., and Horwitz, B. (1994): Network analysis of cortical visual pathways mapped with PET. *Journal of Neuroscience* **14**, 655–666.

Merabet, L.B., Hamilton, R., Schlaug, G., Swisher, J.D., Kiriakapoulos, E.T., Pitskel, N.B., *et al.* (2008): Rapid and reversible recruitment of early visual cortex for touch.*PLOS One* **3**(8), e3046.

Nishitani, N., Schürmann, M., Amunts K., and Hari, R. (2005): Broca's region: From action to language. *Physiology* **20**, 60–69.

Poldrack, R.A. (2006): Can cognitive processes be inferred from neuroimaging data? *Trends in Cognitive Sciences* **10**, 59–63.

Posner, M.I., Petersen, S.E, Fox, P.T., and Raichle, M.E. (1988): Localization of cognitive operations in the human brain. *Science* **240**, 1627–1631.

Pulvermüller, F. (2005): Brain mechanisms linking language and action. *Nature Reviews Neuroscience* **6**, 576–582.

Pulvermüller, F., and Fadiga, L. (2010): Active perception: Sensorimotor circuits as a cortical basis for language. *Nature Reviews Neuroscience* **11**(5), 351–360.

Rhodes, G., Byatt, G., Michie, P.T., and Puce, A. (2004): Is the Fusiform Face Area specialized for faces, individuation, or expert individuation? *Journal of Cognitive Neuroscience* **16**(2), 189–203.

Rigotti, M., Barak, O., Warden, M.R., Wang, X.J., Daw, N.D., Miller, E.K., and Fusi, S. (2013): The importance of mixed selectivity in complex cognitive tasks. *Nature* **497**, 585–590.

Roitman, J.D., and Shadlen, M.N. (2002): Response of neurons in the lateral intraparietal area during a combined visual discrimination reaction time task. *Journal of Neuroscience* **22**, 9475–9489.

Sadato, N., Pascual-Leone, A., Grafmani, J., Ibañez, V., Deiber, M.-P., Dold, G., and Hallett, M. (1996): Activation of the primary visual cortex by Braille reading in blind subjects. *Nature* **380**, 526–528.

Sakai, K., Hikosaka, O., Miyauchi, S., Takino, R., Sasaki, Y., and Pütz, B. (1998): Transition of brain activation from frontal to parietal areas in visuomotor sequence learning. *Journal of Neuroscience* **18**, 1827–1840.

Schall, J.D., and Bichot, N.P. (1998): Neural correlates of visual and motor decision processes. *Current Opinion in Neurobiology* **8**, 211–217.

Schieber, M.H. (2001): Constraints on somatotopic organization in the primary motor cortex. *Journal of Neurophysiology* **86**, 2125–2143.

Schiller, P. (1995): Effect of lesions in visual cortical area V4 on the recognition of transformed objects. *Nature* **376**, 342–344.

Sporns, O. (2011): *Networks in the Brain*, MIT Press, Cambridge.

Steinmetz, P.N., Roy, A., Fitzgerald, P.J., Hsiao, S.S., Johnson, K.O., and Niebur, E. (2000): Attention modulates synchronized neuronal firing in primate somatosensory cortex. *Nature* **404**, 187–190.

Supekar, K., Musen, M., and Menon, V. (2009): Development of large-scale functional brain networks in children. *PLOS Biology* **7**(7), e1000157.

Sur, M., Garraghty, P.E., and Roe, A.W. (1988): Experimentally induced visual projections into auditory thalamus and cortex. *Science* **242**, 1437–1441.

Sur, M., Pallas, S.L., and Roe, A.W. (1990): Cross-modal plasticity in cortical development: differentiation and specification of sensory neocortex. *Trends in Neuroscience* **13**, 227–233.

Takerkart, S., Auzias, G., Thirion, B., Schön, D., and Ralaivola, L. (2012): Graph-based inter-subject classification of local fMRI patterns. In *Machine Learning in Medical Imaging*, ed. by G. Wu, D. Zhang, and L. Zhou, Springer, Berlin, pp. 184–192.

Taylor, D.M., Tillery, S.I.H., and Schwartz, A.B. (2002): Direct cortical control of 3D neuroprosthetic devices. *Science* **296**, 1829–1832.

Thoenissen, D., Zilles, K., and Toni, I. (2002): Differential involvement of parietal and precentral regions in movement preparation and motor intention. *Journal of Neuroscience* **22**, 9024–9034.

Varela, F., Lachaux J.P., Rodriguez E., and Martinerie, J. (2001): The brainweb: Phase synchronization and largescale integration. *Nature Review Neuroscience* **2**(4), 229–239.

Varoquaux, G., and Craddock, R.C. (2013): Learning and comparing functional connectomes across subjects. *NeuroImage* **80**, 405–415.

von Melchner, L., Pallas, L.L., and Sur, M. (2000): Visual behavior mediated by retinal projections directed to the auditory pathway. *Nature* **404**, 871–876.

Yan, C.G., Craddock, R.C., Zuo, X.N., Zang, Y.F., and Milham, M.P. (2013): Standardizing the intrinsic brain: towards robust measurement of inter-individual variation in 1000 functional connectomes. *NeuroImage* **80**, 246–262.

Yang, T., and Shadlen, M.N. (2007): Probabilistic reasoning by neurons. *Nature* **447**, 1075–1080.

Zuo, X.N., Di Martino, A., Kelly, C., Shehzad, Z.E., Gee, D.G., Klein, D.F., *et al.* (2010): The oscillating brain: Complex and reliable. *NeuroImage* **49**, 1432–1445.

17
On the Difference between Persons and Things – Reproducibility in Social Contexts

Kai Vogeley

Abstract. Social interactions fundamentally rely on an adequate understanding of the other, and we ascribe mental states to other persons apparently without effort. This capacity, known as "mentalizing" or "theory of mind," enables us to explain or predict other persons' behavior and constitutes "folk psychology" as a set of psychological rules on how other persons will presumably experience and behave in given situations. However, in our everyday environment we not only encounter persons, but we are also confronted with things – physical objects which "folk physics" regards in terms of natural laws within the framework of Newtonian mechanics. For instance, things reproducibly fall to the ground if we let them go because of gravity. On the contrary, one cannot expect other persons to reliably respond to one's own smile because they will have different reasons to respond or not. While physical things are assumed to behave due to "outer" physical forces in the framework of folk physics, persons behave on the basis of "inner" mental states such as experiences, motivations, or intentions that allow us to predict their behavior in the framework of folk psychology. This fundamental distinction goes back to the canonical work *Psychology of Interpersonal Relations* by Fritz Heider (1958), a key reference with respect to the development of the so-called attribution theory in social psychology. The different concepts of persons and things have consequences for the concepts of reproducibility and predictability in social contexts.

17.1 The Problem of Other Minds and Its Evolutionary Dimension

Reproducibility is one of the fundamental requirements for all types of natural sciences. The results of an experiment should be the same or at least comparable within certain ranges of tolerance if the initial conditions and constraints during the performance of the experiment were the same, independent of the person performing the experiment. Single individual events that cannot be reproduced are not informative in the sense of the empirical sciences (Popper 1959). Close to reproducibility is the concept of repeatability, which refers to the degree of agreement between different measurements within the same experiment in the same experimental environment and conducted by the same experimenter.

Although modern experimental psychology has a self-understanding of an

Reproducibility: Principles, Problems, Practices, and Prospects, First Edition. Edited by Harald Atmanspacher and Sabine Maasen.

experimental empirical science that is comparable to other natural sciences, it is questionable whether reproducibility is possible to the same degree as in the behavioral sciences. This doubt is based on the strong intuition of a difference between persons and things (or physical objects). This difference can be illustrated by empirical evidence from the different fields of experimental social psychology, social neuroscience, and psychopathology.

Philosophically, the problem of other minds is one of the key questions in epistemology and philosophy of mind. It has been neglected for a long time, but now seems to be reanimated, probably stimulated by modern social psychology and social neuroscience (Nagel 1986, Bermudez 1998). This philosophical question is, for instance, explicitly put forward by John Stuart Mill (1889, p. 243):

> By what evidence do I know, or by what considerations am I led to believe, that there exist other sentient creatures; that the walking and speaking figures which I see and hear, have sensations and thoughts, or in other words, possess Minds?

This epistemic gap emphasized by the concept of privileged access to one's own phenomenal experience relates to the psychological question of how we can successfully or adequately attribute mental states such as perceptions, thoughts, emotions, intentions to act, or, briefly, an inner experience to another person that is not in the same way accessible as our own phenomenal consciousness.

Arguing from a naturalistic position as the most plausible premise of natural sciences, subjective mental states that are experienced from a first-person perspective of others are usually defined from a third-person account. The integrated internal representation of the outer world and one's own organism are based on experiences and memories providing reflected responses to the needs of our environment (Vogeley *et al.* 1999). Consciousness can, in this framework, be characterized as the subjective experience of one's own mental states such as perceptions, judgments, thoughts, intentions to act, feelings, or desires.

This issue of perspectivalness corresponds to the distinction of a subjective or private space–time system and an objective or public space–time system (Kuhlenbeck 1982, p. 180). Naturalistically speaking, consciousness is considered "the subjective experience of cognitive function," it can "emerge from the operation of any cognitive function" (Fuster 2003, p. 249) serving our orientation and survival in the world. Cognitive neuroscience focuses on the neural mechanisms of subjective experiences. It is based on the strong intuition that "conscious experience results from the operation and interaction of several functions in complex assemblies of cortical networks" (Fuster 2003, p. 249).

Evolution has provided an account of this deficit. Learning from and adapting to the behavior of others seems to be an indispensable prerequisite for the formation of social groups. The cognitive system of humans in contrast to other species holds particular capacities to process and adapt to complex affordances emerging from our social environment (Tomasello *et al.* 2005, Moll and Tomasello 2007). One can speculate that the unique capacity to create, to process, and to make use of social information that humans share with other conspecifics constitutes a remarkable evolutionary advantage, enabling us to communicate via verbal and nonverbal signals, to make inferences about others' mental states, and to join in complex collaborative efforts (Tomasello 2008). Equipped by nature with unique prerequisites for social cognition, the human cognitive system already in early childhood develops the capability to differentiate self and others (Decety and Chaminade 2003), to infer emotional and cognitive states of other minds (Frith and Frith 2003), to form social impressions, and to adjust actions and communicative behavior accordingly (Decety and Chaminade 2003, Frith and Frith 2003, Vogeley and Roepstorff 2009).

Probably the most interesting and convincing showcase of how we interact with others is gaze behavior or, in brief, social gaze. From early infancy on, the eyes are the primary and most consistent target of visual attention (Haith *et al.* 1977). Despite the development of other tools to navigate the social world (e.g., language), gaze remains a crucial cue system for our understanding of others and serves a variety of social-cognitive functions comprising information-seeking, signaling interpersonal attitudes, regulating the synchronicity of speech during dialogue and interpersonal distance (Argyle *et al.* 1973). Notably, the human eye has a unique morphology: It is characterized by a depigmentated, white sclera contrasting with the dark iris (Kobayashi and Kohshima 2001). This feature might have evolutionarily facilitated the detection of the gaze direction of other individuals (Emery 2000).

17.2 Understanding the Inner Experience of Others

From a psychological viewpoint, the problem of how we can assess and judge upon the inner experience of others, although we do not have direct access to the phenomenal experience of others, is addressed in the so-called attribution theory developed by Fritz Heider (1958). He explicated the differential features of "persons" as opposed to "things" in the sense of physical objects.[1] The aim of Heider's *opus magnum* is "to make explicit the system of concepts that underlies interpersonal behavior" (Heider 1958, p. 12). He is well aware of the fact that

[1] Heider referred to "things" instead of "objects" in order not to confuse the reader with the psychoanalytical terminology according to which objects can also refer to persons.

the understanding of interpersonal behavior cannot be successful without taking into account the inner experience of persons (Heider 1958, p. 1):

> A person reacts to what he thinks the other person is perceiving, feeling, and thinking, in addition to what the other person may be doing. In other words, the presumed events inside the other person's skin usually enter as essential features of the relation.

The main characteristic feature of our interaction with others is that we cannot fully understand the other person. The understanding of others is often enough "unformulated or only vaguely conceived" and "may not be directly evident" (Heider 1958, p. 2). This leads to an inherent ambiguity and uncertainty (Heider 1958, p. 29), which in turn substantially limits our capacity to predict the behavior of other persons.

We are capable of both verbal and nonverbal communication, whereby nonverbal communication is probably the much more powerful registration device of the inner experience of others (Burgoon 1994). In fact, we infer from verbal or nonverbal signals provided by the other person what their inner experience might be at a given moment in order to be able to adequately ascribe mental states to them: "a person reacts to what he thinks the other person is perceiving, feeling, and thinking, in addition to what the other person may be doing" (Heider 1958, p. 1).

Nonverbal cues including facial expressions, gaze behavior, gestures, postures, and body movements have a deep impact on the process and outcome of our communication (Argyle *et al.* 1970, Mehrabian and Wiener 1967, Schneider *et al.* 1979). Nonverbal communication cues influence person perception and construal processes early during social encounters (Willis and Todorov 2006), and a large proportion of social meaning is substantially informed by nonverbal cues (Burgoon 1994, Argyle 1988).

Understanding interpersonal behavior relies on "social perception," i.e., the "perception of such important dispositional and psychological properties of another person as his actions, motives, affects, beliefs, etc." (Heider 1958, p. 58). Nonverbal behavior generally appears to elicit its influence and serves at least four different purposes: (1) modeling and coordination functions in the sense of a "motor contagion" (Blakemore and Frith 2005), (2) discourse functions, (3) dialog functions, and (4) socio-emotional functions (Vogeley and Bente 2010).

As already pointed out, a most relevant nonverbal cue to prepare for imitation and/or coordination before the onset of action is the observed gaze direction of others: "the complexity of feelings and actions that can be understood at a glance is surprisingly great" (Heider 1958, p. 2). However, as a deictic cue, gaze can also be used to direct the attention of another person to an object, "even the direction of a glance may provide a strong hint as to what the person is thinking, feeling, and wishing" (Heider 1958, p. 43). Following another person's gaze teaches us about the attentional focus of this person and thus about her or his inner experience.

Folk psychology thus enables us to adequately communicate and interact with others: "the ordinary person has a great and profound understanding of himself and of other people which ... enables him to interact with others in more or less adaptive ways" (Heider 1958, p. 2). Gaze helps to regulate dyadic encounters, and its coordination can help to establish three-way relations between self, other, and the object world (Argyle and Cook 1976). A particularly interesting phenomenon in this respect is the experience of "joint attention" that is established as soon as a given person follows another individual's gaze to a novel focus of visual attention (Pfeiffer *et al.* 2013). Studies on gaze-based interactions face-to-face and in real-time have been already widely used in different experimental contexts including virtual reality setups and live video feeds of real interaction partners, with and without studying the underlying neural activity (Pfeiffer *et al.* 2013).

These intuitions lead Heider to propose a fundamental difference between what he calls "thing perception" or "nonsocial perception," referring to inanimate physical objects, in contrast to "person perception" or "social perception" referring to other persons (Heider 1958, p. 21). The differentiating and decisive criteria are the following. First, although both inanimate objects or "things" and animate objects or "persons" are "real, solid objects with properties of shape and color ... that occupy certain positions in the environment," the key difference is that persons are "rarely mere manipulanda" but they are "action centers and as such can do something to us." Second, the spontaneous actions that other persons can exert are based on an inner experience of the person, "they can benefit or harm us intentionally," they have intentions, "abilities, wishes and sentiments," "they can act purposefully, and can perceive or watch us" (Heider 1958, p. 21). Third, as gaze does not have an explicit semantic code as, for instance, language-based utterances, looking at another person might be informative, but would still be underdetermined (Heider 1958, p. 77):

> Of course, one may object that the fact that two people look at each other is no guarantee that they really understand each other, or that a real union arises. Both may have unrevealed thoughts, or they may even fight with their glances, in which case there is a struggle and one wants to outstare the other. Nevertheless, there is a peculiar functional closeness and interaction in a mutual glance.

Taken together, Heider argued that in the case of nonverbal communication with other persons we experience a substantially higher degree of unpredictability corresponding to a lower degree of reproducibility (Heider 1958, p. 29): "Probably the constancy in social perception, however, is less perfect than the constancy in thing perception." This account has been very influential and stimulated the development of the so-called attribution theory in social psychology, focusing on the processes that allow us to attribute mental states to others. This attribution is possible on the basis of the actual stimulus (e.g., a smiling face), the context (e.g., a formal or an informal situation), and my knowledge about the person (e.g., a close relative) with whom I interact (Kelley 1967).

17.3 Identifying the Neural Mechanisms of Understanding Others

Social cognitive processes have recently become a key topic in cognitive neuroscience, and social (cognitive) neuroscience has emerged and recently developed into an autonomous scientific discipline (Cacioppo *et al.* 2004, Adolphs 2009). Generally speaking, social neuroscience focuses on processes that are related to the adequate ascription of mental states to others for the purpose of successful communication or interaction between persons.

One very important distinction relates to two different levels of processing social information that can be either implicit or explicit (Frith and Frith 2008). So-called dual-processing accounts propose that verbal descriptions of the behavior of others use an explicit semantic code and are presented in a propositional format. By contrast, nonverbal behavior does not use an explicit semantic code and is presented in a non-propositional format. It is assumed that the two formats have different processing paths and possibly also differential neural mechanisms. In addition, nonverbal communicative cues have a high-dimensional complexity due to the simultaneous presentation of multiple cues (e.g., smile and direct gaze) and a high processual complexity (Evans 2008, Bente *et al.* 2008, Kuzmanovic *et al.* 2011). As a consequence, nonverbal behavior is often produced and decoded without awareness and might, hence, influence impression formation intuitively, but not inferentially (Choi *et al.* 2005).

In other words: Whereas implicit information processing refers to a comparably fast, prereflexive mode that is employed, for instance, during non-verbal behavior, explicit information processing comprises processes in a comparably slow, reflexive, inferential format such as stereotypes or information processing that is based on explicit rules (Lieberman 2007, Barsalou 2008). Empirical studies during the past decade focused on processes of ascribing mental states to others, ranging from "classical" theory-of-mind studies (Vogeley *et al.* 2001) to person perception studies based on mimic behavior (Schilbach *et al.* 2010), gaze behavior (Kuzmanovic *et al.* 2009), or gestures (Georgescu *et al.* 2014a). The creation of virtual characters that can serve as credible artificial humans provide a unique research tool to improve our knowledge about the underlying psychological processes and neural mechanisms (Vogeley and Bente 2010).

As introduced above, social gaze behavior is an extraordinarily interesting showcase of nonverbal communication and has been studied extensively. Emery (2000) has provided a taxonomy of social gaze behavior defining mutual gaze, gaze aversion, gaze-following, joint attention, and shared attention as the core processes. In a study focusing on person perception we were able to distinguish two different subprocesses in the processing of averted and mutual gaze, namely, gaze detection and gaze evaluation by the systematic variation of the experience of being gazed at by virtual characters with different durations in a study employing functional magnetic resonance imaging (fMRI) (Kuzmanovic *et al.* 2009).

According to this study, "gaze detection" – corresponding to the mere perception of being gazed at, irrespective of the duration with which a virtual character gazed at the participant – recruited fusiform and temporoparietal cortices, brain regions known to be responsible for biological motion detection. In contrast to the mere detection, the evaluative component, during which the participants had to make a sympathy rating of the virtual character at different durations of directed gaze, was accompanied by increased neural activity in the medial prefrontal cortex (mPFC). As the detection of social gaze is a necessary requirement for the successful and adequate interpretation of someone's gaze behavior, gaze detection can be interpreted as an early stage of information processing, whereas gaze evaluation can be understood as a late stage of processing social information as shown before in a wealth of related studies focusing on the "meeting of minds" (Amodio and Frith 2006).

In the past, social cognition has mostly been studied from a detached, observational perspective in tasks involving inert social stimuli ("offline social cognition"). This led to a situation in which social cognition was studied in what has been termed "isolation paradigms" without actual social interaction (Becchio *et al.* 2010). Recent claims emphasized that the active engagement with others in interaction ("online social cognition") plays a significant role in understanding other minds (Schilbach *et al.* 2013). The importance of studying behavior and neural activity in truly interactive contexts is particularly important in studies of social gaze, because this always involves two individuals who are engaged with one another face to face and in real time.

Using the phenomenon of joint attention and employing up-to-date eye-tracking and functional neuroimaging methods, it has become possible to successfully induce the experience of a test person of being involved in an ongoing interaction with other persons based on contingent eye movements and gaze behavior of both persons (Wilms *et al.* 2010, Schilbach *et al.* 2010, 2013, Pfeiffer *et al.* 2013, 2014). In this way, the ecological validity of social gaze behavior in real time could be increased in contrast to purely observational paradigms (Schilbach *et al.* 2013).

Employing gaze-contingent social stimuli we investigated the neural correlates[2] of initiating and responding to joint attention in a combined eye-tracking and fMRI experiment (Schilbach *et al.* 2010). Assuming that subjects would participate in a social encounter with another person outside the scanner via gaze behavior, they had the task either to initiate joint attention or to respond to a joint attention initiative by the partner (who was represented by a virtual character and controlled by a computer algorithm throughout the experiment).

The experience of joint attention was associated with increased neural activity in the mPFC besides other brain regions. More specifically, in all cases in which the test person himself/herself was the initiator of the joint attention instantiation, we also found increased neural activity in the ventral striatum, a key component of the reward system (Schilbach *et al.* 2010, Pfeiffer *et al.* 2014). The strength of the blood-oxygen-level-dependent (BOLD) signal in this region correlated with pleasantness ratings in a post-experiment questionnaire (Schilbach *et al.* 2010), which might be interpreted as a neural basis for the intrinsic motivation to share (Tomasello *et al.* 2005).

Already Heider himself dealt with the concept of animacy experience and developed displays of graphically reduced representations of moving objects (Heider and Simmel 1944). Like many other animals, humans are able to detect biological motion in their environment, namely, movement that is performed by biological organisms, irrespective of how this movement is presented. Phenomenally, biological motion relies on a complex perception comprising data about different aspects of the moving objects perceived. This includes, first, the physical properties of the moving object related to weight and size; second, its dependence on the physical environment such as gravity or obstacles; third, its interrelation to the social environment related to approach and avoidance; and fourth, its behavioral capacities, for instance, related to the degree of efficiency during the performance of motor tasks.

The variance in movement patterns leads to the perception of a biological

[2] A critical account of the stability of neural correlates of cognitive states in humans is due to Anderson, this volume.

"being" – often enough a human being who is alive and allows for meaningful inferences (Blake and Shiffrar 2007). In this experimental context, specific movement features have been empirically identified that contribute to the experience of animacy, including self-propelled motion as initiation of movement without an external cause (Leslie 1984), motion contingency based on both spatial and temporal synchrony between objects (Bassili 1976, Johnson 2003, Blakemore *et al.* 2003), or responsiveness to the motion by any component in the environment (Michotte 1946, Abell *et al.* 2000, Castelli *et al.* 2000, Schultz *et al.* 2005).

In another study, we developed a paradigm that allowed us to induce the experience of animacy in a parametric fashion by the systematic variation of movement parameters of two balls presented in animated video sequences. The experience of perceiving animated objects increased with enrichment of the animations by any of the three different movement cues (a break in an otherwise smooth movement trajectory, an approach movement of one object to the other, responsiveness from the addressed object to the actively moving object) or combinations of these movement cues. Results corroborated the involvement of the key structure of the mPFC during the experience of animacy (Santos *et al.* 2010).

17.4 Abduction of the Functional Roles of Neural Networks

Obviously, cognitive functions are not implemented in single regions, but in neural networks, so that the strategy has to be to search for the underlying networks of activated brain regions and not only single brain areas. Social neuroscience has revealed essentially two different systems recruited during social cognitive processes: (1) the "social neural network" (SNN) including essentially the anterior mPFC, the posterior temporal sulcus (pSTS) and temporoparietal junction (TPJ), and the temporal pole (Adolphs 2009) and (2) the so-called human "mirror neuron system" (MNS) covering superior parietal and premotor regions (Rizzolatti and Craighero 2004).

However, it is still an open question that needs more thorough empirical research, what the differential functional roles of both systems could be. One plausible suggestion is that as soon as the attribution of mental states to others is involved, the SNN is activated, whereas the MNS is recruited when a real or virtual motor component is involved, e.g., in actions, simulations, or imaginations thereof (Vogeley *et al.* 2004, Keysers and Gazzola 2007, Wheatley *et al.* 2007). Another suggestion that does not necessarily contradict the first view assumes that the MNS correlates with early stages of social cognition such as the detection of motor expertise and might putatively also underlie the fast processing of "first impressions" in social encounters that are generated on the basis of facial expressions or gestures. In contrast, the SNN might be recruited during comparably "late" stages in the evaluation of socially relevant information.

The real challenge, therefore, is to develop empirically justified hypotheses about the functional roles of the brain regions involved. Obviously, we cannot claim that the mPFC as one of the key regions of social cognition is specific for this cognitive domain because it is not only involved in the processing of self- and other-related information (including theory-of-mind and mentalizing and emotion processing, Mitchell 2009) but also in a number of other cognitive functions (including processing of attention, multitasking and response conflicts, van Overwalle 2009).

This problem has been recognized within the cognitive neuroscience community as the so-called "reverse inference problem" (Poldrack 2006). The only apparent strategy is to search for something like a common ground or a common denominator which all of the cognitive functions under debate share. This abductive hypothesis-generating procedure led Mitchell (2009, p. 249) to the hypothesis that the feature of being "inexact, probabilistic, internally generated" best covers at least the different socio-emotional functions of the mPFC as one key region of the SNN. This can be taken to suggest that social cognition is a "natural kind" and that the underlying processes can be "distinguished from other kinds of cognitive processing."

This proposal shows that the identification of the functional core of a generator of "fuzzy," probabilistic estimates could be a good candidate to explain the capacity of ascribing mental states to oneself and others or the capacity of understanding other persons. Although this abductively identified hypothesis does not formally prove anything, it nicely corresponds with Heider's view, presented earlier, that the behavior of persons as action centers is characterized by inherent ambiguity and uncertainty. Of course, the ambition is that these speculations motivate new research programs.

One of the intellectually most stimulating findings in cognitive neuroscience during the past years has been the description of the so-called "default mode of the brain" (Raichle *et al.* 2001). Driven by the research question what the high and continuous energy consumption of the brain is good for, Raichle and colleagues were able to demonstrate that the mPFC (besides the pSTS/TPJ and the posterior part of the cingulate gyrus) belongs to a network of brain regions that are systematically active during so-called resting states or baseline conditions. These resting states are simply defined by the absence of any external instruction of an experimenter. Loosely speaking, the situation of subjects with respect to their phenomenal experience resembles the situation of white noise.

Every individual subject experiences different topics, possibly even with different degrees of wakefulness, during these so-called resting states.

The corresponding pattern of brain activation is clearly defined in the sense of an attractor when averaged over individuals and over time, irrespective of the imaging modality. This specific pattern of distributed baseline activation has been identified as the so-called "default mode of brain function" or "default mode network" (DMN, Raichle *et al.* 2001). If some cognitive activity requires a higher demand (e.g. instruction by an experimenter in a formal experiment), neural activation "moves," metaphorically speaking, towards the target neuronal network to be recruited, whereas at the same time medial frontal and parietal regions as part of the DMN tend to decrease their activity (Raichle *et al.* 2001).

Intriguingly, the DMN has been observed in humans irrespective of the task they are involved in and the methodology with which brain activity is measured. The DMN is also observed in other mammals (Vincent *et al.* 2007) and can thus be considered a neurobiologically universal building principle of mammalian brains. By now, the DMN has been studied in thousands of empirical studies.

Controlled operational access to the DMN is limited insofar as the intentional induction of a phenomenal experience will disturb and interrupt the resting state. Besides this methodological problem it has been speculated that the DMN might not only reflect a noisy signal, but rather demonstrate, functionally and phenomenally speaking, a "continuous simulation of behavior" or "an inner rehearsal as well as an optimization of cognitive and behavioral serial programs for the individual's future," in short, a state of the "multifaceted self" (Gusnard *et al.* 2001, p. 4263). What appears as a "state of self" on the phenomenal level, appears as DMN on the neural level.

One interesting aspect with respect to the difference of persons and things is that a significant overlap between the SNN and the DMN can be demonstrated in meta-analytical approaches (Buckner *et al.* 2008, Schilbach *et al.* 2012).[3] This convergence supports the hypothesis that we have a disposition for social cognition that is neurobiologically instantiated and potentially disturbed during disturbances of social cognition.

17.5 Psychopathology of the Inner Experience of Others

The aim of psychopathology as an independent scientific endeavor is to describe and understand the inner experience of persons (see also Emrich, this volume).

[3]See Ehm, this volume, for an in-depth discussion of the challenges of meta-analyses from a statistical point of view.

Accordingly, it is not surprising that we find a corresponding distinction between persons and things also in this domain. In his influential textbook *General Psychopathology*, Karl Jaspers (1913, pp. 301f) distinguished two fundamentally different modes of making sense of the experience and behavior of other persons referring to "understanding" (*Verstehen*) and "explaining" (*Erklären*), adopting the terminology by Droysen and Dilthey. While understanding refers to the empathic appreciation of conflicts, hopes, and desires of an individual person, explaining relates to the attempt to consider mental disorders including their neurobiological and genetic prerequisites and, thus, as consequences of impersonal natural laws.

In the world of folk physics, we "attempt to formulate rules" based on causal links between different events (e.g., letting a pen go that subsequently falls to the ground as a consequence of gravity). This is the accepted approach in the natural sciences. But such explanations on the basis of physical forces are neither available nor helpful if we refer to the inner experience of persons. "Psychic events ... by their very nature remain qualitative" (Jaspers 1913, p. 302) – which then formed the ground for Jaspers' (1913, pp. 301ff) proposal of a "psychology of meaning" (*verstehende Psychologie*, Jaspers 1913, 301ff). We have to understand the inner experience of persons "from within" (Jaspers 1913, p. 28), employing what Jaspers called the genetic mode of understanding. This is not only based on "empathy into the content (symbols, forms, images, ideas)" but also on "sharing the experienced phenomena" (Jaspers 1913, p. 311, Vogeley 2013).

Jaspers also noted that the capacity of understanding is limited, it "must be related to the nature of the meaningful and therefore must itself be inconclusive ... a final *terra firma* is never reached" (Jaspers 1913, p. 357). It seems virtually impossible to fully understand the whole experiential space of a person over her or his life time history: "the total psychic constitution in its sequence of different age-levels ... constitute limits to our understanding" (Jaspers 1913, p. 305), and it is, furthermore, limited by cultural influences (Jaspers 1913, pp. 307f).

As our understanding of others may be limited by a lack of data or information we are often forced to "interpret," as we can only ascribe probabilities to meaningful connections between events (Jaspers 1913, p. 305). What can be understood, "moves midway between meaningful objectivity and what cannot be understood" (Jaspers 1913, p. 310). It is a severe complication in psychopathology that we may face psychopathological states that can no longer be understood. It might be difficult if not impossible in single cases to decide upon "what is meaningful and allows empathy and what in its particular way is un-understandable, 'mad' in the literal sense" (Jaspers 1913, pp. 577f).

Generally speaking, psychopathological conditions are defined by characteristic features or norm deviations compared to others in the subjective experience of one's own mental states. We always have to refer to this subjective experiential space when we try to define mental disorders. In essence, mental disorders are norm-deviant disturbances of subjective experiences that are related to any of the following domains: (1) changes in interactive and communicative behavior, (2) inadequate emotional experiences or changes in sharing emotional experiences with others, and (3) inconsistency of subjective experiences or incongruence with experiences of others, leading to a loss of a sense of reality that can be shared by the majority of other persons within the same cultural and traditional background. In other words, one of the constitutive aspects of mental disorders is the fact that they are defined on the basis of norms that are generated or constituted by groups of persons, populations, or social systems (Vogeley and Newen 2009).

This key aspect is related to the capacity of "mentalizing" or the ability to ascribe mental states to others. From the perspective of persons suffering from mental disorders or reporting subclinical exceptional experiences, the interesting cases are the experience of delusions and deficits of communication and interaction disturbances as in autistic disorders. While persons suffering from delusions show an increased tendency of ascribing mental states to others that are often enough related to themselves corresponding to "hypermentalizing," persons with deficits in communication and interaction with others show "hypomentalizing" (Frith 2004).

Most illustrative cases in this context appear in the diagnostic group of high-functioning autism (HFA) or Asperger syndrome (Asperger 1944), suffering from "mindblindness" (Baron-Cohen 1995). Autism spectrum disorders (ASD) are characterized by lifelong and stable deficits in communication and social interaction, while verbal and general learning and memory abilities are independently developed and often fully preserved in case of HFA or Asperger syndrome.

Individuals with ASD have difficulties to adequately process and integrate nonverbal communication cues into their person judgments (Senju *et al.* 2009, Kuzmanovic *et al.* 2011). They avoid the eye region during the visual inspection of faces (Pelphrey *et al.* 2002) and spend significantly less time fixating the eye region of people as compared to non-autistic controls in passive viewing studies involving social scenes (Klin *et al.* 2002). Furthermore, they have difficulties to interpret gaze as a nonverbal cue supporting the disambiguation of social scenes, thereby suggesting a more general problem in using gaze as a tool to infer the mental states of others (Boraston and Blakemore 2007). They do not spontaneously attend to social information, and are thus less able to intuitively interact in social contexts (Klin *et al.* 2003). When confronted with nonverbal signals such as eye gaze, facial expressions, or gestures, individuals with HFA have shown atypical detection (Dratsch *et al.* 2013, Senju *et al.* 2008) and interpretation of such cues (Uljarevic and Hamilton 2013, Baron-Cohen *et al.* 1997).

Generally speaking, individuals with HFA seem to be less affected when processing a task, as compared with typically developed control persons (Schwartz *et al.* 2010, Schilbach *et al.* 2012), and/or they seem to use atypical strategies for social processing (Walsh *et al.* 2014, Kuzmanovic *et al.* 2014). Interestingly, autistic children appear to be able to follow someone's gaze but tend to spend less attention to congruent objects in a gaze-following task (Bedford *et al.* 2012). This suggests that, while core processes of social gaze can be functional, they might be driven by different motives than in non-autistic individuals. On a neural level, we could show that the processing of socially relevant information recruits significantly less of the regions of the SNN, both in a study of gaze detection and evaluation (Georgescu *et al.* 2013) and in a study on animacy experience (Kuzmanovic *et al.* 2014).

Altogether, these results show atypical processing of socially relevant information in ASD both on behavioral and neural levels. This can be interpreted as a decrease in the salience of nonverbal information for individuals with ASD compared to control persons. While the mere perception of nonverbal cues is often comparable to that of control persons, ASD individuals seem to employ different strategies (Georgescu *et al.* 2014b). The psychopathological study of the inner experience of other persons and the specific communication and interaction deficits associated with ASD further illustrate not only the usefulness of, but also the necessity for a distinction between persons and things.

17.6 Conclusions

Summarizing the concepts and empirical results from social psychology, social neuroscience, and psychopathology, it seems justified to accept and support the intuition of Heider (1958) that there is a fundamental difference in processing information related to persons and to things or physical objects. While persons are driven by their inner experiences and intentions to act, objects are fully controlled by external physical forces – at least if the guiding metaphors are embedded in Newtonian mechanics.

This fundamental difference entails a "folk psychology" based on psychological rules dealing with probabilities for how persons behave under given circumstances, and a "folk physics" based on physical forces that explain the behavior of objects. The behavior of persons can be predicted only on the basis of probabilistic estimates: In a given particular instance of a social encounter we can never be sure how a human interactant will react. In contrast, the behavior of physical objects is assumed to be predictable in a deterministic fashion: If we are sufficiently informed about the different forces that influence a given object, we can by necessity predict how it will behave.

While persons are driven by internal reasons and thus exhibit a considerably high degree of unpredictability and low degree of reproducibility, things show a low degree of unpredictability and a high degree of reproducibility. Persons force us to develop concepts such as ambiguity and intentionality, things behave deterministically on the basis of causal laws. It is unclear both conceptually and empirically whether these two poles must be considered exclusive in the relation to each other or whether we should rather conceptualize both as arranged on a continuum or in a relation of balanced homeostasis under normal conditions.

Over the past decades, experimental social psychology has explored and deconstructed the behavior of persons ("top-down") as obviously not completely unpredictable, accidental, or random. There are in fact rules and expectations that have been developed in the framework of attribution theory. On the other hand, we are facing a wealth of newly emerging communication technologies that must be based on a thorough understanding of how we communicate and interact with one another (Vogeley and Bente 2010, Georgescu *et al.* 2014b).

These technologies could be seen as supporting a "reverse engineering" perspective ("bottom-up") on social cognition (Blascovich *et al.* 2002). However, even if we had exhaustive knowledge about the experience and behavior of persons, we would still be left with the explanatory gap between persons and things. It will become an interesting research question for the future whether changes in the social brain networks of the MNS and the SNN relate to models of social cognition that draw upon the notion of predictive coding (Zaki 2013).

What makes low reproducibility and predictability even more complex is the consideration of different sources of variability (see Shiffrin and Chandramouli, this volume). One potentially highly influential source of this kind is the influence of different cultural backgrounds (Vogeley and Roepstorff 2009). A closer look at social cognition and communication from a cultural psychology perspective reveals a considerable variability or diversity regarding the way social information is processed depending on different cultural backgrounds (Kitayama and Cohen 2007) – which also relates to an early insight of Jaspers (1913, pp. 307f).

Culture can be understood as a complex set of competences, practices, and beliefs of groups that shapes and influences the behavior of its members. It is in continuous and dynamic change rather than a rigid body of standardizations of language, habits, or belief systems (Hacking 1999). For instance, nonverbal communication is rooted in human culture and might be a universal tool in communicating with others. However, there is ample evidence for variance on the surface of the expression of nonverbal cues. It is, therefore, important to consider culture as a source of variance. Building on well-established research in the involved disciplines, social and cultural neurosciences aim at broadening our knowledge in the field pursuing an interdisciplinary approach that explores the conditions and implications of diversity from individual and social perspectives.

It is interesting to note that Jaspers, in his psychopathology, extensively emphasized the limitations of our scientific endeavors. This awareness of limitations includes the necessity to take our own viewpoint on the phenomena into account, for instance our own cultural background. In addition, if we cannot find conclusive evidence for a certain chain of events due to a lack of expressions, utterances, or acts, we start to interpret with the consequence that our hypotheses are less robust. In other words, we need to be aware of limitations because what can be understood is not "meaningful objectivity" but has to be located somewhere "midway" between "objectivity" and ignorance as the absence of any insight or knowledge (Jaspers 1913, p. 310).

Accordingly, Jaspers referred to naive, purely neuroscientific accounts of psychopathological phenomena as "brain mythologies" (Jaspers 1913, p. 18). A century ago, the leading neuroscientific explanations for psychic disorders were structural brain abnormalities. Although we now have a much richer and more complex understanding of how the brain works, we are not immune against the mereological fallacy (Bennett and Hacker 2003) that ascribes features to subcomponents of a cognitive system that can only be adequately ascribed to the complete cognitive system or the person. We should be aware of the danger of losing the phenomenal explananda of the inner experience if mental events are partially or completely ignored and reduced to neuroscientific entities.

References

Abell, F., Happé, F., and Frith, U. (2000). Do triangles play tricks? Attribution of mental states to animated shapes in normal and abnormal development. *Cognitive Development* **15**, 1–16.

Adolphs, R. (2009). The social brain: Neural basis of social knowledge. *Annual Review of Psychology* **60**, 693–716.

Amodio, D.M., and Frith, C.D. (2006). Meeting of minds: The medial frontal cortex and social cognition. *Nature Review Neuroscience* **7**, 268–277.

Argyle, M., and Cook, M. (1976): *Gaze and Mutual Gaze*, Cambridge University Press, Cambridge.

Argyle, M. (1988): *Bodily Communication*, Methuen, London.

Argyle, M., Ingham, R., Alkema, F., and McCallin, M. (1973): The different functions of gaze. *Semiotica* **7**, 19–32.

Argyle, M., Salter, V., Nicholson, H., Wiliams, M., and Burgess, P. (1970): The communication of inferior and superior attitudes by verbal and non-verbal signals. *British Journal of Social and Clinical Psychology* **9**(3), 222–231.

Asperger, H. (1944): Die "Autistischen Psychopathen" im Kindesalter. *Archiv für Psychiatrie und Nervenkrankheiten* **117**, 76–136.

Baron-Cohen, S. (1995): *Mindblindness*, MIT Press, Cambridge.

Baron-Cohen, S., Wheelwright, S., and Jolliffe, T. (1997): Is there a "language of the eyes"? Evidence from normal adults, and adults with autism or Asperger syndrome. *Visual Cognition* **4**, 311–331.

Barsalou, L.W. (2008): Grounded cognition. *Annual Review of Psychology* **59**, 617–645.

Bassili, J.N. (1976): Temporal and spatial contingencies in the perception of social events. *Journal of Personality and Social Psychology* **33**, 680–685.

Becchio, C., Sartori, L., and Castiello, U. (2010): Toward you: The social side of actions. *Current Directions in Psychological Science* **19**, 183–188.

Bedford, R., Elsabbagh, M., Gliga, T., Pickles, A., Senju, A., Charman, T., and Johnson, M.H. (2012): Precursors to social and communication difficulties in infants at-risk for autism: Gaze following and attentional engagement. *Journal of Autism and Developmental Disorders* **42**, 2208–2218.

Bennett, M.R., and Hacker, P.M.S. (2003): *Philosophical Foundations of Neuroscience*, Blackwell, Oxford.

Bente, G., Senokozlieva, M., Pennig, S., Al-Issa, A., and Fischer, O. (2008): Deciphering the secret code: A new methodology for the cross-cultural analysis of nonverbal behavior. *Behavioral Research Methods* **40**(1), 269–277.

Bermudez, J.L. (1998): *The Paradox of Self-Consciousness*, MIT Press, Cambridge.

Blake, R., and Shiffrar, M. (2007). Perception of human motion. *Annual Reviews of Psychology* **58**, 47–73.

Blakemore, S.J., and Frith, C. (2005). The role of motor contagion in the prediction of action. *Neuropsychologia* **43**, 260–267.

Blakemore, S.J., Boyer, P., Pachot-Clouard, M., Meltzoff, A., Segebarth, C., and Decety, J. (2003): The detection of contingency and animacy from simple animations in the human brain. *Cerebral Cortex* **13**, 837–844.

Blascovich, J., Loomis, J., Beall, A.C., Swinth, K.R., Hoyt, C.L., and Bailenson, J.N. (2002): Immersive virtual environment technology as a methodological tool for social psychology. *Psychological Inquiry* **13**, 103–124.

Boraston, Z., and Blakemore, S.J. (2007): The application of eye-tracking technology in the study of autism. *Journal of Physiology* **581**, 893–898.

Buckner, R.L., Andrews-Hanna, J.R., and Schacter, D.L. (2008): The brain's default network: Anatomy, function, and relevance to disease. *Annals of the New York Academy of Sciences* **1124**, 1–38.

Burgoon, J.K. (1994). Nonverbal signals. In *Handbook of Interpersonal Communication*, ed. by M.L. Knapp and G.R. Miller, Sage, Thousand Oaks, pp. 229–285).

Cacioppo, J.T., Lorig, T.S., Nusbaum, H.C., and Berntson, G.G. (2004): Social neuroscience: Bridging social and biological systems. In *Handbook of Methods in Social Psychology*, ed. by C. Sansone, C.C. Morf, and A.T. Panter, Sage, Thousand Oaks, pp. 383–404.

Castelli, F., Happe, F., Frith, U., and Frith, C. (2000): Movement and mind: A functional imaging study of perception and interpretation of complex intentional movement patterns. *NeuroImage* **12**, 314–325.

Choi, V.S., Gray, H.M., and Ambady, N. (2005): The glimpsed world: Unintended communication and unintended perception. In *The New Unconscious*, ed. by R.R. Hassin, J.S. Uleman, and J.A. Bargh, Oxford University Press, New York, pp. 309–333.

Decety, J., and Chaminade, T. (2003): When the self represents the other: A new cognitive neuroscience view on psychological identification. *Consciousness and Cognition* **12**, 577–596.

Dratsch, T., Schwartz, C., Yanev, K., Schilbach, L., Vogeley, K., and Bente, G. (2013). Getting a grip on social gaze: Control over others' gaze helps gaze detection in high-functioning autism. *Journal of Autism and Developmental Disorders* **43**, 286–300.

Emery, N.J. (2000): The eyes have it: The neuroethology, function and evolution of social gaze. *Neuroscience and Biobehavioral Reviews* **24**, 581–604.

Evans, J.S. (2008): Dual-processing accounts of reasoning, judgment, and social cognition. *Annual Reviews of Psychology* **59**, 255–278.

Frith, U., and Frith, C.D. (2003): Development and neurophysiology of mentalizing. *Philosophical Transactions of the Royal Society London B* **358**, 459–473.

Frith, C.D. (2004): Schizophrenia and theory of mind. *Psychological Medicine* **34**, 385–389.

Frith, C.D., and Frith, U. (2008): Implicit and explicit processes in social cognition. *Neuron* **60**, 503–510.

Fuster, J.M. (2003): *Cortex and Mind – Unifying Cognition*, Oxford University Press, Oxford.

Georgescu, A.L., Kuzmanovic, B., Roth, D., Bente, G., and Vogeley, K. (2914b): The use of virtual characters to assess and train nonverbal communication in high-functioning autism. *Frontiers in Human Neuroscience* **8**, 807.

Georgescu, A.L., Kuzmanovic, B., Santos, N.S., Tepest, R., Bente, G., Tittgemeyer, M., and Vogeley, K. (2014a): Perceiving nonverbal behavior: Neural correlates of processing movement fluency and contingency in dyadic interactions. *Human Brain Mapping* **35**, 1362–1378.

Georgescu, A.L., Kuzmanovic, B., Schilbach, L., Tepest, R., Kulbida, R., Bente, G., and Vogeley, K. (2013): Neural correlates of "social gaze" processing in high-functioning autism under systematic variation of gaze duration. *NeuroImage: Clinical* **3**, 340–351.

Gusnard, D.A., Akbudak, E., Shulman, G.L., and Raichle, M.E. (2001): Medial prefrontal cortex and self-referential mental activity: Relation to a default mode of brain function. *Proceedings of the National Academy of Sciences of the USA* **98**, 4259–4264.

Hacking, I. (1999): *The Social Construction of What?* Harvard University Press, Cambridge.

Haith, M.M., Bergman, T., and Moore, M.J. (1977): Eye contact and face scanning in early infancy. *Science* **198**, 853–855.

Heider, F. (1958): *The Psychology of Interpersonal Relations*, Wiley, New York.

Heider, F., and Simmel, M. (1944). An experimental study of apparent behavior. *American Journal of Psychology* **57**, 243–249.

Jaspers, K. (1913): *General Psychopathology, 2 Volumes*, transl. by J. Hoenig and M.W. Hamilton, Johns Hopkins University Press, Baltimore 1997.

Johnson, S.C. (2003): Detecting agents. *Philosophical Transactions of the Royal Society London B* **358**, 549–559.

Kelley, H.H. (1967): Attribution theory in social psychology. *Nebraska Symposium on Motivation* **15**, 192–238

Keysers, C., and Gazzola, V. (2007): Integrating simulation and theory of mind: From self to social cognition. *Trends in Cognitive Science* **11**, 194–196.

Kitayama, S., and Cohen, D. (2007): *Handbook of Cultural Psychology*, Guilford Press, New York.

Klin, A., Jones, W., Schultz, R., and Volkmar, F. (2003). The enactive mind, or from actions to cognition: Lessons from autism. *Philosophical Transactions of the Royal Society London B* **358**, 345–360.

Klin, A., Jones, W., Schultz, R., Volkmar, F., and Cohen, D. (2002): Defining and quantifying the social phenotype in autism. *American Journal of Psychiatry* **159**, 895–908.

Kobayashi, H., and Kohshima, S. (2001): Unique morphology of the human eye and its adaptive meaning: Comparative studies on external morphology of the primate eye. *Journal of Human Evolution* **40**, 419–435.

Kuhlenbeck, H. (1982): *The Human Brain and Its Universe*, Karger, Basel.

Kuzmanovic, B., Schilbach, L., Georgescu, A.L., Kockler, H., Santos, N.S., Shah, N.J., Bente, G., Fink, G.R., and Vogeley, K. (2014). Dissociating animacy processing in high-functioning autism: Neural correlates of stimulus properties and subjective ratings. *Social Neuroscience* **9**, 309–325.

Kuzmanovic, B., Schilbach, L., Lehnhardt, F.G., Bente, G., and Vogeley, K. (2011): A matter of words: Impact of verbal and nonverbal information on impression formation in high-functioning autism. *Research in Autism Spectrum Disorders* **5**, 604–613.

Kuzmanovic, B., Georgescu, A.L., Eickhoff, S.B., Shah, N.J., Bente, G., Fink, G.R., and Vogeley, K. (2009): Duration matters: Dissociating neural correlates of detection and evaluation of social gaze. *NeuroImage* **46**, 1154–1163.

Leslie, A.M. (1984): Spatiotemporal continuity and the perception of causality in infants. *Perception* **13**, 287–305.

Lieberman, M. (2007): Social cognitive neuroscience: A review of core processes. *Annual Reviews of Psychology* **58**, 259–289.

Mehrabian, A., and Wiener, M. (1967): Decoding of inconsistent communications. *Journal of Personality and Social Psychology* **6**, 109–114.

Michotte, A. (1946). *La perception de la causalité, etudes psychologiques vi*, Institut superieur de philosophie, Paris.

Mill, J.S. (1889): *An Examination of Sir William Hamilton's Philosophy*, Longmans, London.

Mitchell, J.P. (2009): Social psychology as a natural kind. *Trends in Cognitive Science* **13**(6), 246–251.

Moll, H., and Tomasello, M. (2007): Cooperation and human cognition: The Vygotskian intelligence hypothesis. *Philosophical Transactions of the Royal Society London B* **362**, 639–648.

Nagel, T. (1986): *The View from Nowhere*, Oxford University Press, Oxford.

Pelphrey, K.A., Sasson, N.J., Reznick, J.S., Paul, G., Goldman, B.D., and Piven, J. (2002): Visual scanning of faces in autism. *Journal of Autism and Developmental Disorders* **32**, 249–261.

Pfeiffer, U., Schilbach, L., Timmermans, B., Kuzmanovic, B., Georgescu, A., Bente, G., and Vogeley, K. (2014): Why we interact: On the functional role of the striatum in the subjective experience of social interaction. *NeuroImage* **101**, 124–137.

Pfeiffer, U.J., Vogeley, K., and Schilbach, L. (2013): From gaze cueing to dual eye-tracking: Novel approaches to investigate the neural correlates of gaze in social interaction. *Neuroscience and Biobehavioral Reviews* **37**, 2516–2528.

Poldrack, R.A. (2006): Can cognitive processes be inferred from neuroimaging data? *Trends in Cognitive Science* **10**, 59–63.

Popper, K. (1959): *The Logic of Scientific Discovery*, Routledge, London. First published 1934 in German.

Raichle, M.E., MacLeod, A.M., Snyder, A.Z., Powers, W.J., Gusnard, D.A., and Shulman, G.L. (2001): A default mode of brain function. *Proceedings of the National Academy of Sciences of the USA* **98**, 676–682.

Rizzolatti, G., and Craighero, L. (2004): The mirror-neuron system. *Annual Reviews of Neuroscience* **27**, 169–192.

Saito, D.N., Tanabe, H.C., Izuma, K., Hayashi, M.J., Morito, Y., Komeda, H., Uchiyama, H., Kosaka, H., Okazawa, H., Fujibayashi, Y., *et al.* (2010): "Stay tuned": Inter-individual neural synchronization during mutual gaze and joint attention. *Frontiers in Integrative Neuroscience* **4**, 127.

Santos, N.S., Kuzmanovic, B., David, N., Rotarska-Jagiela, A., Eickhoff, S., Shah, J.N., Fink, G., Bente, G., and Vogeley, K. (2010): Animated brain: A functional neuroimaging study on the parametric induction of animacy experience. *NeuroImage* **53**, 291–302.

Schilbach, L., Wilms, M., Eickhoff, S.B., Romanzetti, S., Tepest, R., Bente, G., and Vogeley, K. (2010): Minds made for sharing: Initiating joint attention recruits reward-related neurocircuitry. *Journal of Cognitive Neuroscience* **22**, 2702–2715.

Schilbach, L., Bzdok, D., Timmermans, B., Vogeley, K., and Eickhoff, S.B. (2012): Introspective minds: Using ALE meta-analyses to study commonalities in the neural correlates of emotional processing, social and unconstrained cognition. *PLOS ONE* **7**, e30920.

Schilbach, L., Timmermans, B., Reddy, V., Costall, A., Bente, G., Schlicht, T., and Vogeley, K. (2013): Toward a second-person neuroscience. *Behavioral and Brain Sciences* **36**, 393–462.

Schneider, D.J., Hastorf, A.H., and Ellsworth, P.C. (1979): *Person Perception*, Addison-Wesley, Reading.

Schultz, J., Friston, K.J., O'Doherty, J., Wolpert, D.M., and Frith, C.D. (2005): Activation in posterior superior temporal sulcus parallels parameter inducing the percept of animacy. *Neuron* **45**, 625–635.

Schwartz, C., Bente, G., Gawronski, A., Schilbach, L., and Vogeley, K. (2010): Responses to nonverbal behavior of dynamic virtual characters in high-functioning autism. *Journal of Autism and Developmental Disorders* **40**(1), 100–111.

Senju, A., Kikuchi, Y., Hasegawa, T., Tojo, Y., and Osanai, H. (2008). Is anyone looking at me? Direct gaze detection in children with and without autism. *Brain and Cognition* **67**, 127–139.

Senju, A., Southgate, V., White, S., and Frith, U. (2009): Mindblind eyes: An absence of spontaneous theory of mind in Asperger syndrome. *Science* **325**, 883–885.

Tomasello M. (2008): *Why We Cooperate*, MIT Press, Cambridge.

Tomasello, M., Carpenter, M., Call, J., Behne, T., and Moll, H. (2005): Understanding and sharing intentions: The origins of cultural cognition. *Behavioral and Brain Sciences* **28**, 675–735.

Uljarevic, M., and Hamilton, A. (2013): Recognition of emotions in autism: A formal meta-analysis. *Journal of Autism and Developmental Disorders* **43**, 1517–1526.

Van Overwalle, F. (2009): Social cognition and the brain: A meta-analysis. *Human Brain Mapping* **30**, 829–858.

Vincent, J.L., Patel, G.H., Fox, M.D., Snyder, A.Z., Baker, J.T., Van Essen, D.C., Zempel, J.M., Snyder, L.H., Corbetta, M., and Raichle, M.E. (2007): Intrinsic functional architecture in the anaesthetized monkey brain. *Nature* **447**, 83–86.

Vogeley, K., and Bente, G. (2010): "Artificial humans": Psychology and neuroscience perspectives on embodiment and nonverbal communication. *Neural Networks* **23**, 1077–1090.

Vogeley, K., Bussfeld, P., Newen, A., Herrmann, S., Happe, F., Falkai, P., Maier, W., Shah, N.J., Fink, G.R., and Zilles, K. (2001): Mind reading: Neural mechanisms of theory of mind and self-perspective. *NeuroImage* **14**, 170–181.

Vogeley, K., Kurthen, M., Falkai, P., and Maier, W. (1999): The prefrontal cortex generates the basic constituents of the self. *Consciousness and Cognition* **8**, 343–363.

Vogeley, K., May, M., Ritzl, A., Falkai, P., Zilles, K., and Fink, G.R. (2004): Neural correlates of first-person perspective as one constituent of human self-consciousness. *Journal of Cognitive Neuroscience* **16**, 817–827.

Vogeley, K., and Newen, A. (2009): Consciousness of oneself and others in relation to mental disorders. In *The Neuropsychology of Mental Illness*, ed. by S. Wood, N. Allen, and C. Pantelis, Cambridge University Press, Cambridge, pp. 408–413.

Vogeley, K., and Roepstorff, A. (2009): Contextualising culture and social cognition. *Trends in Cognitive Science* **13**, 511–516.

Vogeley, K. (2013): A social cognitive perspective on "understanding" and "explaining". *Psychopathology* **46**, 295–300.

Walsh, J.A., Vida, M.D., and Rutherford, M.D. (2014): Strategies for perceiving facial expressions in adults with autism spectrum disorder. *Journal of Autism and Developmental Disorders* **44**, 1018–1026.

Wheatley, T., Milleville, S.C., and Martin, A. (2007): Understanding animate agents: Distinct roles for the social network and mirror system. *Psychological Science* **18**(6), 469–474.

Willis, J., and Todorov, A. (2006): First impressions: Making up your mind after a 100-ms exposure to a face. *Psychological Science* **17**, 592–598.

Wilms, M., Schilbach, L., Pfeiffer, U., Bente, G., Fink, G.R., and Vogeley, K. (2010): It's in your eyes – Using gaze-contingent stimuli to create truly interactive paradigms for social cognitive and affective neuroscience. *Social Cognitive and Affective Neuroscience* **5**, 98–107.

Zaki, J. (2013): Cue integration: A common framework for social cognition and physical perception. *Perspectives on Psychological Science* **8**, 296–312.

PART V
SOCIAL SCIENCES

PART V: SOCIAL SCIENCES
Introductory Remarks

Sabine Maasen and Harald Atmanspacher

The chapters in this part address the issue of reproducibility in the social sciences. And this is almost exactly where their commonality ends. For not only are the social sciences a heterogeneous set of disciplines – they also deal with highly different topics called *social*, ranging from, e.g., individual decisions to political governance. Moreover, they find science and its methodologies thoroughly "infected" by social factors. To complicate things further, the social sciences deeply disagree about what exactly should count as "scientific" in their epistemic culture. Accordingly, the authors in this part tackle highly different issues from highly different angles. Nevertheless, there is no reason to despair: acknowledging *the human factor* or *the social dimension* may indeed help us to both refine research and publication practices, and understand the limitations of reproducibility when applied to society and its diverse objects of study.

The paper by *Zheng Wang, Ohio State University at Columbus, and Jerome Busemeyer, Indiana University at Bloomington,* leads us into novel psychological approaches in the area of decision theory by drawing attention to problems with reproducibility due to so-called *order effects*. Human judgments and decisions are highly sensitive to contexts. More particularly, the authors study how the sequence of questions or tasks in surveys or experiments influences the outcomes. This can be detrimental to their reproducibility, and new ways of understanding such order effects are proposed.

From a conceptual point of view, order effects occur because individuals, being asked a question, rely on knowledge they retrieve from memory related to that question. But if the question is preceded by another one, then the response will be primed by the previous question and the response to it. Wang and Busemeyer present a quantum theoretically inspired model for such question order effects where they follow as a consequence of so-called non-commuting operations, notably measurements. An especially stunning result of their work is a covariance condition expressed by an empirically testable equation that governs *all* order effects of the large body of results from 72 different field and laboratory experiments. This motivates their concluding hypothesis that the quantum account of order effects may provide insight into other decision-related research as well.

Despite such studies and their findings, reproducibility is still a topic that is rarely raised in the social sciences. Whenever it becomes an object of discussion, however, it is not only about quantitative results but, for a large part, about the question of what kind of science the social sciences are. As *Martin Reinhart, Social Sciences at the Humboldt University Berlin,* emphasizes, many of the corresponding debates refer to the split between quantitative and qualitative methodologies.

Long-standing attempts to overcome this rift notwithstanding, it turns out that there is more to his debate than just methodology: it is about *boundary work* (Gieryn 1983), which addresses the demarcation between science and non-science. From the viewpoint of quantitative sociology, qualitative approaches reject reproducibility in any technical or epistemic sense and, thus, are not considered a science. Rather, qualitative approaches insist on a broader notion of reproducibility in the social sphere and advocate the concept of *intersubjective accountability*. In this way, they plead in favor of a more inclusive idea of what "science" in the social sciences stands for.

Based upon empirical material from online discussions on a social science blog, Reinhart shows that and how both reproducibility and non-reproducibility are posited as normative ideals. In such blogs, the opposing factions disagree fundamentally on the value of reproducibility. However, they not only agree on the minor importance of social processes in determining the value of reproducibility but also explicitly distinguish between the social and the epistemic. While one faction holds both the social and the epistemic to be detrimental to the value of reproducibility, the other holds both to be beneficial.

Referring to the notion of "moral economy" by Daston and Galison (2007), Reinhart elucidates that both technical reproducibility and intersubjective accountability are part of a more general normative orientation in science, where "objectivity" is at the heart of the negotiation. In line with Abbot (2004), this can be regarded as one of the numerous oppositions traversing the social sciences, and as a choice that cannot be reconciled but needs to be made.

This applies to the social sciences at large, yet for some more pronouncedly than for others. In public opinion research, for instance, the question arises whether reproducibility is a meaningful criterion at all, especially since its status as a science is subject to ceaseless debates. As the actors in public opinion researchers themselves claim to use scientific methods, *Felix Keller, Humanities and Social Sciences at the University of St. Gallen,* looks for alternative techniques that may produce public trust in their results (Porter 1995). Based upon a historical sketch of modern polling, Keller illustrates how elections figure as the main test of the whole system of public opinion research. The functional analog to reproducibility in experimental science is elections reproducing survey results. Results of elections yield the evidence for the *accuracy* of the predictions due to a poll.

While this procedure might be easily dismissed as flawed in technical or epistemic terms, Keller suggests to change the perspective. From the point of view of a sociologist, polls generate public knowledge, thereby affecting public behavior: Polls may change the real world. Hence, the public opinion reconstructed by surveys may be correct, but as its findings alter the reality it measures, it becomes a component of what it observed and thus changes initial conditions. For this reason, reproduction in any strict sense is impossible – and accuracy is always historical.

From a constructivist angle, this perspective is less sobering as one might expect: The scientific observation of society (e.g., by public opinion research) takes place within the society it describes (Luhmann 1992). There is no external ("Archimedean") point of view from which one can observe society. There are only (scientific) observers who observe observers. From this perspective, any poll "measuring" public opinion generates nothing else than a *possible world* (à la Kripke) according to peoples' political articulation. At the same time, this possible world concurs and interacts with other views, for instance, expressed by media, sociologists, or futurologists. As Keller argues succinctly: Trying to reproduce one of these possible worlds in order to show its accuracy does nothing but generate another one.

The cofunctions of (social) science describing and interacting with society become hardly more apparent than when one looks at the role of *numbers*. Indeed, numbers have assumed an unmistakable power in modern political culture. Opinion polls, surveys, and market research not only quantify and calibrate public judgments, shape citizens, and direct consumers but they also inform health industry as well as agriculture, and virtually any domain of society today (Rose 1991). Statistics is thus not only capable of describing social reality but also impacts it by defining it in terms of numbers.

Theodore Porter, historian at the University of California at Los Angeles, advances the argument that numbers, measures, and statistical reasoning – important elements in the advance of reproducibility – operate in a world full of uncertainties. Porter guides us through notorious problems attached to statistical bias such as taking a census or sampling methods used in surveys. He also points to the numerous difficulties in controlled clinical trials and in translational research (see part IV for more details).

One important message by Porter cannot be stressed enough. Numbers have assumed such a power that they are now subject to distortions and abuse both in the social sphere and in social science. A demonstration of statistical significance may be sufficient to approve or reject new pharmaceuticals and clear hurdles in the subsequent industrial, juridical, and political processes (Porter 1995). And randomized trials may help to produce generalizable results and to resist interest-related manipulation. But efforts to counter misconduct by explicit methodological rules often meet with an increasing incentive to deceive. Therefore, the "technologies of objectivity" ultimately depend on competent interpreters with the knowledge, will, and the authority to defend the spirit of the rules and not merely their formalized application.

Martina Franzen, WZB Berlin Social Science Center, joins in at this point of misconduct and its structural incentives. Science as a social endeavor faces increasing demands to raise the level of reproducibility. While for a long time scientific fraud was couched as a "black sheep phenomenon," leaving the core of science and professional practice almost untouched, this has changed dramatically. Not only has the concept of scientific misbehavior been extended from fabrication, falsification, and plagiarism to "normal misbehavior" (Martinson *et al.* 2005); Journals also have reorganized their role as institutional gatekeepers of scientific integrity.

For a long time, journal editors (and publishers) have maintained that science is based on trust, and a system based on this trust should perform well. For this reason, peer reviewers have not been requested more than essentially qualitative assessments of submissions. Today this has changed: reviewers are asked to identify potential forgers or to foresee possible replication problems. Based on a brief analysis of the so-called STAP saga,[1] Franzen identifies structural problems underlying this example as well as their impact on scientific practice.

In particular, she points to the gap between the media-structured representation mode and the production mode of science (Knorr-Cetina 1981). This gap affects all fields of research – albeit to a different degree. Clearly, experimental research relies on the accurate presentation of reported results. High-level standardizations in reporting are meant to reconcile the two production modes in the scientific paper. The structural gap widens, however, when the editorial programs of the most relevant journals favor high-impact research. For example, authors tend to overstate the level of certainty in their texts to increase the probability of getting their papers published.

On this view, the flaws in scientific reporting can be regarded as an additional factor for problems with reproducibility. Meanwhile, measures to safeguard

[1] "Stimulus-triggered acquisition of pluripotency" was proposed as a new methodology in adult stem cell research which, although published despite of severe peer criticism, soon turn out to be not reproducible.

scientific integrity multiply by new means of communication in the digital age. The editorial control of data ahead-of-print has become one of many routine procedures. Likewise, it has become mandatory to store raw material and share original data.

Summarizing, while there are ways to improve reproducibility within the social sciences in a technical sense, the role of reproducibility in the social sciences is largely contested. Not least due to the importance of (social) science for contemporary societies, additional measures are taken to raise awareness and integrity. Still, given their either limited or possibly ambivalent effects, the articles in this part convey an important message: Reproducibility needs constant scrutiny, also from a social science point of view.

References

Daston, L., and Galison, P. (2007): *Objectivity*, Zone Books, New York.

Gieryn, T.F. (1983): Boundary-work and the demarcation of science from non-science: Strains and interests in professional ideologies of scientists. *American Sociological Review* **48**, 781–795.

Knorr-Cetina, K. (1981): *The Manufacture of Knowledge. An Essay on the Constructivist and Contextual Nature of Knowledge*, Pergamon, Oxford.

Luhmann, N. (1992): Die Selbstbeschreibung der Gesellschaft und die Soziologie. In *Universität als Milieu*, ed. by N. Luhmann, Haux, Bielefeld, pp. 137–146.

Martinson, B.C., Anderson, M.S., and de Vries, R. (2005): Scientists behaving badly. *Nature* **435**, 737–738.

Porter, T.M. (1995): *Trust in Numbers. The Pursuit of Objectivity in Science and Public Life*, Princeton University Press, Princeton.

Rose, N. (1991): Governing by numbers: Figuring out democracy. *Accounting, Organizations and Society* **16**, 673–692.

18
Order Effects in Sequential Judgments and Decisions

Zheng Wang and Jerome Busemeyer

Abstract Question order effects are pervasive in social and behavioral sciences, and they generally cause problems for reproducibility if not being taken into consideration. Although much research has proposed various psychological ideas, such as anchoring and adjustment, to explain the phenomena, we still lack a formal theory of question order effects.

In this chapter, we present a model for question order effects in sequential judgments and decisions inspired by basic ideas of quantum theory. A fundamental feature of quantum theory is that the order of measurements affects the empirically observed statistics. Our quantum model leads to an *a priori*, mathematically precise, parameter-free, and highly testable prediction, called the QQ equality.

We review a series of empirical tests of the model and also compare it to a set of competing cognitive models that are based on classical instead of quantum rules. These empirical findings suggest that the concept of non-commutative measurements, initially used in quantum physics, provides a simple account for measurement order effects in human decision and cognition. Our approach clarifies how problems with reproducibility due to order effects can be constructively turned into their systematic understanding.

18.1 Introduction

Leading researchers in psychology have emphasized the importance of replication for determining whether or not a finding is true (Cohen 1994). Recently, however, an increasing number of psychologists have realized that there are deep problems concerning the replicability of their findings (for instance, Pashler and Wagenmakers (2012) refer to a "crisis of confidence"). Of course there are many reasons for the failure to replicate, including poor scientific and statistical practice (Simmons *et al.* 2011). However, another important reason for the difficulty to reproduce findings in psychology is the fact that human judgments and decisions are highly sensitive to context effects. Without understanding how these context effects work, a replication may fail because of unanticipated changes of context.

The challenges raised by the demands for reproducibility thus provide an incentive and opportunity for development of new contextual theories. A better theoretical understanding about how these context effects operate and moder-

Reproducibility: Principles, Problems, Practices, and Prospects, First Edition. Edited by Harald Atmanspacher and Sabine Maasen.

ate judgments and decisions can lead to improvements in reproducibility. In this chapter, we provide a case study of one important type of context effect, the effect of question orders in survey research, and we examine some possible theoretical explanations for this type of context effect.

Surveys have long served as an indispensable tool for understanding public opinions and guiding policy makers in our society. Given their importance, notable scientific effort has been directed at understanding factors that influence the reproducibility of results obtained from survey questions. One of the major factors affecting reproducibility is the order in which questions are asked (Schuman and Presser 1981).

It is critical to understand why order effects occur. The commonly held explanation is that when a question is asked first, the person must rely on knowledge she or he can retrieve from memory related to the question. But if the question is preceded by another question, then the person will incorporate some of the thoughts retrieved from the previous question into the answer for the second one (Tourangeau *et al.* 2000). The first question changes or disturbs the cognitive context for evaluating the second one. While this is a reasonable conceptual view of the problem, it lacks a rigorous theoretical formulation.

In this chapter, we compare three formal models for question order effects in sequential judgments and decisions in surveys. Two are based on a popular heuristic from judgment research called anchoring and adjustment (Kahneman *et al.* 1982). The third is based on a quantum theory of question order effects that we recently developed (Wang and Busemeyer 2013, Wang *et al.* 2014). The predictions of each model are tested for 72 national survey data sets that examined the effects of question order.

Why apply quantum theory to survey methods? Quantum theory is well known to explain order effects for successive measurements in physics. We will argue that these effects are analogous in many ways to those found in survey research. The first question (or measurement) changes the context for the second question (or measurement), resulting in question order effects due to the non-commutativity of the measurements). Remarkably, Niels Bohr (1958), one of the founding fathers of quantum physics, borrowed the idea of complementarity from William James, one of the founding fathers of psychology (Holton 1970).

Complementarity is an incompatibility relation that prevents answers to different questions from being determined simultaneously. In quantum theory, this means the projectors representing answers to questions are non-commutative, and thus the order of the measurements matters. This is a critical conceptual idea in explaining order effects of measurements in physics. While it is true that quantum concepts have rarely been applied outside of physics, a growing number of researchers are now exploring its usefulness for explaining human judgments (Aerts *et al.* 2013, Busemeyer and Bruza 2012, Khrennikov 2014, Atmanspacher and Römer 2012, Blutner *et al.* 2013, Yukalov and Sornette 2010). In addition, many social and behavioral scientists have pointed out the effects of measurements themselves on measuring belief, attitude, intention, and behavior (Feldman and Lynch 1988, Sharot *et al.* 2010, Ariely and Norton 2008).[1]

18.2 Question Order Effects and QQ Equality

Here is an example of a question order effect based on a Gallup national survey ($N = 945$) reported by Moore (2002). The data set examined a pair of questions on perceptions of racial hostility collected from the "Aggregate of Racial Hostility Poll" during June 27–30, 1996 in the United States. In the poll, respondents were asked "Do you think many or almost all white people dislike blacks?" preceding or following the same question asked about black hostility toward white.

Table 18.1a shows proportions of respondents selecting each combination of answers when the question about white people was asked first, Table 18.1b shows the results when the question about black people was asked first. "Wy" stands for "white yes," "Wn" for "white no," "By" for "black yes," and "Bn" for "black no." The values of $p(WB)$ and $p(BW)$ give the probabilities of the sequential answers, $p(B)$ and $p(W)$ give the summed probabilities in rows and columns, respectively.

Table 18.1a: white-black

	By	Bn	
Wy	$p(WyBy)$ =.40	$p(WyBn)$ =.02	$p(Wy) = .42$
Wn	$p(WnBy)$ =.16	$p(WnBn)$ =.42	$p(Wn)$ =.58
	$p(By) = .56$	$p(Bn)$ =.44	

Table 18.1b: black-white

	By	Bn	
Wy	$p(ByWy) = .40$	$p(BnWy) = .14$	$p(Wy) = .54$
Wn	$p(ByWn) = .06$	$p(BnWn) = .40$	$p(Wn) = .46$
	$p(By) = .46$	$p(Bn) = .54$	

[1]The potentially fruitful role of quantum concepts to understand issues in the social sciences is also addressed by Collins ("experimenters' regress") and Zimmerli ("Merton-sensible systems") in this volume.

According to Table 18.1a (when the question about white people hostility was asked first), "no" was the more popular answer regarding white people hostility and "yes" was the more popular answer regarding black people hostility. However, according to Table 18.1b (when the question about black people hostility was asked first), the opposite pattern occurred. This is an example of a failure to reproduce results caused by the order in which questions are asked. Table 18.1c shows the difference between the summed probabilities in Tables 18.1a and 18.1b. The four entries in Table 18.1c exhibit clear order effects.

Table 18.1c: order effects

	By	Bn	
Wy	0	$-.12$	-.12
Wn	.10	.02	.12
	.10	$-.10$	0

An important property of the order effect exhibited in Table 18.1c is that the sum of the order effects along the minor diagonal is almost zero (actually it is $-.12 + .10 = -.02$), and so is the sum along the major diagonal. In addition, the sum along the major diagonal, $0 + .02 = +.02$, must be equal but opposite in sign to the sum on the minor diagonal, because the sum of both diagonals (which is the sum of all order effects) must equal zero. We define the size of the sum along one of the diagonals as the *q-value*.

If there are no order effects, then the q-value must be zero for obvious reasons. However, if there are order effects, it is a bold prediction that the q-value has to be zero. For example, in Table 18.1c the q-value could have been as large as $.10 + .12 = .22$, if the entries in Table 18.1b changed to $(.38, .14)$ in the first row and $(.26, .22)$ in the second row).

When the q-value is zero, then we say the data set satisfies the *QQ equality*. Later we will show that across the 72 data sets that we analyzed, the q-value remains close to zero and the QQ equality is approximately satisfied.[2] This is an important empirical property of question order effects that any model of the effects needs to explain.

---

[2]The QQ equality actually expresses a *covariance principle* insofar as it governs the covariation of the entries in the matrix of Table 18.1c. For the significance of invariance and covariance principles for the concept of reproducibility, see Tetens, this volume.

18.3 No Order Effect Model and Saturated Model

A model with no question order effects is based on a single theoretical joint probability distribution over binary questions A,B as shown in Table 18.2. This model has three free parameters, because the four joint probabilities sum to one. The joint probabilities in Table 18.2 are derived from three parameters:

$$p(Ay),\ p(By|Ay),\ p(By|An)$$

Table 18.2: joint probabilities for the no order effect model

	By	Bn	
Ay	$p(Ay)p(By\|Ay)$	$p(Ay)p(Bn\|Ay)$	$p(Ay)$
An	$p(An)p(By\|An)$	$p(An)p(Bn\|An)$	$p(An)$
	$p(By)$	$p(Bn)$	

Another model that allows for order effects is the so-called *saturated model*. It uses a separate table for each order (a pair of tables like Table 18.2 but now conditioned on order). For example, the saturated model defines a probability $p_{AB}(AyBy)$ representing "yes" to question A and then "yes" to question B when question A was asked first. This differs from $p_{BA}(AyBy)$, the probability of "yes" to both questions when question B was asked first. The saturated model can perfectly reproduce any pair of tables of observed proportions, and so it is not constrained to satisfy the QQ equality. The two tables of joint probabilities used by the saturated model require six free parameters:

$$p_{AB}(Ay),\ p_{AB}(By|Ay),\ p_{AB}(By|An),\ p_{BA}(By),\ p_{BA}(Ay|By),\ p_{BA}(Ay|Bn)$$

A chi-square test can be used to compare the no order effect model with the saturated model. The null hypothesis states that there is no true difference between the two models, e.g., $p_{AB}(Ay \cap By) = p(Ay \cap By) = p_{BA}(By \cap Ay)$. The two models differ by three parameters. A G^2 statistic is computed from $-2\log$ likelihood using the maximum-likelihood estimates for each model, and a chi-square statistic is based on the difference between G^2 statistics. Using the data from the white-black survey described above, the chi-square statistic is statistically significant ($\chi^2(3) = 73, p < .001$), so we reject the no order effect model in favor of the saturated model.

Wang *et al.* (2014) analyzed a total of 72 data sets, including 66 data sets from all available national field experiments (surveys) during a decade (2001–2011) by the "Pew Research Center" that contained two question orders (A-then-B, and B-then-A), 4 field experiments by Gallup and Survey Research Center, and 2 laboratory experiments. We computed the chi-square statistic from each of these 72 data sets producing 72 chi-square values. If the null hypothesis is the correct model, then these 72 chi-square values should follow a central chi-square distribution with 3 degrees of freedom. Figure 18.1 (left panel) presents a quantile–quantile plot of the observed chi-square quantile against the quantile predicted by the null hypothesis. As can be seen, the observed quantiles are much larger than those expected by the null hypothesis.

Then we divided the chi-square values into 10 categories, with the category bounds defined by the 10 deciles predicted by the null hypothesis, so that the expected frequency within each category should be equal to 7.2 if there are no order effects. Then we used a chi-square test of the difference between the expected and observed frequency distribution to statistically test the null hypothesis. This chi-square test has 9 degrees of freedom. The results for all 72 data sets produced a chi-square lack of fit equal to 38.84, $p < .001$. Therefore, we reject the no order effect model in favor of the saturated model, allowing for order effects.

Among the 72 data sets, 6 were reported in the literature as selected specifically because they contained large order effects. The remaining 66 data sets were not specifically selected to produce order effects. For these 66 data sets, Fig. 18.1 (right panel) presents a quantile–quantile plot of the observed chi-square quantile against the quantile predicted by the null hypothesis that there is no question order effect. It can be seen in this figure that the observed quantiles are much larger than those expected by the null hypothesis (the prediction is the straight line with unit slope).

Again, we divided the chi-square values into 10 categories, with the category bounds defined by the 10 deciles predicted by the null hypothesis, so that the expected frequency within each category should be equal if there are no order effects. Then we used a chi-square test of the difference between the expected and observed frequency distribution to statistically test the null hypothesis. This chi-square test categories) has 9 degrees of freedom. The results for the 66 data sets produced a chi-square lack of fit equal to 30, $p < .004$. Therefore, we reject the no order effect model again in favor of the saturated model.

18.4 The Anchor Adjustment Model

The intuition behind the *anchor adjustment model*[3] is that the answer to the first question produces a bias that carries over to affect the answer to the second

[3]This model was proposed by Junyi Dai in personal communication.

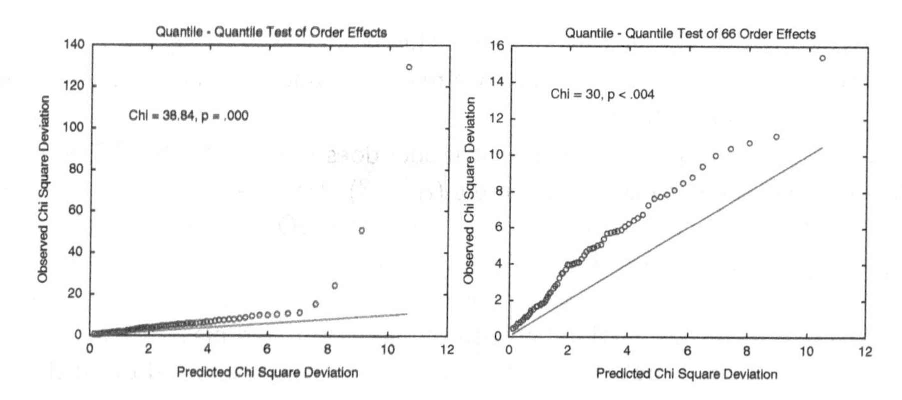

Figure 18.1: Test of the no order effect model. Left panel: All 72 data sets with some specifically selected to have order effects. Right panel: 66 data sets not specifically selected to have order effects. See text for discussion.

question. This bias is represented by a parameter α (for question A first, question B second) or β (for the opposite order). This bias α is added to the implicit conditional probability, $p(By|Ay)$ from Table 18.2, for answering the second question, e.g., $p(By|Ay) + \alpha$. Note that $-\alpha$ must be added to $p(Bn|Ay)$ to produce $p(Bn|Ay) - \alpha$ so that the first row sums to equal $p(Ay)$. Similar rules apply for the second row. The only constraint on the bias is that the adjustment must continue to be a probability, e.g., $0 \leq p(By|Ay) + \alpha \leq 1$.

Tables 18.3a, 18.3b, and 18.3c show the predictions for the joint probabilities conditioned on each order and for the context effects produced by order. Note that in general the model does not have to satisfy the QQ equality unless we assume equal bias, $\alpha = \beta$ for the two orders. The anchor adjustment model has five parameters:

$$\alpha,\ \beta,\ p(Ay),\ p(By|Ay),\ p(By|An)$$

Since the saturated model has six parameters, this leaves one degree of freedom to test the difference between the anchor-adjustment versus saturated models. The null hypothesis states that there is no true difference between the two models. We computed the chi-square statistic for this model comparison using all 72 data sets. The quantile–quantile plot is presented in Fig. 18.2 (left panel). As can be seen, the observed chi-square quantiles are much larger than predicted by the null hypothesis (the prediction line is the straight line with unit slope).

As before, we divided the chi-square values into 10 categories, with the category bounds defined by the 10 deciles predicted by the null hypothesis, so that the expected frequency is equal for each category. Then we used a chi-square test of the difference between the expected and observed frequency distribution to statistically test the null hypothesis. This produced a chi-square lack of fit equal to 40.07, $p < .001$. Therefore, we reject the anchor adjustment model in favor of the saturated model.

In general, the anchor adjustment model does not satisfy the QQ equality when the bias is not equal across orders ($\alpha \neq \beta$). However, if we constrain the model so that it has equal bias ($\alpha = \beta$), then the QQ equality is satisfied. To empirically test this constraint, we estimated the bias from each order for each data set. Figure 18.2 (right panel) provides a scatter plot of the two estimates. As can be seen, the correlation between the two estimated biases equals $-.003$, which disproves the hypothesis of equal bias required by the model to satisfy the QQ equality.

Table 18.3a: anchor-adjust joint probabilities for A–B order

	By	Bn	
Ay	$p(Ay)\,[p(By\|Ay) + \alpha]$	$p(Ay)\,[p(Bn\|Ay) - \alpha]$	$p(Ay)$
An	$p(An)\,[p(By\|An) - \alpha]$	$p(An)\,[p(Bn\|An) + \alpha]$	$p(An)$
	$p_{AB}(By)$	$p_{AB}(Bn)$	1

$$p_{AB}(By) = p(By) + \alpha\,[p(Ay) - p(An)]$$

Table 18.3b: anchor-adjust joint probabilities for B–A order

	By	Bn	
Ay	$p(By)\,[p(Ay\|By) + \beta]$	$p(Bn)\,[p(Ay\|Bn) - \beta]$	$p_{BA}(Ay)$
An	$p(By)\,[p(An\|By) - \beta]$	$p(Bn)\,[p(An\|Bn) + \beta]$	$p_{BA}(An)$
	$p(By)$	$p(Bn)$	1

$$p_{BA}(Ay) = p(Ay) + \beta\,[p(By) - p(Bn)]$$

Table 18.3c: anchor-adjust predicted order effects

	By	Bn
Ay	$\alpha p(Ay) - \beta p(By)$	$-\alpha p(Ay) + \beta p(Bn)$
An	$-\alpha p(An) + \beta p(By)$	$\alpha p(An) - \beta p(Bn)$

QQ equality requires $\alpha = \beta$.

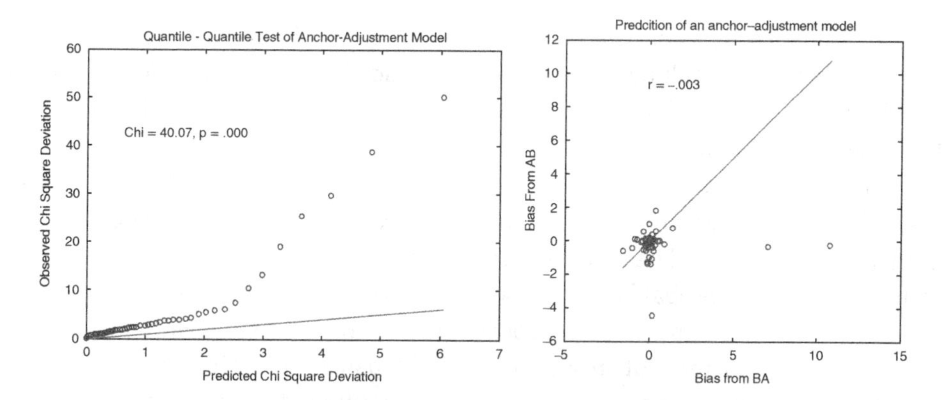

Figure 18.2: Left panel: Quantile plot for the chi-square test of the anchor-adjustment model. Values should fall on a line with unit slope and zero intercept. Right panel: Scatter plot relating the bias parameters obtained from each order. To satisfy the QQ equality, the correlation must equal $+1.0$.

18.5 The Repeat Choice Model

The *repeat choice model*[4] is a variation of the anchor adjustment idea. It assumes that there is some probability to recall and repeat the previous choice, and some probability to make the normal choice conditioned on the first answer. This model includes a parameter α as the probability to recall a previous answer, and a second parameter r as the probability to repeat a choice. Tables 18.4a, 18.4b, and 18.4c show the predictions for the joint probabilities conditioned on each order, and for the context effects produced by order. Altogether this model has five parameters:

$$\alpha,\ r,\ p(Ay),\ p(By|Ay),\ p(By|An)$$

The saturated model has six parameters, so this leaves one degree of freedom to test the difference between the repeat choice versus saturated models. The null hypothesis states that there is no true difference between the two models. We computed the chi-square statistic for this model comparison using all 72 data sets. The quantile–quantile plot is presented in Fig. 18.3 (left panel). As can be seen, the observed chi-square quantiles are much larger than predicted by the null hypothesis (the prediction line is the straight line with unit slope).

As before, we divided the chi-squares into 10 categories, with the category bounds defined by the 10 deciles predicted by the null hypothesis, so that the expected frequency is equal for each category. Then we used a chi-square test

[4]This model was proposed by Bill Batchelder in personal communication.

of the difference between the expected and observed frequency distribution to statistically test the null hypothesis. This produced a chi-square lack of fit equal to 114.37, $p < .001$. Therefore, we reject the repeat choice model in favor of the saturated model.

The repeat choice model is very restricted in its predictions. It must satisfy the QQ equality. Furthermore, it must satisfy another property that can be seen in Table 18.4c. The order effects must be proportional to the difference in marginal choice probabilities. For example, the order effect in the upper left corner of Table 18.4c must be proportional to $p(Ay) - p(By)$. However, as can be seen in the scatter plot in Fig. 18.3 (right panel), the observed correlation between this order effect and the difference between the margins equals $.04$, contrary to the predicted linear correlation.

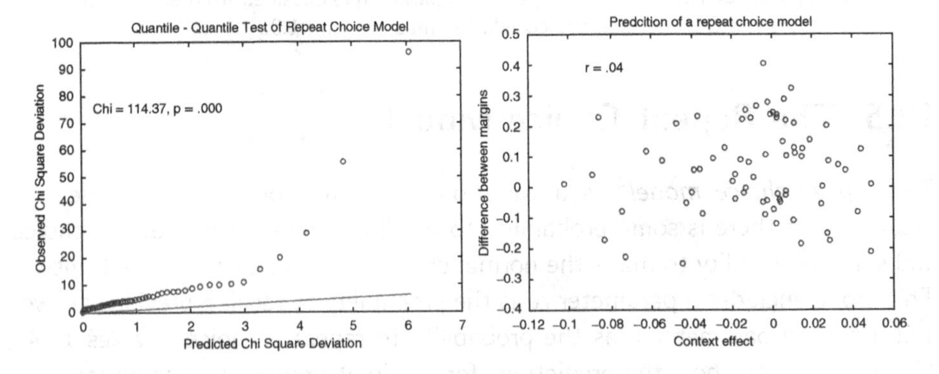

Figure 18.3: Left panel: Quantile plot for the chi-square test of the repeat choice model. Values should fall on a line with unit slope and zero intercept. Right panel: Scatter plot relating the context effect to the difference in marginal choice probabilities. The correlation is predicted to equal $+1.0$.

18.6 The Quantum Model

The *quantum model* is based on different principles. First, it is assumed that the person's beliefs are represented by a vector ψ that lies within an n-dimensional vector space. The answer "yes" to question A is represented as a subspace of this vector space. The projector P_{Ay} projects vectors in the space onto the subspace for the answer "yes" to question A, and $P_{An} = I - P_{Ay}$ is the projector for the "no" answer. The answer "yes" to question B is also represented as a subspace of this vector space. The projector, P_{By} projects vectors in the space onto the subspace for the answer "yes" to question B, and $P_{Bn} = I - P_{By}$ is the projector for the "no" answer.

Table 18.4a: repeat-choice joint probabilities for A–B order

	By	Bn	
Ay	$p(Ay)\,[(1-\alpha)p(By \mid Ay)+\alpha r]$	$p(Ay)\,[(1-\alpha)p(Bn \mid Ay)+\alpha(1-r)]$	$p(Ay)$
An	$p(An)\,[(1-\alpha)p(By \mid An)+\alpha(1-r)]$	$p(An)\,[(1-\alpha)p(Bn \mid An)+\alpha r]$	$p(An)$
	$(1-\alpha)\,p(By)+\alpha\cdot b_{AB}$	$p_{AB}(Bn)$	1

$$b_{AB}=rp(Ay)+(1-r)p(An)$$

Table 18.4b: repeat-choice joint probabilities for B–A order

	By	Bn	
Ay	$p(By)\,[(1-\alpha)p(Ay \mid By)+\alpha r]$	$p(Bn)\,[(1-\alpha)p(Ay \mid Bn)+\alpha(1-r)]$	$(1-\alpha)\,p(Ay)+\alpha\cdot b_{BA}$
An	$p(By)\,[(1-\alpha)p(An \mid By)+\alpha(1-r)]$	$p(Bn)\,[(1-\alpha)p(An \mid Bn)+\alpha r]$	$p_{BA}\,(An)$
	$p(By)$	$p(Bn)$	1

$$b_{BA}=rp(By)+(1-r)p(Bn)$$

Table 18.4c: repeat-choice predicted order effects

	By	Bn
Ay	$\alpha r[p(Ay)-p(By)]$	$\alpha(1-r)[p(Ay)-p(Bn)]$
An	$-\alpha(1-r)[p(Ay)-p(Bn)]$	$-\alpha r[p(Ay)-p(By)]$

The probability of a pair of successive answers is obtained by projecting the belief state ψ onto the subspace for the first answer (P_1) and then projecting the resulting state onto the second answer (P_2), and computing the squared length of the final projection, denoted as $\|P_2P_1\psi\|^2$. This general model produces the predictions shown in Tables 18.5a, 18.5b, and 18.5c.

Table 18.5a: quantum model joint probabilities for A–B order			
	By	Bn	
Ay	$\|P_{By}P_{Ay}\psi\|^2$	$\|P_{Bn}P_{Ay}\psi\|^2$	$\|P_{Ay}\psi\|^2$
An	$\|P_{By}P_{An}\psi\|^2$	$\|P_{Bn}P_{An}\psi\|^2$	$\|P_{An}\psi\|^2$
	$\|P_{By}P_{Ay}\psi\|^2+\|P_{By}P_{An}\psi\|^2$	$p_{AB}(Bn)$	1

Table 18.5b: quantum model joint probabilities for B–A order			
	By	Bn	
Ay	$\|P_{Ay}P_{By}\psi\|^2$	$\|P_{Ay}P_{Bn}\psi\|^2$	$\|P_{Ay}P_{By}\psi\|^2+\|P_{Ay}P_{Bn}\psi\|^2$
An	$\|P_{An}P_{By}\psi\|^2$	$\|P_{An}P_{Bn}\psi\|^2$	$p_{BA}(An)$
	$\|P_{By}\psi\|^2$	$\|P_{Bn}\psi\|^2$	1

Table 18.5c: quantum model predicted order effects		
	By	Bn
Ay	$\|P_{By}P_{Ay}\psi\|^2 - \|P_{Ay}P_{By}\psi\|^2$	$\|P_{Bn}P_{Ay}\psi\|^2 - \|P_{Ay}P_{Bn}\psi\|^2$
An	$\|P_{By}P_{An}\psi\|^2 - \|P_{An}P_{By}\psi\|^2$	$\|P_{Bn}P_{An}\psi\|^2 - \|P_{An}P_{Bn}\psi\|^2$

Table 18.5c shows the constraint that the order effects must sum to zero within each diagonal in the quantum model.[5] In other words, the quantum model requires that the q-values in Table 18.5c are zero, thus satisfying the QQ equality. This prediction places one linear constraint on the theoretical probabilities derived from the quantum model as compared to the saturated model. This leaves one degree of freedom for testing the difference between the quantum and saturated models.

The null hypothesis states that there is no true difference between the quantum model versus the saturated models. Once again, we computed the chi-square statistic for this model comparison using all 72 data sets. The quantile–quantile

[5]This can be theoretically proven from the construction of the quantum probabilities, see Wang and Busemeyer (2013), Wang *et al.* (2014), Busemeyer and Bruza (2012).

plot is presented in Figure 18.4 (left panel). As can be seen, the observed chi-square quantiles agree very closely with the predictions of the null hypothesis (the prediction is the straight line with unit slope).

As before, we divided the chi-squares into 10 categories, with the category bounds defined by the 10 deciles predicted by the null hypothesis, so that the expected frequency is equal for each category. Then we used a chi-square test of the difference between the expected and observed frequency distribution to statistically test the null hypothesis. This produced a chi-square lack of fit equal to 8.53, $p = .48$. Therefore, we do *not* reject the quantum model in favor of the saturated model.

The quantum model predicts that the order effect observed in the upper left corner of Table 18.5c must equal the negative of the context effect observed in the lower right corner of Table 18.5c. Figure 18.4 (right panel) shows the scatter plot of these two order effects. According to the quantum model, the correlation between the two context effects should equal -1; the observed correlation equals $-.82$ which is close to the predicted -1.

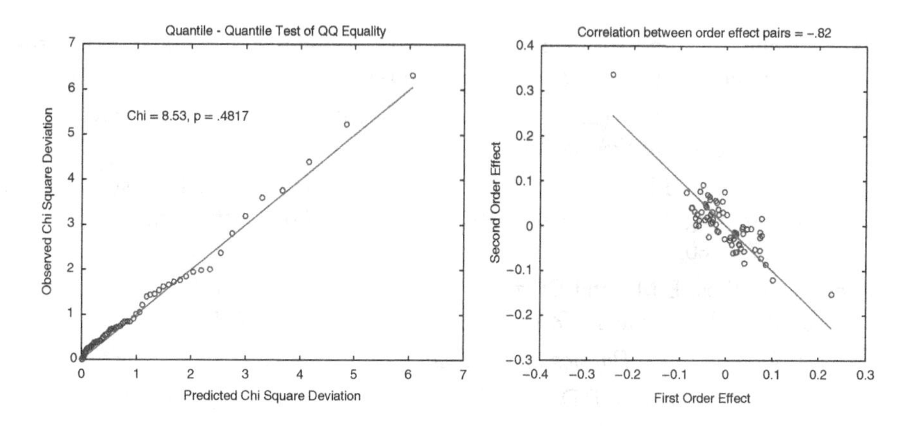

Figure 18.4: Left panel: Quantile plot for the chi-square test of the quantum model. Values should fall on the line with unit slope and zero intercept. Right panel: Scatter plot relating the two order effects in the minor diagonal of Table 18.5c. The observed correlation of $-.82$ is close to the predicted slope of -1.0.

18.7 Concluding Comments

Reproducibility can suffer when unanticipated context effects occur. Context effects can arise in many ways. Order effects of sequential judgments and decisions are one of them. More generally, the questions or stimuli that locally surround a target question can strongly affect the answer to the target question.

Rigorous theories of context effects are needed to understand how the context moderates the findings concerning some target question.

Beyond its obvious role in physics, quantum theory offers a general framework for understanding how context moderates measurement results. Here we have shown that quantum theory makes an *a priori* and precise prediction concerning question order effects, called the QQ equality. It is strongly supported by data from 72 national surveys that included explicit tests for order effects. Two other models, based on heuristic approaches in decision research (anchoring and adjustment), were shown to fail to account for the order effects observed in these surveys.

In quantum theory, order effects are a consequence of non-commuting operations, notably measurements. The same formalism appears to be a powerful natural explanation for order effects concerning survey questions involving sequential decisions and, very likely, for other scenarios in social science as well.

References

Aerts, D., Gabora, L., and Sozzo, S. (2013): Concepts and their dynamics: A quantum-theoretic modeling of human thought. *Topics in Cognitive Science* **5**, 737–773.

Ariely, D., and Norton, M.I. (2008): How actions create – not just reveal – preferences. *Trends in Cognitive Science* **12**, 13–16.

Atmanspacher, H., and Römer, H. (2012): Order effects in sequential measurements of non-commuting psychological observables. *Journal of Mathematical Psychology* **56**(4), 274–280.

Blutner, R., Pothos, E.M., and Bruza, P. (2013): A quantum probability perspective on borderline vagueness. *Topics in Cognitive Science* **5**, 711–736.

Bohr, N. (1958): *Atomic Physics and Human Knowledge*, New York, Wiley.

Busemeyer, J.R., and Bruza, P.D. (2012): *Quantum Models of Cognition and Decision*, Cambridge University Press, Cambridge.

Cohen, J. (1994): The earth is round (p<.05). *American Psychologist* **49**, 997–1003.

Feldman, J.M., and Lynch, J.G. (1988): Self-generated validity and other effects of measurement on belief, attitude, intention, and behavior. *Journal of Applied Psychology* **73**, 421–435.

Holton, G. (1970): The roots of complementarity. *Daedalus* **99**, 1015–1055.

Kahneman, D., Slovic, P., and Tversky, A. (1982): *Judgment Under Uncertainty: Heuristics and Biases*, Cambridge University Press, Cambridge.

Khrennikov, A.Y. (2004): *Information Dynamics in Cognitive, Psychological, Social and Anomalous Phenomena*, Kluwer, Amsterdam.

Moore, D.W. (2002): Measuring new types of question order effects. *Public Opinion Quarterly* **66**, 80–91.

Pashler, H., and Wagenmakers, E.J. (2012): Editors' introduction to the special section on replicability in psychological science: A crisis of confidence. *Perspectives on Psychological Science* **7**, 528–530.

Schuman, H., and Presser, S. (1981): *Questions and Answers in Attitude Surveys: Experiments on Question, Form, Wording, and Content*, Academic Press, New York.

Sharot, T., Velasquez, C.M., and Dolan, R.J. (2010): Do decisions shape preference? Evidence from blind choice. *Psychological Science* **21**, 1231–1235.

Simmons, J.P., Nelson, L.D., and Simonsohn, U. (2011): False-positive psychology: Undisclosed exibility in data collection and analysis allows presenting anything as significant. *Psychological Science* **22**, 1359–1366.

Tourangeau, R., Rips, L.J., and Rasinski, K.A. (2000): *The Psychology of Survey Response*, Cambridge University Press, Cambridge.

Wang, Z., and Busemeyer, J.R. (2013): A quantum question order model supported by empirical tests of an a priori and precise prediction. *Topics in Cognitive Science* **5**, 689–710.

Wang, Z., Solloway, T., Shiffrin, R.M., and Busemeyer, R.J. (2014): Context effects produced by question orders reveal quantum nature of human judgments. *Proceedings of the National Academy of Sciences of the USA* **111**, 9431–9436.

Yukalov, V.I., and Sornette, D. (2011): Decision theory with prospect interference and entanglement. *Theory and Decision* **70**, 283–328.

19
Reproducibility in the Social Sciences

Martin Reinhart

Abstract. The social sciences are a heterogeneous set of disciplines and sometimes reject the normative ideal of reproducibility altogether. Furthermore, reproducibility is a topic that is rarely raised in the social sciences. In this chapter I extend the notion of reproducibility by not just including technical and epistemic but also social and normative aspects. This allows for a discussion of reproducibility that goes beyond the more quantitatively oriented social sciences, showing that reproducibility can be rephrased as intersubjective accountability. From this starting point empirical material from online discussions on social scientific blogs is analyzed where the debate about reproducibility spills over from social psychology to sociology and economics. These represent boundary work, negotiating the difference between the scientific and the non-scientific, and establish a distinction within the social sciences between technical and social aspects pertaining to reproducibility.

Two conclusions are drawn from this case. First, in the social sciences reproducibility as well as non-reproducibility can be normative ideals. Such a dichotomy opens up a space to demand one or the other as a strategic or heuristic tool. Second, reproducibility refers to a more general moral economy of science in which – instead of truth – objectivity and wholeness represent ultimate normative goals. As a normative conclusion: The significance of reproducibility, thus, does not lie in the ability to reproduce certain results but in its contribution to allow science to strive for objectivity and ultimately to not just tell a partial but the whole story.

19.1 Introduction

The social sciences are a heterogeneous set of disciplines. Many properties like object of study, methods, explanatory schemes, publication habits, and so on vary highly between, e.g., social psychology, economics, anthropology, ethnography, sociology, or history. The spectrum seems broader in the social sciences than in the natural sciences where the philosophy of science has been looking for a demarcation criterion to separate science from non-science since Popper (2002) raised the issue 80 years ago – to no avail.[1]

[1] I will exclude the humanities from this discussion for simplicity's sake. However, since I see the social sciences as located between the natural sciences and the humanities, I am assuming that the social sciences are a more diverse set of disciplines as they encompass elements from the natural sciences and the humanities. Finding statements that apply to the social sciences in general have, thus, better chances of being also applicable to the natural sciences and the humanities than the other way around.

Reproducibility: Principles, Problems, Practices, and Prospects, First Edition. Edited by Harald Atmanspacher and Sabine Maasen.

When a demarcation criterion for the natural sciences seems difficult to find because the set of disciplines is too heterogeneous, how can we hope to find something to say in general regarding reproducibility for an even more heterogeneous set of disciplines like the social sciences? First, reproducibility must not be seen as a demarcation criterion. This allows for a more fuzzy definition of the term, that is easier to apply to a diverse set of disciplines. Second, reproducibility must not be seen as a normative concept, but as an actual and discursive practice scientists engage in, by evoking reproducibility to negotiate the boundaries of science.

This allows us to treat instances in the scientific discourse, where reproducibility is appealed to, as boundary work (Gieryn 1983) that can be analyzed empirically. If it is, thus, possible to find something to say in general about reproducibility in the social sciences, chances are that this will also hold for the natural sciences and the humanities.

The main obstacle in reaching general statements about reproducibility that are valid for the whole breadth of the social sciences, however, lies not in their heterogeneity but in the explicit rejection of reproducibility by some of the more qualitatively oriented social sciences. Simon and Goes (2012) wrote:

> In qualitative research, there's no expectation of replication. It is common to see the terms quality, rigor or trustworthiness instead of validity, and dependability, instead of reliability in qualitative studies.

Methods textbooks in the social sciences, especially when focusing on qualitative research, sometimes refer to reproducibility as something unreachable and, thus, not worth striving for (Taylor *et al.* 2005):

> First, there is no widespread agreement about whether there can be any procedures that ensure research and analysis produce the right answers. Second, and this is a problem with qualitative research especially, replication is seldom possible and in most cases doesn't make much sense. When observed or questioned again, respondents in qualitative research will rarely say or do exactly the same things. Whether results have been successfully replicated is always a matter of interpretation.

David Bloor suggested two techniques of validation that are followed in qualitative work to substitute for the quality criterion of reproducibility. The first being triangulation and (Bloor 1997, p. 38)

> the second technique, or rather array of related techniques, judges findings to be valid by demonstrating a correspondence between the analyst's findings and the understanding of members of the collectivity being analysed.

If some part of the social sciences explicitly rejects the idea of reproducibility, should this already be the end of this chapter on reproducibility in the social sciences? Is it not presumptuous to attempt to relate a concept like reproducibility to research fields that seem to set other standards for themselves?

Harry Collins' take (Collins, this volume) on these questions seems clear: Either you accept replication as a relevant criterion or else your work does not qualify as science but as art. Besides this being a very obvious case of boundary work by suggesting replication as a demarcation criterion, two aspects should be kept in mind. First, Collins' starting point are the natural sciences, specifically physics, thus taking up a perspective that might not be best suited to describe and analyze the social sciences. Second, as the quotes from methods textbooks suggest for the qualitative social scientists the problem does not present itself as "replication or nothing," "science or art." Rather, reproducibility can be rejected because other quality criteria are available to take its place. This is good news in that it also indicates that qualitative approaches employ different quality criteria that may have some similarities with the notion of reproducibility.

As a first approximation I will, thus, speak of reproducibility in the technical sense when referring to the moral economy (Daston 1995) of the more quantitative approaches and more generally of reproducibility in the sense of "intersubjective accountability" (Bezzola, this volume) when referring to more qualitative approaches. In this vocabulary, most contributions to this handbook deal with technical reproducibility, especially those on statistical questions. Intersubjective accountability, as a more general concept, can be seen as a rhetorical strategy to convince others of the truth of some statement when no resort to experiments that can be repeated is possible. "Virtual witnessing" (Shapin and Schaffer 1985, pp. 60ff) is a good example of a literary technique to make the reader of a scientific text believe that an experiment is repeatable by giving a very detailed account of it, preferably with pictures.

The term "moral economy" indicates that both technical reproducibility and intersubjective accountability are part of more general normative orientations in science, with objectivity being the most obvious one to relate them to (Daston and Galison 2007). I am, thus, assuming that both belong to the same moral economy of objectivity in the social sciences and I will try to deduce some features of this moral economy form the boundary work surrounding debates about reproducibility.

Since the acceptance or rejection of reproducibility somehow maps onto the distinction between qualitative and quantitative approaches, two strategies present themselves for empirical work. The first would be to investigate boundary work in the qualitative social sciences and look there for technical aspects of reproducibility. The second would be to investigate boundary work in quantitative social sciences and look there for non-technical aspects of reproducibility. As will soon become clear, I will employ the second strategy following a case of boundary work that presented itself more by accident than by systematic selection.

A conceptual problem for both strategies might be that the distinction between qualitative and quantitative is itself the result of boundary work in the social sciences and taking it up as a terminological tool could exaggerate the differences between the two camps. Andrew Abbott's disassembling of the quantitative/qualitative divide by suggesting that it is made up of nine more general distinctions (Abbott 2004) will help to avoid the problem.

The distinction between nomothetic and ideographic approaches may also be helpful. Striving for universal laws within the social (nomothetic) is part of the social sciences as well as striving for precise descriptions of individual cases (ideographic). Both are clearly regulative ideals and not generally reachable goals. Finding laws within the social is limited by the capability of social actors to make sense of the world and to adjust behavior accordingly. Fully doing justice to an individual case by using techniques like thick description can only go as far as others, be they other researchers or the actors that are being described, are able to follow and understand the meaning of such a description. This reminds us that quantitative and qualitative approaches deal with the same problems the social presents to research – its changing and communicative nature.

In the following, I will, first, highlight one discussion about reproducibility in the social sciences by mainly drawing on material from blog posts. Second, I will relate the boundary work that is happening within this material, to different notions of reproducibility and their place in the moral economy of the social sciences. Third, I will argue that reproducibility should not just be seen as a normative aspect within a moral economy but more productively as a heuristic (Abbott 2004) that can be employed in actual social scientific research.

19.2 Reproducibility as a Current Problem in the Social Sciences

One cannot help but notice that reproducibility has become a hot topic in the social sciences also. With social psychology and some cases of scientific misconduct leading the way, a debate on reproducibility is developing through blog posts, news articles, and special issues of academic journals. With many scientists arguing that reproducibility is the gold standard of successful research, it seems peculiar that there is so much discussing going on. If it is the case that reproducibility is one of the main pillars on which the scientific method rests,[2] could we then not expect that any major issue concerning reproducibility should have been cleared long ago? Has the philosophy of science not established the foundation on which working scientists now can argue how reproducibility works and why it is important? Have the history and sociology of science not provided them with analyses on how reproducibility can be achieved and under what conditions research becomes problematic and even fraudulent?

Reproducibility in any of the sciences was only rarely a topic of interest for the news media. It might have popped up while scandalizing one of the spectacular cases of scientific misconduct but besides that, reproducibility was something that only scientists discussed among themselves. When they did, it rarely appeared in books or journal articles but had its place in more informal channels of communication like laboratory talk, conference gossip, or teaching.[3] It is part of what Collins referred to with the metaphor of a ship in a bottle (Collins 1992, pp. 5):

> The ships, our pieces of knowledge about the world, seem so firmly lodged in their bottles of validity that it is hard to conceive that they could ever get out, or that an artful trick was required to get them in.

To stay within the metaphor, even though science is in the business of putting ships in bottles, scientists themselves approach reproducibility as a ship in a bottle, i.e., a piece of knowledge about the scientific method that has always been there and need not be questioned. It seems that reproducibility is seen as so obvious a part of a scientist's habitus that talking about it explicitly is almost a sign of crisis.

Are the current developments then signs of a crisis? Based on the fact that reproducibility has arrived as a topic in news media (The Economist 2013, Johnson 2014, Meyer and Chabris 2014) with scientists struggling to explain publicly why important research is not reproducible and what should be done about this, at least indicates that we are witnessing boundary work (Gieryn 1983), i.e., a debate about where the boundaries of the scientific enterprise are drawn with respect to the question whether the inability to reproduce results is a sign of research that should not be seen as scientific.

---

[2]The term "scientific method" will occur in this text not because I am assuming that science can be easily equated with a specific method, far from it, but because scientists themselves frequently try to legitimize the autonomy of science by recurring to the scientific method. The discourse on reproducibility seems to show this feature prominently.

[3]The field of biomedicine might be an exception as the quality of statistical reasoning and reproducibility have appeared as topics more or less regularly in different fora since the 1970s.

Currently, it seems only possible to speculate on how such a crisis came about exactly. One likely reason is a series of high-profile cases of misconduct – e.g., Diederik Stapel, Hwang Woo-suk, Jan-Hendrik Schön, Ulrich Lichtenthaler, and more – that grabbed media attention and increasingly forced scientists to explain publicly how these were just outliers. The sheer number of cases complicates arguments for the "just-a-few-rotten-apples" theory. Whether the actual number of cases or just media attention has risen is irrelevant for this development.

The 1970s and 1980s also saw a series of alleged and real cases of misconduct that attracted attention and even led to action by political actors in the United States, which was instrumental to the introduction of institutions for detecting, sanctioning, and preventing misconduct like the Office of Research Integrity (Cole *et al.* 1978, Guston 2000). It seems questionable whether a similar development can be expected currently since not just the number of publicized cases has increased but furthermore new forms of governance are in place, intra- and extra-scientific media have multiplied, the Internet speeded up the debate, and larger groups of scientists are now worried that conformity to important scientific norms is declining. In short, pressure from the outside as well as from the inside has risen to put the question on the table how the ship came into the bottle: Why should reproducibility be a cornerstone of the scientific method?

I will not try to reconstruct how the reproducibility debate reached the social sciences, but I assume it is safe to say that social psychology is at the center of this development where the issue is proving to be more than just a news cycle worth of attention.[4] A much publicized letter by Daniel Kahneman (2012), Nobel prize winner in psychology, dated September 2012, called on all researchers in the field to work together to save the reputation of social psychology. He suggested that established scientists should cooperate and reproduce a few of the most important studies in the field.

This suggestion may have backfired, as a recent issue of the journal *Social Psychology* (Nosek and Lakens 2014) reported on 27 attempts to reproduce important findings of which at least 10 failed. Even worse, the area of social priming, which seems the most controversial, scored only one successful replication out of seven.[5] The case of research on social priming is especially relevant

[4]Economics is another field where reproducibility has become a hot topic. See, e.g., the replication wiki in economics: `replication.uni-goettingen.de`.

[5]A more recent many-laboratories effort (Klein *et al.* 2014), which was also a reaction to Kahneman's open letter, yielded the result that 10 out of the 13 most commonly discussed types of social priming have been reproduced consistently across laboratories, yet with considerable variations in effect size.

because explanations for the inability to reproduce, stress the fact that the object of study, humans, is constantly changing because of sociocultural developments.

An effect of social priming may vanish after some time because people have become accustomed to the stimulus or the reaction to the stimulus is culturally changing, i.e., a fad. While many social scientists may see social psychology as only peripheral to the social sciences, such an argument brings possibly the whole of the social sciences into the debate. Jeremy Freese (2014b) explicitly recognized this connection and uses it as a starting point for discussing reproducibility in the social sciences in general (Freese 2014b):[6]

> I worry that a lot of my concern about psychology appears like it's strictly methodological, but a lot of the methodological critique adds up to a dire substantive point that I think sociologists should be extreme [sic] concerned about.

In the following I will analyze a small number of online discussions that represent cases where the debate on reproducibility spills over from social psychology to other social sciences, especially sociology. The sample is very small and consists of five blog posts by four different authors on three different blogs plus the ensuing discussion with a total of 168 comments. The number of comments to each blog post varies considerably with 140, 22, 4, 2, and 0, respectively. The sample is not systematic but has the property that all blog posts are direct reactions to a paper by Jason Mitchell, professor of psychology at Harvard, titled "On the emptiness of failed replications" arguing that failed replications "have no meaningful scientific value" (Mitchell 2014) and, thus, should not be attempted.

Furthermore, all blog posts are directly hyperlinked and some of the authors of the blog posts are also commentators to the other posts. All three blogs are well-established online fora for social scientists with all authors being professors at US universities. Judging from the comments to be analyzed, but also from other posts, the blogs are mainly frequented by working social scientists, whereas two (*scatterplot, Montclair SocioBlog*) address an audience from sociology and one (*Marginal Revolution*) an audience from economics.

A few things can be said to justify such a sample. First, since reproducibility is part of the normative framework of science and is, thus, rarely discussed in the publications of working scientists, more informal channels of communication have to be analyzed. I would suggest that online discussions, such as in

[6]Quotes from blog posts and other online resources will generally have no page numbers as they are not paginated. With the provided URLs and an online search they can, however, be retrieved easily.

blogs, represent a current medium where such topics are addressed. Second, since the discussion is dispersed and, thus, not easy to grasp, a first attempt should be made on a small and admittedly unsystematic sample to generate some exploratory and tentative impressions. Following up on these impressions would then require a more methodical approach. Third, Mitchell's paper can serve as a quasi-stimulus and, with the blog posts all linked to each other, it can be assumed that the material proves coherent enough to not be so many-sided that no patterns are discernible. It goes without saying that the sample can in no way claim representativity for the multifaceted social sciences in general, but aims at working towards some first hypotheses.

19.3 "Reproductions Have No Meaningful Scientific Value"

When Mitchell posted his paper on July 7th, 2014, the debate about reproducibility in psychology seemed to have calmed down with most participants agreeing that replications were necessary and that even failed replications had some value. The paper reignited the debate, because an established psychologist from Harvard took a position that most commentators deemed unscientific. Mitchell's input was not only heard within psychology but also in the social sciences in general with the five blog posts and the following comments from the sample all being reactions within 10 days, most of them within 48 hours.

Why was the paper controversial to the extent that the reactions were so quick and universally negative? All authors of the blog posts are clearly taking position against Mitchell as well as most commentators with some comments even attacking *ad hominem* or questioning the sincerity of the author. "Has anyone looked whether this paper is a hoax?" (Cowen 2014).

On the surface, little of Mitchell's paper seems controversial, as it presents a clear main argument and then discusses three possible rejoinders to this argument. Here is Mitchell's (2014) main argument:

> Recent hand-wringing over failed replications in social psychology is largely pointless, because unsuccessful experiments have no meaningful scientific value.

This argument is supported by the view that experiments are delicate and, thus, so difficult to perform that failure is most likely a consequence of practical mistakes by the reproducing scientists themselves.

The first rejoinder, which also failed replications provide useful information, is dismissed by arguing that performing experiments involves much implicit knowledge that is difficult to transfer to a different researcher or laboratory. This argument is in line with Kahneman's letter asking for replications that involve at least one of the original researchers to ensure that all methodical details are followed. It is also in line, albeit unwittingly, with Collins' experimenters' regress which Freese (2014b) recognizes in one of his blog posts as "the way to a sort of anti-replicationist fundamentalism." Judging from the context, this qualification by Freese is intended to be highly critical maybe even insulting.[7]

The second rejoinder, according to Mitchell, is to argue that effects that are difficult to reproduce must be so flimsy that they are hardly worth studying. Mitchell's response is to point out that many phenomena in social psychology started as effects that were "difficult to obtain" (Mitchell 2014) at first and are now accepted knowledge:

> With twenty years' worth of hindsight, we know that these studies were, in fact, telling us about a highly reliable phenomenon – we just didn't have the right methods for producing it consistently.

Again, this is in line with what we would expect from the experimenters' regress.

The third rejoinder is said to point out that there exists a publication bias for positive results and that failed replications or negative results act as a "counterweight." Mitchell (2014) counters by insisting that replications cannot be used in this way because "the very foundation of science rests on a profound asymmetry between positive and negative claims." Popper's (2002) *Logic of Scientific Discovery* is explicitly mentioned in this context, including the well-known white/black swans example,[8] thus resorting to standard philosophy of science that many scientists should be comfortable with.

Mitchell goes on to explain "why the replication efforts are not science" in his view by adding a social dimension to his arguments. On the one hand, the Popperian philosophy of science is continued by referring to the asymmetry of positive and negative findings. On the other hand, replication studies present themselves as disinterested undertakings, "but are instead motivated by strong prior disbelief in the original findings." According to Mitchell, such a bias is scientifically useless as it attempts to explain away the successes instead of the

[7]Work from the Social Studies of Science (STS) is irregularly mentioned on scatterplot but mostly with negative connotation. The first comment to Freese (2014b) by Hirschman, one of the regular contributors to scatterplot, begins: "How absurd a defense! It's not quite as bad as seeing STS-style soundbites coming from climate change-deniers, but it's in the same direction."

[8]Quine's (1951) *Two Dogmas of Empiricism* and Kuhn's (1970) *Structure of Scientific Revolutions* are also referenced earlier.

failures in detecting a phenomenon. He then takes sides with those researchers who try to perform original work, because (Mitchell 2014)[9]

> these are people who have thrived in a profession that alternates between quiet rejection and blistering criticism, and who have held up admirably under the weight of earlier scientific challenges. ... So we should take note when the targets of replication efforts complain about how they are being treated.

It is implied in the concluding paragraphs of his paper[10] that those who replicate studies must be inferior scientists whose only chance for making it in the brutal world of science is by trying to shoot down the real creators of scientific knowledge by invalidating their work. Mitchell evokes a fragile balance between an epistemic and a social asymmetry. Those with integrity in science face a difficult situation: Not only must they be able to produce positive findings in a world of "quiet rejection and blistering criticism" (epistemic) but they must also defend themselves against those who attempt to denigrate them through unjustified replication efforts (social). In other words, a link between a technical aspect of the scientific method and social norms or customs in science is established, alerting us that reproducibility also has a social dimension besides the technical.

19.4 Reaction from the Blogosphere

The reaction from the blogosphere is immediate and intense, as it is so often. However, even though the analyzed material comes from blogs that tend to maintain a moderate and scientific tone, these specific posts are on the more opinionated and direct end of the spectrum. Most of the comments elaborate on their disagreements with Mitchell's arguments with only a few providing additional conceptual suggestions to show why Mitchell is wrong. From the way the disagreements are phrased it is also possible to conclude where there is common ground between Mitchell and his critics from the social sciences.

[9] In many ways this part of Mitchell's paper echoes Max Weber's (2002) *Science as a Vocation*.

[10] There is a part of Mitchell's paper that is omitted in my summary and analysis. It deals with practical suggestions of how to improve the state of research in social psychology. There are also multiple suggestions in the responses from the blogosphere on this topic but they tend to be disconnected from Mitchell's suggestions and will, thus, not be discussed here. However, a full analysis of the material would have to include these parts as they will allow to draw further inferences on how these scientists view themselves and their disciplines with regard to reproducibility.

The critics are mainly concerned with what they believe is an incorrect way of applying the argument about epistemic asymmetry to the case of reproducibility. According to them, the epistemic asymmetry does not hold in this case. When Mitchell (2014) states:

> When an experiment fails, we can only wallow in uncertainty about whether a phenomenon simply does not exist or, rather, whether we were just a bit too human that time around.

Freese (2014b, comment number 3 by *ljzigerell*) responds:

> But, actually, when an experiment succeeds, we can only wallow in uncertainty about whether a phenomenon exists, or whether a phenomenon appears to exist only because

we were just a bit too human that time around. This objection is thought to apply in general, but can also be developed more specifically for the social sciences. Freese, for example, suggested that experimental findings in the social sciences are not of the form Mitchell presupposes with the example of the white/black swans. Questions are rarely about the existence of a phenomenon, but more often about its frequency or effect size. In such cases "evidence is statistical and demands audience trust" (Freese 2014a), thus rendering positive as well as negative evidence valuable.

Hirschman (2014) extended this argument by proposing to distinguish explicitly between two forms of replication: experimental and statistical. If Mitchell were right, his arguments were more relevant for experimental replication, i.e., in psychology. Most other social sciences deal mainly with the problem of statistical replication and are, thus, less affected. This conceptual distinction is further refined in a comment producing four different kinds of reproducibility: exact reproduction, conceptual reproduction, exact replication, and conceptual replication (Hirschman 2014, comment number 1 by *ljzigerell*).

Reproduction refers to cases where the same data is used, while replications operate with different data. Reproductions or replications are exact when the same analyses are applied, and they are conceptual when different analyses come into play. It seems that the commentators do not have a "theory" of reproducibility at their disposal and are, thus, attempting to develop one on the spot. In this context, another commentator reports on her failed attempts "to come up with more information on how replication works in bioscience" (Freese 2014b, comment number 2 by *olderwoman*) and suspects that there might be a "culture of replication" in certain disciplines. Her call for further references on this topic remains unanswered, suggesting that for this community of social scientists ready access to literature on replication in general and more specifically about theoretical–conceptual or cultural–empirical aspects is lacking.[11]

Returning back to the epistemic asymmetry, Livingston (2014) is not just challenging this part of Mitchell's argument but also what I called the social asymmetry (see above). Mitchell sees the replicators in a position of power by insinuating that illegitimate motives allow to attack knowledge claims through replication with little effort. For Livingston (2014) the situation is the other way around:

> The arrogance Mitchell attributes to the replicators is more common among those who have gotten positive findings. How often do they reflect on their experiments and wonder if it might have been luck or some other element not in their model?

The disagreement seems complete. Mitchell's critics not only find his views on the epistemic and the social dimension of reproducibility lacking but, in both cases, hold the opposite of Mitchell's view to be true.

However, there is an implicit agreement between Mitchell and his critics. By focusing their discussion on the technical or epistemic aspects of reproducibility, they are indicating that the social aspects are of minor value. For Mitchell, the uselessness of replication is first and foremost a consequence of epistemic questions relating to the asymmetry of positive and negative findings. That the replicators are working from an untenable position by invoking their power to discredit the true creators in science is an additional element to the problem, but alone would not suffice.

His critics, respectively, hold replications to be useful because positive and negative findings can be used to strike a balance in the quest for new knowledge. However, even if there were no power differentials between producers and replicators, that would still not render replications unnecessary. Both parties are, thus, subscribing to a scientific model that Todt and Luján (2014) described as "science as arbiter," in which scientific knowledge production is the exclusive domain of cognitive values while noncognitive values only play a role in decision making outside the scientific process. This implies that reproducibility is a purely cognitive, i.e., technical, problem.

In short, Mitchell and his critics disagree fundamentally on the value of reproducibility. Both cite mainly technical, epistemic arguments and use secondary arguments about social processes surrounding replication efforts to reinforce their position. However, they not only agree on the minor importance of social pro-

[11] The author of the blog post could have referred to an earlier publication of his: Freese (2007).

cesses in determining the value of reproducibility but also both find it necessary to distinguish explicitly between the social and the epistemic. Mitchell holds both the social and the epistemic to be detrimental to the value of reproducibility while his critics hold both to be beneficial.

The following two points are relevant for the further discussion. First, even in the domain of quantitative social research, it seems possible to argue in favor as well as against reproducibility. Second, as mentioned in the beginning of this chapter, by looking for non-technical aspects of reproducibility in the quantitative social sciences, it became evident that not just technical but also social aspects are relevant parts of arguments about reproducibility.

19.5 Conclusion

Discourses on reproducibility in the social sciences can be seen as boundary work where either the boundaries between the scientific and the non-scientific or between the qualitative and the quantitative are negotiated. Whether it is the former, the latter, or both depends not only on what we find empirically but also on how broadly we are willing to extend the definition of reproducibility. In contrast to Collins (this volume), I am willing to include those fields into the social sciences that explicitly reject reproducibility as a relevant normative aspiration and I am, thus, extending the definition beyond the purely technical to the social.

Reproducibility encompasses, thus, not only what one commenter identified as exact reproduction, conceptual reproduction, exact replication, and conceptual replication but also what Bloor (1997) subsumed under triangulation and validation. In this sense, reproducibility becomes intersubjective accountability. The empirical material that I reviewed in this chapter represents a case of boundary work from which elements of the boundaries between science and non-science as well as between qualitative and quantitative approaches can be deduced. Even though the case rests firmly within the quantitative domain, from the differences and the similarities between both parties at least some inferences can be drawn regarding both boundaries.

Mitchell's paper shows that it is possible to argue against reproducibility in the technical sense within the quantitative social sciences. This represents a theoretical move Abbott (2004) described as a method of discovery in the social sciences following a heuristic by strategically varying fundamental theoretical dichotomies. Abbott suggested that the social sciences are based on nine fundamental debates, each spanning a continuum between two opposing positions: positivism and interpretivism, analysis and narration, behaviorism and culturalism, individualism and emergentism, realism and constructionism, contextualism and noncontextualism, choice and constraint, conflict and consensus, transcendent and situated knowledge.

For each of these dichotomies, quantitative approaches tend to favor the former and qualitative approaches the latter position. However, by moving along the continuum to the other side of one or multiple dichotomies, new theoretical positions are established and yield new discoveries. Abbott used this model as a tool to describe historical developments in the social sciences (Abbott 2001) and also recommended it as a heuristic for future discovery (Abbott 2004).

I suggest to add a further dichotomy to Abbott's model: reproducibility and non-reproducibility. This is the distinction between thinking you can and should reproduce scientific studies and thinking you cannot and should not. The methodological choice between reproducibility and non-reproducibility "can be made by any kind of analyst at any point" (Abbott 2004, p. 170). However, a move towards non-reproducibility from a quantitative position or from a qualitative position towards reproducibility is more unexpected.

Such choices have higher probabilities for discovering new phenomena because they venture into territory currently not covered by the mainstream of a research field. From the reactions it seems obvious that Mitchell made the unexpected choice as he is confronted with unanimous rejection. However, we should ask whether it is not the other way around. Maybe his critics are those making the unexpected choice by insisting on reproducibility in a context where reproducibility is held high but is not actually practiced? An answer to this question would require more research on the meaning and practice of reproducibility in the social sciences and especially social psychology to show to what extent and on what levels choices between reproducibility and nonreproducibility are made.

In addition, the actors agree that a choice between reproducibility and non-reproducibility should first and foremost be based on technical or epistemic considerations. A social or normative dimension only comes into play in their arguments to reinforce an already held position. This corresponds to the widely held belief in science that the naturalistic fallacy must be avoided. Daston (2014) recently pointed to the complex history of this notion and specifically to its modernity. Strict separation between is and ought, fact and value, nature and nurture represents a development of the 19th and 20th century and replaces earlier meanings that allowed for crossovers between the dichotomies. This is to show that current debates about reproducibility and non-reproducibility are embedded in a historically changing moral economy of science (Daston 1995) that enforces objectivity (Daston and Galison 2007) through the naturalistic fallacy.

Insisting on technical aspects of reproducibility thus appears as a strategic move in scientific boundary work that follows a modern meaning of objectivity drawing mainly on the natural sciences. While insisting on non-reproducibility or on non-technical aspects of reproducibility appears also as a strategic move but as one drawing on a more inclusive meaning of objectivity by stressing intersubjective accountability. With Daston and Galison (2007), it would be possible to argue that these represent different forms of objectivity that researchers are striving for.

However, these different forms of objectivity might be productively seen as part of a larger moral economy surrounding an aspiration towards wholeness (Gouldner 1968). Gouldner also argued that understanding objectivity from the vantage point of "transpersonal replicability" (Gouldner 1968, p. 116) should lead to the realization that there is "something more than the technical machinery of research." This "something more" is conventionally identified in science as the pursuit of truth for its own sake. To Gouldner this seems like obscuring that truth as objectivity is in fact referring to the more general value of wholeness. In this view, science, by striving for reproducibility, objectivity, and truth, is ultimately attempting to tell the whole story.

References

Abbott, A. (2001): *Chaos of Disciplines*, University of Chicago Press, Chicago.

Abbott, A. (2004): *Methods of Discovery: Heuristics for the Social Sciences*, W.W. Norton & Company, New York.

Bloor, M. (1997): Techniques of validation in qualitative research: A critical commentary. In *Context and Method in Qualitative Research*, ed. by G. Miller and R. Dingwall, Sage, London, pp. 37–50.

Cole, S., Rubin, L., and Cole, L.R. (1978): *Peer Review in the National Science Foundation: Phase One of a Study*, National Academy of Sciences, Washington D.C.

Collins, H. (1992): *Changing Order: Replication and Induction in Scientific Practice*, University of Chicago Press, Chicago.

Cowen, T. (2014): How big a deal is replication failure? Available at `marginalrevolution.com/marginalrevolution/2014/07/how-big-a-deal-is-replication-failure.html`.

Daston, L. (1995): The moral economy of science. *Osiris* **10**, 2–24.

Daston, L. (2014): The naturalistic fallacy is modern. *Isis* **105**, 579–587.

Daston, L., and Galison, P. (2007): *Objectivity*, Zone Books, New York.

Freese, J. (2007): Replication standards for quantitative social science. Why not sociology? *Sociological Methods & Research* **36**(2), 153–172.

Freese, J. (2014a): The bigfoot-black swan continuum of behavioral science. *scatterplot*, scatter.wordpress.com/2014/07/09/the-bigfoot-black-swan-continuum-of-behavioral-science/.

Freese, J. (2014b): Why so much psychology? *scatterplot*, scatter.wordpress.com/2014/07/08/why-so-much-psychology/.

Gieryn, T.F. (1983): Boundary-work and the demarcation of science from non-science: Strains and interests in professional ideologies of scientists. *American Sociological Review* **48**, 781–795.

Gouldner, A. (1968): The sociologist as partisan: Sociology and the welfare state. *American Sociologist* **3**(2), 103–116.

Guston, D. (2000): *Between Politics and Science: Assuring the Integrity and Productivity of Research*, Cambridge University Press, Cambridge.

Hirschman, D. (2014): Experimental vs. statistical replication. *scatterplot*, scatter.wordpress.com/2014/07/16/expermental-vs-statistical-replication/.

Johnson, George. (2014): New truths that only one can see. *New York Times*, January 20th, www.nytimes.com/2014/01/21/science/new-truths-that-only-one-can-see.html.

Kahneman, D. (2012): A proposal to deal with questions about priming effects. www.nature.com/polopoly_fs/7.6716.1349271308!/suppinfoFile/Kahneman%20Letter.pdf.

Klein, R.A., Ratliff, K.A., Vianello, M., Adams Jr., R.B., Bahník, S., Bernstein, M.J., Bocian, K., Brandt, M.J., Brooks, B., Brumbaugh, C.C., *et al.* (2014): Investigating variation in replicability: A "many labs" replication project. *Social Psychology* **45**(3), 142–152.

Kuhn, T.S. (1970): *The Structure of Scientific Revolutions*, University of Chicago Press, Chicago.

Livingston, J. (2014): Montclair SocioBlog: Replication and Bullshit. montclairsoci.blogspot.de/2014/07/replication-and-bullshit.html.

Meyer, M.N., and Chabris, C. (2014): Why psychologists' food fight matters. *Slate*, www.slate.com/articles/health_and_science/science/2014/07/replication_controversy_in_psychology_bullying_file_drawer_effect_blog_posts.single.html.

Mitchell, J. (2014): On the emptiness of failed replications. Available at wjh.harvard.edu/~jmitchel/writing/failed_science.htm.

Nosek, B.A., and Lakens, D. (2014): Registered reports: A method to increase the credibility of published results. *Social Psychology* **45**(3), 137–141.

Popper, K.R. (2002): *Logik der Forschung*, Mohr Siebeck, Tübingen.

Quine, W.V.O. (1951): Main trends in recent philosophy: Two dogmas of empiricism. *Philosophical Review* **60**(1), 20–43.

Shapin, S., and Schaffer, S. (1985): *Leviathan and the Air-Pump: Hobbes, Boyle, and the Experimental Life*, Princeton University Press, Princeton.

Simon, M.K., and Goes, J. (2012): *Dissertation and Scholarly Research: Recipes for Success: 2013 Edition*, CreateSpace Independent Publishing Platform, USA.

Taylor, C., Gibbs, G.R., and Lewins, A. (2005): Quality of qualitative analysis. Available at `onlineqda.hud.ac.uk/Intro_QDA/qualitative_analysis.php`.

The Economist (2013): Trouble at the lab. *The Economist*, October 19th, Available at `www.economist.com/news/briefing/21588057-scientists-think-science-self-correcting-alarming-degree-it-not-trouble`.

Todt, O., and Luján, J.L. (2014): Values and decisions: Cognitive and noncognitive values in knowledge generation and decision making. *Science, Technology & Human Values* **39**, 720–743.

Weber, M. (2002): Wissenschaft als Beruf. In *Schriften 1894–1922*, ed. by D. Käsler, Kröner, Stuttgart, pp. 474–511.

20
Accurate But Not Reproducible? The Possible Worlds of Public Opinion Research

Felix Keller

Abstract. In a paper titled "Do the social sciences create phenomena? The example of public opinion research," Osborne and Rose (1999) argued that there are no substantial differences between the construction of scientific objects in the social sciences and in the natural sciences. The existing differences are only gradual, not categorical. Consequently, the same criteria of reproducibility would be applicable to the findings of scientific research and public opinion research.
The thesis presented in this paper is that parallelizing the scientific objects produced by the social sciences and the natural sciences is valid only if the focus is the material techniques of the sciences: the scientific laboratory in the narrow, restricted sense. However, if one recognizes that the context and environment of the scientific objects in the sciences is completely different from those in the social sciences, the parallelization quickly ends. This fact limits the possibility of reproducibility, and alters the definition of the objects themselves.

20.1 Introduction

The criterion of reproducibility shows an ambivalent status in discussions about science: on the one hand, it is regarded as a "general norm of science" (Mittelstrass 1992, p. 61), a "gold standard of sciences";[1] on the other hand, it is seen as difficult to meet even for the hard sciences (Giles 2006).

The question arises whether reproducibility can be a meaningful criterion for public opinion research at all, especially, since their status as a science is debatable if not doubtful compared to "exact" sciences like physics or chemistry (Collins 1994, Kuhn 2012, p. 15). And within the social sciences public opinion research is one of the most controversial areas. It is hard to see how such a strong norm as reproducibility is applicable to public opinion research without the obvious result that the criterion is never met and public opinion research and the social sciences in general are once again seen as the "impossible sciences" (Turner and Turner 1990).

---

[1]See the introductory section to this volume for a general account of the concept, and Steinle's article (this volume) on the historical genesis of norms in the natural sciences.

Reproducibility: Principles, Problems, Practices, and Prospects, First Edition. Edited by Harald Atmanspacher and Sabine Maasen.

However, a careful look at the problems of reproducibility in public opinion polls and the strategies used to develop similar criteria shows that the epistemological situation of public opinion research is so specific that it can enlighten the broader preconditions of reproducibility, at least where social processes are involved. First, I want to discuss the intrinsic idea of public opinion research, and the value given to reproducibility and similar criteria within the discipline. Then I will discuss the status of the knowledge that the discipline creates, with reference to some epistemological reflections, and lastly, I will go back to the scientific status of reproducibility in public opinion research and in the social sciences themselves.

20.2 Reproducibility: A Missing Criterion in Public Opinion Research?

The public is flooded with the tables, numbers, and graphics used in gathering and presenting people's opinions on almost every issue, whether of high or mediocre importance. Public opinion research has perhaps the highest visibility and greatest public presence of all the domains of the social sciences. Moreover, it professes to be just the scientific observation of opinions, a science with its own methods and techniques. However, if reproducibility is a core value of the sciences, then it may be surprising that in the important textbooks representing the current status and core ideas of public opinion research, the question of reproducibility is simply lacking.

In the *SAGE Handbook of Public Opinion Research* (Donsbach and Traugott 2008), the concept is completely missing. The terms "reproducibility" and "reproduction" do not appear in the standard textbook *Polling America: An Encyclopedia of Public Opinion* (Best and Radcliff 2005). It is the same situation in the often-used *Public Opinion: Measuring the American Mind* (Bardes and Oldendick 2012) – here too, the search for a discussion of reproducibility shows no results.

In *Public Opinion Quarterly*, the most influential public opinion research journal, existing since 1937, we find only one article whose main subject is reproducibility: "House Effects and the Reproducibility of Survey Measurements" (Smith 1982). All 76 other articles containing the term "reproducibility" refer to Guttman's concept of a "reproducibility coefficient," used in a specific statistical scaling technique (Guttman 1944), the scalogram analysis, with a different meaning of the word.

Why is the concept of reproducibility missing? Are the observations of public opinion research just not reproducible? Are they, therefore, without scientific validation or objectivity? Are there alternative concepts? And do they have the same functions? In fact, Smith's article on "house effects" (effects created by the survey institutions) is the only one that treats the concept of reproducibility directly. But Smith (1982, p. 55) articulated the same significance of reproducibility as the natural sciences do:

> If a measurement of a particular phenomenon can be replicated by independent observation under like conditions, there is presumptive evidence that the measurement is, within a given degree of exactitude, accurate and reliable.

To investigate the reproducibility of public opinion research, Smith started an experiment asking four survey houses the same questions in "representative" surveys: two university institutes (University of Michigan and University of Chicago) and two of the most important commercial survey agencies (Roper and Gallup). To standardize the experiment as much as possible the questions were worded identically and the researchers tried to "minimize differences in timing and context" (Smith 1982, p. 55).

Differences were found, due to the concept of sampling and due to probability distributions, and that is no surprise (see Stahel, this volume, for a more detailed discussion). However, the differences among the survey houses were quite substantial for all of the answers. Let us look at a typical polling question:[2] Selected individuals were asked if the government should spend more or less money on "halting the rising crime rate." In the four survey waves, the range of agreement that "too much" was spent on crime prevention was between 24.4% and 18.2% in each wave – a difference of more than 6%.[3] Between 72% and 59.7% of respondents wanted the government to spend more money on halting the crime rate – a difference of 12%.[4]

---

[2]The exact wording of the question was: "First I would like to talk with you about some things people think about today. We are faced with many problems in this country, none of which can be solved easily or inexpensively. I'm going to name some of these problems, and for each one I'd like you to tell me whether you think we're spending too much money on it, too little money, or about the right amount. First (READ ITEM A) ... are we spending too much, too little, or about the right amount on (ITEM)? READ EACH ITEM; CODE ONE FOR EACH." Capitalized letters signify interviewer instructions. The answer categories were: "Too Much," "Too Little," "About Right," and "Don't Know."

[3]However, the scales were different: Two waves allowed the answer "Don't Know," two did not. At least in this case, these additional categories do not change the answer patterns and the relative differences systematically (in others they do).

[4]Smith used classical statistical tests to assess whether the differences were hazardous or significant, assuming simple random sampling – although there is probably no modern social survey using a clean random sampling (including this one). Nevertheless, Smith claimed that

So the identical question produced an incoherent pattern of answers among the survey waves, and the differences between the answers were statistically highly significant. Neither Smith, nor a later paper by two important authors in the field, Converse and Traugott (1986), referring to Smith's experiment could explain why this was so. An unknown number and quality of relevant external factors that were not measured may be responsible for the differences. Converse and Traugott summarized them in a "total survey error" which does not focus only on sample errors, but on any source of error.

Smith showed the importance of the reproducibility criterion in public opinion research not only using methodological reasoning, but also based on empirical investigation. There is no argument why reproducibility should not be a criterion as in the hard sciences, so it is very surprising that the scientific community discusses the problems of reproducibility in public opinion research so marginally.[5] At most, reproducibility is discussed in the sense of the related concept "reliability", and almost always in regard either to panel studies or to the stability of scaling techniques.[6] The whole experimental situation of the "house effects" and of what an independent reproduction of the experiment would require (Mittelstrass 1992) are barely ever discussed.

Public opinion research basically seems to ignore this general norm of science but still claims to be a scientific method. This leads to the question whether *alternative* techniques are used for producing scientific-community and public trust in its numbers (Porter 1995), and if so, how similar to reproducibility these techniques are.

Since no science is planned on the paper desks of methodologists and then executed in scientific reality, the best example of a development of alternative

---

the differences have a probability of $p < .0001$ of being accidental (if the "Don't Know" category is excluded, $p = 0.047$; Smith 1982, p. 67).
Recently Nate Silver also estimated house effects of survey institutions, regarding pre-election polls. However, his estimations of the effects are not explained and can, therefore, not be discussed here – see `fivethirtyeight.blogs.nytimes.com/2012/06/22/calculating-house-effects-of-polling-firms/`.

[5]Despite their significance, if reproducibility is in fact a core value of science, the mentioned articles are rarely cited by the community: Smith's paper 23 times, Converse and Traugott's paper 27 times (see `isiknowledge.com/wos`). Other articles discussing reproducibility of the whole survey dispositive (not panel surveys where a person is repeatedly interviewed) could not be found.

[6]More recently, a specific kind of problem with reliability in public opinion research surfaced in so-called question order effects, where the order in which successive questions are asked in a questionnaire or survey makes a difference for the outcomes. A pertinent study of these effects is due to Moore (2002). For approaches to understand them see Wang and Busemeyer (this volume).

concepts for creating trust in numbers is the paradigmatic breakthrough, in Kuhn's sense, by which the technique of polling entered the scientific field. Textbook science focuses on this breakthrough to celebrate the advent of modern scientific polling. But it is not that easy: Parts of the story are not told in the textbooks, though they are of interest concerning the reproducibility and credibility of public opinion research.

20.3 Big Data versus Science: The Breakthrough of Modern Polling

In the United States, where modern polling originated, predictions of election results were part of the political folklore at least since the beginning of the 19th century (Reynié 1998, p. 342; Smith 1990, p. 21f). Almost every occasion where people met – army meetings, festivals, river crossings – was used to gather votes for presidential candidates. The results of these "straw polls," as they were called, were submitted to the local newspapers. Soon the newspapers themselves initiated what they called "probe votes," since the public and the political representatives were curious about the findings (Herbst 2003).

Straw polls were spectacular media performances. Between 1916 and 1926, over 60 straw-poll enterprises were founded. To be competitive with the other forecasters, they tried to optimize their samples by increasing the number of participants in their polls. The obvious philosophy behind this strategy was: the more answers gathered, the greater the validity and credibility of the results. A big data industry *ante nomine* flourished: soon straw polls were collecting tens of thousands of ballots (Bradburn and Sudman 1988, p. 13).

One of the biggest players in this big data game was the New York magazine *The Literary Digest*. In 1936 its database contained 10 million voters (1936a). In the elections of 1920, 1924, 1928, and 1932, its polls were right. "Not only right in the sense that they showed the winner," the magazine wrote (1936b),

> they forecast the *actual popular vote*, with such a small percentage of error [less than 1% in 1932] that newspapers and individuals everywhere heaped such phrases as "uncannily accurate" and "amazingly right" upon us.

A genuine professional experience in building databases with heterogeneous sources was the foundation and basic substance of these polls (Robinson 1932, Chap. IV): gather addresses from almost every source, ask people to describe the different aspects of their political life, summarize the results, and present them as the *volonté générale* of the people.

The practice of linking the results of different polls with elections is exemplary for the discussion of the scientific value of the techniques: a correspondence between a poll and an election means the poll was "unbiased." *Literary Digest* ascribed this quality to its own polling without hesitation: "It has always previously been correct," they boasted, "even its critics admit its value as an index of popular sentiment" (1936a). In 1928 *Literary Digest* took a poll with 18,000,000 ballots. Given the enormous organizational resources needed, even without the help of modern computing techniques, *Literary Digest*'s straw polling was big data analysis in the modern sense. However, whether the forecast would be correct was still in the "lap of the gods" (1936a).

In its last straw poll before the election, *Literary Digest* took a poll of ten million ballots and failed. After receiving more than 2,300,000 ballots, it counted 57% votes for Landon and 43% for Roosevelt. The *American Institute of Public Opinion* predicted 55.7% for Roosevelt and 44.3% for Landon. In the election, Roosevelt received 60.8% and Landon 36.5 %.

George Gallup, based on a sample of only a few thousand probe votes, predicted the right candidate: Roosevelt. The *Literary Digest* promised to wear sackcloth and ashes ("Is our face red!" (1936b)) and in the end had to close its doors. This event has been celebrated as the breakthrough to modern scientific polling. A new star in the heaven of the social sciences had risen (Field 1990, p. 35):

> Even a cursory historical review of the use of the questionnaire survey method in US public opinion research clearly reveals that the so-called "modern era" of polling began in 1936.

By now straw polls are banned in textbooks as examples of wrong, non-scientific sampling. What seemed to be a breakthrough is just a "statistical myth."

20.4 The Birth of a Statistical Myth

Since history is written by the winners, the most popular explanation of the obvious success of the new techniques was what Gallup and his *American Institute of Public Opinion* postulated: the *Digest*'s samples were biased by the criterion of social class. They focused on the upper classes of society, and the workers were underrepresented (Gallup and Rae 1940, p. 42). Surely the election of 1936, in the time of the New Deal, was a special event that redefined the traditional conflict lines and party preferences.

A new "struggle between the haves and the have-nots" replaced the traditional conflict lines. It was believed that the unemployed voted for Roosevelt with a quota of about 80% (Kaufmann 1942, p. 46). In general, the votes for the Democrats originated (Gallup and Rae 1940, p. 48)

> from the ranks of the "ill-fed, ill-clothed, and ill-housed" lower third of the population, the group to which the *Digest*'s sampling plan failed to give full weight.

And why did the *Digest* sample get it wrong? Because the *Digest*, in contrast to other straw-poll institutions, started its sampling process with the purchasers of its own books and magazines, a database containing the names of the reading public: lawyers, architects, physicians, engineers, and members of high-society clubs. Then they extended it with addresses from the telephone directories and the automobile registration files (Robinson 1932, p. 56). So the conclusion was that because the telephone users and automobile owners were seldom in the strata of the "have-nots," the *Digest* database had oversampled the middle and upper classes, and the lower classes were poorly represented.

But it is not that simple. The *Digest*, knowing the rumors about the bias of their samples and aware of the problem, initiated strategies to avoid a sampling bias. And after the fail, the magazine made a great effort to show that its database had indeed included the have-nots, that the idea they were "... not reaching certain strata simply will not hold water" (1936b).

However, there is a factor of the *Digest*'s method (an important difference to Gallup's method) that is of great significance in terms of reproducibility: the *Digest* received its answers anonymously. The use of raw numbers is presented as a quality, an indication of the great authenticity of the polls, that they are not manipulated, that they represent the pure reality (1936a, p. 5):

> These figures are exactly as received from more than one out of any five voters polled in our country – they are neither weighted, adjusted nor interpreted.

But, as a consequence, there was no possibility for a so-called post-stratification: no way to trace back to the respondents and verify which persons and with which characteristics had answered and to check whether there were segments of voters whose answers were inadequate.[7] It was impossible to identify the sampling problems or discover whether certain voter segments had been over- or underrepresented. With no information about the origins of the ballots, the

[7] In the words of *Literary Digest*: "We don't know what proportion went to persons who had voted for Roosevelt in 1932 or what proportion went to persons who had voted for Hoover, because our polls are always secret, and the ballots come back with no signatures, no identifying characteristics of any sort except the post-marks" (1936b, p. 8).

impossibility of tracing them meant the *Literary Digest* could not *reproduce* the problem that it was certainly aware of: Perhaps the bias was not in the sample but in the returns, and this was probably an important cause of the fail (Bryson 1976).

Gallup's technique only counted several thousand ballots. His institute called this a *representative cross section of voters*, or *group sampling*, today known as a quota sample. With this technique the population, the whole country, was segmented along criteria that were *a priori* estimated to be relevant for political opinion making: income level, political preferences, gender, education, occupation, race, and religion – so that "the right proportion of citizens from every walk of life" would be represented in the sample.[8]

Is this the key to the success of modern polling? As mentioned, *Literary Digest*, too, used a form of quota sampling to avoid over-represented or under-represented samples. The difference is that Gallup's new method of sampling established a loop back to the society the sample was drawn from, so the social structure of the sample could be compared with the structure of the population – but compared using only the criteria *already* defined as relevant.

To investigate the causes of *Digest's* failure, the *American Institute of Public Opinion* asked the respondents a series of questions about their previous voting behavior, whether they owned telephones and cars, and whether they had received a *Literary Digest* ballot and sent it back. They were also asked if they had changed their minds after sending the ballot in (compare Squire 1988). The *American Institute of Public Opinion* data could clearly show that the *Literary Digest* polls were inadequate compared to its measurement system: Gallup's survey signalized that 79% of respondents owning no car or phone voted for Roosevelt, whereas 55% of car and phone owners voted for Roosevelt (Squire 1988, p. 130).

These results are the origin of the textbook version of the story. But it is interesting that Gallup's institute did not publish the basis of its calculations "explaining" the fail of *Literary Digest* (Squire 1988, p. 129), even in its compendium of all its questions and findings since its beginnings (Gallup 1972). Why? Simple calculations show that Gallup's data is itself biased: if its replication survey had been as representative as claimed, Roosevelt would have received 66% of the votes instead of 61%, a difference equal to the *Literary Digest* er-

[8] "When the method of group sampling is employed ... care must be taken to see that the national cross section includes the right proportion of citizens from every walk of life. A careful study must first be made of the population structure. By and large, the most important factors which have to be considered are occupational groups, income levels, political preferences, age, education, racial and religious groups" (Gallup 1944, p. 26).

ror (in the other direction). The reason for the bias in Gallup's survey remains unknown. Perhaps, as Squire estimates, it was the psychological effect of respondents wishing they had supported the winners and not wanting to be on the loser's side (Squire 1988, p. 131). But this insight is hypothetical, and other explanations remain speculations.

What is clear is that *Literary Digest*'s survey, and its problems, could not be reproduced even by the new method Gallup provided. Times had changed, the election was over, the voters were either losers or winners, and human memory keeps interacting with historical events. The object of the research had been modified irretrievably: it would take a time machine to reconstruct the failure. So the reasons why *Literary Digest* failed, after years of almost perfect predictions, are not so clear as the textbooks and their celebration of scientific sampling suggest.

There is evidence that the answers must have been biased by a mix of factors, where the very high non-response rates of the big sample also played a role (and hazard, also). And there is evidence that the question why *Literary Digest* failed could never have been answered (Squire 1988). That class-biased sampling led to the fail remains just "a statistical myth," as Bryson (1976) wrote in the *American Statistician* – a myth generated to promote the new sampling techniques.[9]

Furthermore, the *American Institute of Public Opinion*, along with other institutions, experienced a similar disaster some years later in 1948 (Bradburn and Sudman 1988, p. 29). It estimated that Thomas Dewey would receive 49.5% of the vote and Harold Truman would lose with 44.5%. It turned out that Truman received 49.5% and Dewey received 45.15%. Surprisingly, that disaster was not only seen as bad luck, as *Literary Digest* had seen its disaster, but rather as a moment to reflect upon and calibrate the new instruments for measuring political life.

Finally, the discussion of this fail brought in a new scientific norm (and not only as a success in terms of public attention) by which to evaluate survey techniques: the criterion of the accuracy of election predictions. A special analysis of the poll error by the *American Institute of Public Opinion* explained, on the basis of the second questioning of earlier respondents, that the institute had stopped its pre-election interviews too early, at a moment when many people had not yet

[9]The strategies behind the disqualification of *Literary Digest*'s big data method as inadequate can also be regarded as a kind of "boundary work" (a concept that Thomas Gieryn (1983) introduced some time ago: strategies to establish a demarcation line between science and non-science in the emerging field of the quantitative social sciences). See Reinhart's article in this volume for a discussion of boundary work in relation to reproducibility in social science.

decided how they were going to vote; and that the campaign was still going on and people were changing their minds (Gallup 1972, p. 769ff.).

Since the same instruments were used to explain the failure as had produced the failure (sample surveys), the explanation remains a hypothesis. We find the same situation as in the *Literary Digest* disaster. Since the political society as it existed before the election could not be reproduced after the election, the explanation of the fail had to remain *logically* unclear. The question persists in every attempt to replicate a sample: Did the society and its discourses change, or were the samplings inadequate?

20.5 Generating Trust [10]

As a consequence, the prestigious *Social Science Research (SSR) Council*, founded in 1923 to promote social science research, observed with consternation the "wide confusion and misgivings about the reliability of the polls" (SSR 1948, p. 799). The council reclaimed rigid robustness tests of the findings (SSR 1948, p. 622, my italics):

> Good tests of the accuracy of polling methods are important in the development of polling and survey methodology. Without suitable tests there is danger that users of those methods will *drift into an unjustifiably high degree of confidence* in the accuracy of their current methods.

While the council recognized the advent of public opinion research as a ubiquitous tool for producing representations of political opinions, it demanded that the findings be grounded in "real" political life and that a comparison of the poll results with the election results must remain the central criterion of robustness and reliability.

This is called "accuracy." A different test of accuracy, or even a different replication method, is hard to find: the SSR postulated the election forecast as the only way to prove the worth of any polling institution (SSR 1948, p. 602):[11]

[10]The title of this chapter is a reference to Theodore Porter's work *Trust in Numbers* (Porter 1995), which investigates the broader history of statistics to search for the roots of credibility of statistical numbers. See also Porter, this volume.

[11]It is remarkable to observe how the new science was searching for a criterion of stability, called reliability or accuracy. Of special interest is an article by Lucien Warner. He denies the criteria of strict accuracy of election predictions (Warner 1939, pp. 389f.): "But even elections are a far from satisfactory check upon public opinion surveys, for they do not reflect public opinion with complete accuracy. Many factors distort electoral results." Even the weather, he believes, has "a selective influence upon rural versus urban votes." *De facto*, only a census of

> Election returns are not a direct or even a good test of the adequacy of polling on issues like the Marshall plan, civil rights, etc. But elections are useful for testing the adequacy of polling methods for estimating the percentage of the vote going to each candidate from various groups in the population. No better test is now known.

Perhaps it is this document that should be associated with the birth of modern public opinion research, rather than the myth of the breakthrough in 1936.

The argument is that the accuracy of polls in predicting election results is a criterion for judging the reliability and validity of public opinion research. Elections figure, ironically enough, as the main test of the whole system of public opinion research: the researchers observe whether the elections reproduce the survey results. This criterion obviously serves the same function in public opinion research as reproducibility does in the experimental sciences: proof that the phenomenon being measured is not an artifact of the instruments and the experimental devices themselves, but "exists" in the real world, so a result can be "verified with an external source" (Traugott 2008b, p. 408). So far, no other agreed measures or standards exist for evaluating the accuracy of polls (Traugott 2008b, p. 408; Traugott 2011, p. 316).

Due to the rarity of such verifications comparing polls with elections, "accuracy" remains the hard criterion of public opinion research to this day (Gosnell 1937, Sheatsly and Hyman 1953, pp. 480ff, Traugott 2011). Even the publication itself of the differences between predictions and election results has the character of a ritual. For example, the *American National Council of Public Polls* has published its analyses of presidential election polls starting from the origins of modern polling. Every new method of questioning is critically reviewed and measured against other methods.[12] But what does this evidence mean in the broader sense of the reproducibility of the findings that public opinion research presents to the public? What is the point of a science predicting a phenomenon that surely and soon will define itself (and without any doubt): election results?

The answer to the second question is easy. Public opinion research is not primarily interested in predicting elections, probably even in financial terms. Elec-

---

the entire population at the exact same time would assure the reliability of the survey. Since this is not possible for economic reasons, he suggests selecting a "representative community." The original survey "should employ the methods currently accepted" and a "100 per cent canvas of the entire population simultaneously would show if the survey results are replicable." However, he did not recognize that he was only shifting the problem onto another level: Who tests that the selected community is representative? This example shows how arbitrary the definition of criteria of replication are, and perhaps not only in the case of public opinion research.

[12]See recently Graefe (2014), `www.ncpp.org/files/FAQs.pdf`. Independent institutions like Wikipedia also publish results regularly.

tion surveys are only a small part of all the public opinion research being done. Gallup (1939, p. 10) has already formulated the reason why the scientific and public focus is on the election forecasts, in the loaded words:

> The chief usefulness of the sampling survey, however, does not lie in forecasting elections. While such forecasts provide an interesting and legitimate activity, they probably serve no great social purpose. They do, however, offer the public a convincing demonstration of the accuracy of sampling procedures.

This says that if the sampling procedure is correct or accurate in terms of an election forecast, people can be asked almost anything and their answers will accurately represent the whole, the nation. This transfer of "trust in numbers" over to virtually any question that may be asked is the genuine concept of public opinion research to this day (Gallup 1944, p. 26).

There is no room here to seriously discuss all the problems of public opinion research methods. The validity of the answers (do people understand the answers in the same manner? do they answer what they actually think? are the answers the effect of the questions?) and the sampling problems (is the statistical urn well mixed? did all respondents get equal chances to participate?), to name some examples, are all hard problems that have been debated since the beginning of modern polling.[13]

But under the fictional assumption of an ideal world where all methodological problems have been solved, i.e., where sampling bias and validity of questions and answers pose no problems anymore, we can raise some general questions about the reproducibility of scientific and social knowledge where public opinion polls build an instructive case. This is a way to answer our first question: What are the strategies for establishing accuracy in the broader context of scientific reproducibility?

20.6 The Possible Worlds of Public Opinion Research

In the sense of an ideal typical or pure type as postulated by Max Weber (which means: a situation cleared of reality and reduced to its principal idea or logic, so its pure principle and consequences can be studied), we parenthesize all the methodological problems of polling. We assume an ideal world of the pollsters in which to make our reflections on accuracy and reproducibility. Following this postulate, every poll and every poll company creates what analytic philosophers call "a possible world."

[13]See the reconstruction of the debates in my archeology of opinion research (Keller 2001).

For Kripke (1980, pp. 15ff) a possible world is nothing like an exotic world on a distant planet – it is more like a possible state of a situation. Thinking with possible worlds has a lot to do with the probability and possibility of events, with chance and, therefore, with statistics, as Hacking (1975b, Chap. 14) showed in his investigation of the emergence of probability.

Kripke illustrated a possible world with a well-known probability experiment: throwing dice. The dice on a table have 36 possible states, sums of the numbers of dots on each die. Ignoring everything about the world around them, they build 36 possible mini-worlds. Only one of these mini-worlds is actualized by a throw (e.g., 4 and 2 dots), the other 35 worlds remain only possible. But they can become an actual world at the next throw of the dice.

Defining a world with possible states, abstracted from all other aspects, is a very good approach to the idea behind polling: the sides of a throw and number of dots are equivalent to predefined answer categories ("you will vote for whom ...") and distributions of answers. In this sense, it is not surprising that elections and election results were omnipresent in the possible-world discussion. For instance, Kripke chose the example of the 1968 election (Kripke 1980, pp. 40–53 and 115–116) which Nixon won. Kripke discussed a possible world in which Nixon does not win the election and asked (a question of minor interest to us here) if in such a world Nixon is still Nixon (Kripke 1980, p. 44).

Why does the 1968 election figure as the preferred example of a possible world? Because a different election outcome was easily imaginable: the 1968 US elections turned out to be extremely close. Richard Nixon won with 43.4% and was basically equal with Hubert Humphrey, who was supported by 42.7% of the popular vote, while the party-independent George Wallace reached only 13.5%.

A possible world in which Nixon might have lost the elections is of great plausibility, and such possible worlds were explored by the polls, each following the same categories and referring to the same population: before the elections the Gallup Poll measured 43% for Nixon, 42% for Humphrey, and 15% for Wallace. In contrast, the Harris Poll, another big organization, showed Humphrey as winner with 45%, Nixon with only 41%, and Wallace 14%.[14] Given the probability errors inherent to sampling, each of the possible worlds could have been the actual one realized in the elections.

For a pollster, elections, except at their political constitutive moment, are indeed nothing but sample surveys. As Archibald Crossley (1937, p. 25), one of the founding fathers of modern polling, wrote about the *Digest* disaster:

---

[14]See the evaluation made by the National Council on Public Polls (NCPP) at www.ncpp.org/files/1936-2000.pdf.

> The greatest difficulty of all is the fact that the election itself is not a census, but an application of the sampling principle. Every poll is therefore a sample of a sample. There is no guarantee whatever that the 1940 election will be the same kind of sample that the 1936 election was. And the 1936 election probably was a different cross-section of all voters from the cross-section of the 1932 election.

In other words, for pollsters elections and sample surveys both represent possible worlds in an epistemic sense (this is correct, when election results and results of polls both use descriptions like "*the* Americans," an imagined community never counted as a whole, neither in polls nor in elections). It is the power of state, not the pollsters' methodology, that lets the possible world of elections become the "real" one.

Here, a conceptual differentiation is necessary on what "possible" means. As Hacking (1975b, p. 123) insisted, grammatically the term "possible" has two aspects: "possible for" and "possible that." There is an important difference between these two kinds of possibility: polls showed that it was possible *that* Nixon wins and, at the same time, that it was possible *for* Nixon to win. The latter is the possibility in the sense of the Latin *potentia* (power, ability): it signalizes an occasion realizable in the real world. The "possibility *that*" is called epistemic probability (Hacking 1967, p. 148) and concerns the sphere of chance and calculations, a sphere that does not necessarily affect the real world. The range of error ritually mentioned in the published results of polls (usually 3%) is a signifier of this kind of epistemic possibility (even this range is seldom used statistically correctly).[15]

The latter aspect shows, moreover, the political potential: the results of the Harris Poll signalized that it was still possible *for* Humphrey to win the elections (with 45% of supporting answers). In other words, the concept of "possible worlds" includes two aspects (Hacking 1975a, p. 336): the possibility of a state in the sense of its probability, and the possibility of a state regarding its potential to realize as an "actual world."

Crossley's statement above can be reformulated as follows: Since an election is not a census, it is – at an epistemic level – one of a number of conquering possible worlds each containing information about whom the American population will prefer as president. In the case of elections and polls, it is evident that the scientific prestige of the polling institutions is based on the epistemic aspect: the possible worlds generated by polls are structurally similar to the world realized by elections.

Now we are at the root of the strategy of replacing reproducibility with

[15]See Stahel (this volume) concerning the mathematical background of confidence intervals.

accuracy: The accuracy of poll results to an election does indeed mean the surveys are reproduced by another "experiment" with quite a similar structure. But the reflections above show that there is a second aspect of possible worlds affecting the scientific nature of the results: the findings may show that it is possible *for* Humphrey to win. This aspect of the pollster's potential world is inscribed in political life itself, and it transforms the character of the epistemic aspects of the generated knowledge (and its scientific character too).

Here we have a possible difference between the social sciences and the natural sciences. The description of the world changes the world described, but meanwhile, also, the experiment observes the experimenter. And both this description and this observation are using the same techniques: human perception and symbolic communication.

20.7 Looping Effects between Measurement and Measured

Since modern polling began there have been dozens, maybe hundreds, of articles on the effects of public opinion research upon the public, the media, and the political system. These influences are located on at least three levels: the interrogated population, the media system, and the political system and its actors.

On the first level, Gallup and Rae (1940, Chap. 20) already took care of the so-called bandwagon effect: In a circus, the bandwagon carrying the music band leads the caravan of artists and animals. It figures as a metaphor for the leading candidate or party (Marsh 1985) which people tend to applaud because it is the most visible and attractive wagon in the caravan. However, there is also the so-called underdog effect, the tendency to support the losers (Traugott 2008a). Whether results of polls initiate more self-fulfilling or more self-destroying prophecies is difficult to decide (cf. Collins and Zimmerli, both in this volume, concerning similar cases of "Merton-sensible" systems). As a consequence, the impact of polling on individual voting decisions remains unclear (Lang and Lang 1984), although there is strong evidence that polling has effects upon individual perception and behavior (see Moy and Rinke 2012 for a review).

Second, there is also an effect on the media system: creating collective perception and agenda setting (Champagne 1990). Polling results are welcomed by the media as easily consumable "news" (Herbst 2003, pp. 588f), even if the results signalize only a slight difference to a previous state of affairs (sometimes not even exceeding their statistical range of error). By financing studies, parties, and enterprises, all sorts of collective actors can guide the public's collective focus on a real or imagined problem or issue:[16] If people are asked about something, they will answer somehow (Bourdieu 1992), even on completely imaginary issues (Fishkin 1995, p. 82). There is as much research on the effects of public opinion research on the media (Roll and Cantril 1980) as there is on its effects on individuals (Strömbäck 2012).

The close relationship between political actors, pollsters, and media is even described as an "opinion connection" (Cantril and Cantril 1991). Experiencing the strategic uses of public opinion research and techniques that are opaque for outsiders produces a feeling that the media and the public are "being hijacked by campaign strategists, and focused on imagined flaws in survey methods" (Jacobs and Shapiro 2005, p. 638), or even being manipulated.[17] So the judgment of the press about polls is not always friendly (though this does not stop them from publishing the pollsters' results).

Bloomberg Businessweek wrote about the "Dark Art of Polling," about how nobody knows how the pollsters brew their results in their quasi-alchemistic labors.[18] Allan Rivlin from the *National Journal* is more despairing: "First, kill all the pollsters" he titles an article about what he calls the "junk-food polls," and asks "Who's manipulating whom?" Just as the concrete direction of the influences on individual behavior remains unclear, the concrete interlooping effects between the polls and the media also remain unknown, even to the pollsters themselves (Moy and Rinke 2012), and this makes any use of public opinion research a high-risk endeavor, as the following quotation illustrates:[19]

> The body of American public opinion is far too sophisticated and subtle a beast to be prodded in this way. And as if to prove it, on Tuesday the beast bit back.

---

[16] Polls whose only goal is agenda setting are called "push polls." See Medvic (2003), and also NCPP's norm articulated against push polls: `ncpp.org/node/4/#16`. However, the demarcation line between push polls and regular polls is not always easy to identify.

[17] Compare the election of Jacques Chirac as French president (Delpey 1995). However, there seem to be no great national differences in the skeptical attitude toward polls, see an example for Germany by Güllner *et al.* (2007).

[18] "How could a Gallup organization survey published a week before the election show Mitt Romney up by 5 percentage points, while a CBS / New York Times poll from the same period put him 1 point behind President Obama? Even professional poll watchers don't know." Quoted from Peter Coy, November 01, 2012 at `www.businessweek.com/printer/articles/80092-the-dark-art-of-political-polling`.

[19] This quote is due to Frank Luntz in the New York Times of June 11, 2014, after a spectacular false prognosis in the Republican primaries: "Why Polling Fails. Republicans Couldn't Predict Eric Cantor's Loss." The text is available at `www.nytimes.com/2014/06/12/opinion/republicans-couldnt-predict-eric-cantors-loss.html`.

As the third and final point of this short overview, there are direct influences on the public support of the candidates running in the elections. Rising poll support usually generates a monetary push for the candidate's campaign, while declining support dries up the money flow and also the candidate's public presence (Converse and Traugott 1986, p. 1094). In the US system, third party candidates, since they normally receive little support in the early stages of the election campaigns, are given scarce mention in the polls. As a consequence, they disappear from the answer categories step by step, therefore losing public visibility and financial support almost completely (Herbst 2003, p. 589).

In this logic, it is evident that the reputations of the American presidents and candidates for political office and their chances in the elections are not only observed in the polls, but also actively used by the candidates (Holtz-Bacha 2012, p. 272). The use of polls in campaigns is so prominent that in a poll about polls, 42% of the respondents said the pollsters were manipulating their results to show Obama ahead.[20] This poll about election polls is an example of how the possible worlds of polling have established a highly self-referential system, a complex system of looping effects between polls, media, and individuals.

There is no room here to discuss which of the many supposed effects of polling are real or not, and the directions of their influence: the published research about this problem area fills libraries. Since polls generate public knowledge, it is evident that they affect public knowledge and behavior. Here these effects are mentioned only to show the importance of the polls' potential power to change real worlds – the validity of the measurements aside. But what do these evident effects mean with regard to the question of accuracy or reproducibility in general?

20.8 Swarms of Possible Worlds

The *National Council on Public Polls* analyzed the final presidential election polls conducted by the national media dating back more than 50 years.[21] Compared with actual election outcomes, the average error of polls for presidential elections between 1956 and 1996 declined. In terms of criteria of quality, the Council suggests that the polls were getting more accurate over time; that the polls were in fact reproduced in the elections.

This evidence can also be used to support the argument that public opinion

[20]Compare Peter Coy, "The Dark Art of Polling", *Businessweek*, November 01, 2012: `www.businessweek.com/printer/articles/80092-the-dark-art-of-political-polling`.

[21]National Council on Public Polls (NCPP) at `www.ncpp.org/files/1936-2000.pdf`.

research has successfully created its own object: public opinion as measured by its surveys (Blondiaux 1998, p. 71, Osborne and Rose 1999, p. 389), and that it is capable of stabilizing and legitimizing its own phenomena using the elections as another independent experimental system. But here some caution is recommended: phenomena that are stable in time and space, obviously describable, are rare in the sciences, says Hacking (1983, p 227) – even in the natural sciences. Though it seems correct to say that public opinion research talks mainly about "public opinion" today, the conclusion is not necessarily correct that the underlying political articulation is sufficiently established by the tables and numbers of the polls.

After 80 years of research, there is still no consensus about how the phenomena of "public opinion" can exactly be defined, self-contained, and distinguished from other similar phenomena (Holtz-Bacha 2012, pp. 268f). In spite of this fact, public opinion research has successfully transformed the question of how to define public opinion into the question of the interplays between the results of polls and those of elections and to this day that question prevails. However, the election results figure as robust knowledge not because they represent the real expression of public opinion, but because the source of this robustness is state power, legal rules, and even the police who regulate and legitimize the election results. Outside their laboratories, using the concept of accuracy to validate their constructs, the pollsters walk on crutches built of pure governmental techniques.

But more important is that the polls and the elections are not independent of each other. Since the election results temporally follow after the election polls, they can be regarded, to a certain but unknown degree, as an effect of the polls themselves, through the many influence-channels mentioned above.[22] They are never an independent verification of the polls. The higher correspondence between polls and elections observed by the *National Council on Public Polls* could, therefore, be a consequence of the fact that the elections are more and more structured, or even influenced, by the dozens of earlier election polls.

The possible worlds of elections and polls, intermediated by mass media and spin doctors, interact strongly, and stabilize each other. It does not make any sense to regard polls as unique phenomena constructed by lonely and autonomous social scientists in their laboratories. In other words, as soon as an observation of these social realities is fixed and communicated, it becomes a component of what it observed and its original precondition has changed. We find the situation

[22]In this sense, creating, pushing, public opinions in public opinion research is perhaps better understood as an act of agenda setting than as the measurement of an already preformed entity (Holtz-Bacha 2012, pp. 268f). In recent sociological discussion, this is called the performative aspect of polling (Law 2009, Perrin and McFarland 2011, pp. 101f).

of a so-called infinite regress that has no end. So the possible worlds of public opinion surveys may be built correctly in terms of the criteria the survey science sets for itself, but since its findings alter the reality it measures, there is no way to reproduce the *status quo ante*.

This is the destiny of the social sciences, in general, just expressed more clearly in public opinion research than elsewhere: Social scientists' descriptions and measurements are not outside the realities they describe, they form a part of them. All instances, scientists, interviewers, politicians, and media interact on the same level, using the same form of symbolic communication. The scientific observation of society takes place *within* the society it describes (Luhmann 1992). No external (scientific) observation is possible,[23] there are only observers who observe observers.

Or, regarding epistemic techniques of public opinion research, a poll "measuring" a public opinion generates nothing other than a possible world of peoples' political articulation: a world in concurrence and interaction with a whole swarm of other worlds, all pretending to express the political opinions of the people. Trying to reproduce one of these possible worlds in order to show its accuracy does nothing but generate another one. Amazing.

References

— (1936a): Landon, 1,293,669; Roosevelt, 972,897. Final Returns in The Digest's Poll of Ten Million Voters. *The Literary Digest*, 1936, pp. 5-6.

— (1936b): What Went Wrong with the Polls. None of Straw Votes Got Exactly the Right Answer – Why? *The Literary Digest*, 1936, pp. 7-8.

Bardes, B.A., and Oldendick, R.W. (2012): *Public Opinion. Measuring the American Mind*, Lanham, New York.

Best, S.J., and Radcliff, B., eds. (2005): *Polling America. An Encyclopedia of Public Opinion*, Greenwood, Westport.

Blondiaux, L. (1998): *La fabrique de l'opinion. Une histoire sociale des sondages*, Seuil, Paris.

Bourdieu, P. (1992). Meinungsforschung – Eine "Wissenschaft" ohne Wissenschaftler. In P. Bourdieu (Ed.), *Rede und Antwort*, Suhrkamp, Frankfurt, pp. 208–216.

[23]Osborne and Rose (1999, p. 389) argued that there are no substantial differences between the construction of scientific objects in the social sciences and in the sciences. Although there are variations, they should be understood as "matters of degree rather than in terms of a rigid line of demarcation." However, we insist on the fact that in the case of observing society, there is no external observer. There are only observers interconnected by the medium of symbolic communication (as Luhmann argues). See also Collins' chapter in this volume concerning a broader context of this question.

Bradburn, N.N., and Sudman, S. (1988): *Polls and Surveys. Understanding What They Tell Us*, Bass, San Francisco.

Bryson, M.C. (1976): The Literary Digest poll: Making of a statistical myth. *American Statistician* **30**(4), 184–185.

Cantril, A.H., and Cantril, S.D. (1991): *The Opinion Connection: Polling, Politics, and the Press*, CQ Press, Washington DC.

Champagne, P. (1990). *Faire l'opinion. Le nouveu jeu politique*, Minuit, Paris.

Collins, R. (1994): Why the social sciences won't become high-consensus, rapid-discovery science. *Sociological Forum* **9**(2), 155–177.

Converse, P.E., and Traugott, M.W. (1986): Assessing the accuracy of polls and surveys. *Science* **234**, 1094–1098.

Crossley, A.M. (1937): Straw polls in 1936. *Public Opinion Quarterly* **1**, 24–35.

Delpey, R. (1995): *Nicolas Bazire, Edouard Balladur, Nicolas Sarkozy en Examen pour les manipulations des sondages*, Trancher, Paris.

Donsbach, W., and Traugott, M.W. (2008): *The SAGEe Handbook of Public Opinion Research*, SAGE Publications, Los Angeles.

Field, M.D. (1990): Opinion polling in the United States of America. In *The Classics of Polling*, ed. by L. Young, Scarecrow Press, Mettuchen, pp. 34–45.

Fishkin, J.S. (1995): *The Voice of the People. Public Opinion and Democracy*, Yale University Press, New Haven.

Gallup, G., and Rae, S.F. (1940): *The Pulse of Democracy. The Public-Opinion Poll and How It Works*, Simon and Schuster, New York.

Gallup, G.H. (1939): *Public Opinion in a Democracy*, Baker Foundation, Princeton.

Gallup, G.H. (1944): *A Guide to Public Opinion Polls*, Princeton University Press, Princeton.

Gallup, G.H., ed. (1972): *The Gallup Poll. Public Opinion 1935–1971. Volume One 1935–1948*, Random House, New York.

Gieryn, T.F. (1983): Boundary work and the demarcation of science from non-science: Strains and interests in professional ideologies of scientists. *American Sociological Review* **48**, 781–795.

Giles, J. (2006): The trouble with replication. *Nature* **442**, 344–347.

Gosnell, H.F. (1937): How accurate were the polls. *Public Opinion Quarterly* **1**, 97–105.

Graefe, A. (2014): Accuracy of vote expectation surveys in forecasting elections. *Public Opinion Quarterly* **78**, 204–232.

Güllner, M., Hilmer, R., Jung, M., Keller, F., and Lochthofen, S. (2007): Lauter falsche Propheten? In *Macht und Medien. Journalismus in der vernetzten Gesellschaft. Mainzer Tage der Fernseh-Kritik, Band 39*, ed. by P.-C. Hall, Zweites Deutsches Fernsehen, Mainz, pp. 130–156.

Guttman, L. (1944): A basis for scaling qualitative data. *American Sociological Review* **9**(2), 139–150.

Hacking, I. (1967): Possibility. *Philosophical Review* **76**(2), 143–168.

Hacking, I. (1975a): All kinds of possibility. *Philosophical Review*, **84**(3), 321–337.

Hacking, I. (1975b): *The Emergence of Probability. A Philosophical Study of Early Ideas about Probability, Induction and Statistical Inference*, Cambridge University Press, Cambridge.

Hacking, I. (1983): *Representing and Intervening. Introductory Topics in the Philosophy of Natural Science*, Cambridge University Press, Cambridge.

Herbst, S. (2003): Polling in politics and industry. In *The Modern Social Sciences*, ed. by T.M. Porter and D. Ross, Cambridge University Press, Cambridge, pp. 577–590.

Holtz-Bacha, C. (2012): Polls, media and the political system. In *Opinion Polls and the Media Reflecting and Shaping Public Opinion*, ed. by C. Holtz-Bacha and J. Strömbäck, Palgrave, Macmillan, Basingstoke, pp. 267–280.

Jacobs, L.R., and Shapiro, R.Y. (2005): Polling politics, media, and elections campaigns. *Public Opinion Quarterly* **69**(5), 635–641.

Kaufmann, P. (1942): *Wie das Gallup-Institut die öffentliche Meinung ermittelt*, Oesch, Zürich.

Keller, F. (2001): *Archäologie der Meinungsforschung. Mathematik und die Erzählbarkeit des Politischen*, UVK, Konstanz.

Kripke, S.A. (1980): *Naming and Necessity*, Blackwell, Oxford.

Kuhn, T.S. (2012): *The Structure of Scientific Revolutions*, University of Chicago Press, Chicago.

Lang, K., and Lang, G.E. (1984): The impact of polls on public opinion. *Annals of the American Academy of Political and Social Science* **472**, 129–142.

Law, J. (2009): Seeing like a survey. *Cultural Sociology* **3**, 239–256.

Luhmann, N. (1992): Die Selbstbeschreibung der Gesellschaft und die Soziologie. In *Universität als Milieu*, ed. by N. Luhmann, Haux, Bielefeld, pp. 137–146.

Marsh, C. (1985): Back on the bandwagon: The effect of opinion polls on public opinion. *British Journal of Political Science* **15**(1), 51–74.

Medvic, S.K. (2003): Campaign pollsters and polling: Manipulating the voter or taking the electorate's pulse. In *Campaigns and Elections: Issues, Concepts, Cases*, ed. by R.P. Watson and C. Colton, Rienner, Boulder, pp. 31–46.

Mittelstrass, J. (1992): Rationalität und Reproduzierbarkeit. In *Entwicklungen der methodischen Philosophie*, ed. by P. Janich, Suhrkamp, Frankfurt, pp. 54–67.

Moore, D.W. (2002): Measuring new types of question-order effects: Additive and Subtractive. *Public Opinion Quarterly* **66**, 80–91.

Moy, P., and Rinke, E.M. (2012): Attitudinal and behavioral consequences of published opinion polls. In *Opinion Polls and the Media Reflecting and Shaping Public Opinion*, ed. by C. Holtz-Bacha and J. Strömbäck, Palgrave, Macmillan, Basingstoke, pp. 225–245.

Osborne, T., and Rose, N. (1999): Do the social sciences create phenomena? The example of public opinion research. *British Journal of Sociology* **50**, 367–396.

Perrin, A.J., and McFarland, K. (2011): Social theory and public opinion. *Annual Review of Sociology* **37**, 87–107.

Porter, T.M. (1995): *Trust in Numbers. The Pursuit of Objectivity in Science and Public Life*, Princeton University Press, Princeton.

Reynié, D. (1998): *Le triomphe de l'opinion publique. L'espace public français du XVIe au XXe siècle*, Odile Jacob, Paris.

Robinson, C.E. (1932): *Straw Votes. A Study of Political Prediction*, Columbia University Press, New York.

Roll, C.W., and Cantril, A.H. (1980): *Polls, Their Use and Misuse in Politics*, Seven Locks Press, Newport Beach.

Sheatsly, P.B., and Hyman, H.H. (1953): The use of surveys to predict behaviour. *International Social Science Bulletin* **5**, 474–481.

Smith, T.W. (1982): House effects and the reproducibility of survey measurements: A comparison of the 1980 GSS and the 1980 American national election studies. *Public Opinion Quarterly* **46**, 54–68.

Smith, T.W. (1990): The first straw? A study of the origins of election polls. *Public Opinion Quarterly* **54**, 21–36.

Social Science Research (SSR) Council (1948): Report on the analysis of pre-election polls and forecasts. *Public Opinion Quarterly* **12**(4), 585–599.

Squire, P. (1988): Why the 1936 Literary Digest Poll failed. *Public Opinion Quarterly* **52**, 125–133.

Strömbäck, J. (2012): Published opinion polls, strategic party behavior and news management. In *Opinion Polls and the Media Reflecting and Shaping Public Opinion*, ed. by C. Holtz-Bacha and J. Strömbäck, Palgrave, Macmillan, Basingstoke, pp. 263–283.

Traugott, M.W. (2008a): Bandwagon and underdog effects. In *Encyclopedia of Survey Research Methods*, ed. by .P.J. Lavrakas, Sage Publications, Thousand Oaks, pp. 50–52.

Traugott, M.W. (2008b): Validation studies. In *Handbook of Public Opinion Research*, ed. by W. Donsbach and M.W. Traugott, Sage Publications, Thousand Oaks, pp. 408–417.

Traugott, M.W. (2011): The accuracy of opinion polling and its relation to its future. In *Oxford Handbook of American Public Opinion and the Media*, ed. by R.Y. Shapiro and L.R. Jacobs, Oxford University Press, Oxford, pp. 316–331.

Turner, S.P., and Turner, J.H. (1990): *The Impossible Science. An Institutional Analysis of American Sociology*, SAGE Publications, Newbury Park.

Warner, L. (1939): The reliability of public opinion surveys. *Public Opinion Quarterly* **3**, 376–390.

21
Depending on Numbers

Theodore M. Porter

Abstract. Reproducibility in modern science is often defined in terms of probability and error. These are reasonably straightforward when it is a question of measuring an identical quantity according to a well-defined protocol. But reproducibility is more typically about finding new ways to measure, and often about objects subject to variation, such as living organisms. Even physical science works much harder to extend its findings than to repeat them, while in medicine and psychology few results can be expected to replicate with high precision.
Under such circumstances, reproducibility is not a rigorous standard, but more often a loose one that depends on expertise and interpretation. Moreover, science that matters will often be caught up in human interests. The effort to repress these by means of rigorous quantification succeeds only some of the time and can easily create more problems than it solves. The perspective of history offers ways of understanding this dynamic, and even perhaps of adapting the ideal of reproducibility to the demands of practical and public science.

21.1 Introduction

Many of the most vexing and controversial problems of reproducibility in science are about numbers. This was by no means an expected outcome of the rise of statistics as a form of empirical investigation and quantitative reasoning. In its 19th-century form, as a science of quantities relevant to the state, statistical counts and measures seemed to be as straightforwardly factual as knowledge could be, to the point that the Statistical Society of London tried out the idea of excluding opinion from their meetings.

In the more mathematical form of error theory, statistics grew up as a set of probability-based methods for managing measurement variation. From about the 1920s and 1930s, this kind of statistics gained wide acceptance as the proper basis of scientific inference, allied to techniques of randomized sampling and experimental design. Statistics was brought forward as a neutral and authoritative way to settle or to avoid controversy. Especially in therapeutic medicine and the human sciences, statistical methods have become virtually obligatory for some very important categories of empirical investigation.

The adjective "evidence-based," which by now extends well beyond medicine,

---

Reproducibility: Principles, Problems, Practices, and Prospects, First Edition. Edited by Harald Atmanspacher and Sabine Maasen.

is defined by a set of statistical techniques. A technology of reasoning like this one, promising to reduce debate to calculation, is bound to be challenged, and statistics has been the site of bitter debates. Even among experts who agree on methods of taking measurements and reducing data, reproducing results with the degree of precision predicted by basic statistical models is difficult, and even uncommon (Stigler 1986, Daston 1988, Porter 1986, Krüger *et al.* 1987a,b, Hacking 1990).

21.2 Statistical Error

"Statistical error" is more or less synonymous with sampling error and is calculated using basic formulas of probability theory. Perhaps the most common and most damaging mistake of statistics is to suppose that this is the only kind of error that matters. Trained statisticians do not assume that error is random and independent. Rather, their professional competence takes in a set of techniques, typically involving some form of structured randomization, to create data sets that can be treated as random samples. Often they transform the data to bring it into line with a standard frequency curve such as the normal distribution.

These tools have become indispensable to controlled experimentation and to survey samples and have a role now in many or most publications in the natural and social sciences. If we wished to determine whether this "many" is indeed "most," I can think of no way of proceeding except to try to define a sampling technique that would be appropriately random. In practice, it is often not possible to measure or sample in a way that excludes bias, which in statistics is defined simply as a difference between the expected value of the estimation process and the true value or true average of a population. Such bias is ubiquitous even in measures of basic physical quantities.

A well-known instance is provided by the astronomical unit, defined as the mean distance from the Earth to the Sun. There is very little ambiguity in the quantity itself, and astronomers over the centuries have devoted some of their best measuring techniques to the task. It is known with impressive accuracy. In 1972, Youden of what was then called the American Bureau of Standards presented a table including 15 expert measurements of this quantity over the preceding eight decades, with error estimates. Each new value on his list lies outside the limits of probable error given by the previous one (Youden 1972, Gigerenzer *et al.* 1989, pp. 82f). This was for him no counsel of despair, but a plea for methods to assess and manage this kind of bias.

We must recognize that most improvements in accuracy come with new instruments and techniques, perhaps even new methods of measurement, which alter without eliminating the bias. The measurement of physical constants is one of the uncommon scientific problems for which a scientist can make a reputation by reproducing (and improving) a measurement that has been done before, and for that reason it makes visible what otherwise might not be easily noticed. Even as measured values become more and more accurate, discrepancies remain.[1] Here we find bias where we might expect it least. In fact we should not be surprised. The narrowing of measurement variation, synonymous with an increase of precision, makes visible ever more minute sources of measurement bias.

Statistical bias does not imply bias in the colloquial sense, a misrepresentation due to interest or prejudice, perhaps involving outright dishonesty. This latter kind of bias seems to have an intrinsic interest to the human species. The hope to prevent it is sometimes used as an argument against relying on samples at all, on the theory that only a complete count will eliminate all opportunity for self-interested manipulation.

But things are often not as they seem. We can in most cases reliably count the number of people sitting around a dinner table, e.g., Jesus and his twelve disciples, one of them with a little accumulation of silver coins and a furtive look. Taking a census is not much like this. In the United States, each of the last few decennial censuses has provoked a debate about whether to use sampling methods to estimate the uncounted population. The US Constitution calls for a census every 10 years, and determining what should qualify as a census is in the end a legal question, and not, perhaps, a simple one.

If we are thinking instead about accuracy and reproducibility, we must recognize that it remains impossible to count every member of a population of 300 million. The result of a complete census depends on the efforts made to track down those who do not send in forms as well as on removing those who have been registered more than once. Canadian census-takers in 1861 were paid in proportion to the number of persons they registered, inspiring a 110% effort that generated an appreciable overcount (Curtis 2001).

The undercount in a "complete" enumeration in the United States is perhaps on the order of 10 million, and the missing ones include a higher density of poor, urban, ethnic minorities, and especially (it sounds like a tautology) the undocumented. They are concentrated in the biggest states, which tend to be Democratic states. Since all of this is well known to the Congress, arguments that appear to be about fair, objective, reliable procedures of counting or sampling are also, and perhaps mainly, about political impact and the allocation of federal funds (Anderson and Fienberg 1999).

---

[1]See also Stahel, this volume, for the early history of measurements of the speed of light.

For practical purposes, sampling is generally more manageable, and it was in order to get quick results more cheaply that governments and then businesses began developing systematic survey techniques. Sampling can be used to identify inaccuracies in a census, and may well be more accurate on its own than a complete count. But sampling, too, presents thorny problems of reproducibility.

As a basis for political prediction, passive sampling with no systematic effort to represent different types of persons in the population was discredited in the 1936 US presidential election, when a prominent poll by the *Literary Digest* magazine anticipated a massive loss for Franklin Roosevelt. The election itself gave him a landslide victory. Keller shows (this volume) that the *Literary Digest* writers were well aware of the dangers of an unrepresentative sample, and never put faith merely in large numbers of straw ballots, as their critics claimed.

At the head of the critics was the George Gallup polling firm, which got the outcome right in 1936 and claimed vindication for their scientific methods of sampling. These never included the use of chance to select interviewees, but only an effort to identify the most consequential dimensions of voter heterogeneity and to balance these in their sample population. In 1948, the Gallup firm was hoist on a related petard. Their incorrect prediction of a loss by the incumbent Truman may have had less to do with flawed techniques for sampling the population than to shifts of voter sentiment after the last poll, hence a mismatch between the population surveyed and the population at the time of the vote.

Identifying the relevant population is not straightforward in political polling, which aims to survey the people who will actually turn out and vote, not simply those who are entitled to vote. Perhaps for this reason, perhaps for some other, the last population polled and the one that turned out on election day were different. Heraclitus had a good sense of the general problem: you can never survey the same population twice.

The failure of the Gallup survey was acutely embarrassing for American social science, which immediately mobilized to control the damage. After all, as Keller points out, since most surveys correspond at best indirectly to anything that happens in the world outside, election results provide practically the only empirical evidence that surveys really work. Hence, a failure to call elections correctly was potentially very damaging to this fundamental tool of social research. But the difficulties are not limited to social science. Almost nowhere is sampling unproblematic, and failures of experimental replication have been as vexing as those in survey results.

Historically, the procedure for clinical trials in medicine has multiple sources. They are often credited to the influence of Ronald Aylmer ("R.A.") Fisher, who by the 1930s had achieved great prominence in the application of statistics to medicine, and to so many of the human and biological sciences. But Karl Pearson and his students, especially Udny Yule and Major Greenwood, had already worked out experimental methods for the statistical comparison of experimental and control populations by the 1910s.

More interestingly, the problem was very much on the agenda of physicians in Britain and elsewhere even before statisticians like Pearson had shown much interest in health and medicine. In few, if any, disciplines did formal methods of statistical design and analysis arrive simply as imports from mathematics. And in every field of application, they had to be adapted to specific problems of the substantive disciplines involved. Not only the basic experimental design but also the techniques for recruiting patients and assigning them to experimental and control groups amounted to much more than minor adjustments to a basic methodology worked out by mathematicians like Fisher.

Indeed, agricultural experiment stations were already notable site of statistical experimentation when Fisher began develop developing techniques for measuring the effects of different fertilizers or irrigation regimes on crop growth at the Rothamsted Experimental Station. Disciplines like medicine and psychology had their own reasons for testing therapies and even for randomizing. Other prominent sites of experimental designs involving statistical analysis, sometimes even incorporating techniques of randomization, included psychical research and educational psychology (Hacking 1988, Dehue 1997, Porter 2004, Chalmers 2005).

The logic of the controlled clinical trial is deceptively simple. Patients suffering from, let us say, tuberculosis, are divided at random into treatment and control groups. Tossing a coin or consulting a random number table provides some assurance against conscious or unconscious bias by an experimenter who, even apart from any financial interests in the outcome, may be eager to demonstrate that medical research has given rise to something that really works.

Since the patient groups are in principle indistinguishable, a difference in the pattern of outcomes should be the result of the different medical regimes. Of course the numbers must be sufficiently large that an effect, which may be small, can be distinguished from random noise. Indeed, the official methodology requires the doctors to decide in advance how small an effect they want to be able to detect – the *power* of the test – as well as what the experimental dosage should be. Perhaps they will complicate the experiment by applying more than one dosage, each to a different group of patients.

They try to neutralize the effect of the beliefs of doctors and of the hopes and expectation of patients by giving the control patients a placebo and by keeping patients and doctors alike in the dark about who is getting the treatment. Yet the patients may still confuse the outcome by dropping out of the experiment.

If an appreciable fraction of the patients become missing data, the treatment and control groups may no longer be comparable. The experimenters cannot assume that the decision to withdraw is independent of the results of treatment thus far. There are of course ways of dealing with disappearing patients, but none are entirely satisfactory (Marks 1997).

We could go on in this vein. Some of the most important statistical trials of the postwar era were not real experiments at all, but quasi-experiments. The most famous of these involved smoking and cancer. It would take years to have useful results from a controlled trial, if such an experiment were even possible. Really, it is not, since you could never get people to commit to a lifetime of smoking or not smoking, or smoking then quitting at a moment dictated by the experimental design.

If tobacco really proved to be carcinogenic, the experiment itself could lead to thousands of unnecessary cancers and even deaths, which would scarcely be to the credit of scientific medicine. In place of this, the medical scientists tried to identify populations that were similar in every respect except that some smoked and others did not. Some of the earliest of these comparisons were based on data from populations of physicians. Rather unexpectedly, it turned out that the increased disease incidence from smoking was not limited to lung cancer, but included a whole range of conditions, including heart disease as well as other cancers.

Such results might be interpreted as showing the wide-ranging deleterious health effects of smoking, or they might be explained by supposing that control and treatment groups were not really comparable at all. Some medical and statistical experts did make this latter argument. Did they make it honestly? The cigarette companies, as is well known, began very early to buy off experts in order to cover up or distort the medical results of smoking. One of the most prominent of the skeptics was the preeminent statistician Fisher, whose research, as it happens, was funded in part by tobacco money. The companies were willing to spend so much on this campaign that they even bought off a considerable number of historians (Brandt 2007, Proctor 2011).

Qualitatively, at least, the tobacco results were eminently reproducible. That is, elevated rates of cancer and death appeared consistently when smokers were compared with nonsmokers, and this conclusion was supported in certain respects by other kinds of evidence. Quantitatively it is much more complicated. Preventive medicine offers few interventions whose positive effects are even on the same order of magnitude as the negative ones of cigarette smoking. Once people began looking seriously, the medical effects of tobacco, and especially of manufactured, mass-distributed cigarettes were hard to overlook, even though smoking acts over a long time period and involves behaviors that can change unpredictably and that are not accessible in detail to researchers.

Meanwhile much else in the world is changing, and not in the same way for different categories of people. We can scarcely expect a precise match of numbers between past and future experience for phenomena like these. In practice, the effects of smoking, like so many in the world, are estimated with models and debated by economic and political actors as well as scientific ones.

21.3 Translation

The "blockbuster" drug, a creation of the postwar era, was measured by the revenue it brought in and not by its contribution to human health. There have been fewer of these in the past few decades, despite the reiterated promises of molecular genetics. Pharmaceutical companies, having seen their revenues and stock values expand hugely over the late 20th century, are troubled by the relative stasis so far in the 21st.

One by-product of searching their instrumental souls has been a recent efflorescence of writing on replication and reproducibility. Company spokespersons lament that so many studies are not reproducible and do not lead to products that show benefits in humans. They typically do not mean to confess that something is fishy with trials on human subjects. Rather, a drug has been shown to produce a certain physiological effect on experimental animals, an effect that is supposed to be desirable for some category of humans.

But when the trials are extended to humans, the corresponding effect does not appear, or proves not to be beneficial. Sometimes the failure of translation is between one population of humans, perhaps made up of carefully selected persons in a highly controlled setting, and a diverse population of subjects who may differ widely in bodily characteristics or in behavior.

Here, once again, "reproducibility" is not the same as repetition of the same experiment under similarly controlled conditions. Instead, it is a different river. There have been plenty of problems replicating experiments that show genetic causes or pharmaceutical therapies for human illness. Here the company laments pertain to conditions for which there is no special problem in producing the effect reliably in laboratory mice.

The failure takes place at the point of transition from a kind of basic to applied research, which in many cases corresponds to the transition from laboratory animals to suffering humans. In a university laboratory, at least in the days before patents became an important criterion for promotion, such an outcome might be disappointing, but would not be deemed a failure, because they had at least demonstrated something about physiological response in mice, which might someday contribute to a more generalized biological knowledge.

But the difficulty of reproducing animal effects in humans certainly affects university science too. The massive postwar expansion of biology was achieved to a large degree with medical money. It was a kind of marriage, consecrated by a fusion of the two partners into a hybrid "biomedicine" (Gaudillière 2002). The master field for biomedicine has been genetics, and projects such as the human genome initiative were funded on the basis of optimistic promises of medical breakthroughs (Mirowski 2011, Comfort 2012).

These have been slow in coming, as the narrowed drug "pipeline" attests, and the support for medical research has of late been shifting toward applied work on very specific diseases or risk factors. The animal experiments performed by pharmaceutical companies are typically focused in this way. But their work, too, begins in the highly artificial world of purified reagents and cloned, patented experimental mice. These facilitate research at the beginning, but heighten the difficulties of translation, of reproducing results in human bodies (Löwy 1996).

The reasons for experimenting on mice or other non-human organisms may include more rapid results associated with quicker generation times, and certainly involve the possibility of more complete control as well as larger sample sizes. We have not yet bred populations of genetically identical humans. The crucial advantage of the animals is that the ethics of experimentation are much more relaxed. There need be no expectation that the treatment should benefit the animals, no need to avoid subjecting the animal to risk. Indeed, many thousands of mice may be treated and then sacrificed to see what effect a substance has had on a gland or an organ, as a basis merely for clarifying the possibilities of treatment.

The ideal candidate for a risky human experiment was at one time a criminal sentenced to death, who seemed to have nothing to lose. Physicians also sometimes performed heroic experiments on themselves, and still do, whereas medical ethics now clearly prohibits dangerous experiments on prisoners, including condemned criminals, who lack now the standing to give informed consent. Right into the postwar era, deadly medical experiments were sometimes performed without real consent on the poor and on racial minorities, and the role of guinea pig is clearly not always a privilege to be cherished.

But it came for a time to seem that way, at least in some contexts. During the middle decades of the 20th century, the preferred subjects of human experimentation in the United States were immune to the complications of childhood development or old age, and had no risk of becoming pregnant. They were healthy white males. A satirical song written by Malvina Reynolds in 1962, and performed most famously by Pete Seeger, gives us "doctors and lawyers and business executives" in houses "all made out of ticky-tacky, and they all look just the same." This is not quite the cloned laboratory mouse, but the suppression of variability was welcomed by medical research, and it might have remained acceptable for longer were it not that the maladies of white male professionals absorbed so much of the medical research funding.

The neglect of diseases that particularly affected women, above all breast cancer, became an important cause for the women's movement in the 1970s. The preference for reducing complexity among subjects of research gradually gave way to an ideal or representativeness, and even of representation in a political sense. This ethic was applied first of all to biological sex, and before long to ethnoracial groups.

The emergence of AIDS in the 1980s opened the way for sexual preference as a category that should be considered. By the onset of the new millennium, the recruitment of subjects for therapeutic experiments had become so complicated that specialist firms grew up with the capacity to advertise to diverse categories of potential subject and to persuade them to participate. Apart from therapies that encourage the hope of a miracle cure for a dread disease, where there is often fierce competition to be accepted into an experiment, the research subjects must typically be paid to participate (Epstein 2007). Reproducibility has become more complicated, a result that had passed the statistical test of validity for the white men might not hold for women or children.

There are many other dimensions of difference that will matter from time to time, including age, social class, occupation, ethnicity, and other medical conditions. Most human variation does not reduce to any of these. But therapeutic research, according to the current ideal, is based on statistics and applies to groups rather than individuals. Although much variation is averaged away, the general results obtained for a diverse population may not hold for every subset, and certainly not for every individual.

Conversely, there is now precedent for clinical trials that have failed the test of significance for the entire population of subjects but achieved it for one group. Often it is not easy to determine if such differences are real. But ambiguity in statistics brings opportunity. For a pharmaceutical company operating in the United States, approval for a single group can be almost as good as approval for everyone, since once a product is allowed on the market, physicians are permitted to rely (often irresponsibly) on their presumed expertise to prescribe it to other kinds of people and even for other conditions.

21.4 Statistical and Substantive Significance

The basic idea of statistical significance did not begin with Fisher. Joseph Fourier in 1823 proposed a standard of certainty of 1 in 20,000 as sufficient to take a thing as valid. Two decades later, the medical statistician Jules Gavarret was so discouraged by the number of cases required to reach this standard that he lost heart (Matthews 1995). Karl Pearson checked his comparisons of anthropometric and medical measurements for degree of confidence that the difference was nonzero, and sometimes spoke of those with a low probability of occurring by chance as significant.

But it was in fields like psychology of education and psychical research as well as agriculture where the demonstration to a high probability that an effect would not occur uncaused became standard procedure. Showing that a certain result is sufficiently unlikely to have arisen merely by chance serves, among other things, as a standard of reproducibility. Here again, we do not find the statistician providing something brand new that transformed substantive fields from outside, but a clearer articulation using more sophisticated mathematics for what physicians and psychologists were already trying to do.

Karl Pearson, who went a long way toward creating statistics as a mathematical field, found himself much in demand across a range of fields, though he was also much reviled. Fisher, too, engaged in fierce controversies, but more often with other statisticians such as Pearson himself and Jerzy Neyman rather than with social and medical scientists. Fisher's incorporation of the significance test into a general program of experimental design was a huge success across a wide range of fields. It was valued not least for supplying a rigorously impersonal standard of a demonstrated fact.

A solid fact was one that emerged as significant at the probability level of 0.05, the magical 5%. This meant, in probability terms, that if there were no real effect, an experiment would have only a 5% chance of deviating so far from what he called the "null hypothesis." It requires other, highly questionable assumptions to turn the statement around and conclude with 95% certainty that the effect is real and points to some causal factor.

What Fisher and Neyman (in alliance with Karl Pearson's son Egon) fought bitterly about, textbook writers in a range of disciplines fused into a methodology, mostly Fisherian, that they called, simply, "statistics" (Gigerenzer *et al.* 1989). Significance of this sort became *de facto* a necessary if not quite sufficient standard of publishability for empirical findings. This reflected a supposition that a low p-value (< 0.05) implied a high probability that the effect did not appear merely by some statistical fluke and, hence, could be reproduced.

From the beginning, there were critics of Fisher's program, which often was followed virtually as a recipe. Many have pointed out that it emphasizes the wrong thing, measuring a probability value rather than an outcome or effect. An effect that passes the p-test, showing significance at the level of 0.05, even if everything is done correctly, may still be too small to matter. Statistical significance, as has been pointed out again and again, is not the same thing as substantive significance (Ziliak and McCloskey 2008).

Using a massive data base such as a national census, almost every form of variation will pass the test of significance, though many will not involve a direct causal relationship, and many of those that do will be too small to matter. To be sure, it is important for science to be confident that an apparent effect is not merely a random fluctuation. A few categories of causal relationship, as in psychical research, would, if their findings held up, have great substantive significance even if the effect strength is very small.

This would generally not be true in medicine, still less in agriculture. One might, however, understand the p-value, with the assurance it provides that such results will not occur very often purely by chance, as one criterion for deciding whether it is worth exploring further. In these follow-up studies, the scientist might take more trouble to determine the magnitude of the effect and, at the same time, to explore more deeply the mechanisms involved.

One more consideration, perhaps the most important of all, is the interactions of the effect with other variables which, for humans, might include other treatments and genetic or bodily traits, as well as various conditions of life. By refining their techniques, teachers, physicians, or psychics might be able to increase the effect or to specify more tightly the domain of its proper application, to the point that it acquires a high practical worth.

In scientific publications, however, "significant" results are more often presented as factual nuggets. To blame Fisher for all the misuses of statistics associated with this fixation on significance testing is not entirely fair. He sometimes described the calculation of significance as primarily a basis for deciding whether to continue a line of investigation. But he also made strong claims to the effect that an experiment had no more purpose than to give the facts an opportunity to refute the null hypothesis. And this is how it has often functioned.

The ideal of reproducibility may remain purely hypothetical. Other researchers rarely regard it as worthwhile to see whether an experiment and its findings can be replicated. They will get little credit if they succeed, and may have their competence or their objectivity called into question if their results diverge from the original study. Irreproducible results seem most often to disappear silently, not because they fail in a crucial experiment but as a result of the inability (or lack of interest) of other researchers to continue a line of investigation.

The disinclination of scientists to repeat reported experimental work and to bolster or call into doubt its validity occasions much public hand-wringing. Indeed, there are sometimes reasons to worry, especially, if a result has only loose or indefinite relations to other empirical findings and if it is not tightly interwoven with models or theoretical predictions. These are the circumstances in which fraud will be most difficult to detect, and that are typical for fields like pharmacology, where it can be most profitable.

In other areas of science, reproduction and testing occur mainly in other ways. As Collins argues (this volume) duplicating an experiment from another laboratory is far from straightforward even in the most rigorous areas of science. More typically, scientists do not repeat experimental but try to incorporate novel findings or techniques into new researches aimed at new results. This form of reproducibility – putting a result to work to generate something new – may be the one that functions most effectively in experimental science (Hacking 1983). It depends, however, on well-developed ways of linking experiments to other empirical results, often by way of theoretical expectations. Such sciences, most of the time, depend only a little on formal statistical inference.

The logic of statistical inference, with its requirement of a sufficiently low p-value as gatekeeper to knowledge, was developed primarily to validate results that cannot be screened off from their complex circumstances, and not as a strategy for building a structure of a general theory. Modern research in psychology and medical therapeutics rely mainly on statistical meta-analysis to combine related studies, and in this way to test the validity of particular experiments. They always face the perplexing question of whether different experiments are really similar enough to be combined in this way.

The reproduction of the conditions of an experiment must first be put into question before outcomes can be compared to decide if the indicated cause or effect is "reproducible." Meta-analysis, we must note, includes tools to test the homogeneity of a group of studies, to evaluate the quality of the data, and even to seek out biases, in particular the "publication bias," that investigations are unlikely to be published if they do not reach statistical significance (see Ehm, this volume, for more details). At this level, there is much evidence that affiliations of researchers matter, that who pays the piper can indeed often call the tune.

As a form of reproducibility, meta-analysis in statistics works to accumulate results, often with practical importance. This should include an attempt to sort out results that appear contradictory. The significance test usually underlies the meta-analysis, defining a minimum standard of reliability by which a paper earns the right to be treated as comparable to other empirical studies of the question at issue.

It makes diverse papers commensurable and perhaps interchangeable, and licenses the meta-analysis to use the research paper as datum without having to call into question every aspect of the experiment. This is consistent with the public use of statistical tests, which are certified as appropriately solid and impersonal by their data and machinery of analysis. Statistical design and testing demand data that can be fed into a rigorous form of quantitative analysis, one that leaves as little space as possible for loose or self-interested reasoning.

The exclusion or suppression of data from studies that do not achieve statistical significance, however, undermines the meta-analysis. Determining what kinds of questions are appropriate for this kind of cumulative analysis become a very subtle question. On a wide range of questions, such as decisions to approve or reject new pharmaceuticals, a demonstration of statistical significance may be sufficient by itself to get over an important hurdle in the decision process (Porter 1995).

But here we encounter an irony. When numbers acquire such power, the temptation to distort or subvert them becomes intense. Often it is irresistible. The social psychologist Donald Campbell, a distinguished quantifier, remarked that to invest automatic power in a numerical indicator leads almost inevitably to its corruption. What is too readily accepted as rigorous will be turned to putty. Suspicion cannot be put to rest so easily (Porter 2012).

21.5 Irreproducible Numbers

The alliance of quantification and deception is often about measures and estimates applied to unique objects or situations, and so may not involve a straightforward notion of reproducibility. The pharmaceutical companies that are vexed by a failure of animal studies to work out in human populations need not presume dishonesty or incompetence in the conduct of the experiment. They had some grounds for hoping that a result on mice would apply also to humans, and the failure of the human trials shows that some unanticipated interspecies difference has come into play.

Inconsistencies in the outcomes of two different trials on humans would seem to be more incriminating. But these might reflect explicit purposes of the experiment, to make comparisons between men and women, old and young, healthy and sick, carriers and non-carriers of some allele. Even when the trial is designed to be representative of a diverse population, it may be hard to match one experiment with another. We cannot presume fraud, but perhaps only a flawed design, if the results do not match, not even when the evidence of a p-value suggests that the difference would be highly unlikely to arise from pure chance. There are too many circumstances that make it difficult to match human populations. Charges of fraud typically rest on more flagrant misbehavior, such as the fabrication or suppression of data. And yet, the whole point of a clinical trial, an educational test, or an experiment in social psychology is to establish results of general validity, results that will hold good when extended to the population at large.

The experimental technology of randomized trials was designed specifically to supply generalizable results and to resist interested manipulation. But efforts to rein in fudging and cheating by means of explicit methodological rules can scarcely prevail in the face of a sufficiently strong incentive to deceive. The technologies of objectivity depend on competent interpreters with the knowledge, will, and the authority to defend the spirit of the rules and not merely their explicit content.

The vulnerability of the randomized experiment has been clearly explained. The first problem is publication bias, the suppression of unsuccessful trials. Unless there is an explicit and enforceable requirement to report all results that might someday be used for regulatory approval, it will be possible for a company to sponsor repeated trials until at last the desired result appears. A dozen trials on average will include one that achieves the required level of significance purely by chance.

Similar temptations confront academic researchers seeking only to publish exciting results. But this kind of rule can only be enforced in a very highly regulated research endeavor. Do we require every experiment in every field to be carried to completion, prohibit researchers from changing their minds, outlaw scientific work by anyone who has not registered with the proper authorizes? In science, the question to be asked is not always obvious in advance, yet adjusting the question offers many opportunities to get the desired result from random noise.

In pharmaceutical research, it is often not self-evident whether a new drug should be compared simply to a placebo, or to some existing treatment. A less effective drug might still have value for patients who do not tolerate the preferred treatment. A comprehensive experimental trial might consider two or more dosages, perhaps in combination with other medications or in different treatment regimes, and the researcher might even analyze the outcomes separately for different classes of patients.

None of these ways of proceeding is inherently objectionable. On the contrary, it accords with the statistical aspiration – articulated quite specifically by

Fisher – to maximize the information from an experiment. But the effect is to turn a single experiment into many experiments, providing abundant opportunities for one or more to meet the standard of significance (Healy 2012).

Of late such moves have become controversial. The tools of meta-analysis can be used to estimate the prevalence of unreported trials. Critics argue that every trial should be registered in advance, and the data made available for public inspection, whether or not the company intends to use it in an application for regulatory approval. To the extent that these moves are successful, they might bring drug regulation up to a standard of disclosure that has been recognized, e.g., in parapsychology for decades.

Social psychology, which has never been subject to such withering doubt as the experimental study of psychical effects, has also not had its statistical techniques regulated to the same degree. Neither does psychology confront pressures and temptations like those facing a big pharmaceutical company, which must maintain profits by keeping the pipeline of new patents flowing. Yet striking results, demonstrated with adequate statistical significance, make academic careers, perhaps especially in psychology, and cemeteries are crowded with the unidentified remains of psychological experiments that never achieved statistical significance.

John Ioannidis (2005), a well-known critic of ordinary practices of significance testing as a basis for reliable findings in medical therapeutics, has recently extended this skepticism to a wave of experiments in social psychology on subconscious "priming."[2] Priming in its various forms has often been highly newsworthy, all the more so when a finding seems most strikingly improbable. The suggestion that some at least of the phenomena of priming may be artifacts of dubious statistical and experimental practices was picked up recently by general news outlets, creating a scandal and stimulating prominent appeals to close at last these familiar loopholes in the methodology of the statistical experiment.

To be sure, some of the most forceful objections do not specifically involve the statistical aspects of experimental design (Bartlett 2013). Most critical commentators point to careerism, combined perhaps with a touch of the herd mentality, as the reason for so many results they find incredible. This would be in contrast to the cynical business practices involving ghost authorship as well as suppressed data that have tainted so much regulatory testing of pharmaceuticals.

---

[2]Stimulated by an open e-mail in which Nobel prize winner Daniel Kahneman recently urged social psychologists to restore the shaken credibility of their field, a large-scale study of direct replications concerning 13 different kinds of social priming were conducted in 36 independent samples and settings. This heroic effort led to a 51-authors paper (Klein *et al.* 2014) which reports that 10 out of 13 kinds of social priming have been reproduced consistently, yet with considerable variations in effect size, across laboratories.

But a moment of web searching turns up entrepreneurs trying to make their fortune peddling the magic of psychological priming as well as secrets of brain science to improve scores of classroom students on tests now in use to evaluate teaching effectiveness. We should not underestimate the commercialization of contemporary science.

The hope for enacting rules of reproducibility itself is undermined somewhat by its inescapable ambiguities. The results of astronomical measurements could be reliably reproduced, but the bounds or error did not apply reliably to measurements with new instruments and different observers. Clinical trials, if successful, lead to new pharmaceutical treatments on a wider range of patients, which are to be monitored in the next step of the process. This involves extending, and not merely replicating, an experimental result.

The comparison is made more difficult by the inherent variability of the subjects of therapeutic or social–psychological experiments, quite in contrast to the astronomical unit. Priming, as it appears, did not inspire much effort to replicate specific experiments until its results were called into question. Its credibility, which many still defend, depends less on the absolute solidity of any particular finding than on the great multitude of situations in which priming has been detected.

Whole families of experiments may in the end be discredited as suffering from the same kinds of flaws. Which is to say that every satisfactory demonstration of priming supports the others, while failures to replicate particular experiments are taken to cast doubt on all. We have, then, two levels of replication which, however, are not at all separable. And yet it could happen that when the dust has settled, some instances will be vindicated and others discredited. A notorious instance of fraud in priming research has been damaging to the whole field, yet proves rather little by itself.

21.6 Reproducing Calculations

We know from legal documents that pharmaceutical companies can be utterly unsentimental about a thing with so little cash value as truth. Yet they exploit ambiguity rather than advancing outright lies. Statistical inference is about reducing ambiguity to a minimum for the sake of true or valid knowledge. Since it is not easy to expel ambiguity from the world, it must be managed. There is always a temptation to suppose that the truth lies on the side of what is more rigorous. Sometimes, however, the ambiguities are merely driven underground by reliance on rules of quantification. In many situations, the methods for reducing ambiguity may function as conventions more than as roads to truth. Especially in the hands of researchers who lack any deep training in the field, statistics can be pointlessly formulaic.

This preference for what can be tightly specified over what may seem most reasonable is also a technique of reproducibility. Here it is not a question of repeating an experiment and achieving the same results but of being able to follow instructions and to perform calculations reliably (compare Bailey *et al.*, this volume). Rules of procedure are most needed in situations where many actors would prefer to evade such constraints. In this respect, science may be compared to other professional activities, such as accounting. Tax systems are exemplary sites of the exploitation of ambiguity, which persists in the face of every effort to clarify and make rigorous. Every hydra-head that Hercules can sever, every suppressed accounting trick, is replaced by new ones.

In science and scholarship, whose practitioners in most cases are trying to understand one another and to reach agreement, the situation is usually not so desperate. But scientists too are human, and there is no consensus on the rules for determining who is a competent scientist. Honest and competent (if fallible) science has its enemies, many of whom come dressed as if they were the real scientists.

The great line from the files of tobacco litigation, "doubt is our product," is only one of the tricks they use to sow confusion. It has been equally effective to insist on absolute rigor and unambiguous demonstration, to be utterly literal-minded, and to refuse to accept a line of reasoning until every point has been purged of all uncertainties. Then, and only then, they say, does a claim deserve the name of "sound science," and whatever falls short in the least measure is mere propaganda, to be condemned as politics masquerading as science.

Public discussions of climate change are often derailed by the pretense of insisting on the highest standards (see Feulner, this volume). And yet the most respectable scientists can be wrong and are not necessarily free of selfish interests. We think of the chemists who lined up to dismiss the allegations of the science writer Rachel Carson's regarding DDT.

Querulous objections to little lapses of rigor and reproducibility are less often at stake in discussions within the established community of research disciplines than they are in more public debates. But we must recognize that the opponents of science are drawing on a rhetoric that is not far from the language that scientists sometimes use in teaching or addressing the public. Numbers, measures, and statistical reasoning are important elements in the advance of reproducibility, but they cannot isolate us from the bigger world, with its real uncertainties as well as its oceans of exploitable ambiguity.

Numbers, too, depend for their effectiveness on trust and discernment.

Statistics has a key role in providing a foundation for such discernment. But there is no secret algorithm by which, in the business of pursuing knowledge, we can cease to be human.

References

Anderson, M., and Fienberg, S. (1999): *Who Counts? The Politics of Census-Taking in Contemporary America*, Russell Sage, New York.

Bartlett, T. (2013): Power of suggestion. *Chronicle of Higher Education*, January 30, 2013.

Brandt, A. (2007): *The Cigarette Century: The Rise, Fall, and Deadly Persistence of the Product that Defined America*, Basic Books, New York.

Chalmers, I. (2005): Statistical theory was not the reason that randomization was used in the British Medical Research Council's clinical trial of streptomycin for pulmonary tuberculosis. In *Body Counts: Medical Quantification in Historical and Sociological Perspectives*, ed. by G. Jorland, A. Opinel, and G. Weisz, McGill-Queens University Press, Montreal, pp. 309–334.

Comfort, N. (2012): *The Science of Human Perfection: How Genes Became the Heart of American Medicine*, Yale University Press, New Haven.

Curtis, B. (2001): *The Politics of Population: State Formation, Statistics, and the Census of Canada, 1840–1875*, University of Toronto Press, Toronto.

Daston, L. (1987): *Classical Probability in the Enlightenment*, Princeton University Press, Princeton.

Dehue, T. (1997): Deception, efficiency, and random groups: Psychology and the gradual origination of the random group design. *Isis* **88**, 653–673.

Epstein, S. (2007): *Inclusion: The Politics of Difference in Medical Research*, University of Chicago Press, Chicago.

Gaudillière, J.-P. (2002): *L'Invention de la biomédecine: la France, l'Amérique, et la production des saviors du vivant après 1945*, La Découverte, Paris.

Gigerenzer, G., Swijtink, Z., Porter, T., Daston, L., Beatty, J., and Krüger, L. (1989): *The Empire of Chance: How Probability Changed Science and Everyday life*, Cambridge University Press, Cambridge.

Hacking, I. (1983): *Representing and Intervening*, Cambridge University Press, Cambridge.

Hacking, I. (1988): Telepathy: Origins of randomization in experimental design. *Isis* **79**, 427–451.

Hacking, I. (1990): *The Taming of Chance*, Cambridge University Press, Cambridge.

Healy, D. (2012): *Pharmageddon*, University of California Press, Berkeley.

Ioannides, J. (2005): Why most published research findings are false. *PLOS Medicine* **2**(8), e124.

Klein, R.A., Ratliff, K.A., Vianello, M., Adams Jr., R.B., Bahník, S., Bernstein, M.J., Bocian, K., Brandt, M.J., Brooks, B., Brumbaugh, C.C., *et al.* (2014): Investigating variation in replicability: A "many labs" replication project. *Social Psychology* **45**(3), 142–152.

Krüger, L., Daston, L., and Heidelberger, M., eds. (1987a): *The Probabilistic Revolution, Vol. 1*, MIT Press, Cambridge.

Krüger, L., Gigerenzer, G., and Morgan, M.S., eds. (1987b): *The Probabilistic Revolution, Vol. 2*, MIT Press, Cambridge.

Löwy, I. (1996): *Between Bench and Bedside: Science, Healing and Interleukin-2 in a Cancer Ward*, Harvard University Press, Cambridge.

Marks, H. (1997): *The Progress of Experiment: Science and Therapeutic Reform in the United States, 1900–1990*, Cambridge University Press, Cambridge.

Matthews, J. (1995): *Mathematics and the Quest for Medical Certainty*, Princeton University Press, Princeton.

Mirowski, P. (2011): *Science-Mart: Privatizing American Science*, Harvard University Press, Cambridge.

Porter, T. (1986): *The Rise of Statistical Thinking*, Princeton University Press, Princeton.

Porter, T. (1995): *Trust in Numbers: The Pursuit of Objectivity in Science and Public Life*, Princeton University Press, Princeton.

Porter, T. (2004): *Karl Pearson: The Scientific Life in a Statistical Age*, Princeton University Press, Princeton.

Porter, T. (2012): Thin description: Surface and depth in science and science studies. *Osiris* **27**, 209–226.

Proctor, R. (2011): *Golden Holocaust: Origins of the Cigarette Catastrophe and the Case for Abolition*, University of California Press, Berkeley.

Stigler, S. (1986): *The History of Statistics: The Measurement of Uncertainty before 1900*, Harvard University Press, Cambridge.

Youden, W. (1972): Enduring values. *Technometrics* **14**, 1–11.

Ziliak, S., and McCloskey, D. (2008): *The Cult of Significance: How the Standard Error Costs Us Jobs, Justice, and Lives*, University of Michigan Press, Ann Arbor.

22
Science Between Trust and Control: Non-Reproducibility in Scholarly Publishing

Martina Franzen

Abstract. In the past few years, the revelation that certain claims in prominent scientific papers were not reproducible has sparked a discussion linking reproducibility and scientific integrity. The public discourse on scientific reproducibility points to the general constraints that science faces with regard to the scientific paper. Reporting research is not equivalent to doing research. Inconsistencies of reporting is one factor for the lack of reproducibility. To raise the level of reproducibility, journals have introduced new editorial measures to ensure the accuracy of research ahead of print. Authors are now requested to make their protocols, data, and materials available. The argument developed here is that the emerging transparency regime generates a structural tension between trust and control that might even hinder scientific development.

22.1 Introduction

Alarming numbers of non-reproducible studies published in top-tier journals (Begley and Ellis 2012, Prinz *et al.* 2011) together with high-profile cases of scientific fraud has put the issue of non-reproducibility at the center of a recent discussion on accountability in science and scholarly publishing. The increased pressure to publish, the publication bias towards positive findings or poor reporting of experimental details are among the factors explaining reproducibility problems (see also Bailey *et al.*, this volume).

This chapter, however, tackles the issue of reproducibility from a communications theory perspective. This perspective understands reproducibility as a communicative act rather than an ontological state. In contrast to other constructivist approaches that are concerned with "doing science" (Collins 1985, and in this volume) this contribution is concerned with the issue of "reporting science." And just as there are technical constraints to overcome when "doing science," there are also structural constraints on the act of "reporting science" that might impede the replication of research results after publication.

I follow the theoretical and practical *dictum* that research results only become completely scientific when they are published (Ziman 1969, p. 319, see

Reproducibility: Principles, Problems, Practices, and Prospects, First Edition. Edited by Harald Atmanspacher and Sabine Maasen.

also Price 1981, Stichweh 1987) and thus use the scientific paper as a reference point for studying reproducibility within scholarly communication. The argument developed is that the new transparency measures that seek to increase the level of reproducibility in science mask a structural dilemma of output control that might even hinder scientific progress. Before I start the analysis, it is necessary to consider the sociological aspects of the reproducibility concept more closely.

22.2 Reproducibility as the Touchstone for Distinguishing Science from Non-Science

Scientific knowledge production is based on the publication and reception of published research findings. The founding of scientific journals and the invention of the experimental article provided the basis for science to flourish in the 19th century. Science evolves by publishing new findings that trigger follow-up research, which in turn refers to previous research results. Thus, publications allow for scientific reproduction.

From a temporal perspective, it is important to note that reproducibility is an *a posteriori* concept. In other words, the knowledge claims presented in scientific publications can subsequently either be confirmed or refuted by the scientific community. In practice, however, the most likely outcome is for new findings to simply be ignored. Instead, the level of awareness of scientific findings increases to a greater extent when they are associated with recognizable names and issues rather than due to their "truthiness."

This makes clear that communicative connectivity and reproducibility are not identical. Topical relevance is detached from the robustness of single findings.[1] Nevertheless, the findings' utility is coupled with their reproducibility, because this enables further development. To establish scientific facts is the path along which science can progress, and along which negative findings can guide the way. In sum, the successful replication of experimental results is pivotal for translating new scientific findings into scientific instruments and technologies.

Refuting knowledge claims for scientific reasons, prior to or after publication, is inherent to the professional identity of science. This has been characterized by the Mertonian norm of organized skepticism (Merton 1942). Non-reproducibility, however, relates not only to the original message but also to the act of reproduction. If an attempt to reproduce experimental results fails, it would be premature to assume that the primary claim was untenable or that the original data were falsified.

[1] It further implies that scientific evidence cannot be captured by the level of citation counts.

Problems with reproducibility occur for a variety of reasons. In addition to errors in the original research, failed attempts at experimental reproduction may occur due to technical constraints (apparatus, material, or equipment), data constraints (if, for instance, raw data are not available), or personal and social constraints (e.g., inexperienced researchers or missing tacit knowledge). There are general methodological constraints on the reproduction of experimental results, which cause Collins (1985, and in this volume) to understand reproducibility as a chimera, stemming from the notion of what he calls the "experimenters' regress."

Furthermore, in addition to understanding reproducibility merely as a scientific norm facing technical constraints, it is worth considering its social dimension. Scientific practice is not about identically replicating previous findings. As scientific achievements are measured by originality, scientists are encouraged to produce new results rather than to reproduce the well known: "A chef cannot make a reputation for himself by demonstrating bad recipes" (Broad and Wade 1982). As long as published findings do not directly affect a scientist's own field of research there is no practical need to test the reliability or efficiency of another scientist's experimental results.

As a self-regulated system, science relies on its self-correcting nature. Systemic trust is perpetuated by the institution of pre- and post-publication peer review. Hirschauer (2010) defined journal peer review as a means of ensuring the reciprocal accountability of judgments among peers. Journal peer review is concerned with evaluating manuscripts with regard to their technical soundness, the empirical support for their conclusions, their novelty, their comprehensibility, and their appropriateness for the journal in question. Giving proof of the reproducibility of "firsts" is, therefore, beyond the scope of prepublication peer review.

Concerning the factual dimension of reproducibility, it is important to note that scholarly publications are not "scientific" per se. A scientific paper is merely a *knowledge offer* that enters into scholarly communication and informs scientific knowledge production. The scientific community in areas such as molecular biology sticks to the general rule that published findings must be replicated by at least three independent research teams before they are believed to be scientifically useful. The fact that scientific claims can later turn out to be wrong is part of the uncertain process of searching for newness, which is the defining feature of modern science. The (self-)description of science as a self-correcting system assumes that irreproducible findings simply vanish from scientific communication over time. But if replication is not part of routine scientific practice (Collins 1985), what is the emerging discourse of irreproducible research about?

In the following, I will explore the public discourse on reproducibility problems as well as the measures journals have taken to increase the level of reproducibility of their published record. To put the role of scholarly journals in context, I will start the analysis by reconstructing the so-called STAP saga, which provides valuable insights into how new claims are revealed as non-reproducible and the impact of these revelations on science.

22.3 Contested Claims: The Story behind STAP

The STAP saga is an intriguing example of how replication problems get into the public sphere. The acronym STAP stands for "stimulus-triggered acquisition of pluripotency," a new methodology published by a Japanese team at the RIKEN Center for Developmental Biology in January 2014. In two accompanying *Nature* papers the authors claimed to have found a new method for stimulating the differentiation capacity of adult stem cells by using external stressors, such as a bacterial toxin or a weak acid bath. Their remarkable success in reprogramming adult stem cells into every cell type in such an easy manner was highly promoted and received worldwide attention. Stem cell research seemed to have moved a step closer to the clinic.[2]

But it took only a few days before rumors suggested that one image might have been manipulated. Allegations were posted anonymously on blogs and online platforms such as PubPeer, which are dedicated to informal discussion on scientific papers. In addition, a few science bloggers eagerly followed the suspicions and created new ones. Among them was the academic stem cell researcher Paul Knoepfler, who runs the Knoepfler Lab Stem Cell Blog on `www.ipscell.com`. In the context of the release of STAP, Knoepfler raised the level of skepticism by conducting a poll, updated weekly, on public perceptions of the validity of STAP, which showed decreasing support levels of STAP.

Notable critics, who declared the results non-reproducible, appeared on the scene. Their criticisms were covered by journal news sections and the general media. In response to the widespread allegations that data may have been trimmed or cooked, the RIKEN Center for Developmental Biology in Kobe, Japan, announced a formal investigation to test the results. The committee report provided proof of flaws in image duplications and serious misconduct in respect to data fabrication and plagiarism: Parts of the method section turned out to be plagiarized. RIKEN responded by immediately announcing they would take preventive measures, "to ensure that misconduct does not occur again."

[2]See `hsci.harvard.edu/bringing-ips-cells-closer-clinic`.

These measures were included in a press release accompanying the investigation report on April 1, 2014. In July of the same year, the authors eventually agreed to retract the papers. As the junior scientist and lead author Haruko Obokata was found guilty of research misconduct, the incident also blemished the reputation of the senior authors. In the aftermath, Yoshiki Sasai, vice director of the RIKEN institute, co-author of both papers and supervisor of Obokata, committed suicide.

The STAP saga shares characteristics with earlier, highly publicized cases of misconduct, particularly in stem cell science (Franzen 2012), though it stands out in the following respects. In the *temporal dimension*, we can observe an accelerated process of debunking published scientific claims.[3] The papers began to be questioned immediately after the release, and this process spread through blogs, social, and general media, and science magazines. Only two weeks after the papers had been released, the RIKEN institute set up a formal investigation committee to assess the accusations. The committee's report was published six weeks later and gave proof of misconduct. It took a lot of time, however, to reach an authors' agreement on how to proceed with the papers. Only five months after their initial release the original articles were editorially retracted.

In the *factual dimension*, the central claim to have reprogrammed adult stem cells for pluripotency challenged a central paradigm of developmental biology. Although this claim resembles various similar claims made in a series of controversial papers in this area, for STAP there was a complete lack of appropriate methodological descriptions to substantiate the findings. STAP had the appearance of "a magical approach," as one of the referees of the earlier version stated (cf. Vogel and Normile 2014) – or simply of pure fiction.

In the *social dimension*, we can observe the rise and the fall of a science star, the 30-year-old Japanese scientist Haruko Obokata, who reminds us of Hwang Woo Suk, once celebrated as the "king of cloning" before he became an "unscrupulous fraudster". In contrast to the latter, however, the STAP case exhibits an unaccustomed sharpness in the way the lead author was personally blamed. The official statement from RIKEN's investigation report includes the following note (Research Paper Investigative Committee 2014, p. 9, my emphasis):

> Given the poor quality of her laboratory notes it has become clearly evident that it will be extremely difficult for anyone else to accurately trace or understand her experiments, and this, too, is considered a serious obstacle to healthy information exchange. Dr. Obokata's actions and sloppy data management lead us to the conclusion *that she sorely lacks, not only a sense of research ethics, but also integrity and humility as a scientific researcher.*

[3] The quicker time-to-retraction seems to mirror a general trend (cf. Steen *et al.* 2013).

Such personal affronts are rather unusual for the professional academic context. The harsh judgment might be explained by conflicting cultural norms that may have added to the tragic denouement of the STAP saga as well. It moreover indicates that much of the issue of non-reproducibility is negotiated in the social dimension rather than in the factual dimension.

The public investigation uncovered further that an earlier version of the STAP papers had been rejected by the three top journals *Science*, *Nature*, and *Cell* a year earlier, as reported in a news article in *Science* (Normile and Vogel 2014). This new piece of information in the STAP saga led to speculation in science blogs about the reasons for its earlier rejection and later acceptance by *Nature*. Two months after the official retraction of the two papers in *Nature*, the blog Retraction Watch released *Science's* review reports of the former paper version.[4]

This "leak" of confidential referee reports of a rejected paper expresses a new kind of dynamic in the public control of science. The documents confirm that the referees were fully aware of the potential flaws in the paper. Although the referees agreed on the groundbreaking potential of the proposed methodology, they did not see the claim well supported, as exemplified by the following statement of one referee for *Science* (cited in Retraction Watch September 10, 2014):

> There are several issues that I consider should be clarified beyond doubt because of the potential revolutionary nature of the observations.

One day after the release of the *Science* reviews, members of *Science's* news team published the initial correspondence between its competing journal *Nature* and Obokata, dated 4 April 2013 (Vogel and Normile 2014). We learn that *Nature's* referees were equally unconvinced of the evidence of the findings. One of the three *Nature* referees stated (cited in Vogel and Normile 2014):

> Of paramount importance for the legitimacy of this paper is that the authors provide a full step by step account of their method such that the community can rapidly validate the reproducibility of the findings. The present method description is minimal and key elements are not properly defined.

Instead of outright rejection, however, the *Nature* editor did encourage the lead author to submit a revised paper (cited in Vogel and Normile 2014):

> Should further experimental data allow you to address these criticisms, we would be happy to look at a revised manuscript.

[4]Retraction Watch offers an informal repository on retractions of scientific papers and operates under the slogan "tracking retractions as a window into the scientific process."

The number and seriousness of inconsistencies raised by the numerous referees makes it rather unlikely that the authors could have overcome all the doubts on the scientific reliability of STAP in their final versions – nevertheless, it was published by the high-impact journal *Nature* and received worldwide recognition. It is thus an intriguing case of a "media conflict in science," reflecting the conflicting values of the epistemic robustness and newsworthiness of extraordinary claims inherent in the scholarly publication system (Franzen 2012). In this sense, reproducibility problems in science can be considered a symptom of the structural constraints on scholarly communication due to the internal logic of scholarly publishing. It is thus necessary to further explore why reporting research is not equivalent to doing research.

22.4 The Structural Gap between the Production and Representation of Scientific Facts

Science relies on the published record to facilitate scientific progress that takes new ideas into account. One of the key structural dilemmas of journal peer review is that it is based on textual knowledge for judging the reliability of empirical findings. By referring to scientific publications we can only look at the front-stage component of research, the representational side. The backstage, i.e., the knowledge production component, however, remains in the dark.

The integration of these two processes of research and its further interpretation into a scientific manuscript (Luhmann 1990) has two implications for scholarly communication. First, the presentation of results always exceeds what has been actually investigated. Knorr-Cetina (1981, p. 130) speaks of a gap between "the dynamics of the research process and the literary dramatics of the paper," in what she called the "double mode of production."

Based on the ethnographic approach and in line with Latour and Woolgar (1979) she identified numerous literary strategies used to persuade the reader of the actual scientific relevance of the presented work. While the laboratory studies aimed to show that science is structured by non-scientific factors that undermine the proclaimed epistemic authority of science, the distinction between the production and the representation mode of science can also help us to analytically distinguish science from non-science on the basis of its formal communication (Esposito 2005).

Moreover, a media theoretical perspective points to the selective nature of science reporting. Scientific publications cannot fully represent the workflow of knowledge production. Any scientific publication must adapt to fit the structural requirements of language (binarity), writing (sequentiality), and distribution media (intersubjective comprehensibility), which structure journalistic print forms as well (compare Franzen 2011, Chap. 2). In addition, scientific results are presented in somewhat rigid formats due to editorial standardization. This formal representation can restrict the possibility of reproduction, e.g., if relevant details are left out in order to adhere to space limitations or to produce a well-rounded paper.

Thus, the structural divergence between the representation mode of knowledge and the production mode of knowledge is one factor explaining the difficulty of replication. Obviously, the gap is even greater in the experimental sciences (than in other empirical sciences), in which the textual knowledge is always the result of a translation process of technical measurements and individual observations and perceptions. It is only feasible to verify claims based on the written text in disciplines such as mathematics, in which new results do not refer to the empirical world. Mathematical proof integrates both the production and the presentation side of knowledge (Heintz 2000).

In the empirical and experimental sciences, however, one has to be convinced of the scientific relevance based on how well the authors present the results. In order to judge whether the conclusions drawn in a paper are fully justified, the reader and the referee in particular must first of all trust the author that the findings presented were produced according to the methodological standards of "good science."

One factor explaining the problem of reproducibility in "doing science" lies in the constraints of "reporting science." These partly lie in the technical difficulties of fitting research results into editorial formats that do not necessarily facilitate replication. At the very least, detailed protocols are needed to reconstruct the claims that are made. In the past decades, however, the presentation of the workflow has been systematically reduced in favor of highlighting the outcomes (Berkenkotter and Huckin 1995).

But when detailed protocols are missing, it becomes even more difficult to interpret the results and to identify potential misinterpretations in reviewing a paper. When it comes to replicating other findings, not even the methodological description is sufficient, since tacit knowledge is needed (Polanyi 1966, Collins 1985 and in this volume). In some areas of research access to the original material such as cell lines or particles is necessary to reproduce results, preferrably under the supervision of the researchers who authored the original paper.

However, it is not possible to achieve certainty of the reliability of the proposed outcome by reading the manuscript. The ultimate proof lies only in the (re)production of the experiments. Some journals commission a preverification of results prior to publication in cases of high-risk research such as primate cloning to prevent "unwarranted speculation and controversy" (*Nature* Editorial 2007, p. 458) but this is far from being a routine editorial practice. Moreover, it contradicts the genuine function of a journal article of providing new but still tentative knowledge, and to open it up for contestation by its peers before it can acquire textbook status (Fleck 1981). The demand to certify scientific knowledge claims acquired an organizational shape in the recently founded Reproducibility Initiative, which offers a "certificate of reproducibility" for single papers.

22.5 The Increasing Awareness of Reproducibility Problems

The overall rate of false claims in the scholarly literature is unknown. What we can observe, however, is an increasing awareness of peer-review failures, reproducibility problems, and scientific misconduct in recent decades. Highly publicized cases of fraud together with the indication of high percentages of non-reproducibility of published findings, particularly in preclinical research (Begley and Ellis 2012), have increased the level of public scrutiny of science.

With the emergence of electronic communication technologies and the advent of Web 2.0, new media platforms and weblogs have been established in the last couple of years in order to monitor scientific activities and to red-flag any transgressions. Among these is the blog Retraction Watch, which encourages an investigation of questionable scientific behavior and editorial misuse in order to build up an informal repository of failures in the scientific publication record. It was launched in 2010 and now has a readership of 34,664 followers (September 2014).

Some stories attract widespread attention. Adam Marcus and Ivan Oransky, who run the blog, started to compile annual ranking lists of the "most memorable retractions," which also caters to media interest (see *The Scientist* "Top Ten Retractions 2013" from December 30, 2013). Retraction Watch collaborates with the science exchange weblog PubPeer to gain new insights into internal controversies that are being monitored.

Along these lines, journals and their publishing authors are facing a growing ex-post scrutiny of published research. The scholarly publication system, particularly in biomedicine, is increasingly being checked by (anonymous) peers investigating and reporting on malpractice in science by using new forms of public communication through the Web 2.0. While mass media communication was based on organizational structures and the journalistic profession, the use of social media is no longer restricted to organizations or professional roles. Any kind of content can be published online by anyone. The message does not need to be authorized; the Internet offers space for anonymous postings. The shift from print to electronic distribution technology, therefore, has implications for the public discourse on science.

So far, public allegations against scientific papers had to pass through the filters of news organizations. Journalists broke stories of questionable research practices or fraudulent papers, and whistleblowers had to work hand-in-hand with journalists to substantiate their allegations in order to make them public. Science journalists felt obliged to act as watchdogs for science when pointing to scientific misbehavior (Stollorz 2008). Now, public allegations against scientific papers can be made directly – even anonymously – and this seems to be happening with growing intensity. Moreover, media coverage maximizes the public dissemination of such information. While genuine whistleblower sites such as `science-fraud.org` are being taken down by lawsuits, one operator puts his (former) blog success in numbers (Brookes, cited in Pain 2014):

> Out of 275 papers discussed, there have so far been 16 retractions and 47 corrections. It's good, but it's not enough; there are many potential problems still unresolved.

A look into one open post-publication platform called `pubpeer.com` reveals the reasons for increased post-publication scrutiny: the erosion of trust among (young) scientists, especially in the life sciences. In an opinion piece of July 27, 2014, the blog operators reflected on the recent developments in post-publication peer review as follow:

> When we created PubPeer, we expected to facilitate public, on-the-record discussions about the finer points of experimental design and interpretation, similar to the conversations we all have in our journal clubs. As PubPeer developed, and especially once we enabled anonymous posting, we were shocked at the number of comments pointing out much more fundamental problems in papers, involving very questionable research practices and rather obvious misconduct. [...] We have come to believe that these comments are symptomatic of a deep malaise: modern science operates in an environment where questionable practices and misconduct can be winning strategies.

Participants in those blogs seem to share an eagerness to make violations of scientific integrity publicly known to enable a solid and fair assessment of scientific merits. Against this backdrop, journals are facing increased pressure to safeguard scientific quality prior to and after publication in order to restore trust in science to enable scientific progress and to improve the translation of research into practice (Laine 2007).

While displaying errors and mistakes in published literature used to be a rare exception, retractions of scholarly papers have exponentially increased in the past decade (Van Noorden 2011). It would be premature, however, to interpret these new figures as a rise in fraudulent behavior in science in the same time period. Causal explanations of this kind are difficult and they are even more difficult due to the notorious fuzziness of the concept of retraction as practiced in scholarly publishing.

As empirical studies have concluded, only one in four retractions comes under the banner of fraud (Nath *et al.* 2006, Steen 2011). There are just eight different reasons for a paper to be retracted, from journal error to outright data fabrication, as a content analysis of retraction notes highlighted (Steen 2011). Retraction counts would, thus, overrate the actual incidents of misconduct. In turn, the rate of research misconduct or irreproducibility would be systematically underrated by solely referring to retractions associated with reported cases.

What we can indeed confirm is an increased awareness of scientific integrity on multiple levels, for which the rising retraction rates are only one indicator. A key aspect of the discourse on reproducibility relates to the responsibility of scholarly journals to take on a gate-keeping role for science.

22.6 The New Transparency: Bridging the Gap in Scholarly Publishing

Journals walk a fine line, not only in manuscript selection but also in their interpretation of what kind of measurements should be applied to controversial papers. Highly publicized cases of scientific fraud in the past two decades have increased public pressure on peer-reviewed journals to ensure the quality of the scientific record. As famous fraudsters like Jan Hendrik Schön, Diederik Alexander Stapel, and Anil Potti have managed to publish fabricated data in journals of any rank, raising the standards of publishing has become a central topic in high-ranking journals.

With respect to the editorial handling of manuscripts to safeguard research quality, guidelines for publication ethics have been established. Formal guidelines give advice on many managerial aspects to increase both the level of fairness and of scientific quality in scholarly publishing. The bulk of journals in the biomedical and many other journals in the hard sciences are now committed to the code of conduct by COPE (The Committee on Publication Ethics) or other similar ones. Committees such as COPE, originally founded as self-help groups for coping with incidences of research misconduct, have since developed a life of their own and have acquired a mission to ensure the reliability of scientific findings.

In the light of public concerns on the peer-review crisis, journals have begun to request detailed information about author contributions, ethics committee approvals, or conflict of interest statements, but most notably they have intensified the checking of data ahead of print by the use of different measures to control the robustness of results ahead of print:

Scanning: randomly or systematically scanning for text plagiarism and image duplications,

Testing: reanalysis of statistical data by (in-house) statisticians,

Reporting: disclosure of all relevant methodological elements of the experimental design,

Witnessing: authors' confirmation requested that more than one person in the laboratory has reproduced the results,

Verification: prior-to-publication replication of the experimental results by an independent research team commissioned by the journal.

Papers are now subjected to routine scanning or random spot checking for image and figure duplications or manipulations, although specific journal policies still vary. In addition some journals are using scanning services to systematically detect text plagiarism. Authors are prompted by many journals to provide the primary data the paper is based on and to give access to the original protocols. Thus, the structural constraints on an adequate representation of scientific findings within publications are in a sense counterbalanced by the expansion of the idea of a scientific paper. In the digital age, tight space restrictions can be overcome. Authors are now requested to include more background information in the supporting online material that accompanies the scientific paper (Fig. 22.1).[5]

Such additional material is considered to improve the reliability of the prepublication peer review. The only way, however, to give proof of the veracity of results is their independent replication through empirical work. To safeguard against possible biases and to protect their role, some journals are now requesting a factsheet of statistical, methodological, and ethical details to be taken into consideration in the decision making process. *Nature*, for instance, has recently

[5]Despite the fact that a full representation of the research process is unattainable, the disclosure of all relevant data can meet with reluctance on part of the authors. As far as scientists are honored for originality, there is a general tendency for keeping relevant data aside to allow them for follow-up publications. The structural tension between disclosure and secrecy is nicely illustrated by the following quote of an interviewed physicist by Collins (1974, p. 176): "What you publish in an article is always enough to show that you've done it, but never enough to enable anyone else to do it. If they can do it then they know as much as you do." Thus, the new dictum of sharing data meets different concerns (Haeussler 2011).

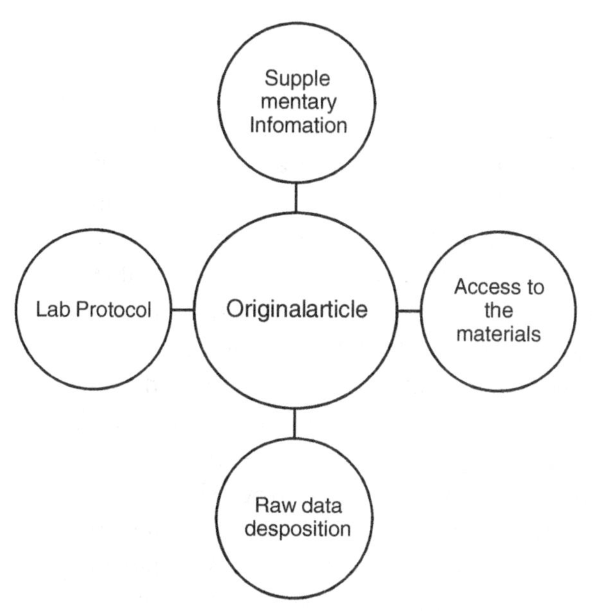

Figure 22.1: Types of accompanying materials for an original paper

introduced a specific Reporting Checklist for life sciences articles to guard against the lack of reproducibility.

Nature officially announced its new transparency guidelines to enhance reproducibility in 2013 (*Nature* Editorial 2013), and its American counterpart *Science* set up new guidelines in 2014 (McNutt 2014). Both have also modified their General Information for Authors to make clear that manuscripts are spot-checked for image manipulations and text plagiarism and that requests for materials, further experiments, laboratory protocols, or any primary data necessary to verify the conclusions may be required prior to acceptance. *Nature's* statement reads as follows (*Nature* Editorial 2013):

> To ease the interpretation and improve the reliability of published results we will more systematically ensure that key methodological details are reported, and we will give more space to methods sections. We will examine statistics more closely and encourage authors to be transparent, for example by including their raw data.

The barriers to getting papers on experimental data published are currently substantial.[6] However, if journals do not stick to their own policies, they compromise their credibility. This was apparently the case in the STAP saga where *Nature*

[6]Researchers complain that in the high-ranked journals, reviewers' requests that further experiments be conducted only impede the publication process, and they feel in danger of being scooped (Lawrence 2005).

was put to the test in the public debate. Editors from competing journal titles argued that the image manipulation spotted in one of the two papers could have been easily detected by using common software tools that were not used (Cyranowski 2014). Although many editors claim the "beautification" of images has become common practice (Pulverer 2014), image manipulation remains a first hint of questionable behavior. Once image irregularities in papers are publicly alleged, deeper investigations and more revelations follow.

Part of the editorial work is to decide what measures need to be taken if scientific findings turn out to be false. As no uniform guidelines for disclosing failures exist in the scholarly literature yet, journals decide on a case-by-case basis, a practice that is critically monitored by Retraction Watch. The retraction is only one way to display identified errors – there are others such as corrigendum. Public pressure from ongoing controversies has, however, contributed to increasing the level of withdrawals in order to clean up the published record. The accepted reason for a retraction is if data manipulation and unintentional errors affect the conclusions of a manuscript (Kennedy 2006). But here, we also find exceptions to the rule, as the following statement on the recent retraction of another controversial stem cell paper illustrates (*Nature* retraction note on Conrad *et al.* 2008, 21 August 2014):

> The new data have brought to light that the original conclusions are not as robust as presented in the original paper. Nature does not dispute the main claim that the cells are pluripotent to some level, but the level of proof of pluripotency shown is not in line with regular criteria for such papers in Nature. Consequently, the authors have agreed to retract their manuscript.

Against this backdrop, it is worth to note that paper retractions do not indicate that the results are invalid.[7] Invalid results, moreover, are either interpreted as "honest error" or "fraud," distinguished by the "intention to deceive." Although individual intentions are difficult to grasp, it makes a difference to which category non-reproducible studies are attributed. Irreproducibility of research is regarded as slowing scientific progress, but that it can be attributed either to "honest error" or to "fraud" entails its genuine social dimension. The label of non-reproducibility, often associated and formally marked by retractions, can have important consequences for the related authors and comes down to the issue of social exclusion from science.

[7] Plagiarism is an intriguing example of a common retraction type that is not concerned with invalidate results.

22.7 Conclusions

The problem of reproducibility mirrors the general constraints science faces due to its use of publications. The circulation of knowledge is bound to the materiality of the printed or electronic paper. Publications can only give a selective picture of the relevant pieces of information needed for the reproduction of the results.

The structural gap between the media-structured representation and the production mode of science affects all fields of research – but to a different degree. Experimental research in particular relies on the accurate presentation of results in the reporting. If doing the research and writing the paper imply different modes of production (Knorr-Cetina 1981), high-level standardizations in reporting should help to bridge the gap between the two production modes in the scientific paper.[8]

The structural gap widens when the editorial programs of the most relevant journals favor high-impact research. Authors tend not only to overstate the level of certainty in their textual explanations to increase the probability of getting their papers published. In textual explanations, there is much room left to raise the level of certainty by trimming the data or exaggerating the overall implications, in so far as the presentation of data always exceeds what has been investigated. Inconsistencies in reporting science can be one factor for the lack of reproducibility.

The implementation of new measurements in scholarly publishing to make the invisible visible highlights that the proof of validity of the claimed findings has become a main concern for reviewers and editors alike. If we consider the collective experiences of editors, reviewers, and peers, the trimming of data has become routine behavior. Tentative contradictions are ruled out in order to produce a well-rounded paper that can be placed into the pages of the high-impact journals that promise career advances (Lawrence 2003). Since notifications of reproducibility problems, particularly in the biomedical field, have been widely discussed in the general media as well as in online platforms, the authority of peer-reviewed journals has been tarnished.

Irrespective of the immanent reasons for the failure of reproducibility (Collins 1985), the recent debate on irreproducible findings is triggered by the normative expectations of good scientific practice. In this context, science as a social endeavor is faced with an increased public pressure to raise the level of reproducibility. High-profile cases of scientific fraud and a series of irreproducible

[8]In the nonempirical sciences and especially in the humanities, the two modes of production are more closely coupled.

research papers published and detected in eminent journals has apparently tarnished the credibility of science in the past decades.

Every single revelation of inconsistencies in the literature, however, erodes the journal's traditional gatekeeping role. For a long time, journal editors have insisted that science is based on trust. For this reason, they have convincingly argued that it would be too much to ask of a peer reviewer to identify potential forgers or to foresee any replication problems. Scientific fraud was in this sense framed as a black sheep phenomenon – leaving the core of science and professional academic practice nearly untouched (Franzen *et al.* 2007).

By extending the concept of scientific misbehavior from fraud (FFP: fabrication, falsification, plagiarism) to mundane transgressions of "normal misbehavior" in science (Martinson *et al.* 2005), the perspective has changed. The call to action has been referred back to the journals as the institutional gatekeepers of reliable scientific information. Journals are forced to take responsibility in order to avoid losing their legitimacy.

Meanwhile, the editorial control of data ahead-of-print constitutes a routine behavior, accompanied by the technical request to store raw material and share the original data in the realm of the open science movement. Although the standards of publication have been formally raised in the past decade, the revelations of reproducibility problems continued to increase. The emergence of the transparency regime, however, contributes to the erosion of social trust. Trust has been replaced by new measures of organizational control that puts scholarly communication into the realm of overbureaucratization.

As the general constraints of scientific papers in "telling the truth" are well known, the increasing trend towards revealing inconsistencies in the scientific literature cannot be explained in the factual dimension but in the social dimension of the meaning of science. The meaning of the scientific paper has shifted from being a means of communication to the "discovery itself" (Price 1981, p. 3). When, in the persistent evaluation culture, publications (instead of the acquisition of knowledge) are equated with reputation gains and even financial benefits (Adam 2002), this implies an inversion of means and ends that affects the motivational structure of science in two directions. On the one hand, the pressure to publish has increased in regards to quantity and journal rank. On the other hand, the post-publication scrutiny is on the rise in regards to the overall awareness of scientific misconduct.

The emergence of the reproducibility discourse can thus be explained by the perceived misdirection of science, concerning the evaluatory regime limited to publication output, facilitated and quickened by new means of communication in the digital age.

References

Adam, D. (2002): The counting house. *Nature* **415**, 726–729.

Begley, C.G., and Ellis, L.M. (2012): Drug development: Raise standards for preclinical cancer research. *Nature* **483**, 531–533.

Berkenkotter, C., and Huckin, T.N. (1995): *Genre Knowledge in Disciplinary Communication: Cognition/Culture/Power*, Lawrence Erlbaum, Hillsdale.

Broad, W.J., and Wade, N. (1982): *Betrayers of the Truth: Fraud and Deceit in the Halls of Science*, Simon and Schuster, New York.

Collins, H.M. (1974): The TEA set: Tacit knowledge and scientific networks. *Science Studies* **4**, 165–185.

Collins, H.M. (1985): *Changing Order. Replication and Induction in Scientific Practice*, Sage, Beverly Hills.

Conrad, S., Renninger, M., Hennenlotter, J., Wiesner, T., Just, L., Bonin, M., *et al.* (2008): Generation of pluripotent stem cells from adult human testis. *Nature* **456**, 344–349.

Cyranoski, D. (2014): Research integrity: Cell-induced stress. *Nature* **511**, 140–143.

Esposito, E. (2005): Die Darstellung der Wahrheit und ihre Probleme. *Soziale Systeme* **11**, 166–175.

Fleck, L. (1981): *Genesis and Development of a Scientific Fact*, University of Chicago Press, Chicago.

Franzen, M. (2011): *Breaking News. Wissenschaftliche Zeitschriften im Kampf um Aufmerksamkeit*, Nomos, Baden-Baden.

Franzen, M. (2012): Making science news: The press relations of scientific journals and implications for scholarly communication. In *The Sciences' Media Connection – Public Communication and its Repercussions*, ed. by S. Rödder, M. Franzen, and P. Weingart, Springer, Dordrecht, pp. 333–352.

Franzen, M., Rödder, S., and Weingart, P. (2007): Fraud: Causes and culprits as perceived by science and the media. Institutional changes, rather than individual motivations, encourage misconduct. *EMBO Reports* **8**(1), p. 3–7.

Heintz, B. (2000): *Die Innenwelt der Mathematik. Zur Kultur und Praxis einer beweisenden Disziplin*, Springer, Wien.

Haeussler, C. (2011): Information-sharing in academia and the industry: A comparative study. *Research Policy* **40**(1), 105–122.

Hirschauer, S. (2010): Editorial judgments: A praxeology of "voting" in peer review. *Social Studies of Science* **40**(1), 71–103.

Kennedy, D. (2006): Responding to fraud. *Science* **314**, 1353.

Knorr-Cetina, K. (1981): *The Manufacture of Knowledge. An Essay on the Constructivist and Contextual Nature of Science*, Pergamon, Oxford.

Laine, C. (2007): Reproducible research: Moving toward research the public can really trust. *Annals of Internal Medicine* **146**(6), 450–453.

Latour, B., and Woolgar, S. (1979): *Laboratory Life*, Princeton University Press, Princeton.

Lawrence, P.A. (2003): The politics of publication. *Nature* **422**, 259–261.

Luhmann, N. (1990): *Die Wissenschaft der Gesellschaft*, Suhrkamp, Frankfurt.

Martinson, B.C., Anderson, M.S., and de Vries, R. (2005): Scientists behaving badly. *Nature* **435**, 737–738.

McNutt, M. (2014): Reproducibility. *Science* **343**, 229.

Merton, R.K. (1942): Science and technology in a democratic order. *Journal of Legal and Political Sociology* **1**, 115–126.

Nature Editorial (2006): Let's replicate. *Nature* **442**, 330.

Nature Editorial (2007): Replicator review. *Nature* **450**, 457–458.

Nature Editorial (2013): Announcement: Reducing our irreproducibility. *Nature* **496**, 398.

Nature Retraction (2014): Generation of pluripotent stem cells from adult human testis. *Nature* **512**, 338.

Normile, D., and Vogel, G. (2014): STAP cells succumb to pressure. *Science* **344**, 1215–1216.

Obokata, H., Wakayama, T., Sasai, Y., Kojima, K., Vacanti, M.P., Niwa, H., *et al.* (2014): Stimulus-triggered fate conversion of somatic cells into pluripotency. *Nature* **505**, 641–647.

Pain, E. (2014): Paul Brookes: Surviving as an outed whistleblower. *Science Careers*, `sciencecareers.sciencemag.org/career_magazine/previous_issues/articles/2014_03_10/caredit.a1400061`.

Polanyi, M. (1966): *The Tacit Dimension*, Doubleday, New York.

Price, D.J. de Solla (1981): The development and structure of the biomedical literature. In *Coping with the Biomedical Literature*, ed. by K.S. Warren, Praeger Publications, New York, pp. 3–16.

Prinz, F., Schlange, T., and Asadullah, K. (2011): Believe it or not: How much can we rely on published data on potential drug targets? *Nature Reviews Drug Discovery* **10**(9), 712.

Pulverer, B. (2014): STAP dance. *EMBO Journal* **33**, 1285–1286.

Research Paper Investigative Committee (2014): Report on STAP Cell Research Paper Investigation, March 31, 2014, `www3.riken.jp/stap/e/f1document1.pdf`.

Steen, R.G. (2011): Retractions in the scientific literature: Is the incidence of research fraud increasing? *Journal of Medical Ethics* **37**(4), 249–253.

Steen, R.G., Casadevall, A., and Fang, F.C. (2013): Why has the number of scientific retractions increased? *PLOS ONE* **8**(7), e68397.

Stichweh, R. (1987): Die Autopoiesis der Wissenschaft. In *Theorie als Passion*, ed. by D. Baecker *et al.*, Suhrkamp, Frankfurt, pp. 447–481.

Stollorz, V. (2008): Ist der Platz zwischen allen Stühlen der richtige Ort? Essay über die Frage, was Wissenschaftsjournalismus heute soll. In *WissensWelten: Wissenschaftsjournalismus in Theorie und Praxis*, ed. by H. Hettwer *et al.*, Bertelsmann Stiftung, Gütersloh, pp. 566–582.

van Noorden, R. (2011): Science publishing: The trouble with retractions. *Nature* **478**, 26–28.

Vogel, G., and Normile, D. (2014): EXCLUSIVE: Nature reviewers not persuaded by initial STAP stem cell papers. *Science Insider* of 11 September 2014. `news.sciencemag.org/asiapacific/2014/09/exclusive-nature-reviewers-not-persuaded-initial-stap-stem-cell-papers`.

Ziman, J.M. (1969): Information, communication, knowledge. *Nature* **224**, 318–324.

PART VI
WIDER PERSPECTIVES

PART VI: WIDER PERSPECTIVES
Introductory Remarks

Sabine Maasen and Harald Atmanspacher

The final part of the handbook convenes articles that venture into wider perspectives concerning principles, practices, problems, as well as prospects with regard to the issue of reproducibility. In fact they are "wider" in three ways. First, they explore further domains of thought and practice, such as literature, philosophy of time and psychoanalysis, which provide us with notions and dimensions of reproducibility that remain unconsidered in other contributions to this handbook. Second, in order to cope with the variety of reproducibilities required in interdisciplinary settings, it is important to inquire into the epistemic and ontological question of how to identify system properties relevant for reproduction across disciplines. Third, science and technology take place in the wider context of society – a knowledge society utterly depending on science and technology. This urges us to reconcile constant innovation with reliable knowledge characteristic of contemporary knowledge societies.

Taken together, these perspectives provide deliberations that might help us to understand the notoriety of reproducibility as a "problem." The chapters in this part emphasize, albeit in highly distinct ways, the paramount importance of the counterparts of reproducibility and its neighboring concepts such as repetition. Whoever talks about reproduction also talks about producing not only the same but variance as well. In domains as different as literature, philosophy of time, psychoanalysis, stability of explanation across levels of description in interdisciplinary research, or science policy, reproduction cannot properly be thought of and/or practiced without considering or even integrating its opposite.

The general idea of considering both the same and the other is perhaps best visible in literary studies: As early as in 1963, Wellek and Warren stated that when it comes to entertain the reader "the totally familiar and repetitive pattern is boring." however, "the totally novel form will be unintelligible" (Wellek and Warren 1963, p. 235). In a similar vein, Steve Neale argued later that pleasure is indeed derived from "repetition and difference" (Neale 1980, p. 48). Radicalizing these points of view for the practice and analysis of literature, *Ladina Bezzola Lambert, Department of English at the University of Basel,* emphasizes that reproducibility as a generalized norm in science needs to be reconsidered when it comes to the social sciences and humanities.

Literature (as well as social media) depend on their readers' context, be it personal, situational, cultural, or historical. Even more fundamentally, literature rests on language that, whenever used, is reproduced with a difference. This holds true even where reproducibility is explicitly aimed at. For instance, the most scientific branches of linguistics, following de Saussure, do not support reproducibility but rather a culturally contextualized norm of *intersubjective comprehensibility* (also required by so-called qualitative approaches in the social sciences). Likewise, literary theory highlights the concept of mimesis: the skillful imitation of an artistic role model. It is revealed as an artistic practice that, at the same time, respects the past and self-assertively accomplishes a genuine literary artifact. Seemingly perfect imitation and expertly crafted innovation go hand in hand.

Furthermore, literary translation indicates forcefully that, even when done most painstakingly, it can never reproduce the original. Whether using the strategy of domestication (faithful reproduction) or foreignization (making particularities of language and context apparent), a translation will remain derivative. This is not least due to the fact that cultural context matters – or the writer (imitator or translator), the reader, or the literary scholar. This again points to the limited extent of reproducibility within literature and across cultural domains as well as to intriguing interactions between all elements of the literary network.

Hinderk Emrich, Psychiatric Clinic at the Medical School of Hannover, reframes the notion of reproducibility within yet another context. He inquires into the dialectic context of remembering and forgetting as conceptualized by psychoanalysis, based upon insights from the philosophy of time and practiced, most prominently, in trauma therapy. This move implies an important shift from earlier accounts that draw a link between *psychoanalysis* and reproducibility. For instance, Popper (1972) famously considered psychoanalysis a pseudoscience because of its impossibility to be falsified, thereby presupposing a nomothetical understanding of science including, *inter alia*, the independence of the subject, objective measurability, and reproducibility.

By contrast, Emrich starts from a pronouncedly different angle, namely, from the notion of repetition itself. In actual fact, this offers two kinds of interpretation, one simply reiterating a procedure, another one "requesting" something or even "taking" it back. In the philosophy of time and in neuropsychology, phenomena of remembering play a key role: here, the act of remembering is conceived as a constructive mental act. The actualization of the past within the process of remembering induces experiential contents that result in emotional manifestations of the past, shaping, if not determining, the remembering person's present.

This is particularly important in psychoanalytic trauma treatment. In contrast to splitting, repression, or denial, this approach relies on repetition as "taking back" and proceeds by "actualizing deactualization." The traumatic event must first be activated and recognized before it can be resolved. The corresponding coping process thus implies a constitutive dialectic of the urgent need to forget and the necessity to remember. In this respect, co-affection is of highest significance when trying to remember the forgotten or to forget the remembered. The repetition of a trauma ultimately helps "to take it back" by way of mimesis, i.e., the capacity to experience reality from the perspective of the other, or rather the otherness of his or her co-affected self.

On a methodological note, *Harald Atmanspacher, Collegium Helveticum Zurich,* addresses the particular difficulties of reproducibility in interdisciplinary research, already within the natural sciences. While scientists from different disciplines usually agree about the possibility of discovering "true" aspects of nature, and about what sort of demonstration will count as valid, they differ as to what counts as relevant to be reproduced in their respective discipline.

Atmanspacher proposes to distinguish the *relevance* of attributes of the system in question as depending on the level of description targeted by individual disciplines. As a rule, different descriptive levels go along with different degrees of *granularity*: Lower-level descriptions address systems in terms of micro-properties, while higher-level descriptions address macro-properties. For interdisciplinary research it is thus important to identify the criteria for describing the relevant attributes of the system for the involved disciplines. For example, temperature is a relevant property at a thermodynamical description, while the individual molecules are relevant at a mechanical level.

Under the label of "contextual emergence," introduced by Bishop and Atmanspacher (2006), he suggests an empirically applicable procedure to construct level-specific criteria for relevant attributes across disciplines and then to identify adequate interlevel relations. This procedure is explicitly designed to avoid both unwarranted reductionism and relativism. Instead, and following Quine's plea for "ontological relativity" and Putnam's "pragmatic realism," the idea is that once features are identified as relevant for a proper explanation of an observation, one has also identified the features that are relevant for its reproduction.

The final chapter puts the issue of reproducibility in science and technology in the context of contemporary *knowledge societies* that are characterized by a constant quest for innovation. Given that innovation privileges newness and, by implication, the not-yet reproducible, *Sabine Maasen, Center for Technology in Society at the Technical University Munich*, asks how, if at all, these opposing requirements can be reconciled in scientific practice. Radicalizing a position held by Rheinberger (1997), she identifies both reproducibility and innovation as representing a regulative dual that cannot be met in a straightforward fashion, but needs to be orchestrated both epistemically and socially.

For quite some time, this has been safeguarded exclusively by the scientists' general compliance with the so-called Mertonian norms of a scientific ethos. They refer to a set of implicit rules that operate mostly implicitly and are thought to establish trust among scientists and enforce the reliability of knowledge production. However, at times of accelerated innovation, carefully arranged empirical settings emerge that complement and challenge these norms. Prototypical examples are the plea for "social robustness of science" and schemes of "responsible research and innovation." The corresponding procedures render the implicit norms explicit and, hence, can be regarded as instances of institutionalized reflexivity in science and as increased societal "responsiveness" of science.

However, these practices of responsiveness may turn out to be a double-edged sword: While they are designed to effectively balance the need for innovation and reliable knowledge, the underlying "pressure of practice" (Carrier 2011) effectively threatens the production of reproducible knowledge – an ambivalence that will most certainly yield more sound science programs.

Summarizing, the prospects entailed by the diversity of these perspectives do have a least one common denominator: the joint conviction that reproducibility comes in degrees and aims at identifying and balancing internal and external challenges. Inevitably though, it is precisely due to constant challenges by internal (e.g., epistemic or technical) or external (e.g., political) factors that the reflection on reproducibility is bound to reproduce itself (albeit with differences)!

References

Bishop, R., and Atmanspacher, H. (2006): Contextual emergence in the description of properties. *Foundations of Physics* **36**, 1753–1777.

Carrier, M. (2011): Knowledge, politics, and commerce. Science under the pressure of practice. In *Science in the Context of Application*, ed. by M. Carrier and A. Nordmann, Springer, Dordrecht, p. 11–30.

Neale, S. (1995): Questions of genre. In *Approaches to Media*, ed. by O. Boyd-Barrett and C. Newbold, Arnold, London, pp. 460–472.

Popper, K. (1972): *Conjectures and Refutations. The Growth of Scientific Knowledge*, Routledge, London.

Rheinberger, H.-J. (1997): *Toward a History of Epistemic Things: Synthesizing Proteins in the Test Tube*, Stanford University Press, Stanford.

Wellek, R., and Austin, W. (1963): *Theory of Literature*, Penguin, Harmondsworth, Chapter 17.

23
Repetition with a Difference: Reproducibility in Literature Studies

Ladina Bezzola Lambert

Abstract. Reproducibility is commonly defined as a general scientific norm established to assure control over scientific claims. In the experimental sciences, reproducibility more specifically indicates the requirement for experiments to be repeatable with the same results irrespective of place and time (Mittelstrass 1992, p. 61). The inquiry into the significance of reproducibility in literature studies shows that this "general scientific norm" requires reframing and renegotiating as in the processes of literary production and reception, time and place cannot be blinded out, but rather prove essential factors in the production of meaning. In consequence, replication is never total.
The application of the concept of reproducibility moreover needs to be extended from the critical engagement with literary texts to literary production itself and to its medium, language. Language, as the linguist Ferdinand de Saussure argued, is based on difference, but language also creates difference. As words are repeated, their meanings change. The essay illustrates this principle in connection with a number of literary and critical practices where reproducibility is an acknowledged aim.

Rose is a rose is a rose is a rose.
(Gertrude Stein)

23.1 Introduction

Language is in many ways a prototypically reproductive medium: words are meaningful as a consequence of their continual requotation within a linguistic community. Yet their meaning is inflected (and, hence, renegotiated) by the specific contexts in which words are used. Written language reproduces oral speech, making it reproducible in other places and times. Yet language thus resituated acquires new meanings as it meets with different audiences in different cultural contexts. Writing practiced as a form of art, commonly termed "literature," seeks to defy the reproducibility of its medium by using language in an ostensibly original way. Yet an artistic use of language evidently remains tied to the laws of intelligibility and cultural communities and to recognizable genres and modes that at once facilitate and influence reception. Its links to a literary and

Reproducibility: Principles, Problems, Practices, and Prospects, First Edition. Edited by Harald Atmanspacher and Sabine Maasen.

more broadly cultural tradition moreover simultaneously offset and give meaning to the notion of originality, which evidently depends on the definition of a benchmark.

While language, writing, and literature are thus closely associated with reproduction, they defy stabilization and control. As social media, their use is context-dependent and inflected by historical processes. They cannot be treated as stable, neutral entities. In consequence, reproducibility and repetition can only be productive concepts in the context of literature studies if, rather than aiming at non-ambiguity and identity, they allow for variability and multiplicity. Language that is reproduced is always reproduced with a difference.

With considerable foreshortening, we may take for granted that extant texts are perfectly reproducible in later times and other places,[1] but when it comes to reproducing their significance, we are facing quite a different challenge: Meaning production is always determined by the individual reader's personal history and particular experience of the world. With the disappearance of the original cultural context in which a literary text originated and that, to some extent, channeled its reception, the factors that diversify reception increase.

The following observations are structured in four parts: after an introductory section concerned with language and signification in relation to reproduction and difference (Section 23.2), the focus is set on three practices associated with literature and literature studies where reproducibility plays a central role: literature as imitation of reality or earlier literature (Section 23.3); literary translation as a special reproductive technique in the domain of literature (Section 23.4); reproducibility in the hermeneutic process of reception (Section 23.5). The strong interrelations of these three practices will become evident in their discussion.

23.3 Language and Difference

The discontinuities of reproductive practices in the field of literature and literature studies are rooted in the fact that meaning is not intrinsic to language, does not inhere in a literary text; meaning is rather assigned to a text by the reader and critic. It, therefore, cannot, strictly speaking, be reproduced from extant writing, but always needs to be produced anew from the reader's ever shifting cultural and personal vantage point.

[1] Nevertheless, it evidently makes a difference whether a text is handwritten on a piece of vellum or paper, printed in a book, or downloaded in digitalized form on a tablet. It also makes a considerable difference whether a text is printed as facsimile copy of the original publication with the original spelling retained or in modern, standardized form. In recent scholarship, such issues have received increasing attention.

From the point of view of linguistics, this Heraclitean principle may be traced back to Ferdinand de Saussure's theory of language as developed in his *Cours de linguistique générale* (1916), the text that generally represents the starting point of modern linguistics and, from there, of the semiotic approach to a wide array of cultural phenomena. Saussure described the discipline of *semiology* (today more commonly referred to as semiotics) as "a science which studies the role of signs as part of social life" (de Saussure 1983, p. 15f).[2]

He first of all defines language as a social product and, secondly, as a system based on relations of difference. Accordingly, particular words develop from a particular society's needs, rather than resulting from a need to name pre-existing concepts. Moreover, the relation between the two parts that in Saussurian theory constitute the sign – the sound image (*signifiant*) and the concept it denotes (*signifié*) – is arbitrary and not fully identical among different speakers of the language, whose individual life experiences and cultural backgrounds lay the basis for different conceptual patterns.

No sign means by itself; it only conveys meaning through its relation to other signs. To quote a classical example in illustration: the meanings associated with the word "house" change with the addition of words like "palace," "hovel," "hut," "cabin," and so on. The value of a sign (*valeur*) is thus not substantially, but differentially defined. This entails that any addition to or elimination from the overall system of signs affects the larger conceptual framework that language constitutes. Language change thus also involves more wide-ranging conceptual change, a fact that directly affects the reproducibility of meaning.

The differential principle enabled the fruitful transfer of Saussurian theory to all forms of communication and it explains the theory's seminal role in the development of the humanities in the 20th century. Literature studies, in particular, took an emphatically scientific turn in its wake. Nevertheless, even those branches of literature studies which, under the influence of Saussurian theory, strongly aspired to the ideal of an "objective science," like formalism and structuralism, do not support a strict distinction between "right/reproducible" and "wrong/irreproducible" but allow for several – if, clearly, not any number – of parallel explanations. Rather than reproducibility in the strict sense, the criterion is the intersubjective comprehensibility (*Nachvollziehbarkeit*) of claims, a criterion that is evidently tied to a shared culture.

This hermeneutic ambivalence marks the genesis of Saussurian linguistic theory itself. The *Cours* is not actually de Saussure's own work; it is a text

---

[2]Saussurian linguistic theory had an interdisciplinary orientation from the beginning, and it is no coincidence that the term *semiotics* originates from medical discourse, where it referred to the interpretation of bodily symptoms.

based on lecture notes by two of his students who had not personally attended the lecture course, but relied on notes taken by colleagues. The foundational text of Saussurian linguistics may then be portrayed as an attempt at two removes to reproduce the gist of Saussure's lectures in written form. This is worth pointing out here because the *Cours'* tortuous genesis results from what may be described as Saussure's quite pathological aversion against formulating his ideas in writing. This aversion seems to have sprung from a strong sense of failure to reproduce "the author's conclusive thoughts," his awareness that both the system of language and history rendered such an endeavor impossible.

Saussure's project in the *Cours* to characterize the basic logic of "language in general" further conflicted with his personal propensity for a historical approach to specific linguistic phenomena focusing on idiosyncrasies and linguistic development. The incompatibility between the two approaches may be said to lie at the basis of the concept of reproducibility in literature studies. While, from a generalist standpoint, reproduction focuses on *stasis* and claims an objective, neutral point of observation, within the domain of communication, it involves change in the very process of reproduction. In communication, repetition destabilizes rather than stabilizes meaning.

The famous quotation from the poem *Sacred Emily* by Gertrude Stein (1874–1946) in the epigraph may be used to illustrate this claim: *Rose is a rose is a rose is a rose.* As the reader of the repetitive assertion proceeds from one rose to the next, the perspective on the word "rose" and what it denominates changes. As Stein herself claimed in a comment on this line, an effect of the insistent repetition may be that of overcoming the hollow routine which, over the centuries, has weakened a formerly powerful symbol to a merely denominational concept. The repetition, she claimed, had the effect of bringing the rose back to life in that it conveyed to the reader a sense of "continued presence" and "pure being."[3]

In the quoted line, repetition thus challenges the role of referential language, the authority of words to refer to things, but then powerfully recreates the proximity of things in language. The repetition also destabilizes meaning in that it brings about a proliferation of the concepts the word could denote: the actual flower called rose, a woman's name Rose, the word "rose," the focused engagement with the idea of a rose, the concept of the rose, and its central role in the

[3]Stein's famous reply on being asked about the line in a lecture: "Now listen! I'm no fool. I know that in daily life we don't go around saying 'is a ... is a ... is a ...' Yes, I'm no fool; but I think that in that line the rose is red for the first time in English poetry for a hundred years" (Stein 1947, p. iv).

debate about universals.[4] The effect then is not just the critique or affirmation of the concept of denomination, but the destabilizing of reference.

The quotation from Stein's poem has had a significant afterlife. The line's fame has lent it a monadic character. It has detached it from its original context in *Sacred Emily* (1913) to a degree that the very fact that the line was part of a long poem has been erased in the popular imagination. This brings up another crucial point about reproduction in a language context. Since quotation, the reproduction of words elsewhere, is always partial reproduction, it brings about a change of context that also affects the meaning of words quoted. In consequence, not even faithful quotations are strict reproductions.

Through reproduction, Stein's line has assumed a life of its own also as parody. Parody, one of the principal literary modes concerned with the concept of reproducibility, depends on the fundamental incompatibility of imitation and model and on partial reproduction. Linda Hutcheon has aptly defined it with the phrase "repetition with a difference" (Hutcheon 1985), which I have used to characterize the concept of reproducibility for literature studies. Parody may involve different types of imitation, ranging from irreverent or critical to playful or, more generally, self-conscious imitations of a model. Structurally, parody depends on the reader's recognition of the text as an imitation of a model he or she is familiar with. However, the interpretation of parody may depend on additional intertextual links that may or may not be available to the reader, but which, to contribute to the parodic effect, need to be intentional on the side of the author.

Ernest Hemingway's parody of Stein's line is a case in point: "a stone is a stein is a rock is a boulder is a pebble" first appears to be merely a playful parody of Stein's original line based on the semantic meaning of her name. However, when related to two different contexts, mutually incompatible meanings become available. On the one hand, Hemingway's range of words with related meanings (stone, stein, rock, bolder, pebble) suggests a reference to Saussure's theory that the semantic meaning of words is based on difference. Referred to Stein's original model, this first intertextual link then gives testimony to the interpretive diversity (rather than the tautology) produced by Stein's repetitions and acknowledges her authority as a writer. The reference thus lends Hemingway's parody a complimentary inflection. The opposite is the case with the second intertextual link, the burlesque Latin lesson in William Shakespeare's *The Merry Wives of Win-*

[4]The philosophical debate about universals concerned the question whether general concepts like "man" or "rose" have real existence or whether they are mental constructions. In the Middle Ages, the "name of the rose" was an often used example in the debate. This is also the motivation for Umberto Eco's novel with this title.

sor, which Hemingway's semantic use of Stein's name and the "pebble" strongly suggest to be intentional (Shakespeare 1997, 1272 — 4.1.26-31):

> [Schoolmaster] Evans: What is *"lapis"*, William?
> William: A stone.
> Evans: And what is "a stone", William?
> William: A pebble.
> Evans: No, it is *"lapis"*. I pray you, remember in your brain.
> William: *"Lapis"*.

The reference to Shakespeare's pedantic schoolmaster gives Hemingway's parody of Stein a decidedly less flattering inflection, as it places her repetitions in the same category as the tautological rant of Shakespeare's pedantic schoolmaster. This shows that how we interpret parody strongly depends on the contexts or cotexts we share (or believe to share) with the author, and also that the meaning of Hemingway's line is not carved in stone. It rather oscillates between the models the line may or may not invoke and depends on how the parody is seen to relate to these models.

A more stable and reverential relation to a literary model may be found in the ancient practice of *imitatio*, the skillful imitation of an artistic role model, and in literary translation, the two domains to which we now turn.

23.3 Mimesis, Imitatio, and Parody

Consideration of the role of reproducibility in literature and literature studies requires a reference to the theory of *mimesis*, one of the oldest and most fundamental concepts in literary theory. A term of wide-ranging aesthetic, psychological and social meanings (compare Emrich, this volume), "mimesis" may here be narrowed down to a definition of art as fundamentally imitative, as a representation of something in the real world. Most forms of Western art beyond realism set art in relation to reality, even though, as Roland Barthes among others has pointed out, "the real" is itself a "conceptual category." It is posited as that which just is, beyond meaning, while its copy in art that produces a realistic effect involves an act of interpretation (Barthes 1986, p. 148, Potolsky 2006, p. 99).

Such an imitative definition of art raises the question whether the imitation produced in and through art represents a copy of the world "as it is" or whether it rather "defines art as a self-contained 'heterocosm' that simulates the familiar world, and in effect copies our ways of knowing and understanding things" (Halliwell 2002, p. 5, Potolsky 2006, p. 3). These two possibilities follow the different characterizations of mimesis given by Plato and Aristotle, who, according to the classicist Stephen Halliwell, might be said to stand at the outset of the two major strands of Western thought characterizing the nature of art and its use of imitation.

Plato, who first introduced the concept of mimesis in his dialogue *Republic*, described art as an illusion copying reality outside. The copy is illusory and untruthful on account of its shallowness; it is, moreover, dangerous because it lures the human mind away from an engagement with truth. Where Plato held that art "merely" imitates what is, even as he warned of the deceptive and manipulative potential of such an imitation, Aristotle adopted and radically revised his teacher's theory of mimesis in his *Poetics*, one of the founding texts of literary theory.

Rather than denouncing artistic imitation as removed from truth and demagogical, as Plato had done, Aristotle emphasized the formative role of imitation for human cognition and psychology. In his view, art does not just copy the real world, but offers a condensed and structured imitation of reality in the form of a unified, conclusive action that conveys the human condition in a meaningful way. Against Plato's presentation of art as a form of deception, Aristotle characterized art as the product of both rational and ethical artistic choice and as congruent with the norms of human cognition (Potolsky 2006, pp. 36ff).

A second crucial distinction to be made in relation to the theory of mimesis concerns the question whether art imitates the real world or other art. The concept of art as a reproduction of reality underlying the Greek term "mimesis" was supplemented by the Roman version of mimesis termed *imitatio*, which referred to the imitation of artistic role models. First developed from word by word translations of Greek texts into Latin, the practice gradually developed into a more self-assertive creative mode that today we might more properly term adaptation. In this literary or more broadly artistic practice, Greek works of art were translated, emulated, and competed with in an attempt to link up to a revered cultural tradition and, in a second step, to achieve cultural maturity through imitation.

The seminal importance of this practice for the history of art cannot be exaggerated. As Matthew Potolsky points out, *imitatio* constitutes an artistic practice that from the height of the Roman empire until the late 18th century essentially determined cultural production in the West and did not then disappear from view either. Beside its cultivation in the Roman empire, it had its peak in Renaissance culture, which, through the imitation of classical art, strove to confer authority on vernacular cultures through the imitation of classical works of art, which in this process only came to assume the position of classics. The practice depended on a conception of authorship as imitation: the skillful imitation of earlier writers was seen as the key to literary success.

At the same time, authorial status depended on the ability to make something new out of old models (Potolsky 2006, p. 50, Greene 1982). In the practice of *imitatio*, imitation as mere copying or crude translation was emphatically distinguished from more ambitious and competitive emulation (*aemulatio*). The 12th-century author Pierre de Bois offered an effective summary of this particular combination of emulation and competition, in an image often quoted in the context of the empirical sciences to characterize the structure of scientific progress: He defended the backward looking engagement with ancient literature with the famous image of "dwarfs mounted on the shoulders of giants; with their assistance we can see further than they can" (quoted in Potolsky 2006, p. 59). In its original context, the image suggested that artistic production without imitation is unthinkable, but that imitation necessarily involves adaptation and change.

The paradigmatic example of literary *imitatio* as an artistic practice that is reverential toward the past and self-assertive in terms of its own artistic abilities may be traced from Vergil's adaptation of the Homeric epics in the Aeneid to Dante's Christian adaptation of the Virgilian model in the *Divine Comedy*, John Milton's appropriation of Catholic Christian epic in the context of English protestant culture in *Paradise Lost*, and James Joyce's return to and deviation from Homer in *Ulysses*, which reproduces Odysseus' years of wandering through the course of a single day in modern Dublin, the 16th of June 1904, as experienced by the "wandering Jew" Leopold Bloom.

In terms of its relation to the concept of reproducibility, Joyce's *Ulysses*, published in 1922, provides an instructive contrast to another modernist landmark of literature in English published in the same year: T.S. Eliot's long poem *The Waste Land*. Notoriously, both works make eclectic use of cultural allusions and have sent generations of readers and critics on a research quest in an attempt to identify a plethora of sources.

Yet they differ fundamentally in the attitude they display toward the elements of culture, both remote and contemporary, which they reproduce. *Ulysses* is characterized by a belief in the act of cultural translation as fruitful metamorphosis, in the possibility of telling the story of the wily Ulysses and his heroic adventures again, on a much smaller scale and with an eye focused on the minutiae of daily life. In this new context, the effect of the Homeric echoes ranges from the comic to the profound.

By way of contrast, *The Waste Land*, as its title suggests, uses the cultural remnants as debris that has become utterly meaningless for latter-day humanity.

Thus, Eliot adopts the beginning of Chaucer's *Canterbury Tales* (written in the late 14th century) for the opening of his poem but replays the promising note of Chaucer's forward-looking verses in a chopped and ominous tone. The wanderlust provoked by Chaucer's sweet April showers and soft breezes is aborted in a stifling deadlock:

Whan that Aprill with his shoures soote The droghte of March hath perced to the roote, And bathed every veyne in swich licour Of which vertu engerdred is the flour; Whan Zephirus eek with his sweete breeth Inspired hath in every hold and heath The tendre croppes, and yonge sonne Hath in the Ram his half cours yronne, And smale foweles maken melodye, That slepen al the niyght with open ye (So priketh hem nature in hir corages), Thanne longen folk to goon on pilgrimages (Chaucer 1988, 23, ll. 112)	When in April the sweet showers fall And pierce the drought of March to the root, and all The veins are bathed in liquor of such power As brings about the engendering of the flower, When also Zephyrus with his sweet breath Exhales an air in every grove and heath Upon the tender shoots, and the young sun His half-course in the sign of the Ram has run, And the small fowl are making melody That sleep away the night with open eye (So nature pricks them and their heart engages) Then people long to go on pilgrimages. (Chaucer 1986, 19)

April is the cruellest month, breeding
Lilacs out of the dead land, mixing
memory and desire, stirring
Dull roots out of spring rain.
(Eliot 1970, 1000, ll. 14)

Eliot's imitation of the Chaucerian model is a parody in the literal sense of that word, a "counter-song." This form of repetition with a difference does not ridicule its model but allows it to stand firm as an enduring work of English literature by the first poet buried in the poets' corner in Westminster Abbey; while it invokes its model with reverence, it insists on the incompatibility of the different cultural moments, expressing doubt about the earlier poet's faith in an unencumbered act of poetic creation, inspiration as a spring breeze, as it were.

By the same token, Eliot commits himself to what might be called a poetics of cultural encumbrance, together with the poem's exalted dedicatee, Eliot's contemporary Ezra Pound, who is addressed with the headline *"Il miglior fabbro"* ("the better blacksmith"). The quotation from Dante's *Purgatory* imitates the great Tuscan's reverential gesture to the 12th-century troubadour Arnaut Daniel, thereby subscribing to a poetry that is not a convivial pastime, but the product of hard work.

23.4 Literary Translation: Domesticating versus Foreignizing

As we have seen, the literary imitation of role models corresponds to a form of free translation or adaptation that initially had developed out of the effort to make the texts of an earlier, highly civilized foreign culture accessible and, in a second step, aimed to exploit them for the cultivation of an indigenous literary culture. The cultural rivalry characterizing *imitatio* promotes the imitation's emancipatory detachment from the model with a view to lending it artistic authority of its own. In the wider context of literary translation, the liberties taken in this process may, however, also be said to depend on a particular conception of faithfulness toward the model. Accordingly, different theories of translation disagree as to what faithful reproduction involves.

For interlingual translations, i.e., translations from one language into another,[5] reproducibility is a central concern even if, given the particularities of individual languages, total reproduction is evidently impossible and it remains subject to debate how the level of reproduction may be maximized. In the case of literary translation, the situation is more accentuated because, as Walter Benjamin observed in an important essay on translation, literature (and art more generally) is not about transmitting information, it is not meaning *for* the reader and, therefore, does not communicate.

By contrast, the translation of a literary text is produced to communicate to readers who do not understand the original. For Benjamin (2004, p. 75), this "seems to be the only conceivable reason for saying 'the same thing' repeatedly." This points to "a divergence of their standing in the realm of art" and makes clear that a translation can never assume the same position as an original; it will always, to some degree, remain derivative. Moreover, the aspiration to reproduce a given literary text as fully as possible faces the challenging task of transferring it from one cultural context to another. In translation theory, there are two basic approaches to this undertaking: domesticating and foreignizing translation.[6]

---

[5]Roman Jakobson distinguishes between (1) intersemiotic translation, which involves an original formulated in language and a translation/adaption that is not linguistic; (2) intralingual translation (translation between different registers of one language); and (3) interlingual translation (translation from one language into another). See Boase-Beier (2011, p. 4).

[6]Translation theory as systematic discipline is of relatively recent date. Commentary on translation in older texts is offered in the form of passing remarks, not as a systematic argument (Venuti 2004, p. 13). In the 16th and 17th centuries, which beside translations from Greek and Latin saw many translations from one modern vernacular into another, competing views about the right mode of translation begin to enter into dialogue with each other and to compete also in terms their ambitions to win cultural authority.

In the first case, the translation aims to exert its force in its native culture as a literary text in its own right. Rather than serving at the table of the great, it claims a seat at this table, as an autonomous work of art akin to the original. This mode of translation leans toward the practice of *imitatio* or overlaps with it. The difference between the two is essentially a difference in purpose: Its artistic ambition notwithstanding, the domesticating translation is not produced in an emancipatory gesture to compete with the original but to reproduce it faithfully in a work of art that can measure up to that original. It typically belongs to a cultural phase where the target language is felt to have already achieved cultural maturity, such as the literary culture of 17th-century France.

As Lawrence Venuti observes, the French translator Nicolas Perrot d'Ablancourt (1606–1664) is exemplary for the domesticating approach. It is his maxim that "diverse times require not only different words, but different thoughts" (Venuti 2004, p. 16). Translation is thus characterized as a form of interpretation with an acute sense of historical difference. D'Ablancourt initiated a tradition of translation that was soon to be called *"les belles infidèles,"* beautiful and unfaithful, as, following Voltaire's *bonmot*, good translations were compared to attractive women. Viewed from a different angle, the *"belles infidèles"* may also be considered as faithful reproductions of the original for the way they attempt to preserve the unity of the impression the work made on its original readers.

In an English context, Alexander Pope's widely popular 18th-century translations of Homer's epic into *heroic couplets* (rhyming iambic pentameter) present a classic example of this approach to translation. A generation before him, John Dryden (1631–1700), another influential English translator and translation theorist, had criticized imitation for disappointing those "inquisitive to know an Authour's thoughts", shrewdly observing that "'tis not always that a man will be contented to have a Present made him, when he expects the payment of a Debt" (quoted in Venuti 2004, pp. 38f).

In Germany, in the late 18th and early 19th century, there is a countertrend to domesticating translations, which are explicitly criticized as products of French self-sufficiency. Accordingly, Gottfried Herder (1744–1803) and Friedrich Schleiermacher (1768–1834) promoted a foreignizing mode of translation that does not appropriate the original but rather attempts to guide the reader to that work in a way that makes the particularities of its language apparent in the target language. Schleiermacher offered the most elaborate formulation of this approach. First, he emphasized the central role of translation not only between different languages but also to enable communication between different historical moments, different linguistic registers and social classes, even between one person's changing attitudes over the course of time. He then emphatically distinguished between pragmatic translation or interpreting "in the area of business", which is highly conventionalized, and literary translation, which is concerned with a use of language that lies beyond convention and is strongly determined by an "author's own particular way of seeing and drawing connections."

For Schleiermacher, language as "self-expression" or expression of "independent thought" was the distinguishing feature of literary art (Schleiermacher 2004. pp. 44ff). This use of language must be understood in two different senses: in terms of the thoroughly irrational "genius of language" and in terms of "the speaker himself, as a use of language that can only have emerged out of, and be explained as a product of, his particular being."

Language, he argued, does not express thought and meaning transparently, but shapes them according to linguistic structures and cultural traditions. On the other hand, "every free-thinking, intellectually independent individual shapes the language in his turn" and for this reason alone deserves to be "heard outside his immediate sphere of activity" (Schleiermacher 2004, pp. 46f). This active interference with linguistic convention may become invisible as the language develops over time, but it is also what the skilled translator aims to make visible in translation.

An adept translator needs to understand an author's particular relationship to the language he writes in at a particular historic moment and try to convey it in translation. This is best done in a style that makes a writer's linguistic idiosyncrasies palpable through a foreignizing style and, in so doing, contributes to the development of the target language by detaching it from conventional usage. Foreignizing translation is here imagined as a nationalist practice that may contribute to the development of German language and literature, emancipating it from foreign (mainly French) cultural domination.

The mode of translation Schleiermacher advocated thus shares the emancipatory impetus of the ancient tradition of *imitatio*. It requires from the translator a personal intellectual contribution, which is essentially an act of reading and interpretation that defines what essentially constitutes the original work. As personal interpretations, translations of the same text evidently can take diverse forms; there is not one definitive reproduction, but each has only relative and subjective value (Venuti 2004, p. 19).

Walter Benjamin revived Schleiermacher's notion of defamiliarizing translation and the assumption that "language is not so much communicative as constitutive in its representation of thought and reality" (Venuti 2004, p. 71). Translation ought not to reproduce "sense" (*das Gemeinte*), but the "mode of signification" *(die Art des Meinens)*, the author's particular way of reaching beyond linguistic conventions. The quality of a literary translation then depends less on a correct rendering of semantic content than on the degree to which the poetic quality of the original is recognizable in the target language. Changing parameters, such as the temporal distance to the context in which the work originated, are seen as offering an opportunity to react to the kind of changes to which the original itself is subjected (Benjamin 1991, pp. 12f, Benjamin 2004, p. 77): "Es gibt eine *Nachreife* auch der festgelegten Worte" ("even words with fixed meaning can undergo a *maturing process*").

Such a *Nachreife* finds expression in translation. A translation thus participates in the "afterlife" (*Überleben*) of the foreign text, enacting an interpretation that is informed by a history of reception. This interpretation does more than transmit messages; it recreates the values that accrued to the foreign text over time (Benjamin 2004, p. 71). The process of reception combined with the changes that languages undergo over time render literary works unstable. Translation adopts the concept of reproducibility not by striving for similarity with the original "in its ultimate essence" (*ihrem letzten Wesen nach*), but by combining linguistic transmission with epistemological analysis, a process that is inevitably tied to a particular moment in the history and culture of the target language (Benjamin 2004, p. 77).

In one of his short stories, Jorge Luis Borges has carried the notion of reproducibility of a cultural moment and the *Nachreife* of translation to their logical extreme. The story reports how a French author, a certain Pierre Menard, ventures to recreate *Don Quixote*: "His admirable ambition was to produce a number of pages which coincided – word for word and line for line – with those of Miguel de Cervantes." His first, historical method to achieve this – "Learn Spanish, return to Catholicism, fight against the Moor or Turk, forget the history of Europe from 1602 to 1918 – be Miguel de Cervantes" – he soon abandons in favor of the much more demanding task of "continuing to be Pierre Menard and coming to the Quixote *through the experiences of Pierre Menard*" (Borges 1998, p. 91). To generate the same results independent of time and place – the guiding principle of the natural sciences – has here been scrupulously transferred to the sphere of art and reduced to absurdity.

However, in the literary context, repetition brings about difference through the change in cultural context. As the narrator observes full of admiration: "The Cervantes text and the Menard text are verbally identical, but the second is almost infinitely richer" (Borges 1998, p. 94). Menard's singular experiment illustrates the thesis that it is more likely for two works of art than for two readings to turn out identical. The new text contains a *Nachreife* which, if not outwardly apparent in the letters that make up the text, can be intuited by readers aware of the circumstances in which the two texts were produced. The "difference" between the Cervantes text and Menard's recreations of short passages from it is a cultural value that is implied in the reader's historic consciousness.

23.5 Reproducing Cultural Significance

Shifting cultural perspectives and their impact on the reproducibility of meaning concern not only the translator and imitator; they also concern the literary scholar who approaches texts from remote historical contexts, aiming to understand the conditions of their production. Given the cultural determinedness of meaning production, the issue is whether it is possible to recreate the cultural field in which a text originated in sufficient complexity and detail to be able to recreate a historically accurate understanding of its significance for an original audience.

At issue is also the status of literary texts within a wider cultural context as well as whether and to what extent a work of art may to some degree be liberated from the ideological constraints of the culture in which it was produced so as actively to transform that culture.[7] In hindsight, the broader question is whether different cultural perspectives can at all be made compatible and whether intercultural communication is generally possible. Such questions touch on the debates between two different versions of historicism and those between a historicist and a formalist approach to literature.

An older, Hegelian, positivist historicism believed in the possibility of reproducing the conditions under which a literary text had been produced in sufficiently accurate and ideologically neutral form to enable a reading that might be made useful in the service of enlightenment and historical progress. A newer, more skeptical, relativist historicism negates the freedom of access, emphasizing the impossibility to step out of one's proper cultural sphere and judge in a neutral manner. This "new historicism," as this second approach has come to be called in the domain of literary criticism, is also a critical response to the formalist approach to literature of the 1940s and 1950s, with which it nevertheless retains structural affinities.

The so-called "new criticism" blocked out consideration of the historical context of a literary work, focusing exclusively on its internal dynamics of repetitions and contrasts, such as irony and paradoxes, images, sounds, and rhythms. It considered literary works as self-contained, self-referential aesthetic objects and as cultural icons of timeless aesthetic value. The new historicism, by contrast, considers literary works as "signs of contingent social practices" rather

[7]This connects to Schleiermacher's claim that a proper author is only partly tied to constraints of the language, while also able to shape it.

than as sources of numinous authority, and literature more generally as just one of many different practices interrelated with each other in the wider nexus of culture (Greenblatt 1988, p. 5).

While the new historicism thus reactivates the historicist focus on cultural context, it complicates the possibilities of access to it. It also takes a more egalitarian approach to culture. The British Marxist theorist and literary critic Raymond Williams defined culture as a "whole way of life" as opposed to the artistic productions of the social elite (Williams 1960, p. xiv). Stephen Greenblatt, one of the main representatives of the new historicist approach to culture, adapted and reformulated Williams' definition by defining culture as a "network of negotiations" between a broad range of social practices and material exchanges (Greenblatt 1995, pp. 228f).

Culture is thus the product of a dynamic exchange between different areas of social life and the discourses connected with them. Literature immediately participates in this exchange, absorbing the various discourses that characterize a particular cultural network. Literary texts are then considered to be privileged objects of investigation only in as much as they are defined as sites of concentrated social energy, concentrated not least because of the social attention bestowed on them. Reconnecting literary texts with other extant fragments of its culture may allow us partly to recover the particular cultural texture that produced them.

The relevance of the concept of reproducibility in the new historicist approach to literature becomes clear when we relate this definition of culture to the theory of ideology connected to it. According to the Marxist theorist Louis Althusser, ideology represents the imaginary relationship of individuals to their real conditions of existence. Ideology is a fictional representation of material circumstances, a form of narrative by which individuals are interpellated by ideology and directly turned into its subjects. Their very constitution as "subjects" is a consequence of their being "subjected" to ideology or protagonists in a particular narrative. For Althusser (1971), there is no outside ideology, and its subjects have no access to their real conditions of existence.

According to Louis Montrose (together with Greenblatt a leading new historicist critic), Althusser's theory appeals to a double-edged process of subjectification: On the one hand, culture produces individuals who (within the frame of a given narrative) are capable of agency; on the other hand, it produces social networks and cultural codes that elude the individual's understanding and control. This suggests that the relation between collective structure and subjectivity is not oppositional, but dynamic: both are products of mutual exchange (Montrose 1989, p. 21).

The issue of the agency of subjects brings up the role of literature. The debate is whether literature, as fictional representation, can offer a counter-discourse to the fictional representation of ideology, one that has the power to interfere with ideology's domineering construction of reality and, potentially, promote change, or whether literary works irredeemably reproduce dominant ideology, thus serving as an instrument for the solidification of power. This latter view is the position that Greenblatt declares to be his own, although his textual criticism contradicts this claim. For such a rigid instrumental function does not square with the dynamic conception of culture he adopted from Williams, which in Greenblatt's criticism comes to life in the artful juxtaposition of texts. Neither does it exclude the possibility of willful human agency, even if such agency may theoretically be limited to the dominant social order (Bezzola 2010).

It, therefore, seems more appropriate to say that the new historicism aims to show both how literary texts represent dominant forms of social conduct and thus perpetuate them, as well as how they may actively influence such social forms through the medium of representation. This opens up possibilities for literature as a social tool of intervention beyond indoctrination. It also suggests that reading a literary text of a remote culture may offer glimpses of how the dominant narratives of ideology are challenged in the very process of their reproduction and re-presentation in literary fictions. While acknowledging this, it is important to add that the social relations reproduced in the fictional framework of a literary text and the imbrications of literary works in the social world of their making (including patronage systems, censorship, the marketplace, and other factors) often makes it difficult to distinguish between (blind) reproduction and critique.

For the literary scholar approaching texts from a remote period and culture, the crucial issue is how to break out of the ideological prison-house of his or her own culture and gain access to a literary artifact from a different one. Here the impossibility, self-consciously posited by the new historicism, of reproducing a foreign culture in adequate complexity joins with the hope that the blind spot of ideology characterizing each culture may be overcome through the temporal and cultural distance involved in the act of reading. Put more simply, the possibility that an outsider may identify cultural patterns that the natives of that culture are blind to. Another hope is that the inevitable loss of textual data over time may be made up by a new manageability of extant materials and the greater likelihood of the "serendipitous find" of significant connections.

Louis Montrose (1989, p. 20) has phrased the critical premise of the new historicism in chiastic terms as positing "the historicity of texts and the textuality of history." The phrase serves as a reminder that all texts have a history and that history is construed not only exclusively from texts but only from those texts that have survived into the present, which are then artificially tied together in a narrative. The new historicist mode of historical reconstruction insists on the fragmentary character of its materials, on the lacunae gaping between them, and on the fact that meaning extrapolated from revisited texts and contexts requires constant remaking. In the wake of poststructuralism, the new historicism assumes a skeptical attitude toward the grand metanarratives of historiography, arguing that these are produced artificially through an arbitrary selection and concatenation of data.

The same objections may, however, be raised against the powerful narratives of the skilful new historicist: Although these narratives typically do not spell out in a grand narrative what the combination of the materials they assemble signifies, their allegedly fortuitous juxtapositions suggest that they produce significant correlations, or correlations that – from the viewpoint of the receiving culture – appear self-evident. It proves extremely difficult to defy the power of narrative, particularly of those narratives that go without saying.

23.6 Conclusions

Our brief survey brings up a range of issues that merit consideration also in connection with the natural and empirical sciences, where reproducibility is a central methodological requirement. They primarily concern the cultural identity of meaning as well as the element of difference, which a more adequate definition of reproducibility may need to make allowance for.

The first issue is the awareness that cultural contexts matter, not only in terms of their material constraints on the factual, but in terms of what, within a specific cultural setting, is thinkable or perceptible (see the contributions in part I of this volume). Secondly, meaning is human meaning rather than facts speaking for themselves. Meaning is the product of an act of interpretation from a specific point of view; its production involves the (arbitrary) selection and combination of materials and always depends on a limited store of information available. Accordingly, meaning is always meaning *for* someone. Its production is never disinterested but rather governed by a particular agenda and subject to ideological constraints to which the producers of meaning tend to be blind, taking them to be factual constraints.

This brings up a third issue: the awareness of what inevitably gets lost or altered in translation. Any translation aiming to capture the actual spirit rather than the dead letter of an original builds on the realization that this original is the product of choices and constraints not available in the new context. In the empirical sciences, the transfer of insights produced within a specific context to a different context may confront similar difficulties. In a foreignizing approach to translation and in historicist criticism, the aim is to bridge a cultural gap without belying its existence. This entails acknowledging the radical artificiality of human categories of knowledge as well as the act of transgression which their transcultural combination necessitates. A critical awareness of the cultural specificity of meaning and its limited degree of reproducibility across cultural domains may render the cross-pollination between different cultural settings and disciplines fruitful.

Acknowledgments

This essay was originally planned to be written in collaboration with Alexandra Kleihues (Collegium Helveticum, Zurich), who has had to cut back on the project. But the essay still draws on first collaborative blueprints and indispensable conversations on the topic. Thank you, Alexandra!

References

Althusser, L. (1971): Ideology and ideological state apparatuses. In *Lenin and Philosophy and Other Essays*, transl. by B. Brewster. Monthly Review Press, New York, pp. 121–176.

Barthes, R. (1986): *The Rustle of Language*, transl. by R. Howard. Berkeley University Press, Berkeley.

Benjamin, W. (2004): The task of the translator: An introduction to the translation of Baudelaire's *Tableaux Parisiens*. In *The Translation Studies Reader*, ed. by L. Venuti, Routledge, London, pp. 75–85.

Benjamin, W. (1991): Die Aufgabe des Übersetzers. In *Gesammelte Schriften, Vol. IV.1*, ed. by T. Rexroth. Suhrkamp, Frankfurt.

Bezzola Lambert, L. (2010): The new historicism: Stephen Greenblatt. In *Literaturtheorien des 20. Jahrhunderts*, ed. by U. Schmid. Reclam, Tübingen, pp. 377–392.

Boase-Beier, J. (2011): *A Critical Introduction to Translation Studies*, Continuum, London.

Borges, J.L. (1998): Pierre Menard, Author of the Quixote. In *Collected Fictions*, transl. by A. Hurley. Viking, New York, pp. 88–95.

Chaucer, G. (1988): *The Riverside Chaucer*, ed. by L.D. Benson, Oxford University Press, Oxford.

Chaucer, G. (1986): *The Canterbury Tales*, transl. by N. Coghill. Penguin, London.

Eliot, T.S. (1970): The waste land. In *The Norton Anthology of Poetry*, W.W. Norton, New York, pp. 10001012.

Greenblatt, S. (1995): Culture. In *Critical Terms for Literature Study*, ed. by F. Lentricchia and T. McLaughlin, University of Chicago Press, Chicago, pp. 225–232.

Greenblatt, S. (1988): *Shakespearean Negotiations: The Circulation of Social Energy in Renaissance England*, University of California Press, Berkeley.

Greene, T.M. (1982): *The Light in Troy: Imitation and Discovery in Renaissance Poetry*, Yale University Press, New Haven.

Halliwell, S. (2002): *The Aesthetics of Mimesis*, Princeton University Press, Princeton.

Mittelstrass, J. (1992): Rationalität und Reproduzierbarkeit. In *Entwicklungen der methodischen Philosophie*, ed. by P. Janich, Suhrkamp, Frankfurt, pp. 54–67.

Montrose, L. (1989): The poetics and politics of culture. In *The New Historicism*, ed. by H.A. Veeser. Routledge, London, pp. 15–36.

Hutcheon, L. (1985): *A Theory of Parody*, Methuen, New York.

Potolsky, M. (2006): *Mimesis*, Routledge, London.

de Saussure, F. (1983): *Course in General Linguistics*, transl. by R. Harris, Duckworth, London.

Schleiermacher, F. (2004): On the different methods of translating. In *Translation Studies Reader*, ed. by L. Venuti, Routledge, London, pp. 43–68.

Shakespeare, W. (1997): *The Norton Shakespeare*, ed. by S. Greenblatt, W.W. Norton, New York.

Stein, G. (1947): *Four in America*, Yale University Press, New Haven.

Venuti, L., ed. (2004): *Translation Studies Reader*, Routledge, London.

Williams, R. (1960): *Culture and Society: 1780–1950*, Doubleday, Garden City.

24
Repetition Impossible: Co-Affection by Mimesis and Self-Mimesis

Hinderk Emrich

Abstract. The notion of "repetition" not only refers to behavioral phenomena but also to "deeper" – within an understanding in terms of the philosophy of time – and different kinds of subjects' apprehension of their own past. These kinds of apprehension are reconstructions as well as manifestations of a "real past," in the sense of a past impacting the present. They are related to the concepts of co-affection and, furthermore, to mimetic experiences and the notion of self-mimesis as elaborated by René Girard.

24.1 Introduction

The notion of "repetition" offers two kinds of interpretation. On the one hand, there is pure repetition in the sense of repeating the same procedure over and over again. The other meaning of "repetition" derives from the separate parts "re" and "petition" and implies the meaning of "requesting back," for instance, something that had occurred in the past.[1] For the second interpretation, of course the question arises as to whether a "real" manifestation of taking back the past is possible at all.

The two types of understanding repetition open a field of ambiguity, within which a borderland, a greater spectrum of kinds of recall, memorizing, and repetition is established. For instance, when a theater director asks an actor to repeat a sequence of words or a gesture, already the simple fact that this is requested implies a change of the situation. There are expectations and hypotheses from the part of the actor: What does the director really want me to do? What was wrong or questionable or should be different? A new point of view is formed, leading to the appearance of another subtext, another meaning, another context of the sentence, and so on.

However, this is not only true for the manifestation of semantic material

[1]The second meaning becomes even more evident in German language where "wiederholen," the translation of "to repeat," means "to take something back" rather than merely requesting it back.

Reproducibility: Principles, Problems, Practices, and Prospects, First Edition. Edited by Harald Atmanspacher and Sabine Maasen.

within learning and performing. As a matter of fact, it holds also for physical processes like the motion of a pendulum. The theory of nonlinear dynamical complex systems tells us that precise repetitions are practically impossible, since no system can be isolated from the totality of the universe and, therefore, not be prepared with the same initial conditions. From a philosophical perspective, the mathematician and philosopher Whitehead (1929) argued that the physical world is permanently creating itself again and again, as it were, and any of these processes have the meaning of a new reality in ever and ever new manifestations.

A deeper insight into the problem of repetition in our everyday life may be derived from philosophical and neuropsychological theories of recall and memory. The ambiguity of the term repetition in its twofold interpretation – processes of elementary repetition and semantic phenomena of the past – is represented by the very complex processes of recalling and memorizing, of emotional experiences within the past. We are coping with the phenomena of remembering which, as the philosopher Theunissen (2001) enunciated within a Hegelian account of memory, can be seen as transferring the content of a recall into the inner world of the memorizing person.[2]

24.2 Repetition within the Philosophy of Time

What is our philosophical understanding of time? Time does not represent an object of our existence. In the understanding of Kant, it belongs to the universal categories of experience: an *a priori* condition for the possibility of experience. Time is an inconceivable background of our being. Why is time philosophically so relevant?

This has to do with the fact that time is an irreducible frame of our human condition, our being-in-the-world. One of the important subjective aspects of time can be seen in the statement "take your time," famously suggested by Wittgenstein (1980, p. 80) as the way in which philosophers should salute each other. Any relevant events in our lives – birth, maturation, acquiring identity, success, or suffering – are undeniably phenomena of time.

In this regard, Theunissen (1991, pp. 218ff) speaks of the "dominance of time upon human beings," and states: "Time dominates us human beings in the same way as objects." And, therefore, he claims, there is both a necessity of "resistance against the dominance of time" and of "suffering from the dominance of time." Theunissen's *opus magnum* with the title *Negative Theology of Time*

[2]Again, this becomes clearer in an etymological analysis of the German translation of remembering, "erinnern," which means literally to turn someting "outside in," and to bear it "in mind."

is in fact an account of the ways in which the experience of time is altered in psychopathological disorders.[3]

From the point of view of phenomenology – and also from results of brain research – we know that memory is different from playing back a tape recording or running a CD-ROM. The act of remembering is an act of the constructive mind, enabled by neuropsychological processes within the brain. The act of remembering does not necessarily relate to what has actually taken place; it has a peculiar, intrinsically metaphorical quality of an "as if."

The actualization of the past within the process of remembering induces experiential contents in the sense that it appears to me "as if" it would be now that something was occurring in the past. Realistic memorizing represents a process of actual imagination and recall of the emotional manifestations of the past, molding and determining the present. What we experienced, and what we now are, is our reality in the sense of experiential qualities of the past. Our existence is always the manifestation of processes of meaning-generating constructs from the past within ourselves.

A radical, very lucid example of such phenomena has been created by Marcel Proust (2002) in his famous novel *In Search of Lost Time*. In his interpretation of Proust's way of remembering and coping with the past, Theunissen pointed out that this way of recall does not only refer to the conscious facts of the past but also to the memory traces of the "subliminal." The highly elucidative "Combray Chapter" in the novel has two major implications. The first is Proust's insight into the processes of remembering, induced by an "external trigger" in the sense of the gustatory "madeleine experience." The second aspect, to be discussed in detail below, has to do with the fascinating fact that the "madeleine experience" is, apparently, embedded into three types of contexts related to the phenomenon of "co-affection."

Here is the "madeleine experience" by itself, apparently more than a simple repetition (Proust 2002, p. 47):

> I raised to my lips a spoonful of the tea in which I had soaked a morsel of the cake. No sooner had the warm liquid mixed with the crumbs touched my palate than a shudder ran through me and I stopped, intent upon the extraordinary thing that was happening to me. An exquisite pleasure had invaded my senses, something isolated, detached, with no suggestion of its origin. And at once the vicissitudes of life had become indifferent to me, its disasters innocuous, its brevity illusory – this new sensation having

[3]Vogeley, this volume, discusses psychopathological disorders in the different context of social neuroscience with respect to deviations in interactive and communicative behavior or inconsistencies of emotional experiences, e.g., in individuals with autism spectrum disorders.

> had on me the effect, which love has, of filling me with a precious essence; or rather this essence was not in me, it *was* me. I had ceased now to feel mediocre, contingent, mortal. Whence could it have come to me, this all-powerful joy?

Proust's impressive observation of the spontaneous emergence of (constructive) memory traces from the subliminal, from the subconscious, indicates strong psychotherapeutic implications as represented by Freud's concepts of remembering, repetition, and "working through." Repetition here is one of the most important entrances into any form of a dialogue between the subject – e.g., a patient with a traumatic past – and the so-called "trauma-related zones of complexes."

24.3 Re-Presenting Forgetting

Following Theunissen's concept of "resistance to the dominance of time," we will now deal with "active dis-acknowledgment" as a type of forgetting process. Dis-acknowledgment here means forgetting as a creative interpretive act of representing. It involves using one's imagination to access the path that lies behind, while including all options which were not chosen and by bringing to life the corresponding subliminal "experiences." One might call this the "Proustian method," which Theunissen (1991, p. 62, translation HE) described as follows:

> Proust is able to recall "regained" time by shattering it. Memory, which functions as an organ of reconciliation, is spontaneous in that it leaps out from the solid context of occurrences and opens itself to the unique in which the subject perceives the internal subjective world which he has continually suppressed on the basis of what appeared to be external. In that it tears open the surface of memory in which we normally drift around.

What is this concept of internalization and how does it relate to a theory of forgetting? The following experience of a philosopher, a friend of mine, may be illuminating. Asked how one can regain something that has been literally "lost in time," he recounted the following anecdote. He and his family were booked on a holiday flight to Karachi the next day, but when he came to look for the airline tickets he could not find them. The family searched the whole house, without success. My friend decided to meditate and suddenly his inner eye saw the title of a particular book. Although he thought this was an absurd and irritating result of the meditation, he nevertheless decided to take the book down from the shelf. Astonishingly, the tickets were between the pages.

His explanation was that a visiting friend had been browsing in his library the day before. The tickets had been lying on the desk and the visitor must have used them as a bookmark before replacing the book on the shelf with the tickets in it. My friend must have subliminally seen the book on the desk and his unconscious must have "known" that the tickets were in the book. However, in order to inform him about this "forgotten material," his unconscious had to take a roundabout route via meditation and by presenting the title of the book. It could not otherwise have "communicated" the knowledge it had at its disposal.

This anecdote indicates how it is possible for perceptions succumbed to forgetting (e.g., because of "selective attention") to be reactivated via an indirect route. Bringing subliminal options of the past back to life, options that have not been lived (or only in an incipient form), means to apply the Proustian method, the search for lost time. This does not only imply that truth is, in Hegel's sense, a result of the actual past, but that it also includes all those pasts which the path actually chosen has passed by.

This way of recalling past subliminal options and possible lives is a process of making time relative. Theunissen observed that this kind of recall is blocked for depressive patients. It is a process of making the subject autonomous, a process whose energy is derived from re-presenting libido. It, therefore, lives within the totality of its own life, a remarkable blend of possibility and reality – and this type of living would seem to be the only adequate form of forgetting. It is a method of re-presenting remembering that de-actualizes what has passed by and allows us to integrate what is to be forgotten. It relativizes the dominance of time within life, so that the projection of the past into the present can be set to rest.

If this kind of re-presenting forgetting does not occur, if the processing of past traumas through memory does not take place, there is the risk of a "conspiracy of silence." This can often be observed in the psychotherapy of victims of rape, where there is a subconscious inhibition to reimagine – to repeat – the humiliation and to allow it to become painful within the transference–countertransference relationship.

The method of actualizing the de-actualization of what has passed by is the only appropriate type of forgetting in cases of severe trauma. This is because it can be assumed that after the re-presenting has taken effect, conflicting components of the self can be reconciled. This "Nirvana principle," a Freudian term characterizing the drive for the least amount of tension, yields an obliteration of the semantic contents of the traumatized area within the self. This liberates creativity.

To give an example from psychoanalytic practice, a patient who had been involved in a relationship which executed a "curse" on her further life succeeded in removing the curse. This became possible when she went back to the roots of the curse *ab ovo* in her imagination and discovered sides of her self which were free of the curse. In contrast to splitting, repression, and denial, this "transformational forgetting" makes psychological forces available for future developments. While ritualized memory can be seen as a type of unintended and indeed unsuitable, unsuccessful forgetting, proper forgetting constitutes a type of re-presenting within which time is de-actualized and made relative.

The process of "actualizing de-actualization" sets free – in Freudian terms – libido and "Nirvana energy" so that they can become effective simultaneously or successively. What has to be "extinguished" must first be activated and recognized. Nietzsche's "active inhibitive capacity" can only be applied to something which is made present, which has made its appearance in consciousness.

Actualizing de-actualization should not be seen as a one-off event. Rather, it is as a cyclical, process-like, reiterative progression of remembering and forgetting, of libido and "Nirvana principle." To put it differently, only a problem which can be solved can also be disbanded. To this extent, a type of forgetting that is able to dissolve the dominance of time always consists of two phases: actualization through memory and the subsequent "Nirvana"-related de-actualization and rejection. If this is correct, the tension of inner conflicts can be discharged by means of re-presenting forgetting.

24.4 Repetition, Co-Affection and Trauma: Identity and Coping with the Past

In his novel *Fiasco*, the great writer and Nobel-prize winner Imre Kertész (2011) described the impossibility of a traumatized person to reestablish a coherent personal identity (p. 109):

> I looked about. Everything around me was seething and bubbling, a chirping twitter of voices from all sides, as if carried by invisible telegraph wires on invisible telegraph poles; ideas, offers, plans, and holes jumped across like flashing electric discharges from one head to another. Yes, somehow I had been left out of this vast global metabolism of mass production and consumption, and at that moment I grasped that this was what had decided my fate. I am not a consumer, and I am not consumable.

And another quotation (p. 77):

> I didn't have anything to think about. Yet slowly something nevertheless was taking shape inside me. If I distinguish it from the mild dizziness caused by walking and from other contingent impression, I discover a definable feeling. I suppose my state of affairs was materialising in it. It would

> be hard for me to put it into words – and that's exactly the point: it settles itself in spaces that lie outside of words. It cannot be couched in an assertion, nor in a bald negation either. I cannot say that I don't exist as that is not true. The only word with which I could express my state, not to speak of my activity, does not exist. I might approximate it by saying something like "I amn't". Yes, that's the right verb, one that would convey by existence and at the same time denote the negative quality of that existence – if, as I say, there were such a verb. But there isn't. I could say, a bit ruefully, that I have lost my verb.

Fiasco represents a metaphor for the structure of our existence. Our life represents a "fiasco" insofar as we always appear to be situated within an outer court of being, but not within the existence of being itself. From the traumatic experiences of the protagonist in the novel, it raises a radical metaphor for life itself: the fiasco of the inconceivability and impenetrability of existence (p. 70):

> Destiny – since that's its nature – would have robbed me of any future which was definitive and thus could be contemplated. It would have bogged me down in the moment, dipped me in failure as in a cauldron of pitch: whether I would be cooked in it or petrified hardly matters. I was not circumspect enough, however. All that happened was that an idea was shattered; that idea – myself as a product of my creative imagination, if I may put it that way – no longer exists, that's all there is to it.

With respect to the dichotomy between language and life itself, we always live in an unredeemed situation, we live in the fiasco that our language, our terms, our thoughts, our pure reason, our spirits cannot describe ourselves and cannot lead us to salvation. (This, unfortunately, may also be true for our therapeutic work with patients.) The fiasco has to do with the insight that our life normally is in the foreground, not in the background of being. It is not lived as an experience of the foundations that would yield safety and substance. In this regard the novel by Kertész emphasizes the central topic of autobiographic writing:[4] representing the topics referring to questions of one's own identity.

Within a world, within a life that is absorbed by categories and terms, within a world of fiction, the question arises how an "initiation" may bring an individual into contact with his existence. Traumatization destroys the alliance between cognition and being, between thoughts and vivid impressions. In this peculiar, dangerous, and often catastrophic psychological situation, the continuum of

[4] In his work *Time and Narrative*, the French philosopher Ricoeur (1988) explicated that human beings are unable to manifest some type of absolute, "metaphysical" identity. Instead, individuals are able to manifest themselves as a "narrative" identity which, in a specific type of self-reference, is molded by the way of telling their own biography.

self-being in a coherent personal subjective–objective world is fragmented. As a consequence, the only solution for the traumatized person is autobiographic writing: a form of repetition.

In describing his attempts and repeated reattempts to write his novel, the traumatized author within the novel could only experience and describe the concretizations of his immediate surroundings, sometimes with details of utmost banality. This approach to concreteness and precision apparently has to do with the inevitable necessity to approach – within the immediate surroundings of the author – some type of "holding," some type of stability and thus to overcome the extreme difficulty to remember the traumatic experiences of the past.

Trauma destroys the relation between categories of life, language, cognitions, thoughts, and streams of consciousness, on the one hand, and the relation between those and life itself on the other. It destroys the normally existing coherence, the normal rules of parallelism. After this destruction the person is helpless and requires a redeeming concreteness within his or her surroundings. This new experience of concreteness reestablishes the lost coherence between sensual perceptions and feelings and the related cognitions, ideas, fantasies, and ways of understanding.

24.5 The Dialectics of Remembering and Forgetting

The literary example by Kertész makes the fragility of personal identity after traumatization obvious, its paralyzed constitution, its loneliness and alienation, and its fragmentary character. This entails the necessity to become stabilized by the concreteness of surroundings as the only type of reality that may yield stability. It also entails the urge to describe aspects of a possible stabilization of personal identity of the traumatized person in terms of coping with the past – a trauma-loaded past that transmits flashbacks, by dissociated states of consciousness, to the self of the subject.

As a response to this a coping process is induced, which contains a subtle inner dialectics between the necessity and wishfulness of *forgetting* and the existentially founded tendency to *remember.* Remembering and forgetting establish an unresolvable cycle of interrelated processes that characterize the severe fate and the burdens of the traumatized person. On the one hand, the trauma is indescribable and unspeakable and cannot be forgotten. On the other hand, it has to be deactivated to dissolve its powerful detrimental effects. How may this be possible? Central aspects are represented by creative processes of transfiguration, enabled by an integration of the subliminal, and establishing a new coherence between cognition and concreteness.

24.6 Co-Affection and Memorizing Recall

Interpersonal and intrapersonal affective involvements, influences, and experiences are of highest importance regarding the internal dialogues between our subliminal unconscous memory traces of the past and the representations within our manifest consciousness. In Proust's novel, this has the following structure.

Firstly, the spontaneous emergence of the "madeleine experience" in its whole intensity and vitality is induced by a narrative about the singularity and unequivocality of the presence of the mother, especially in one single night. The mother occupies and dominates the whole semantic field of the Combray representation within the conscious mind (Proust 2002, p. 30):

> My agony was soothed; I let myself be borne upon the current of this gentle night on which I had my mother by my side. I knew that such a night could not be repeated; that the strongest desire I had in the world, namely, to keep my mother in my room through the sad hours of darkness, ran too much counter to general requirements and to the wishes of others for such a concession as had been granted me this evening to be anything but a rare and casual exception.

The second interpersonal aspect within these topics of co-affection is the representation of Aunt Leonie (Proust 2002, p. 55):

> And suddenly the memory revealed itself. The taste was that of the little piece of madeleine which on Sunday mornings at Combray (because on those mornings I did not go out before mass), when I went to say good morning to her in her bedroom , my aunt Léonie used to give me, dipping it first in her own cup of tea or tisane. The sight of the little madeleine had recalled nothing to my mind before I tasted it.

The third dimension of co-affection within the madeleine experience is – as a matter of fact – the narrator himself. It is really the intimate, flourishing, and unexpectedly emerging world of emotions of pleasure, which came up due to the taste of the cookie and tea within the madeleine experience. The author asks what it means and where it comes from (Proust, p. 33):

> So in that moment all the flowers in our garden and in M. Swann's park, and the water-lilies on the Vivonne and the good folk of the village and their little dwellings and the parish church and the whole of Combray and its surroundings, taking shape and solidity, sprang into being, town and gardens alike, from my cup of tea.

The phenomenon of co-affection may be described within a concept elaborated by Girard (1977): *mimesis* as a process of identifying oneself not only with the behavior but also with the intentions of the other in his/her otherhood. Further elaborations of mimesis are due to Gebauer and Wulf (1992), who proposed and developed the important concept of "self-mimesis."

24.7 Mimesis

One of the great modern figures in the theory of mimesis is René Girard. In various works, notably *Violence and the Sacred* (Girard 1977), he developed a theory of mimetic occurrences. In Girard's thinking, the main point of the oedipal drama is not the oedipal aspect as such – the desire for the mother in the Freudian sense – but the mimetic rivalry with the father. What does this mean? In a marvelous passage, Peter Handke (1994) showed how much children in particular live by models that they act out on the basis of assimilation and imitation, even to the point of utterly schematic self-stylizations:

> For hours, already, two very young couples (14, 15 years old) are moving on one spot in the sand, and all their movements (embracing, hitting, clinging to each other) are always executed only as feints, as fleeting, as ephemeral intimacies. They spend their whole day, even in their shouting, talking, and looking, solely with such little, insinuating rituals, in a strange montage of karate, porno, and adventure films – pirates carrying women away; a woman is putting her foot on the man's neck; one man walking on another's stomach; someone, pretending to be dead, is brought to life with slaps on the face –, while the girls usually stand there (but also actually play-act standing there), intervening at most with the tips of their fingers or "lasciviously" bending backwards before the boys' howls (or playfully and quickly are gliding down the bodies of the boys). Whenever this ever rapidly changing course of events (rapid as with a team of acrobats) threatens to become naively tender and to turn normal, one of the boys immediately screws up his face, lets out a karate yell (more a snarl), and transforms the "menacing tenderness" into a feint of violence or enslavement (or he plays the slave himself, gesturing utmost resignation). An eerie quiet descends upon these changing images only when one of them has to wipe the sand out of his eyes – a strange beauty, too – but then he falls to his knees again, throws his head back, and yells with bared teeth; or shakes a girl back and forth by the neck or drags the girl through the sand by the arm; or the women come running to separate the men who are fighting, whereupon one of the boys has a "thorn removed" by a girl.

Though this text passage focuses more on the teenagers' creation of anonymity and self-alienation, it points out an aspect of mimetic action coursing in many different facets through the cultural history of all humanity. Gebauer and Wulf (1992) documented how vast a spectrum there is in the history of mimesis, ranging from origins in the psychomotor system of dancers to the constructive role it plays in the upbringing of young people. Mimesis is a creative performance of what has been observed of outstanding human beings; the goal is to become like them. This extends as far as to mimesis as a representation of political power, as a medium of theater, as self-formation, and as the "mimetic constitution of social reality."

In one of his earlier works, *Deceit, Desire and the Novel*, Girard (1961) described a discovery central to the entire development of his theory. Namely, he was able to substantiate the phenomenon of "mimetic desire" in 19th- and 20th-century novels from Stendhal to Flaubert to Dostoyevsky and Proust (and finally Camus).[5] The novels he examined share a peculiar basic structure. As Gebauer and Wulf (1992) wrote, the main characters, according to Girard, desire someone else, and their feelings are directed toward a third person, a "mediator." Analyzing Stendhal, Girard (1961, p. 61, translation HE) said:

> In most cases of Stendhalian desire, the mediator himself desires the object, or could desire it. This desire, whether real or assumed, is precisely what makes the object infinitely desirable in the eyes of the subject.

The preferred victim of this mechanism is the one who is possessed by vanity (Girard 1961, p. 20, translation HE):

> For a vain person to desire an object, it is sufficient to convince him that this object is desired by a third party who enjoys a certain prestige. In this respect the mediator is a rival.

Desires, needs, preferences, and yearnings are thus described as something not genuinely given, something brought forth internally by drives, but taken on from a third party because of mimetic dependence. The mediator in Stendhal is a character said to be real, a person different from the hero of the novel. This protagonist knows that he depends on someone else. For Girard, this mimetic interlacing of desire is a product of dynamic social relations. According to Gebauer and Wulf (1992, p. 328, translation HE):

> The competition from others who are considered models for the actor lend the object the character of the uniqueness. ... The mediator has as little need to possess a special value as does the desired object. It is sufficient for the mediator to have prestige. ... Satisfaction can arise from successes over the rival. Achieving the goal, however, cannot bring lasting happiness. Having once actually achieved it, the hero is ultimately left only with disillusionment, the feeling of a void.

---

[5]For a detailed account of mimesis in literature theory in general see the contribution by Bezzola Lambert, this volume.

Girard referred to the novels he analyzed as "anti-romantic" because they undercut what he calls the "romantic" conviction "that the persons exercise autonomous and self-determined choices over their goals and that those goals possess inherent value." According to him, both fundamental persuasions, the autonomy of the subject and the inherent value of that person's goals, are transcended in the novels he examined.

Gebauer and Wulf (1992) expanded on this analysis, expressing Girard's discovery in a more analytical language. Characterizing the superpersonal "internal medium," they wrote (p. 330, translation HE):

> The internal medium is not easy to discover, for messages conveyed by it are generally held to be intersubjective reality, contingent on persons. Girard finds it by scrutinizing the role of the mediator. The mediator shows the fictional protagonist, a special world symbolically consisting in mental images, affects, ascribed values, (and) nomenclature that are different from the usual ones. It is an inner world; the mediator lets them be seen from the perspective of the person belonging to them and emotionally involved in them. It springs from speech, from stories, by means of clues, descriptions, names. The hero thinks that the inner world conveyed to him is the social reality. Actually, it is produced in the subjective view of the mediator and can be perceived only from that person's perspective. It is the model world on which the hero organizes his or her own world. The hero's reality is intended to be like that one. Philosophically speaking, model worlds provide interpretations of the subject's world. Every individual customarily formulates interpretations of the world. The ones at issue here are of a special nature. They are mimetically generated under the influence of a medium, and they determine action. The medium is the predominant thing. It is like a film containing the hero's ideals and dream stories, like situations, constellations, (and) projections by which the individual is attracted.

In Girard's theory of religious sacrifice, this "anti-mimetic novel" plays a key role. It is concerned here with both rescuing and superseding the theory of sacrifice that Freud developed in *Totem and Taboo.* Girard (1977) himself summarized his intentions like this (p. 169):

> As an interpretative tool the concept of mimetic rivalry is far more serviceable than the freudian complex. By eliminating the conscious patricide-incest desire it does away with the cumbersome necessity of the desire's subsequent repression.

According to Girard, Freud fails to get to the essence of his system because, at the critical juncture, he "completely forgets the ... father identification" and concentrates solely on the driving desires directed to the mother.

As a prime example of a preanalytic work of art, in which Girard's mimetic rivalry models the central action, is the theater play *Blunt or the Guest* published by Moritz in 1780. Blunt, the father, like Hiob in Girard's analysis, has fallen from high social status to the poorest social circumstances. During this social demise, he has repudiated his son, William, who "blossomed like a rose ... in his Hussar attire" – an image mimetically imitating the father. Living in abject poverty and being too proud to accept help from his brother, the father dreams one night that he will come into "immense wealth" through the "sacrifice of the stranger" – whereupon he stabs his son to death, who lay sleeping unrecognized in bed.

The cruelty of this archaic act of killing the first-born son, antecedent of the oedipal complex, compels Moritz to offer an alternative, made possible by "blessing imagination": the mother intervenes, awakening (i.e., enlightening) both the father and the son. The result is a "humanized variant" in which mimetic occurrences henceforth win out over archaic filial murder, albeit at the cost of the father's utter moral and social deprivation.

With Girard, Moritz's version can be interpreted to mean that the son's described process of identity formation can be arrived at only through several complicated and dangerous stages of rivalry and battle with the father. This conception can be harmonized with the thesis that the crux of Girard's conception of the "sacred" as the "founding power" lies precisely in the fact that it symbolizes processes of identity formation and the concomitant external and internal "sacrifices."

The image of the double (the German "Doppelgänger"), which is likewise set up in Blunt and which would seem crucial for a theory of self-mimesis, thus means that one's imaginary world suddenly shows up in the form of something apart or separate (as symbolized by Blunt's "guest"). Mimetic rivalry and triangulation thereby appear as requirements through which a psychological event ultimately arises as "self-mimesis" in the first place – just as in Freud's theory the oedipal nature of the psyche is an essential condition.

But there are negative examples of mimetic triangulation as well. One is the case of a patient, an important film actor about fifty years old, who reported having noticed more and more since the beginning of his artistic development that a kind of "veil" was drawn over his entire life. He said that reality was somehow estranged from him, detached, removed, that he had the feeling to not really relate to other people or things, and that he even did not really feel alive, not really present. – Now this changes completely when the cameras start rolling. He then becomes totally alive and present, things and people become animated, the veil is lifted and true reality, a feeling of existence, arises.

The patient said that he experienced this kind of "illness" as oppressive and uncanny. However, faced with the alternative of living a "more real life", which might be possible through psychotherapy but might also cause the "awakenings" due to filming to cease, he decided strictly against that option.

Another case concerns the syndrome of infantile autism, which has to do with extremely solitary or "veiled" subjects. In some instances, this veil has been lifted a little in recent years through "facilitated communication." An especially notable case is described in the texts by an autistic 20-year-old with the name of Birger Sellin. Here are a few passages impressively illustrating the impossibility for the patient to step out of his psychological condition into interpersonal reality (Sellin 1993, p. 124, translation HE):

> An old simple point of view
> a stepping out of yourself
> a crate from which I rise
> that would be a dream like everyone dreams it
> but I don't see any way out of this person crate
> this important letter isn't enough either
> a way out of destroys my old security
> I'm afraid of it ...
> simply tearing out fear would be the rescue like an
> effective sure unutterable miracle

The "interrogation" of the suprapersonal domain of art in the subjective context (of the actor) can be condensed to a general interpretation of psychological events. That which is psychological is seen as that which is individual, as the special within the general, as the space of freedom.

But how is such space for freedom constituted, and how is it used? The basic problem of oedipal, or mimetic, triangulation is the otherness within each person, which comes as an outsider to the imaginary world, as a "guest" (in Blunt). Like in Bischof's (1989) *Oedipus Riddle*, it is decoded and deciphered as a dichotomy between autonomy and relatedness. Second-order mimetic rivalry occurs when the rival strangers in the imaginary world produce different reality options that signify spaces of freedom in the self. In other words, the possibility of "perspective mimesis," the ability to experience reality from the perspective of the other, creates freedom. It makes human beings human.

Perspective mimesis and mimetic rivalry are indispensable parts of human experience from the outset. Co-affection manifests itself by the "otherness of the self." There is a semantic field within the subject which – though it remains partially unknown – is the most intimate and truly intricate and affectionate representation of inner reality. The subject is (mimetically) co-affected by his or her unconscious self. And this co-affection entails a transfiguration of reality.

References

Bischof, N. (1989): *The Oedipus Riddle*, Piper, München. Originally published in 1985 in German.

Gebauer, G., and Wulf, C. (1992): *Mimesis: Kultur – Kunst – Gesellschaft*, Rowohlt, Hamburg.

Girard, R. (1961): *Mensonge romantique et vérité romanesque*, Grasset, Paris. English translation: *Deceit, Desire and the Novel: Self and Other in Literary Structure*, Johns Hopkins University Press, Baltimore 1966.

Girard, R. (1977): *Violence and the Sacred*, transl. by P. Gregory, Johns Hopkins University Press, Baltimore. Originally published 1972 in French.

Handke, P. (1994): Excerpt of a theater program at "Schaubühne Berlin": "The hour we did not know about one another". Originally in German.

Kertész, I. (2011): *Fiasco*, transl. by T. Wilkinson, Melville House, New York. Originally published in 1988 in Hungarian.

Proust, M. (2002): *In Search of Lost Time, Vol. 1)*, ed. by C. Prendergast, Allen Lane, London. Originally published 1913–1027 in French.

Ricoeur, P. (1988): *Time and Narrative*, transl. by K. Blamey and D. Pellauer, University of Chicago Press, Chicago. Originally published 1983–1985 in French.

Sellin, B. (1993): *Ich will kein inmich mehr sein. Botschaften aus einem autistischen Kerker*, Kiepenheuer & Witsch, Köln.

Theunissen, M. (1991): *Negative Theologie der Zeit*, Suhrkamp, Frankfurt.

Theunissen, M. (2001): *Reichweite und Grenzen der Erinnerung*, Mohr, Tübingen.

Whitehead, A.N. (1929): *Process and Reality*, Macmillan, New York.

Wittgenstein, L. (1980): *Culture and Value*, ed. by G.H. von Wright, transl. by P. Winch, Basil Blackwell, Oxford.

25
Relevance Criteria for Reproducibility: The Contextual Emergence of Granularity

Harald Atmanspacher

Abstract. Reproducibility is a particularly difficult issue in interdisciplinary research where the results to be reproduced typically refer to more than one single level of description of the system considered. In such cases it is mandatory to distinguish the relevant attributes or observables of the system, depending on its description. Usually, different descriptive levels go along with different degrees of granularity. While "lower-level" descriptions address systems in terms of micro-properties (position, momentum, etc.), other, more global, macro-properties are more suitably taken into account for "higher-level" descriptions.
The transformation between descriptive levels and their associated granularities is possible by an interlevel relation called "contextual emergence." It yields a formally sound and empirically applicable procedure to construct level-specific criteria for relevant observables across disciplines. Relevance criteria merged with contextual emergence challenge the old idea of one fundamental ontology from which everything else derives. At the same time, the scheme of contextual emergence is specific enough to resist the backlash into a relativist patchwork of unconnected model fragments.

25.1 Introduction

When Willie Sutton was in prison, a priest who was trying to reform him asked him why he robbed banks. "Well," Sutton replied, "that's where the money is." Garfinkel (1981, p. 21), who introduces his chapter on "explanatory relativity" with this example, argued that the palpable misfit of question and answer arises because Sutton and the priest have different sets of alternatives in mind, different "contrasts" as it were. The priest wants to know why Sutton goes robbing rather than leading an honest life, Sutton explains why he robs banks rather than gas stations or grocery stores.

Another example of explanatory relativity is illustrated by van Fraassen (1980, p. 125, quoting Hanson): "Consider how the cause of death might have been set out by a physician as 'multiple haemorrhage', by the barrister as 'negligence on the part of the driver', by a carriage-builder as 'a defect in the brakeblock construction', by a civic planner as 'the presence of tall shrubbery at that turning'." All these different ways to "explain" the death of one particu-

Reproducibility: Principles, Problems, Practices, and Prospects, First Edition. Edited by Harald Atmanspacher and Sabine Maasen.

lar person in one particular incident are examples of different relevance criteria, corresponding to different contrast classes.[1]

On the accounts of Garfinkel and van Fraassen (and others), explanations are not only relationships between theories and facts; they are three-place relations between theories, facts, and contexts. Relevance criteria as well as contrast classes are determined by contexts that have to be selected, and are not themselves part of a scientific explanation.[2] This assertion is an essential piece of van Fraassen's anti-realist stance of "constructive empiricism." Science tries to explain the structure of the phenomena in the world (possibly all of them), but it does not determine which parts of that structure are salient in particular situations.

Explanatory relativity backed up by relevance criteria can vitally serve the discussion of reproducibility across scientific disciplines. Features that are relevant for a proper *explanation* of some observation should have a high potential to be also relevant for the robust *reproduction* of that observation. But which properties of systems and their descriptions may be promising candidates for the application of such relevance criteria? One option to highlight relevance criteria is to consider the *granularity* (coarseness) of a description, which usually changes across disciplines.

25 Contrast Classes, Coarse Grains, Partition Cells

In order to illustrate the basic concepts with a simple example, let us consider a set of eight elements: three bats (B), three cats (C), and two vats (V). They can be grouped by properties, e.g., bats and cats are mammals, and their number (in this example) can be divided by three; cats and vats are unable to fly and weigh more than a pound. These groupings are *partitions* into contrast classes, into which elements fall if they are equivalent with respect to one of the properties. Partitions may have different granularity. If one partition is a refinement of another one, its cells (or grains) are finer, otherwise they are coarser.

[1]On p. 127, footnote 34, van Fraassen (1980) acknowledged that this "idea was independently developed by ... Alan Garfinkel in *Explanation and Individuals* (Yale University Press, forthcoming)". Obviously, this note refers to a preliminary version of Garfinkel (1981).

[2]The importance of context in determining relevance is also emphasized in "relevance theory," a pragmatic theory of communication formulated by Sperber and Wilson (1986). Relevance theory aims to explain how the receiver of a message infers the meaning intended by the sender based on the evidence provided in the associative *context* of the message.
The significance of context is already indicated in Weaver's commentary to Shannon's purely syntactic theory of communication (Shannon and Weaver 1949). It is an essential part of the concept of pragmatic information proposed by von Weizsäcker (1974).

How relevant a chosen partition is depends on the context of the selected properties and is not universally given. For instance, the fact that the class of mammals and the class of elements divisible by three are identical is purely accidental, and the fact that the class of elements unable to fly and the class of being heavier than a pound does not at all mean that nothing heavier than a pound cannot fly. Although the classification in our example may be statistically significant beyond any reasonable doubt, it is not substantive (cf. Porter, this volume).

Now imagine yet another possibility for the above set to be partitioned, namely, into contrast classes (f) and (m) with, say, four elements in (f) and two in (m), which differ from the contrast classes (B), (C), and (V). And imagine a property for classification that looks rather quaint: the property that the sum of elements of any class in {(B), (C), (V)} and of any class in {(f), (m)} is uneven. A grouping according to this property yields f cats (4+3), m cats (2+3), f bats (4+3), and m bats (2+3). (Groupings with the even number of two vats are even because the numbers of (f) and (m) are even too.)

Technically speaking, this can be represented as a *product partition* of the form $X \vee Y = Z$ for the factor partitions X and Y:

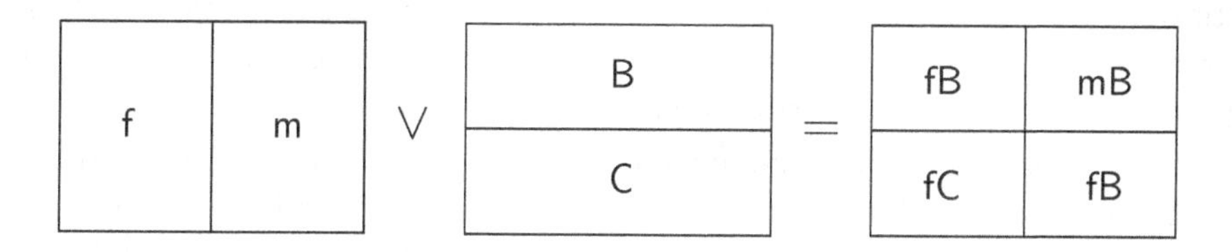

What about the *relevance* of the "quaint" property of unevenness? The product partition, which is a refinement of both f/m and B/C, shows that the result of applying the corresponding grouping matches with the well-known fact that bats and cats can be female or male, and vats have no sexes. This is a very relevant classification for living beings, although the unevenness property giving rise to it is entirely accidental.

By definition, properties are *invariant* within the same class, or partition cell. More precisely, one may change some parameter of a system S *without* changing a property that is invariant under that parameter change. For instance, all female living beings are female, irrespective of changes between species.

A property F is called invariant under the transformation π if $F(S) = F^{(\pi)}(S)$. Transformations under which properties are invariant are *symmetry* operations (e.g., inversion or rotation).[3] In order to identify invariances or sym-

[3]See Tetens, this volume, for the signifiance of invariances and symmetries in physical laws.

metries, one needs distinguishable parameters for the transformation π! Without a precisely defined change, it is pointless to speak of invariances or symmetries.

An *equivalence class* with respect to an element a of a set X is the subset $[a]$ of all elements $x \in X$ satisfying an equivalence relation denoted as "$\sim$": $[a] = \{x \in X | a \sim x\}$. A property F defines an equivalence class if and only if $F(x_i) = F(x_j)$ for $x_i \sim x_j$. In this case, F is an invariant of $[a]$. Equivalence classes can be represented as cells A_i of a partition of X which cover X completely, $\bigcup A_i = X$, and do not intersect, $A_i \cap A_j = 0$ for $A_i \neq A_j$.[4]

There are two extremal partitions: (1) the *identity partition*, for which each cell A_k contains exactly one element $A_k = x_k$, and (2) the *trivial partition*, for which all $x \in X$ are equivalent. Between these extremes, we have the general case of *finite partitions*. As mentioned above, the product partition of two finite partitions is a refinement of its factors.

25.3 Two Examples

25.3.1 Temperature

Partitions into equivalence classes are a key to mereological relations between parts of a system and systems as a whole. For instance, a rather simplistic instance for a fine-grained partition would be the number of particles in a vat. The vat as a whole would allow for characterizations of the temperature, the viscosity, or the scent of its content (or other "global" properties). This example points to the much-discussed example of relations between a mechanical and a thermodynamical description of a system.

Consider a liter of gas at a particular temperature T containing some 10^{23} gas molecules with mass m moving around with different velocities v. The distribution of those velocities (actually, their absolute values) can be calculated from their equations of motion and turns out to be the well-known canonical Maxwell–Boltzmann distribution

$$p(v) = 4\pi \left(\frac{m}{2\pi kT}\right)^{3/2} v^2 \exp\left(-\frac{mv^2}{2kT}\right), \qquad (25.1)$$

where k is the Boltzmann constant. The dependence of this distribution on the mid term v^2 implies that the distribution approaches zero if the velocity approaches zero and the third, exponential term lets the distribution decrease

[4]In the terminology of clustering, partitions are crisp clusters (see Estivill-Castro, this volume). By contrast, fuzzy clusters do not satisfy the two conditions for the union and intersection of cells.

toward zero for large velocities. The mean squared velocity of the particles turns out to be

$$v^2 = \frac{3kT}{m}, \tag{25.2}$$

which gives rise to a mean kinetic energy:

$$E_{\text{kin}} = \frac{3}{2}kT. \tag{25.3}$$

This equation is also known as a *bridge law* (Nagel 1961) relating mechanical kinetic energy and thermodynamical temperature.

This bridge law, and the way it has been derived, suggests that the macro-property temperature can be reduced to the micro-property of particle velocity. Temperature would then be "nothing else" than the mean kinetic energy of particles in an ensemble. Rhetorical figures like this have been used to establish a picture of reductive methodologies as prime drivers of progress in physics and science in general. But things are much less innocent upon closer inspection. In fact, the correlation between thermodynamical and mechanical properties, heuristically known since the mid-19th century, was rigorously understood not before the work of Haag *et al.* (1974) and Takesaki (1970); see also Section 25.4.

It is important to realize that, although the bridge law above gives correct quantitative information about the numbers at both sides of the equation sign, the observables on both sides are fundamentally different conceptually. Technically speaking they belong to algebras with very different properties and are *coextensive but not identical.* Contemporary informed accounts of bridge laws are much more careful and tend to avoid premature claims as to reduction as the overarching regulative principle in science (see, e.g., Chibbaro *et al.* 2014).

Concerning the issue of granularity, the example of mechanics and thermodynamics illustrates a relation between the two extremal partitions mentioned above. At the mechanical level, each molecule is considered with its own individual velocity (and other micro-properties). This means that *each individual molecule* is represented pointwise in the corresponding state space, leading to an *identity partition* where every state is a singleton. Such a description is as fine-grained as it could be.

On the other hand, the velocity *distribution* of the molecules in a statistical description, which serves as a basis for thermodynamics, refers to an object (the distribution) which covers the entire system of *all molecules together.* In contrast to the mechanical description, this represents the *trivial partition*, which is as coarse-grained as it could be. Within one coarse grain, differences between individual molecules are irrelevant by definition. Only the treatment of the ensemble as a whole gives rise to system properties such as temperature.

Concerning the issue of reproducibility, it would be absurd to reproduce a particular thermodynamical property (e.g., a particular temperature) of a system by preparing the states of individual molecules such that their velocities together yield a Maxwell–Boltzmann distribution. Obviously, the relevant level for reproducing temperature is the thermal level.

25.3.2 Chirality

Another interesting example for descriptions based on different partitions is the relation between basic quantum mechanics and properties of matter at the molecular level. The classic tool of choice to calculate the energy eigenvalues of quantum systems, including molecules, is the solution of the time-independent Schrödinger equation

$$H\psi(r_e, r_k) = E \cdot \psi(r_e, r_k)\,, \tag{25.4}$$

where H is the molecular Hamilton operator, ψ is the state of the system, r_e are the positions of the electrons, r_k are the positions of the nuclei, and E are the energy levels of the molecule. But the bad news is that an exact solution of this Schrödinger equation (25.4) is unattainable for the in-principle reason that nuclei and electrons are generally entangled.

However, there are ways to circumvent this objection and look for convenient approximative solutions, For instance, one can treat the system as if the nuclei were fixed in space and only the electrons are moving. Under this condition, the Schrödinger equation yields the discrete electronic energy levels of the molecule. A more sophisticated approximation, originally due to Born and Oppenheimer (1927), is that the nuclei move slowly as compared to the electrons, because their mass M is much greater than the electron mass m, $m/M \ll 1$. In this case, one obtains solutions including the motion of the nuclei as well, i.e., their vibrational and rotational energy levels.

Through the fixed or slowly varying position of the molecular nuclei, both approximations define a nuclear structure which in turn determines the shape (the "gestalt") of the molecule.[5] The emergent classical property of having a shape entails that molecules can have all kinds of symmetries, among them the discrete transformation of mirror (inversion) symmetry. If two molecular species

[5] Actually the notion of an "approximation" is somewhat misleading. It misses the fact that the molecular shape that it defines gives rise to the emergence of new properties that do not follow from the Schrödinger equation alone.

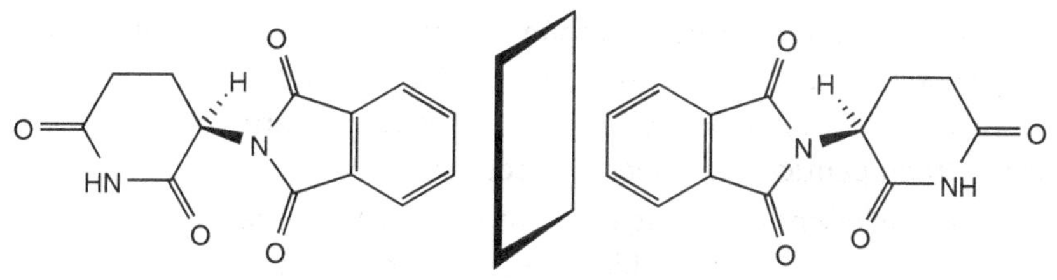

Figure 25.1: Two chiral versions of the molecule thalidomide, also known under the brand name Contergan® (also Softenon®); left: R-thalidomide acts as a sedative, right: S-thalidomide is acutely teratogenic and caused thousands of babies with deformed extremities in the 1950s.

are *not invariant under reflection* with respect to an axis or a center, they are called *chiral* molecules or *enantiomers*. They have identical energy eigenvalues (solutions of the Schrödinger equation), but are distinct concerning the classical property of chirality.[6] There are no quantum superpositions of chiral states of molecular systems.

Chiral molecules have many properties that do not follow from their first-principle quantum treatment as in Eq. (25.4). For instance, optically active substances (such as many sugars and aminoacids) rotate the polarization of light. Chiral molecules in naturally occurring substances exhibit a preference for one of the chiral versions, for instance, D-glucose (right) and L-aminoacids (left), which is called homochirality.[7] As of today, neither its origin nor its purpose is finally understood (Meierhenrich 2008).

The biochemical consequences of chiral molecules can be dramatic. While L-aminoacids, as the building blocks of peptides and proteins, belong to the basis of life on earth, many D-aminoacids are toxic in living organisms. The multifaceted history of thalidomide (see Fig. 25.1), originally produced and distributed as a sedative, is another case in point for the entirely different impact that different enantiomers of the same molecule can have in biological environments (Stephens and Brynner 2001).

In order to reproduce the effects of chiral molecules in living organisms, it is clearly insufficient to reproduce their energy spectrum. The descriptive level of the Schrödinger equation (25.4) does not include the classical observable of chirality. The relevant context for its proper consideration is the biochemistry of

---

[6]Chirality was discovered by Pasteur in 1848, the term chirality is due to Lord Kelvin in 1904. It derives from the Greek word $\chi\epsilon\iota\rho$ for "hand," since human hands are a most illustrative example of a chiral pair.

[7]R (rectus) or D (dexter) stand for the right-handed versions of a chiral molecule, L (levo) or S (sinister) stand for the left-handed versions.

chiral molecules in their natural environment, for instance, their interaction with (chiral) receptors, enzymes, transporters, or DNA. Chiral drugs, i.e., biomedical applications of enantiomers, typically have different pharmacokinetic, pharmacodynamic, therapeutic, and adverse effect profiles.

The example of chirality is again of interest for the general issue of partitioning. The bipartition due to two chiral variants emerges from a more fundamental (quantum mechanical) description in which the corresponding distinction is not available. In other words, the more fundamental description is, with respect to chirality, coarser than the less fundamental description. This differs from the case of temperature where the more fundamental mechanical description requires the finest possible identity partition while the less fundamental thermal description leads to the coarsest possible trivial partition.

Within a hierarchy of descriptions, the choice of partitions can be a very powerful tool to identify rigorous relations between descriptive levels. Yet, these relations would not be of much interest in science if this choice were arbitrary, so that each choice might lead to different interlevel relations. But, as the next sections will indicate, choices can be understood as depending on contexts that are characteristic for *particular* descriptive levels. This means that interlevel relations (e.g., bridge laws) can be formulated, which reflect how the *particular* context of a description, implemented as a partition within a coarser level, gives rise to emergent properties to which the context pertains. The formal framework accomplishing this task has been denoted as *contextual emergence*.

25.4 Contextual Emergence

Contextual emergence was introduced by Bishop and Atmanspacher (2006)[8] as an alternative to a number of approaches to address relations between levels of description in science, so-called interlevel relations. Briefly speaking, it is more flexible than plain reduction on the one hand, where a fundamental description is assumed to "fix everything," and, on the other hand, not as arbitrary as a radical emergence where "anything goes."

The basic goal is to establish a well-defined interlevel relation between a "lower" (L) level and a "higher" (H) level description of a system. This is done by a two-step procedure that leads in a systematic and formal way (1) from an *individual* description L_i to a *statistical* description L_s and (2) from L_s to an individual description H_i. This scheme can in principle be iterated across any

[8]See also Atmanspacher and beim Graben (2009) for more details and, in particular, for applications.

connected set of descriptions, so that it is applicable to any situation that can be formulated precisely enough to be a sensible subject of scientific investigation.

The essential goal of step (1) is the identification of equivalence classes of individual states that are indistinguishable with respect to a particular ensemble property. This step implements the multiple realizability of statistical states in L_s by individual states in L_i. The equivalence classes at L can be regarded as cells of a partition. Each cell is the support of a (probability) distribution representing a statistical state, encoding limited knowledge about individual states.

The essential goal of step (2) is the assignment of *individual* states at level H to *statistical* states at level L. This cannot be done without additional information about the desired level-H description. In other words, it requires the choice of a *context* setting the framework for the set of *observables* (properties) at level H that is to be constructed from level L. The chosen context provides constraints that can be implemented as *stability criteria* at level L. It is crucial that such stability conditions cannot be specified without knowledge about the context at level H. In this sense the context yields a top-down constraint or downward confinement (sometimes misleadingly called downward "causation").

The notion of stability induced by context is of paramount significance for contextual emergence. Roughly speaking, stability refers to the fact that some system is robust under (small) perturbations. For example, (small) perturbations of a homeostatic or an equilibrium state are damped out by the dynamics, and the initial state will be asymptotically retained. The more complicated notion of a stable partition of a state space is based on the idea of coarse-grained states, i.e., cells of a partition whose boundaries are robustly maintained under the dynamics. In this way, the partition cells define equivalence classes that do not change with time (which would be highly undesirable).

The technical tools to construct such partitions belong to the theory of ergodic systems and symbolic dynamics (Lind and Madcus 1995). This is not the place to introduce these fields in detail. But the basic idea is to define symbols (as higher-level states) based on generating (or Markov) partitions in the state space of the lower-level description. This construction is applicable to all situations, including systems far from thermal equilibrium, which are well-enough defined to enable a state-space representation of their dynamics. For explicit examples of such constructions, theoretically and empiricallly, see Atmanspacher and beim Graben (2007) and Allefeld *et al.* (2009).

The mentioned stability criteria guarantee that the statistical states of L_s are based on a robust partition so that the *emergent observables* in H_i are well defined. Implementing a contingent context at level H as a stability criterion in L_i yields a proper partitioning for L_s. In this way, the lower-level state space is endowed with a new, *contextual topology*. From a different perspective, the context selected at level H decides which details in L_i are relevant and which are irrelevant for individual states in H_i. Differences among all those individual states at L_i that fall into the same equivalence class at L_s are irrelevant for the chosen context. In this sense, the stability condition determining the contextual partition at L_s is also a *relevance criterion*.

This interplay of context and stability across levels of description is the core of contextual emergence. Its proper implementation requires an appropriate definition of individual and statistical states at these levels. This means in particular that it would not be possible to construct emergent observables in H_i from L_i directly, without the intermediate step to L_s. And it would be equally impossible to construct these emergent observables without the downward confinement arising from higher-level contextual constraints.

In this spirit, bottom-up and top-down strategies are interlocked with one another in such a way that the construction of contextually emergent observables is *self-consistent*. Higher-level contexts are required to implement lower-level stability conditions leading to proper lower-level partitions, which in turn are needed to define those lower-level statistical states that are coextensional (not necessarily identical!) with higher-level individual states and associated observables. Paraphrasing this self-consistency, the scheme of contextual emergence is "top down all the way from the bottom up."

The framework of contextual emergence has been shown to be applicable to a number of examples from the sciences. Paradigmatic case studies are: (1) the emergence of thermodynamic observables such as temperature from a mechanical description (cf. Section 25.3.1; Bishop and Atmanspacher 2006), (2) the emergence of hydrodynamic features such as Rayleigh–Bénard convection from many-particle theory (Bishop 2008), and (3) the emergence of molecular observables such as chirality from a quantum mechanical description (cf. Section 25.3.2; Primas 1998, Bishop and Atmanspacher 2006).

If descriptions at L and H are well established, as it is the case in these examples, formally precise interlevel relations can be straightforwardly set up. The situation becomes more challenging, though, when no such established descriptions are available, e.g., in cognitive neuroscience or consciousness studies, where relations between neural and mental descriptions are considered. Even there, contextual emergence has been proven viable for the construction of emergent mental states (Atmanspacher and beim Graben 2007, Allefeld et al. 2009) and an informed discussion of the problem of mental causation (Harbecke and Atmanspacher 2011).

25.5 Ontological Relativity: Beyond Fundamentalism and Relativism

A network of descriptive levels of varying degrees of granularity raises the question of whether descriptions with finer grains are more "fundamental" than those with coarser grains. The majority of scientists and philosophers of science in the past have tended to answer this question affirmatively. As a consequence, there would be one *fundamental* ontology, preferentially that of elementary particle physics, to which the terms at all other descriptive levels can be reduced.

But this reductive credo has also produced critical assessments and alternative proposals. A philosophical precursor of trends against a fundamental ontology is Quine's (1969) "ontological relativity" (carrying Garfinkel's "explanatory relativity" into ontology). Quine argued that if there is one ontology that fulfills a given descriptive theory, then there is more than one. In other words, it makes no sense to say what the objects of a theory are, beyond saying how to interpret or reinterpret that theory in another theory.

For Quine, any question as to the "quiddity" (the "whatness") of a thing is meaningless unless a conceptual scheme is specified relative to which that thing is discussed. The inscrutability of reference (in combination with his semantic holism) is the issue that necessitates ontological relativity. The key motif behind it is to allow ontological significance for any descriptive level, from elementary particles to icecubes, bricks, and tables, and further to thoughts, intentions, volitions, and actions.

Quine proposed that a "most appropriate" ontology should be preferred for the interpretation of a theory, thus demanding "ontological commitment." This leaves us with the challenge of how "most appropriate" should be defined, and how corresponding descriptive frameworks are to be identified. Here is where the idea of relevance criteria becomes significant. For a particular system in a given context, a "most appropriate" framework is the one that provides the "most relevant" granularity. And the referents of this descriptive framework are those which Quine wants us to be ontologically committed to.

Putnam (1981, 1987) developed a related kind of ontological relativity, first called "internal realism," later sometimes modified to "pragmatic realism." Ontological (sometimes conceptual) relativity in Putnam's internal realism differs from Quine's usage of the term in an important detail. While Quine's ontological relativity is due to the impossibility of a uniquely fixed relationship of our concepts to objects to which those concepts refer, Putnam's position is more radical insofar as he questions that we know at all what we mean when we speak of such objects.

On the basis of these philosophical approaches, Atmanspacher and Kronz (1999) suggested how to apply Quine's ideas to concrete scientific descriptions, their relationships with one another, and with their referents. One and the same descriptive framework can be construed as either ontic or epistemic, depending on which other framework it is related to: bricks and tables will be regarded as ontic by an architect, but they will be considered highly epistemic from the perspective of a solid-state physicist.

This farewell to the centuries-old conviction of an absolute fundamental ontology (usually that of basic physics) is still opposed to modern mainstream thinking today. But in times in which fundamentalism – in science and elsewhere – appears increasingly tenuous, ontological relativity offers itself as a viable alternative for more adequate and more balanced frameworks of thinking. And, using the scientifically tailored version of ontological relativity, it is not merely a conceptual idea but can be applied for an informed discussion of concrete scenarios in the sciences.

Coupled with an ontological commitment that becomes explicit in relevance criteria, the relativity of ontology must not be confused with dropping ontology altogether. The "tyranny of relativism" (as some have called it) can be avoided by identifying relevance criteria to select proper context-specific descriptions from less proper ones. The resulting picture is more subtle and more flexible than an overly bold reductive fundamentalism, and yet it is more restrictive and specific than a patchwork of arbitrarily connected model fragments.

Acknowledgments

Parts of this contribution are based on material in an earlier publication by Atmanspacher, Bezzola Lambert, Folkers, and Schubiger (2014). I am grateful to the coauthors of that article for the permission to re-use that material.

References

Allefeld, C., Atmanspacher, H., and Wackermann, J. (2009): Mental states as macrostates emerging from EEG dynamics. *Chaos* **19**, 015102.

Atmanspacher, H., Bezzola Lambert, L., Folkers, G., and Schubiger, P.A. (2014): Relevance relations for the concept of reproducibility. *Journal of the Royal Society Interface* **11**, 20131030.

Atmanspacher, H., and beim Graben, P. (2007): Contextual emergence of mental states from neurodynamics. *Chaos and Complexity Letters* **2**, 151–168.

Atmanspacher, H., and beim Graben, P. (2009): Contextual emergence. *Scholarpedia* 4(3), 7997.

Atmanspacher, H., and Kronz, F. (1999): Relative onticity. In *On Quanta, Mind and Matter*, ed. by H. Atmanspacher, A. Amann, U. Müller-Herold, Kluwer, Dordrecht, pp. 273–294.

Bishop, R. (2008): Downward causation in fluid convection. *Synthese* **160**, 229–248.

Bishop, R., and Atmanspacher, H. (2006): Contextual emergence in the description of properties. *Foundations of Physics* **36**, 1753–1777.

Born, M., and Oppenheimer, R. (1927): Zur Quantentheorie der Molekeln. *Annalen der Physik* **389**, 457–484.

Chibbaro, S., Rondoni, L., and Vulpiani, A. (2014): *Reductionism, Emergence and Levels of Reality*, Springer, Berlin.

Garfinkel, A. (1981): *Forms of Explanation*, Yale University Press, New Haven.

Haag, R., Kastler, D., and Trych-Pohlmeyer, E.B. (1974): Stability and equilibrium states. *Communications in Mathematical Physics* **38**, 173–193.

Harbecke, J., and Atmanspacher, H. (2012): Horizontal and vertical determination of mental and neural states. *Journal of Theoretical and Philosophical Psychology* **32**(3),161–179.

Lind, D.A., and Marcus, B.A. (1995): *An Introduction to Symbolic Dynamics and Coding*, Cambridge University Press, Cambridge.

Meierhenrich, U.J. (2008): *Amino Acids and the Asymmetry of Life*, Springer, Berlin.

Nagel, E. (1961): *The Structure of Science: Problems in the Logic of Scientific Explanation*, Routledge, London.

Primas, H. (1998): Emergence in the exact sciences. *Acta Polytechnica Scandinavica* **91**, 83–98.

Putnam, H. (1981): *Reason, Truth and History*, Cambridge University Press, Cambridge.

Putnam, H. (1987): *The Many Faces of Realism*, Open Court, La Salle.

Quine, W.V.O. (1969): Ontological relativity. In *Ontological Relativity and Other Essays*, Columbia University Press, New York, pp. 26–68.

Shannon, C.E., and Weaver, W. (1949): *The Mathematical Theory of Communication*, University of Illinois Press, Urbana.

Sperber, D., and Wilson, D. (1986): *Relevance: Communication and Cognition*, Blackwell, Oxford.

Stephens, T., and Brynner, R. (2001): *Dark Remedy – The Impact of Thalidomide and Its Revival as a Vital Medicine*, Perseus, Cambridge.

Takesaki, M. (1970): Disjointness of the KMS states of different temperatures. *Communications in Mathematical Physics* **17**, 33–41.

van Fraassen, B. (1980): *The Scientific Image*, Clarendon, Oxford.

von Weizsäcker, E.U. (1974): Erstmaligkeit und Bestätigung als Komponenten der pragmatischen Information. In *Offene Systeme I*, ed. by E. von Weizsäcker, Klett-Cotta, Stuttgart, pp. 83–113.

26
The Quest for Reproducibility Viewed in the Context of Innovation Societies

Sabine Maasen

Abstract. This chapter reframes the issue of reproducibility in science and technology within the context of contemporary knowledge societies that are characterized by a constant quest for innovation. Given that innovation privileges novelty and, by implication, the not-yet-reproducible, the first question is how, if at all, these opposing requirements can be reconciled in scientific practice. Being ideal-typical goals they represent a regulative dual that cannot be met in a straightforward fashion but need to be orchestrated both epistemically and socially. Secondly, in times of accelerated innovation, carefully arranged empirical settings emerge that complement and challenge these norms. "Social robustness of science" and "responsible research and innovation" are prototypic examples. As instances of institutionalized reflexivity in science they may eventually change the normative structure of science.

26.1 Introduction

Innovation, so it seems, is ubiquitous: innovation is discussed in the scientific and technical literature, in the social sciences such as history, sociology, management, economics, and in the humanities and arts. Innovation is also a central idea in popular discourses, in the media as well as in the economy and in public policy. That is to say, innovation is not only a phenomenon to be studied from various angles but also the icon of modern knowledge societies, a universal remedy for resolving real-world problems in particular. For Helga Nowotny, it is nothing less than the definition of our epoch: the never-ending fascination and constant quest for innovation (Nowotny 2006, Godin 2008, p. 5).

Adopting a social constructivist stance, innovation, however, does not exist as such. Rather, it is constructed through discourses and practices attributing something as innovative (Godin 2008, p. 7). Most prominently, innovation is framed as a break with the past by way of acknowledging both continuities and discontinuities (Godin 2008, p. 8). Or, as Braun-Thürman (2005) has it: From a systems theoretical point of view, innovation takes neither recourse to triggering factors nor to normative factors but rather to an observation that

---

Reproducibility: Principles, Problems, Practices, and Prospects, First Edition. Edited by Harald Atmanspacher and Sabine Maasen.

something denotes a difference to what hitherto counted as familiar, normal, just reproduced. Whenever this observation is communicated, others can connect to it: confirm it, contest it, or debate it. Thus, on this general level, innovation is conceived as an attribution. Moreover, it is regularly attached to evaluative judgments, albeit often ambivalent ones: to progress and prosperity, as well as to uncertainty and risk, respectively.

The increasing importance and ambiguity of innovation for contemporary societies does not leave science and technology unaffected. On the one hand, new knowledge is an important driver of economic growth and social development. On the other hand, multiple pressures on science, technology, and innovation entail the risk that knowledge production might be practiced in inappropriate ways, among others, by disregarding the norm of producing knowledge that is sound as well as responsible with respect to societal needs and concerns. Ever-more often, this situation calls for action. When, for instance, UNESCO inquired into its role for assisting its member states in building knowledge societies based on peace and sustainable development, crucial elements were (UNESCO 2012)

> to elaborate and promote ethical norms and principles and to embed ethical practices in the institutions of science and technology at all levels. It is further necessary to connect such ethical norms, principles and practices to the institutional design of science, technology and innovation systems.

Policy documents of this type abound on all scales (institutional, national, transnational) and address different aspects such as particular scientific domains (e.g., biotechnology, big data), specific practices (e.g., reporting principles), or the systems of science and technology at large (e.g., by national funding agencies). More often than not they result in codes of conduct or best practices detailing norms and procedures, such as new training modules on enhancing reproducibility and transparency of research findings (Collins and Tabak 2014). Moreover, since about two decades, intensive research and debate emerged in Science and Technology Studies (STS) as to how increased pressure on innovativeness could be reconciled with producing reproducible or "sound" science, respectively.

Two of the many notions will be pointed out: While one strand of research focuses on the concept of "socially robust knowledge" (Gibbons *et al.* 1994, Nowotny *et al.* 2001), another strand of research, or, for some, a further development of the former, is concerned with "responsible research and innovation" (e.g., Owen *et al.* 2012). It not only addresses the need to reflect on the (right) impact of science but also to institutionalize appropriate practices such "that deliberation and reflection can be coupled to action (i.e., responsiveness)" (Owen *et al.* 2012, p. 755).

Interestingly enough, in the sphere of STS as in science policy, we do not hear so much (or noticeably less) about "reproducibility" (cf. Collins, this volume), yet much more about "robustness," "responsibility," and "responsiveness" of science and technology. Unless it concerns issues of genuine misconduct (above all fraud and plagiarism), the latter three terms have gained increasing relevance, particularly within the European Commission's Science in Society program, framed within the EU Horizon 2020 initiative.

The hypothesis of this paper is that in innovation societies the quest for and procedures to secure reliable knowledge emerges in response to the increased pressure to innovate. Robustness, responsibility, and responsiveness are designed to counterbalance the concomitant risks of science being either epistemically flawed or insignificant, if not problematic for politics, industry, or civil society. In this perspective, robustness, responsibility, and responsiveness might be regarded as accompanying measures extending the concept of "sound science" beyond the existing intra-scientific norm of reproducibility into the sociopolitical domain. Science and technology are no longer systems governing themselves exclusively according to self-set standards (such as reproducibility) but rather are co-governed and currently challenged by external actors and their norms (such as safety, utility, and profitability). As the precarious balance of innovation and robustness in science has become severely threatened by the ubiquitous quest for the new in our society, efforts increase to address the soundness of science in ever-more differentiated ways.

In order to set the stage, the paper will start with a brief genealogical account of innovation to show the recency of its contemporary meaning and importance. Thereafter, the notion of sound science in innovation societies will be explained as the reframing of intra-scientific norms, postulated by Merton in the 1940s. Within this set of norms, reproducibility and innovation appear as paradoxical goals that have become the object of ongoing socio-epistemic balancing acts. An additional challenge arises when the scientific ethos meets with intensified emphasis on innovation in knowledge societies. This is the moment when the set of self-regulating norms is increasingly complemented by explicit efforts at enforcing social robustness of science as well as responsible research and innovation. Ultimately, these efforts may be regarded as not only challenging and complementing intra-scientific norms by external norms but also changing the former.

26.2 A Genealogical Sketch of "Innovation"

Innovation has not always been that prominent, neither in society nor in science. Neither has it been conceived as positively as today – ambivalences notwithstanding. Well into the 18th century innovation was received predominantly negatively. For example, a "novator" was thought of a person to be deeply mistrusted. In politics, change was seen as deviance; in religion as heresy. Also in economics, when inventors ("projectors") started to make money from their inventions, "they became objects of satire by many authors because of insufficient science, bad management and fraud" (Godin 2008, p. 24).

Likewise, innovation was only late in becoming an object for theory building. In sociology, it was Gabriel Tarde (1890, 1895, 1898, 1902) who distinguished statics from dynamics and was interested in explaining social change (or social evolution). Although Tarde frequently referred to the term innovation (and a cluster of synonyms such as creation, originality) as novelty, he never defined it (Godin 2008, p. 26). Only in the early 1920s, authors like Ogburn and Gilfillan started looking at inventions, above all technological inventions, as causes of cultural change or social change. Documented quantitatively by Hart from the 1930s to the 1960s (Hart 1931, 1957, 1959), Ogburn (1922, p. 4) observed the growth and acceleration of material culture as a social process that needs to be planned and controlled. He noticed a "cultural lag," more precisely, an increasing gap between the material culture (technology) and the rest of culture (adaptive culture) due to inertia and lack of social adaptation. On this account, innovation was not conceived outright negatively, yet risky and in need of constant monitoring.

In the domain of economics, it was evolutionary economics, notably Schumpeter, who introduced and used innovation as a category. In his view, capitalism equals creative destruction: disturbance of existing structures, unceasing novelty, and change (Schumpeter 1928, 1942, 1947). Moreover, inventions and innovations needed to be distinguished: while he conceived of invention as an act of intellectual creativity, innovation was an economic decision, e.g., a firm applying or adopting an invention. In this vein, innovation was increasingly received as technological innovation and as commercialized innovation. In the course of this development, innovation studies no longer focused solely on individuals but also on collectives, e.g., on organizational innovation (Godin 2008, p. 39). Corresponding studies referred to institutional or political innovation in schools (Mort 1967) or government agencies (e.g., Chapin 1928, McVoy 1940, Walker 1969, Mohr 1969).

This shift in the meaning of innovation as both important and indeed something to be induced and monitored became ubiquitous in society and was consequently applied to science as well. After World War I (Mees 1920), the management of research activities gained increasing attention because of their contribution to economic performance. In this context, (technological) innovation came to be regarded as output stemming from investments in research and development. Therefore, scientific and technological discoveries and inventions became a genuine subject of study by psychologists (Pelz and Andrew 1966, Myers and Marquis 1969), well before science and technology studies (Mulkay 1972a, b) and innovation research (for an overview see Fagerberg *et al.* 2013) adopted this issue from multiple perspectives. It is particularly due to management of technology (Anthony and Day 1952) and research evaluation (Rubenstein 1957), though, that innovation became studied in terms of efficiency (Godin 2008, p. 40).

It should not go unnoticed that the importance of innovation in both society and science played out because innovation, at the same time, was primarily a policy-driven concept. The researchers involved frequently were consultants to governments. They provided expert advice on many different policy issues. "What was called science policy in the 1960s became science and technology policy in the 1970s, then innovation policy in the 1990s" (Godin 2008, p. 41). As Godin correctly points out, however, there has never been a "policy for science" period (see also Elzinga and Jamison 1995). Instead, there was a "science for policy" program, by which public research and universities were urged to contribute to technological innovation (Godin 2007). "From its very beginning, science policy, whether implicit or explicit, was constructed as a means to achieve social, economic and political goals" (Godin 2008, p. 42).

In the course of this genealogy, innovation came to be predominantly defined as useful innovation. In the eye of the sociologist, when used and adopted, an invention becomes an innovation; in the eye of the economist, invention is innovation when commercialized. Here, the focus of both sociologists' and economists' theories was technological innovation. In public representations, in policy, and in social studies, innovation came to be identified with technological innovation. Previous meanings or predecessors of the concept such as invention, ingenuity, or imagination, were all subsumed under "innovation" (Godin 2008, p. 46). In this spirit, creative abilities were placed in the service of organizations, industrial development, and economic growth.

As mentioned at the beginning, this shift in meaning corresponds to a shift in orientation of contemporary societies as knowledge societies. As their socio-economic reproduction predominantly rests on science and technology, their government programs increase public spending for scientific research as well as for, e.g., IT infrastructure, and efficient patent systems. In knowledge societies, the state tries to coordinate and foster interactions between the government, universities, and the private sector in order to promote innovative activities. Innovations are deliberate interventions designed to initiate and establish future developments concerning technology, economics, and social practices. Acting as a "descriptive umbrella notion" (Howaldt and Schwarz 2010), the term innovation today covers a series of complex and interrelated governance changes aimed at ensuring systemic competitiveness in a globalized world society. Thus, the triangle of technology, innovation, and society is of central concern.

Summarizing, and irrespective of the precise definition, innovation has become part and parcel of characterizing and governing modern knowledge societies. Innovation is the key attribution – including "innovation speak" – and (according to Vinsel 2014)

> key to societal improvement, including politicians reaching for appealing rhetoric and numerous groups opportunistically looking for handouts (e.g., university administrators; academic and industry scientists and engineers; firms, from large corporations to small startups; bureaucrats in government labs).

It is all the more important that we trust in science and technology to produce not only innovative but also "sound science." Yet, what is "sound science"?

26.3 Reframing Scientific Ethos: Sound Science

Following the recommendations of the Society of Environmental Toxicology and Chemistry (SETAC 1999, p. 1),

> sound science can be described as organized investigations and observations conducted by qualified personnel using documented methods and leading to verifiable results and conclusions.

It implies that a set of data, facts, or conclusions of a scientific nature are supported by studies that follow the standards of the scientific method. The latter describe investigational attributes and practices such as the formulation of a testable hypothesis, the use of systematic and well-documented experimental or analytical methods (e.g., adequate sample sizes, appropriate control experiments), the application of appropriate data analysis tools (e.g., statistics and mathematical models) to the data, and conclusions that provide an answer to the hypothesis and that are supported by the results. The scientific method is broadly applicable to a wide range of investigations, ranging from the so-called "hard sciences" to the humanities. In addition, further aspects have to be taken care of: notably, the acquisition, storage, management, and sharing of data as well as the communication of scientific knowledge and information (SETAC 1999).

The quest for sound science has been voiced ever since science has acquired the status of a genuine subsystem operating upon very specific practices and norms (e.g., "method") – a subsystem of society that increasingly interacts with, yet is markedly distinct from, e.g., economic or political subsystems. While on the one hand, science thus became somewhat opaque to the rest of society, on the other hand, it helped to establish a particular scientific ethos, i.e., a set of rules that are supposed to establish trust among scientists and guarantee the reliability of the knowledge created in the process.

This ethos has been given its most influential account by Merton (1957) who defined it in terms of four basic norms. Dating back to the establishment of the academies in England and France in the 17th century, the norms of science can be understood as ideal-typical behavioral patterns that evolved into a set of institutional imperatives. The ethos "is that affectively toned complex of values and norms which is held to be binding on the man of science" which, although not codified, can be inferred from, above all, the "moral indignation directed toward contraventions of the ethos" (Merton 1957, pp. 551f):

- Communism ("communality") refers to the norm that the findings of science "are a product of social collaboration and are assigned to the community" (Merton 1957, p. 556). The scientist's claim to intellectual property is constrained to recognition and esteem by the members of his community.
- Universalism is the principle that truth claims are "subjected to pre-established impersonal criteria" (Merton 1957, p. 553) irrespective of the social attributes of their protagonists, e.g., nationality, race, class, or religion.
- Disinterestedness rests upon the public and testable nature of science. It contributes to the integrity of scientists by sanctioning and by the accountability to their peers to resist the temptation of using improper means to their own advantage.
- Organized scepticism is "both a methodologic and an institutional mandate," i.e., the scrutiny of claims and beliefs on the basis of empirical and logical criteria (Merton 1957, p. 560).

Taken together, these norms constitute a system of communication uniquely geared toward producing knowledge that may be considered true. Truth in science, however, means knowledge being reliable but always open for further analysis and debate. Universalism guarantees, in principle, accessibility by all scientists and, at the same time, prevents interference of types of knowledge (political, religious, ethnic) other than those accepted as belonging to science itself. Communism subjects all knowledge to open communication. Proprietary interests of scientists are framed as gaining recognition exclusively by obtaining priority of discovery. Disinterestedness distinguishes science from the professions in that it has no clientele: The primary clients of science are scientists themselves. Fraud or plagiarism may lead to temporary advantages, at best. Finally, organized skepticism is the flip side of that norm. Institutionalized in the peer review system of journals and funding agencies, it requires the impersonal scrutiny of any truth claim as a general principle of scientific communication.

It has often been countered that scientists do not behave in accordance with the norms and that these are subject to historical change, as can be shown in cases of patenting or commodification of science, not least driven by the contemporary imperative of innovation. While it cannot be denied that these norms change over time, their existence, however, is not necessarily reflected in actual behavior. Rather, Merton's theoretical construction of the norms is a complex combination of different elements (Weingart 2012):

> first, patterns of attitudes that are expressed in internalized (but not necessarily explicit) reactions to violations of the norms, in the awareness that one's own actions or those of others are breaching a code; second, patterns of sanctions, i.e. mechanisms institutionalized in science that sanction, positively or negatively, certain behaviors like plagiarism (negatively) or the open exchange of information (positively).

The problems concerning reproducibility and the plethora of debates and activities meant to solve them are cases in point. It is precisely because poor reproducibility violates the Mertonian norms (above all the norm of universalism), that the community of scientists, the higher education system at large as well as funding agencies, and science political bodies are taking important steps in analyzing the problem of irreproducible scientific results in its complexity and finding workable solutions.

With this amendment, the Mertonian norms can be regarded as an analytical scheme that helps to explain the unique status of science as a set of methods to produce certified knowledge and to accumulate that knowledge. It also entails a set of, albeit changing, cultural values that govern scientific activities – such as politeness (Shapin 2003) or, today, increasing competitiveness (see Franzen, this volume). The central assumption is that scientific knowledge has to be and is only certified if it has passed the scrutiny of peers. For it is only peers that are competent to test truth claims. Only if such truth claims are accepted as such by the collective of scientists can they be considered – for the time being – as secure and reliable, and by implication, reproducible knowledge (Weingart

2012). There is an inner epistemic tension, however, in producing reproducible knowledge.

26.4 Reproducibility and Innovation: A Regulative Dual

As Collins (1985, p. 19) put it, "replicability, in a manner of speaking, is the Supreme Court of the scientific system," and he continued:

> It corresponds to what the sociologist Robert Merton (1945) called the "norm of universality". Anybody, irrespective of who or what they are, in principle ought to be able to check for themselves through their own experiments that a scientific claim is valid.

What is more, this ideal should at least theoretically be applicable (Collins 1985, p. 19):

> Replication is the scientifically institutionalized counterpart of the stability of perception. It is just that with scientific phenomena one looks through a complex instrument called an experiment. Thus the acceptance of replicability can and should act as a demarcation criterion for objective knowledge. Public agreement to the existence of a new concept implies that its reproducibility can be confidently affirmed even if, as a matter of fact, it is never tested.

But Collins also pointed out that, in practice, the ability to reproduce a given experiment cannot be separated from personal skills and expertise of the persons involved (see also Collins, this volume). As a consequence, there is plenty of room for insecurity and, hence, the need for trust, even among closest peers – and even more so in interdisciplinary contexts where standards of meeting the gold standard of good science vary considerably among the disciplines involved.

As importantly, science is not only about producing certified and, hence, reproducible knowledge. It is also about producing new knowledge. In this spirit, Ziman (1984) added another norm to the Mertonian ensemble, namely, originality, according to which science is the discovery of the unknown. All scientific work must be novel, continually adding to or correcting the existing body of scientific knowledge.

On this view, scientific practice is basically all about balancing two basic tasks: producing new knowledge (innovation!) and certified knowledge (via reproducibility!). Both tasks are clearly paradoxical as innovative knowledge leads into yet unknown territory, not yet or not fully reproducible. Certified knowledge can be trusted in, yet, by implication, fails to bring about something new. In this vein, Rheinberger adds another perspective to the problem of reproducibility by tying it together with innovation as flip sides of the same coin. His prototypical example is the scientific experiment (Rheinberger 1997, p. 3, my italics):

> experimental systems must be capable of *differential reproduction* ... in order to behave as devices for producing scientific novelties that are beyond our present knowledge, that is, to behave as generator[s] of surprises. Difference and reproduction are two sides of one coin; their interplay accounts for the shifts and displacements within the investigative process.

On a general note, he concluded that novelty has its roots in reproducibility, building the grounds for the scientific process as such (Rheinberger 1997, p. 75):

> All innovation, in the end and in a very basic sense, is the result of such reproduction. Reproduction, far from being simply a matter of securing appropriate and reproducible boundary conditions for the experiment, characterizes scientific activity as a material process of generating, transmitting, accumulating, and changing information.

On both accounts, Collins' and Rheinberger's, reproducibility in science is thus hard to attain. While Rheinberger conceives of it as a balancing act by inherent epistemic necessity, Collins offers no rational solution to the paradox constituted by the dual norm of innovation and reproduction. His central argument about experimental practice revolves around the thesis that facts can only be generated by adequate instruments. However, whether or not instruments are adequate can only be recognized if they produce facts (experimenters' regress). For Collins, scientific controversies thus cannot be closed by the "facts" because there are no criteria independent of the outcome of the experiment that could help to decide whether or not an experimental setting is indeed adequate.

Godin and Gingras countered this argument with an argumentation approach. It suggests that knowledge consists of "opinions" that have survived objections (Perelman 1963, p. 117). From this angle, scientific statements become knowledge only if they are neither ignored nor contradicted (Ziman 1968). Scientific communities submit experiments and calculations to critical investigation (Popper 1962). Godin and Gingras (2002, p. 150) concluded:

> Collins' empirical descriptions and analysis of scientific debates make perfect sense when resituated within the dynamic of a scientific field constrained by rules which are themselves the product of a past history of experimentation, calculation and argumentation.

Epistemically speaking, this is because reproducibility and originality are what Kant considers "regulative ideas": Unlike "constitutive ideas," regulative ideas (Briesen 2013, p. 3)

> are of purely regulative use, i.e., they are used to formulate principles from which instructions can be derived regarding how to proceed in our projects of systematizing our cognitions.

In actual fact, as all science must strive for both reproducibility and originality, they may more accurately be regarded as "regulative dual." The norm of reproducibility, while still effective in governing scientific behavior, in principle (Collins and Tabak 2014), is not easy to attain on the grounds of this inherent tension – a tension which results in a complicated socio-epistemic game of experimentation and argumentation, always balancing reproducibility and originality.

An additional challenge arises when the scientific ethos meets with intensified emphasis on innovation in knowledge society. This is the moment when the scientific ethos as a largely implicit set of self-regulating norms becomes increasingly complemented by explicit efforts at enforcing "sound science."

26.5 Making the Implicit Explicit I: Social Robustness of Science

The socio-epistemic act of balancing reproducibility and innovation in scientific practice has become a matter of increasing concern between science and society as well. As the role of scientific evidence in knowledge societies is not only to support economic productivity but also to legitimate political decisions (cf. Porter, this volume), this has a number of repercussions on science itself. To mention but three of them: first, science is increasingly influenced by politics. The use of scientific expertise by politicians and other groups, such as non-governmental organizations, involves science in political decision making and installs scientists within political groups. Second, science is increasingly commercialized. Scientific knowledge is becoming a commodity that can be traded almost like any other product. Third, science and the media are increasingly coupled as scientists use the media in order to raise public support for their research.[1]

All these trends point to a loss of distance between science and society, which has been the most important basis for the production of reliable knowledge. Most visibly, scientific reports, studies, and expert commissions are now part of our political and social landscape. Indeed, the status of scientific knowledge has changed; it is no longer cloaked in the traditional signs of scientific expertise (Innerarity 2012, p. 3) but has become subject to various efforts at

[1]The mass media, in turn, observe the sciences in terms of their "newsworthiness" (Weingart 2002) – according to this logic, "breakthroughs" are of equal news value as scandals such as fraud or lack of reproducibility.

democratization (Funtowicz and Ravetz 1923, Brown *et al.* 2004, Maasen and Weingart 2006).

Contemporary approaches in socio-political analysis have thus suggested the need for seriously rethinking the "social contract" between science and society (Gibbons *et al.* 1994, Latour 2004, Elam and Bertilsson 2002). In this framework, it does no longer suffice to produce knowledge that is both certified, hence reproducible, and innovative. In addition, society does want to have a say in what counts as "sound science" and what kinds of innovation are worth pursuing. This has led to debates that culminate in two prototypic concepts: social robustness of science and responsible innovation.

The concept of science aiming at social robustness has been developed by Gibbons *et al.* (1994). The authors distinguish between an "old" and a "new" mode of knowledge production. While the old paradigm of scientific discovery (mode 1) was characterized by the hegemony of disciplinary science and driven by the autonomy of scientists and their host institutions, the universities, the new paradigm of knowledge production (mode 2) was said to be socially distributed, application oriented, transdisciplinary, and subject to multiple account-abilities – not only by peers but also by extra-scientific stakeholders, such as the state, users, and industrial partners.

This hypothesis was neither new nor unique; it was preceded by the so-called "finalization science" (Böhme *et al.* 1983). Its distinctive claim was that all disciplines follow three phases (explorative, paradigmatic, post-paradigmatic), the latter of which being the one, where "finalization" may occur. In this mature stage of a discipline, it becomes open to orientation in accordance with external (e.g., political) objectives.

Similar notions with slightly different accentuation are "strategic research" (Irvine and Martin 1984), "strategic science" (Rip 2004), or "post-normal science" (Funtowicz and Ravetz 1993). The latter's most striking characteristic is public participation, either by engaging stakeholders in decision-making processes or in the quality assessment of scientific knowledge production ("extended peer review", Funtowicz 2001). Ziman (2000) advanced the notion of post-academic science that, *inter alia*, emphasizes the utility of knowledge produced, the increased competition for funds and the links between science and industry. The closer coupling of industry, university, and government is key to the "triple helix model" by Etzkovitz and Leydesdorff (1998).

Together, these and other notions such as the enterprise university (Marginson and Considine 2000), academic capitalism (Slaughter and Leslie 1997), as well as a number of variants of innovation systems approaches stress the numerous ways in which science is encircled by governing actors, institutions, and organizations. They all see to it that "the right kind of science is being done" by way of addressing not only intra-scientific quality criteria of producing reproducible science but also extra-scientific ones (such as participation, competitiveness, utility, etc.) of producing "sound science." In so doing, science and technology today not only adhere to robustness but also responsibility and responsiveness.

26.6 Making the Implicit Explicit II: Responsible Research and Innovation

When it comes to responsible innovation, what's new in this concept? While innovation has shaped modern society from its very inception, it is currently indeed gaining new dimensions: innovation is becoming increasingly reflexive, heterogeneously distributed, and ubiquitous. Reflexivity implies more than the intentional transformation of routine actions; it also refers to the transformation of social practices based on continuously (re-)produced knowledge about innovation. Yet, exactly how specific technologies ultimately contribute to the resolution of societal issues and problems often depends on how these technologies interact, as well as on societal factors such as political or economic conditions, public acceptance, and ethical considerations. To this end, concepts like "responsible research and innovation" and offerings like "Technology, Innovation and Society Programs" address the need for systematic research of the many and varied interconnections between technology, innovation, and society.

Small wonder that, with respect to responsible innovation, a similar story can be told as with respect to social robustness of science. Although the term responsible innovation has become increasingly fashionable in recent years (notably in European policy circles, cf. Owen *et al.* 2012), it is known for more than a decade (e.g., Hellstrom 2003, Guston 2004, von Schomberg 2011). And the debate over responsible research also has a genealogy: it stretches from variants of technology assessment to ELSI programs (ethical, legal, and social implications of science and technology).

Firmly anchored, e.g., in the EU Framework Program, it fosters what Guston calls collaborative assurance: scientists are taking societal demands on their research more self-consciously in their own hand. In some domains, responsible innovation has led to Codes of Conduct, an example being the Code for Responsible Nanosciences and Nanotechnologies Research (European Commission 2008). Based upon many established approaches such as upstream engagement (Wilsdon *et al.* 2005), constructive technology assessment (Schot and Rip 1996), or anticipatory governance (Barben *et al.* 2008), the idea of responsible research and innovation fosters the reflexivity of scientists, broadens the scope of strategic choices, and embeds extra-scientific concerns into scientific practice.

The corresponding field of activity is called Responsible Research and Innovation (RRI). RRI describes a research, development, or innovation process that takes its possible effects and potential impacts on society into account. It has been defined (von Schomberg 2013) as

> a transparent, interactive process by which societal actors and innovators become mutually responsive to each other with a view to the (ethical) acceptability, sustainability and societal desirability of the innovation process and its marketable products in order to allow a proper embedding of scientific and technological advances in our society.

Ultimately, it puts more pressure on science to produce not only true but also relevant and trustworthy knowledge. Alongside with, and partly substituting the notion of socially robust knowledge, RRI science develops forms of knowledge production that are designed to increase responsibility by way of addressing four tasks:

1. *Anticipation.* Anticipation disposes researchers and organizations to ask "what if" questions (Ravetz 1997) to consider contingency, what is known, what is likely, what is plausible, and what is possible. Anticipation aims at resilience.
2. *Reflexivity.* Reflexivity, at the level of institutional practice, means critically considering one's own activities, commitments, and assumptions, being aware of the limits of knowledge (codes of conduct, moratoriums, and the adoption of standards).
3. *Inclusion.* This task emphasizes the importance of public dialogue in "opening up" (Stirling 2008) and framing issues that challenge entrenched assumptions and commitments (Lövbrand *et al.* 2011).
4. *Responsiveness.* Its two aspects relate to the two meanings of the word *respond* – to react and to answer (Pellizzoni 2004). Responsiveness involves responding to new knowledge as this emerges and to emerging perspectives, views, and norms.

While social robustness initiated first steps in accountability beyond the scope of researchers, agendas in the framework of RRI advance a set of explicit intuitions and procedures as summarized by von Schomberg (2011, p. 53):

- organizing collective co-responsibility: codes of conduct for research and innovation,
- ensuring market accountability: standards, certification, and accreditation schemes and labels,

- ethics: a "design" factor of technology in order to increase social–ethical reflexivity in research practices,
- feedback with policymakers: models for responsible governance,
- public debate: moderating "policy pull" and "technology push."

At this point one may ask: What difference does it make to follow an RRI scheme as opposed to just follow intra-scientific quality criteria in the framework of epistemic discussions in science? In general, RRI schemes can be characterized as reflections triggered by controversies arising from the acquisition of new scientific knowledge. During such reflexive processes, scientific methods and the fundamental understanding of the nature of the subject matter become subjects of dispute. In these cases, researchers of different scientific disciplines are mutually challenged in terms of which discipline can claim to offer the best answer to a given problem. Recent examples of epistemic discussions in science include the debates between molecular biologists and ecologists on the risks of genetically modified organisms, and the debate on climate change as either being induced by human interventions or as caused by natural cycles. Typically, epistemic discussions provoke public debate way before any scientific answer on the issue can be expected. Still, such debates provide a significant challenge for developing reasonable public policy. Plausible epistemic approaches on the acquisition of knowledge in science are associated with problem definitions which, in turn, frame policy approaches – albeit often only implicitly.

By way of an example, the Socio-Technical Integration Research (STIR) is an experiment in responsible innovation (Fisher and Mahajan 2006). It convened social scientists and scholars in the humanities in over 25 laboratories on three continents. The goal was to test 'midstream modulation' as an integral part of science and engineering research. So-called midstream modulation processes allow for the incremental adjustment of innovation through regulation, or upstream, during policy and priority setting. To this end, the STIR program uses a protocol that scrutinizes social and ethical values midstream, that is, as decisions by scientists and innovators are being made in real time.

The protocol takes the collaborators through an iterative set of questions by social scientists and philosophers probing their capacities for responsible innovation: more particularly, it explores the capacity to reflect upon opportunities (reflexive capacity), considerations (deliberative capacity), alternatives (responsive capacity), and outcomes (anticipatory capacity). As a result of using the protocol regularly during laboratory decision making, STIR engagements have shown 'productive research disruptions'. Eventually, they have led to changes in research direction, experimental design, safety and environmental practices, and public outreach (responsiveness). These changes are voluntary, as the laboratory researchers opt freely whether or not, and if so, how to make them. They are integrative, as they cover both social values and technical options. Last but not least, they are collaborative.

Summarizing, one may conclude that in times of accelerated and ubiquitous innovation the quest for certified, hence reproducible, knowledge is supplemented by concepts and programs that attempt to render scientific knowledge socially robust and innovation responsible as well as responsive to societal demands (Kaldewey 2013, p. 3). In the eyes of the protagonists, RRI may provide an opportunity for reshaping research practices, research agendas, and research policy as well as funding governance (Stahl 2013, p. 208)

> to include a new conception of responsibility that extends beyond conventional, although important, notions associated with ensuring "experimental reproducibility" when working in the laboratory and with following research process guidelines.

From this angle, reproducibility, however conceived and practiced, today does not suffice any longer. Given its importance for contemporary knowledge societies and their future, accountability of science extends from the laboratory or the study well into the economy, politics, and civil society. Producing sound science and responsible innovation not only complements the scientific aim of producing reliable knowledge but also challenges and may eventually change the corresponding Mertonian norms.

26.7 Mertonian Norms Challenged Anew: Institutional Reflexivity and Responsiveness

Seen from a different angle, one may conceive of robustness and responsibility as instances of institutional(ized) reflexivity. According to Anthony Giddens (1990), institutional reflexivity refers to the fundamental structural dimension of modern society in which the regularized use of knowledge in society is a constitutive element in its organization and transformation. In its original definition, institutional reflexivity refers to the fact that ever-more segments of social life, and material relations with nature, are susceptible to continuous revision in the wake of new information and knowledge. Notably science and technology provide us with ever-new knowledge and devices that require constant adaptation by way of learning. It is characteristic of modern knowledge societies that its members (individuals or organizations) as well as society at large are thought of as always in need of justification and revisable, i.e., "reflexive" within the framework of reflexive modernization (see, e.g., also Beck and Bonss 2001). What is true for society at large has become true for science and technology as well: in response to societal demands, codes of conduct and procedures emerged that are explicitly designed to reflect upon and promote sound science and responsible innovation. These codes and procedures, hence, institutionalize ongoing reflexivity in science and technology.

This trend is often couched in terms of "responsiveness." Science is expected to be responsive to societal demands, and science policy is expected to ensure that science actually takes this responsibility. On a critical note, it is important to hint at the implicit suspicion that science in itself is deemed not to be responsively enough, and that a certain external pressure is necessary to achieve it. This notion is based upon the long-standing, albeit unconvincing model of science as an "ivory tower" – set apart from society and its expectations, and pursuing nothing but intra-scientific agendas.

At this point, one may get back to Merton and his ideas as to the normative structure of science. In the 1940s, his set of institutional norms governing science as a specific and specialized system of communication within society stressed the autonomy of scientists *vis-à-vis* the political struggles of their time. Nowadays, we have to reconsider the norms that are constitutive for the modern system of science. Indeed, one should keep in mind that the normative structure of science is in constant flux and challenged by other values. In particular, conceiving of science as *either* autonomous (independent from society) *or* heteronomous (utterly entangled with society) does not go far enough. By contrast, Kaldewey (2013, p. 10) suggested to think

> about whether there may be a complementary responsive structure of science, i.e. a set of norms and values that determines the ways science reacts proactively to problems in its societal and natural environments.

And yes, there are candidates: responsible innovation, for example, is explicitly designed as the scientists' proactive attempt to address societal demands on their research. In this line of reasoning, current science policy discourses (e.g., on the "grand challenges"), however, should be hypothesized as highly ambivalent. They both support the novel norm of responsiveness and erode it by way of enforcing a "pressure of practice" (Carrier 2011) to the point of threatening the production of reproducible science.

While this hypothesis still awaits empirical clarification, this much seems to be clear: (institutional) reflexivity in doing sound science is here to stay.

References

Barben, D., Fisher, E., Selin, C., and Guston, D.H. (2008): Anticipatory governance of nanotechnology: Foresight, engagement, and integration. In *Handbook of Science and Technology Studies*, ed. by E.J. Hacket, O. Amsterdamska, M. Lynch, and J. Wajcman, MIT Press, Cambridge, pp. 979–1000.

Beck, U., and Bonss, W., eds. (2001): *Die Modernisierung der Moderne*, Suhrkamp, Frankfurt.

Böhme, G., van den Daele, W., and Krohn, W. (1983): *Finalization in Science: The Social Orientation of Scientific Progress*, Reidel, Dordrecht.

Braun-Thürmann, H. (2005): *Innovation*, Transcript, Bielefeld.

Briesen, J. (2013): Is Kant (W)right? On Kant's regulative ideas and Wright's entitlements. *Kant-Yearbook* **5**, 1–23.

Carrier, M. (2011): Knowledge, politics, and commerce. Science under the pressure of practice. In *Science in the Context of Application*, ed. by M. Carrier and A. Nordmann, Springer, Dordrecht, pp. 11–30.

Chapin, F.S. (1928): *Cultural Change*, Century, New York.

Collins, H.M. (1985): *Changing Order: Replication and Induction in Scientific Practice*, University of Chicago Press, Chicago.

Collins, F.S., and Tabak, L.A. (2014): Policy: NIH plans to enhance reproducibility. *Nature* **505**, 612–613.

Elam, M., and Bertilsson, M. (2002): Consuming, engaging and confronting science: The emerging dimensions of scientific citizenship. *European Journal of Social Theory* **6**(2), 233–225.

Elzinga, A., and Jamison, A. (1995): Changing policy agenda in science and technology. In *Handbook of Science and Technology Studies*, ed. by S. Jasanoff, G.E. Markle, J.C. Peterson and T. Pinch, Sage, Thousand Oaks, pp. 572–597.

Etzkowitz, H., and Leydesdorff, L. (1998): The endless transition: A "triple helix" of university industry government relations. *Minerva* **36**, 203–208.

European Commission (2008): European Commission's Code of Conduct for Responsible Nanotechnologies Research. Available at `ec.europa.eu/research/science- society/document_library/pdf_06/nanocode-apr09_en.pdf`.

Fagerberg, J., Martin, B.R., and Andersen, E.S. (2013): *Innovation Studies. Evolution and Future Challenges*, Oxford University Press, Oxford.

Fisher, E., and Mahajan, R.L. (2006): Midstream modulation of nanotechnology research in an academic laboratory. In *Proceedings of the ASME International Mechanical Engineering Congress and Exposition 2006*, pp. 189–195.

Funtowicz, S., and Ravetz, J.R. (1992): Three types of risk assessment and the emergence of post-normal science. In *Social Theories of Risk*, ed. by S. Krimsky and D. Golding, Praeger, New York, pp. 251–273.

Funtowicz, S. (2001): Peer review and quality control. In *International Encyclopaedia of the Social and Behavioural Sciences*, Elsevier, Oxford, pp. 11179–11183.

Gibbons, M., Limoges, C., Nowotny, H., Schwartzman, S., Scott, P., and Trow, M. (1994): *The New Production of Knowledge*, Sage Publications, London.

Giddens, A. (1990): *The Consequences of Modernity*, Polity Press, Oxford.

Godin, B., and Gingras, Y. (2002): The experimenters' regress: From skepticism to argumentation. *Studies in History and Philosophy of Science* **33**, 37–152.

Godin, B. (2007): National innovation system: The system approach in historical perspective. Available at `www.csiic.ca/PDF/Godin_36.pdf`.

Godin, B. (2008): Innovation: History of a category. Available at `www.csiic.ca/PDF/IntellectualNo1.pdf`.

Hart, H. (1931): *The Technique of Social Progress*, Henry Holt, New York.

Hart, H. (1957): Acceleration in social change. In *Technology and Social Change*, ed. by F.R. Allen, Appleton, New York, pp. 27–55.

Hart, H. (1959): Social theory and social change. In *Symposium on Sociological Theory*, ed. by L. Gross, Harper and Row, New York, pp. 196–238.

Hellstrom, T. (2003): Systemic innovation and risk: Technology assessment and the challenge of responsible innovation. *Technology in Society* **25**, 369–384.

Hicks, J.R. (1932): *The Theory of Wages*, Macmillan, London.

Howaldt, J., and Schwarz, M. (2010): Social innovation: Concepts, research fields and international trends. Sozialforschungsstelle Dortmund. Available at `www.internationalmonitoring.com/fileadmin/Downloads/Trendstudien/IMO%20Trendstudie_Howaldt_englisch_Final%20ds.pdf`.

Hutter, M., Knoblauch, H., Rammert, W., and Windele, A. (2001): Innovation society today: The reflexive creation of novelty. Technical University Technology Studies Working Paper No. 4, p. 1.

Innerarity, D. (2012): Power and knowledge: The politics of the knowledge society. *European Journal of Social Theory* **16**, 3–16.

Irvine, J., and Martin, B.R. (1984): *Foresight in Science: Picking the Winners*, Frances Pinter Publishers, London.

Kaldewey, D. (2013): ECRC paper, science dynamics and research systems: The role of research for meeting societal challenges, Madrid, 8–9 April 2013.

Latour, B. (2004): *Politics of Nature: How to Bring the Sciences into Democracy*, Harvard University Press, Cambridge.

Lövbrand, E., Pielke, S., and Beck, S. (2011): A democracy paradox in studies of science and technology. *Science, Technology & Human Values* **36**, 474–496.

Maasen, S., and Weingart, P., eds. (2005): *Democratization of Expertise? Exploring Novel Forms of Scientific Advice in Political Decision-Making*, Springer, Dordrecht.

Marginson, S., and Considine, M. (2000): *The Enterprise University: Power, Governance and Reinvention in Australia*, Cambridge University Press, Cambridge.

McVoy, E.C. (1940): Patterns of diffusion in the United States. *American Sociological Review* **5**(2), 219–227.

Merton, R.K. (1957): Science and democratic social structure. In *Social Structure and Social Theory*, rev. ed. by R.K. Merton, Free Press, Glencoe, pp. 550–561.

Mohr, L.B. (1969): Determinants of innovation in organizations. *American Political Science Review* **63**(1), 111–126.

Mort, P.R. (1964): Studies in educational innovation from the institute of administrative research. In *Innovation in Education*, ed. by M.B. Miles, Columbia University Teachers College Press, New York, pp. 317–328.

Mulkay, M.J. (1972a): Conformity and innovation in science. *Sociological Review Monograph Series* **18**, pp. 5–23.

Mulkay, M.J. (1972b): *The Social Process of Innovation: A Study in the Sociology of Science*, Macmillan, London.

Myers, S., and Marquis, D.G. (1969): *Successful Industrial Innovations: A Study of Factors Underlying Innovation in Selected Firms*, National Science Foundation, Washington.

Nowotny, H., ed. (2006): *Cultures of Technology and the Quest For Innovation*, Berghahn, New York.

Nowotny, H., Scott, P., and Gibbons, M., eds. (2001): *Re-Thinking Science. Knowledge and the Public in an Age of Uncertainty*, Polity Press, Oxford.

Ogburn, W.F. (1922): *Social Change with Respect to Culture and Original Nature*, Viking Press, New York.

Owen, R., Macnaghten, P., and Stilgoe, J. (2012): Responsible research and innovation: From science in society to science for society, with society. *Science and Public Policy* **39**, 751–760.

Pellizzoni, L. (2004): Responsibility and environmental governance. *Environmental Politics* **13**, 541–565.

Pelz, D.C., and Andrews, F.M. (1966): *Scientists in Organizations: Productive Climate for Research and Development*, Wiley, New York.

Perelman, C. (1963): Self-evidence and proof. In *The Idea of Justice and the Problem of Argument*, ed. by C. Perelman, Routledge, London, pp. 109 –124.

Popper, K.R. (1962): *Conjectures and Refutations*, Harper and Row, New York.

Ravetz, J.R. (1997): The science of "what-if?" *Futures* **29**, 533–539.

Rheinberger, H.-J. (1997): *Toward a History of Epistemic Things: Synthesizing Proteins in the Test Tube*, Stanford University Press, Stanford.

Rip, A. (2004): Strategic research, Post-modern universities and research training. *Higher Education Policy* **17**(2), 153–166.

Schumpeter, J.A. (1928): The instability of capitalism. *Economic Journal* **38**, 361–386.

Schumpeter, J.A. (1942): The process of creative destruction. In *Capitalism, Socialism and Democracy*, Harper, New York, pp. 81–86.

Schumpeter, J.A. (1947): The creative response in economic history. *Journal of Economic History* **7**, 149–159.

Schot, J., and Rip, A. (1996): The past and future of constructivist technology assessment. *Technological Forecasting and Social Change* **54**, 251–268.

SETAC (1999): Sound science. Available at httpc.ymcdn.com/sites/www.setac.org/resource/resmgr/publications_and_resources/setac_tip_soundsci.pdf.

Shapin, S. (2003): The image of the man of science. In *Cambridge History of Science 4, Eighteenth-Century Science*, ed. by R. Porter, Cambridge University Press, Cambridge, pp. 159–183.

Slaughter, S.G., and Leslie, G. (1997): Academic capitalism, managed professionals, and supply-side higher education. *Social Text* **51**, 9–38.

Stahl, B.C., Eden, G., and Jirotka, M. (2013): Responsible research and innovation in information and communication technology: Identifying and engaging with the ethical implications of ICTs 199. In *Responsible Innovation: Managing the Responsible Emergence of Science and Innovation in Society*, ed. by R. Owen, J. Bessant, and M. Heintz, Wiley, London, pp. 199–218.

Stirling, A. (2008): "Opening up" and "closing down": Power, participation, and pluralism in the social appraisal of technology. *Science, Technology & Human Values* **33**, 262–294.

Tarde, G. (1902): L'invention, moteur de l'évolution sociale. *Revue internationale de sociologie* **95**, 561–574.

Tarde, G. (1898): *Les lois sociales: esquisse d'une sociologie*, Edition Institut Synthélabo, Le Plessis-Robinson, reprinted in 1999.

Tarde, G. (1897): *L'opposition universelle: Essai d'une théorie des contraires*, Edition Institut Synthélabo, Le Plessis-Robinson, reprinted in 1999.

Tarde, G. (1895): *La logique sociale*, Edition Institut Synthélabo, Le Plessis-Robinson, reprinted in 1999.

Tarde, G. (1890): *Les lois de l'imitation*, Edition Seuil, Paris, reprinted in 2001.

UNESCO (2013): Towards knowledge societies for peace and sustainable development. Available at www.unesco.org/new/fileadmin/MULTIMEDIA/HQ/CI/CI/pdf/wsis/WSIS_10_Event/wsis10_outcomes_en.pdf.

Vinsel, L. (2014): How to give up the I-word. Available at culturedigitally.org/2014/09/how-to-give-up-the-i-word-pt-1/.

von Schomberg, R. (2011): Prospects for technology assessment in a framework of responsible research and innovation. In *Technikfolgen abschätzen lehren: Bildungspotenziale transdisziplinärer Methoden*, ed. by M. Dusseldorp and R. Beecroft, VS, Wiesbaden, pp. 39–61.

von Schomberg, R. (2013): A vision of responsible innovation. In *Responsible Innovation: Managing the Responsible Emergence of Science and Innovation in Society*, ed. by R. Owen, J. Bessant and M. Heintz, Wiley, London, pp. 51–74.

Walker, J.L. (1969): The diffusion of innovations among the American states. *American Political Science Review* **63**, 880–899.

Weingart, P. (2002): The moment of truth for science: The consequences of the "knowledge society" for society and science. Available at onlinelibrary.wiley.com/enhanced/doi/10.1093/embo-reports/kvf165/.

Weingart, P. (2012): Openness in science. Available at whataboutscience.com/2012/06/13/guestpost-openess-in-science-by-peter-weingart/.

Ziman, J. (1968): *Public Knowledge: The Social Dimension of Science*, Cambridge University Press, Cambridge.

Ziman, J. (1984): *An Introduction to Science Studies: The Philosophical and Social Aspects of Science and Technology*, Cambridge University Press, Cambridge.

Ziman, J. (2000): *Real Science: What It Is, and What it Means*, Cambridge University Press, Cambridge.

Index